Pollution, Atmosphere and Carbon Balance

Pollution, Atmosphere and Carbon Balance

Edited by Bernie Goldman

SYRAWOOD
PUBLISHING HOUSE

New York

Published by Syrawood Publishing House,
750 Third Avenue, 9th Floor,
New York, NY 10017, USA
www.syrawoodpublishinghouse.com

Pollution, Atmosphere and Carbon Balance
Edited by Bernie Goldman

International Standard Book Number: 978-1-68286-547-7 (Hardback)

Cataloging-in-Publication Data

Pollution, atmosphere and carbon balance / edited by Bernie Goldman.
    p. cm.
Includes bibliographical references and index.
ISBN 978-1-68286-547-7
1. Pollution. 2. Air--Pollution. 3. Atmosphere . 4. Carbon dioxide mitigation.
I. Goldman, Bernie.
TD174 .P65 2018
363.73--dc23

TABLE OF CONTENTS

Preface..VII

Chapter 1   **Quantifying and understanding carbon storage and sequestration within the
Eastern Arc Mountains of Tanzania, a tropical biodiversity hotspot**.. 1
Simon Willcock, Oliver L Phillips, Philip J Platts, Andrew Balmford, Neil D Burgess,
Jon C Lovett, Antje Ahrends, Julian Bayliss, Nike Doggart, Kathryn Doody,
Eibleis Fanning, Jonathan MH Green, Jaclyn Hall, Kim L Howell, Rob Marchant,
Andrew R Marshall, Boniface Mbilinyi, Pantaleon KT Munishi, Nisha Owen,
Ruth D Swetnam, Elmer J Topp-Jorgensen and Simon L Lewis

Chapter 2   **Modelling forest carbon stock changes as affected by harvest and natural
disturbances.I. Comparison with countries' estimates for forest management**....................... 18
Roberto Pilli, Giacomo Grassi, Werner A. Kurz, Raúl Abad Viñas and
Nuria Hue Guerrero

Chapter 3   **Projecting the spatiotemporal carbon dynamics of the Greater Yellowstone
Ecosystem from 2006 to 2050**... 36
Shengli Huang, Shuguang Liu, Jinxun Liu, Devendra Dahal, Claudia Young,
Brian Davis, Terry L Sohl, Todd J Hawbaker, Ben Sleeter and Zhiliang Zhu

Chapter 4   **Spatial distribution of temporal dynamics in anthropogenic fires in miombo
savanna woodlands of Tanzania**.. 51
Beatrice Tarimo, Øystein B Dick, Terje Gobakken and Ørjan Totland

Chapter 5   **A framework for estimating forest disturbance intensity from successive remotely
sensed biomass maps: moving beyond average biomass loss estimates**............................... 66
T. C. Hill, C. M. Ryan and M. Williams

Chapter 6   **These are the days of lasers in the jungle**.. 75
Joseph Mascaro, Gregory P Asner, Stuart Davies, Alex Dehgan and Sassan Saatchi

Chapter 7   **Advancing reference emission levels in subnational and national REDD+
initiatives: a CLASlite approach**... 78
Florian Reimer, Gregory P Asner and Shijo Joseph

Chapter 8   **Time-series maps of aboveground carbon stocks in the forests of central Sumatra**............ 89
Rajesh Bahadur Thapa, Takeshi Motohka, Manabu Watanabe and
Masanobu Shimada

Chapter 9   **Allometric equations for estimating belowground biomass of *Androstachys
johnsonii* Prain**... 102
Tarquinio Mateus Magalhães

Chapter 10   **Effects of harvest, fire, and pest/pathogen disturbances on the West Cascades ecoregion carbon balance**...117
David P Turner, William D Ritts, Robert E Kennedy, Andrew N Gray and Zhiqiang Yang

Chapter 11   **Integrating forest inventory and analysis data into a LIDAR-based carbon monitoring system**...129
Kristofer D Johnson, Richard Birdsey, Andrew O Finley, Anu Swantaran, Ralph Dubayah, Craig Wayson and Rachel Riemann

Chapter 12   **Rapid forest carbon assessments of oceanic islands: a case study of the Hawaiian archipelago**...140
Gregory P. Asner, Sinan Sousan, David E. Knapp, Paul C. Selmants, Roberta E. Martin, R. Flint Hughes and Christian P. Giardina

Chapter 13   **An assessment of the carbon stocks and sodicity tolerance of disturbed *Melaleuca* forests in Southern Vietnam**..153
Da B Tran, Tho V Hoang and Paul Dargusch

Chapter 14   **Analysis of biophysical and anthropogenic variables and their relation to the regional spatial variation of aboveground biomass illustrated for North and East Kalimantan, Borneo**...167
Carina Van der Laan, Pita A Verweij, Marcela J Quiñones and André PC Faaij

Chapter 15   **Choice of satellite imagery and attribution of changes to disturbance type strongly affects forest carbon balance estimates**..179
Vanessa S. Mascorro, Nicholas C. Coops, Werner A. Kurz and Marcela Olguín

Chapter 16   **EU mitigation potential of harvested wood products**...194
Roberto Pilli, Giulia Fiorese and Giacomo Grassi

Permissions

List of Contributors

Index

PREFACE

The purpose of the book is to provide a glimpse into the dynamics and to present opinions and studies of some of the scientists engaged in the development of new ideas in the field from very different standpoints. This book will prove useful to students and researchers owing to its high content quality.

The main cause for the increase in pollution is human activities. It is generally produced from sources such as industries, deforestation, vehicle emission, etc. Technologies have been developed in recent times to remove greenhouse gas emissions from the Earth's atmosphere. The topics included in this book on pollution with respect to atmosphere and carbon balance are of utmost significance and bound to provide incredible insights to readers. Scientists and researchers actively engaged in this field will find this book full of crucial and unexplored concepts.

At the end, I would like to appreciate all the efforts made by the authors in completing their chapters professionally. I express my deepest gratitude to all of them for contributing to this book by sharing their valuable works. A special thanks to my family and friends for their constant support in this journey.

Editor

Quantifying and understanding carbon storage and sequestration within the Eastern Arc Mountains of Tanzania, a tropical biodiversity hotspot

Simon Willcock[1,2]*, Oliver L Phillips[1], Philip J Platts[3], Andrew Balmford[4], Neil D Burgess[5,6], Jon C Lovett[1], Antje Ahrends[7], Julian Bayliss[4], Nike Doggart[8], Kathryn Doody[9], Eibleis Fanning[10], Jonathan MH Green[11], Jaclyn Hall[12], Kim L Howell[13], Rob Marchant[3], Andrew R Marshall[3,14], Boniface Mbilinyi[15], Pantaleon KT Munishi[15], Nisha Owen[10,16], Ruth D Swetnam[17], Elmer J Topp-Jorgensen[18] and Simon L Lewis[1,19]

Abstract

Background: The carbon stored in vegetation varies across tropical landscapes due to a complex mix of climatic and edaphic variables, as well as direct human interventions such as deforestation and forest degradation. Mapping and monitoring this variation is essential if policy developments such as REDD+ (Reducing Emissions from Deforestation and Forest Degradation) are to be known to have succeeded or failed.

Results: We produce a map of carbon storage across the watershed of the Tanzanian Eastern Arc Mountains (33.9 million ha) using 1,611 forest inventory plots, and correlations with associated climate, soil and disturbance data. As expected, tropical forest stores more carbon per hectare (182 Mg C ha^{-1}) than woody savanna (51 Mg C ha^{-1}). However, woody savanna is the largest aggregate carbon store, with 0.49 Pg C over 9.6 million ha. We estimate the whole landscape stores 1.3 Pg C, significantly higher than most previous estimates for the region. The 95% Confidence Interval for this method (0.9 to 3.2 Pg C) is larger than simpler look-up table methods (1.5 to 1.6 Pg C), suggesting simpler methods may underestimate uncertainty. Using a small number of inventory plots with two censuses ($n = 43$) to assess changes in carbon storage, and applying the same mapping procedures, we found that carbon storage in the tree-dominated ecosystems has decreased, though not significantly, at a mean rate of 1.47 Mg C ha^{-1} yr^{-1} (c. 2% of the stocks of carbon per year).

Conclusions: The most influential variables on carbon storage in the region are anthropogenic, particularly historical logging, as noted by the largest coefficient of explanatory variable on the response variable. Of the non-anthropogenic factors, a negative correlation with air temperature and a positive correlation with water availability dominate, having smaller p-values than historical logging but also smaller influence. High carbon storage is typically found far from the commercial capital, in locations with a low monthly temperature range, without a strong dry season, and in areas that have not suffered from historical logging. The results imply that policy interventions could retain carbon stored in vegetation and likely successfully slow or reverse carbon emissions.

Keywords: Eastern Arc Mountains; Tanzania; IPCC Tier 3; REDD+; Forest; Disturbance; Degradation; Ecosystem service

* Correspondence: S.P.Willcock@soton.ac.uk
[1]School of Geography, University of Leeds, Leeds LS2 9JT, UK
[2]School of Biological Sciences, University of Southampton, Southampton SO17 1BJ, UK
Full list of author information is available at the end of the article

Background

Tropical forests are globally significant ecosystems; accounting for ~50% of global forest area [1], storing ~ 45% of all carbon in terrestrial vegetation [2-4], maintaining high biodiversity [5], and providing ecosystem services, such as timber, non-timber forest products [6], and climate change mitigation [7,8]. However, within the last few decades, vast areas of tropical forests have been converted to other land-uses or degraded. For example, between 1990 and 1997, 4.4-7.2 million hectares of humid tropical forest were converted each year and an additional 1.6-3.0 million hectares of forest were visibly degraded [9]. This process increased in the early 2000s, with an estimated 5.1-5.7 million hectares of humid tropical forest (and 3.5-4.7 million hectares of dry tropical forest) deforested per year between 2000 and 2005 [10]. The gradual and sustained reduction in forest quality and quantity has resulted in substantial emissions of CO_2 [11]. Globally, deforestation and forest degradation accounted for 6-20% of anthropogenic GHG emissions in the 1990s and early 2000s [12-14]. Tropical regions make a substantial contribution to this, emitting 0.7-1.5 Pg C yr^{-1} between 1990 and 1999 [9,15-17] and 0.71.5 Pg C yr^{-1} between 2000 and 2007 [13,16-18]. These processes also impact the future potential of forests to remove carbon from the atmosphere [7,19,20].

Recently, attempts to mitigate increasing anthropogenic CO_2 emissions through reducing emissions from degradation and deforestation (REDD+) have been instigated [21]. The REDD+ programme is aimed at contributing to a reduction in greenhouse emissions whilst providing economic incentives for better management and protection of forests. This policy has been widely welcomed and may provide a financial incentive to significantly reduce carbon emissions [22,23], although the equity and justice issues surrounding the impact on local livelihoods are actively debated [24,25]. Key technical issues for the successful implementation of REDD+ include (but are not limited to) the accuracy of monitoring systems, preventing leakage and establishing accurate historical baselines. Thus, the success of REDD+, in part, rests on robust scientific information on the magnitude and extent of carbon storage in tropical regions and how it changes over time [26].

The Intergovernmental Panel on Climate Change (IPCC) provide a three "Tier" system through which carbon stocks and emissions can be reported, each with a different level of methodological complexity and accuracy. Tier 1 is the simplest method, using global default values obtained from the IPCC literature [27,28]. The intermediate Tier 2 level improves on Tier 1 by using country specific data. Tier 3 is the most rigorous approach, using local forest inventory data, focusing on the direct measurement of trees, repeated over a time series [27-29]. Here we

develop a Tier 3 methodology for the Eastern Arc Mountains (EAM) watershed area.

The estimates become progressively more robust from Tier 1 to 3 due to changes in two main systematic errors [29]. The first, completeness, refers to the number of IPCC carbon pools that are included, with studies including all five pools (aboveground live, litter, coarse wood debris [CWD], belowground and soil carbon) considered complete. The second, representativeness, derives from the substantial natural variability in the carbon stored across landscapes, even within a biome or country [30]. The aboveground biomass of a forest within a landscape may differ considerably from global default (Tier 1) values or even from country-specific (Tier 2) values. For example, in the Peruvian Amazon, data from the Los Amigos Conservation Concession [31] were shown not to be representative of forests nationally. Nearby forests situated to the north and south of this local study are estimated to contain 20-35% less carbon per unit area [32], suggesting that Los Amigos Conservation Concession is an area of locally high biomass. Since Tier 3 methods account for variation observed within biomes and countries, the representativeness of the carbon estimates is higher than those associated with Tier 1 and 2 methodologies [32,33].

However, Tier 3 methods are more expensive [34,35] and some nations may lack the capacity to adopt such methods [36]. Whilst, in some cases, the capability to apply Tier 3 guidelines is being rapidly developed, multi-temporal inventory data and data on historical carbon stock changes can take several decades to accrue [37,38]. It is expected that REDD+ requirements will allow data provisions from several tiers in a single report. Highly variable and/or substantial carbon pools should be estimated using Tier 3 methodology (e.g. forest aboveground live carbon [ALC]), whilst Tier 1 or Tier 2 methodology may be sufficient for smaller carbon pools (e.g. CWD) or carbon poor land cover categories (e.g. bare ground).

In Tier 3 methods, in order to extrapolate from plot data, it is necessary to develop correlations with remotely sensed data to scale to the study area or country-wide estimates. Generally, carbon storage is either estimated via statistical correlation with electromagnetic properties, ground-truthed by volumetric measurements, such as diameter at breast height (DBH), which are converted to biomass estimates using allometric equations. A variety of remotely sensed data sources have been employed for carbon mapping and these can be aggregated into four groups: photographic imagery, RADAR, LiDAR, and ancillary geographic information systems (GIS) data (see Additional file 1: SI1 for an evaluation of each method). Here, we use ancillary GIS data as such data have three main advantages: 1) wide availability, often free of charge; 2) a suitable resolution (e.g. 90 m [39]); and

3) correlations with these ancillary GIS data may indicate which variables directly affect carbon storage. Developing an understanding of how these variables influence carbon storage is vital for accurate scenarios of future emissions.

Here, we correlate carbon storage estimates from tree inventory plots ($n = 1,611$, median size = 0.1 ha) with data on climatic (e.g. temperature, precipitation, and solar radiation), edaphic (e.g. soil water holding capacity and soil fertility) and proxy variables for direct human interventions (e.g. governance type, distance from the main economic demand centres, population pressure, and historical logging), and variables that derive from climate-human interactions (e.g. burnt area index) for the Tanzanian watershed of the Eastern Arc Mountains (hereafter, EAM [40]), which covers 33.9 million ha (Figure 1; see Swetnam et al (2011) [41] for further details). We develop Tier 3 type correlation equations to estimate the total ALC stored across the forested and wooded land cover categories, an advancement on previous Tier 2 estimates for the region presented in Willcock et al (2012) [42]. Additionally, we investigate the most influential correlates of spatial differences in carbon storage and how these result from changes in either species composition affecting wood density (specific gravity) or the number of large trees present. Lastly, a smaller number of inventory plots (n = 43, median size 0.1 ha) have two censuses, and by applying the same mapping procedures, we assess changes in carbon storage over time, providing a first-order estimate of sequestration across the region.

Results
Carbon stocks
Utilising 1,611 plots and scaling to the 33.9 million ha study area we estimate that 1.32 (95% confidence

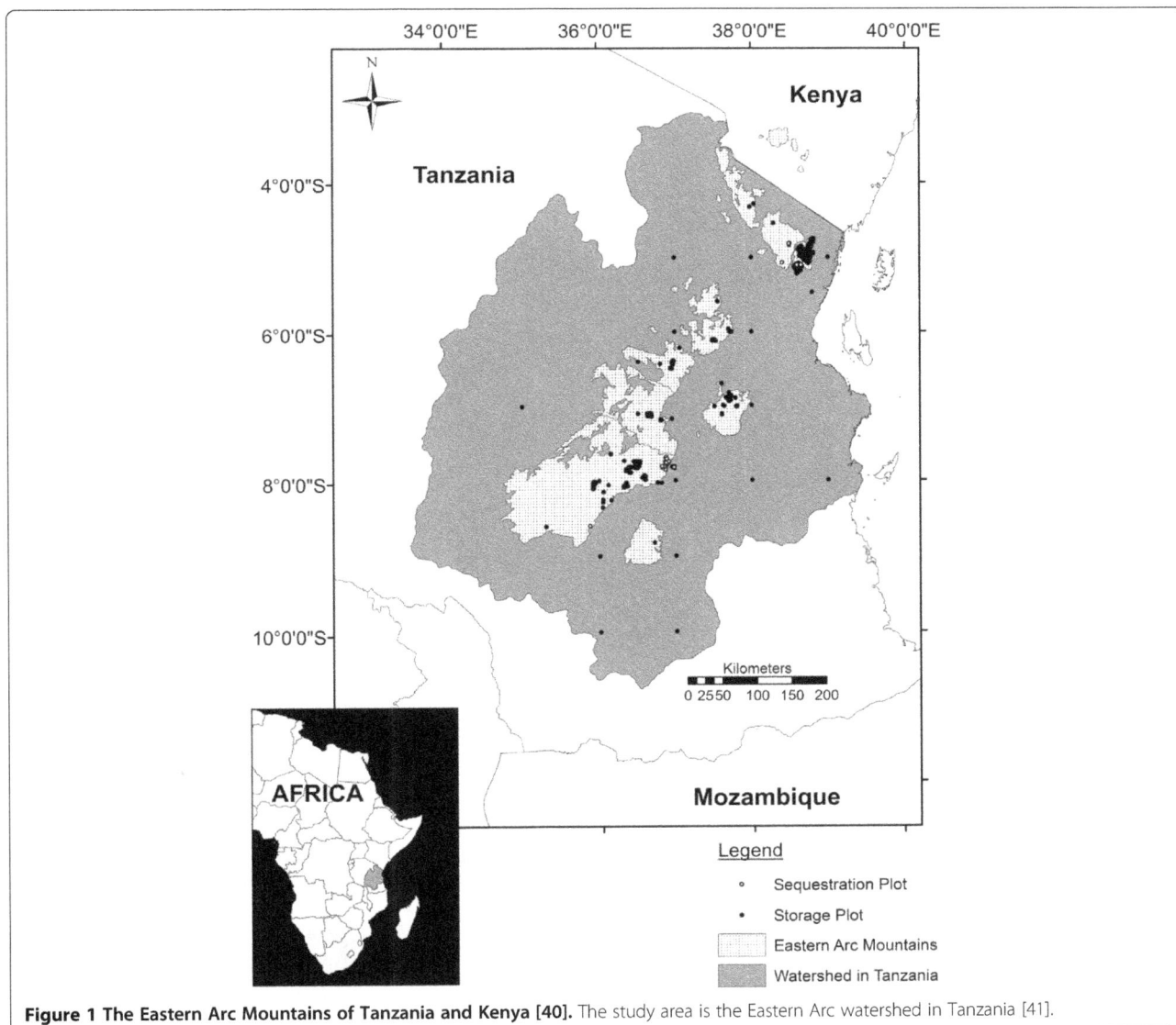

Figure 1 The Eastern Arc Mountains of Tanzania and Kenya [40]. The study area is the Eastern Arc watershed in Tanzania [41].

interval [CI] ranges from 0.89 to 3.16) Pg C was stored in the aboveground live vegetation in the year 2000 (Figure 2; Table 1). Woodland and bushland contributed most to the amount of stored aboveground live carbon (ALC) in the study region, with open woodland storing the most ALC (0.49 [0.47 to 1.60] Pg C over 9.6 million ha); followed by bushland (0.29 [0.15 to 0.51] Pg C over 5.0 million ha) and closed woodland (0.18 [0.13 to 0.61] Pg C over 1.8 million ha).

Best estimate values from our methodology, per unit area, in each land cover class, are given in Table 2. Forest contained the greatest ALC per unit area, with highest values in sub-montane forest (189 [95 to 588] Mg ha^{-1}), followed by lowland (182 [152- to 360] Mg ha^{-1}), upper montane (166 [69 to 533] Mg ha^{-1}), montane (130 [62 to 702] Mg ha^{-1}), and forest mosaic (121 [55 to 485] Mg ha^{-1}). Woodlands held less ALC than forests, with closed woodland storing 100 (70 to 331) Mg ha^{-1} and open woodland storing 51 (38 to 165) Mg ha^{-1} (Table 2),

but more than the landscape average of 39 (26 to 93) Mg ha^{-1}.

Our sequestration model suggests that the landscape may be losing 0.05 (-0.07 to 0.26) Pg C yr^{-1} (mean net flux to atmosphere of 1.47 [-2.13 to 7.75] Mg C ha^{-1} yr^{-1}). Of the 12.3 million ha of tree-dominated land in our study area, only 1.4% (0.17 million ha) shows a carbon decrease over the entire 95% CI range and only 0.8% (0.10 million ha) a definite carbon increase (Figure 3). The locations showing net carbon uptake are in the Udzungwa mountains, while the locations with net reductions in carbon storage are mainly in the Pare and Usambara mountains.

Links between carbon stock and influential variables

The variables that influence carbon storage and sequestration may be inferred from relationships within the correlation models. Forward selection results are presented in the following paragraphs as these best indicate causal relationships [43-45]. In general, backward models were in

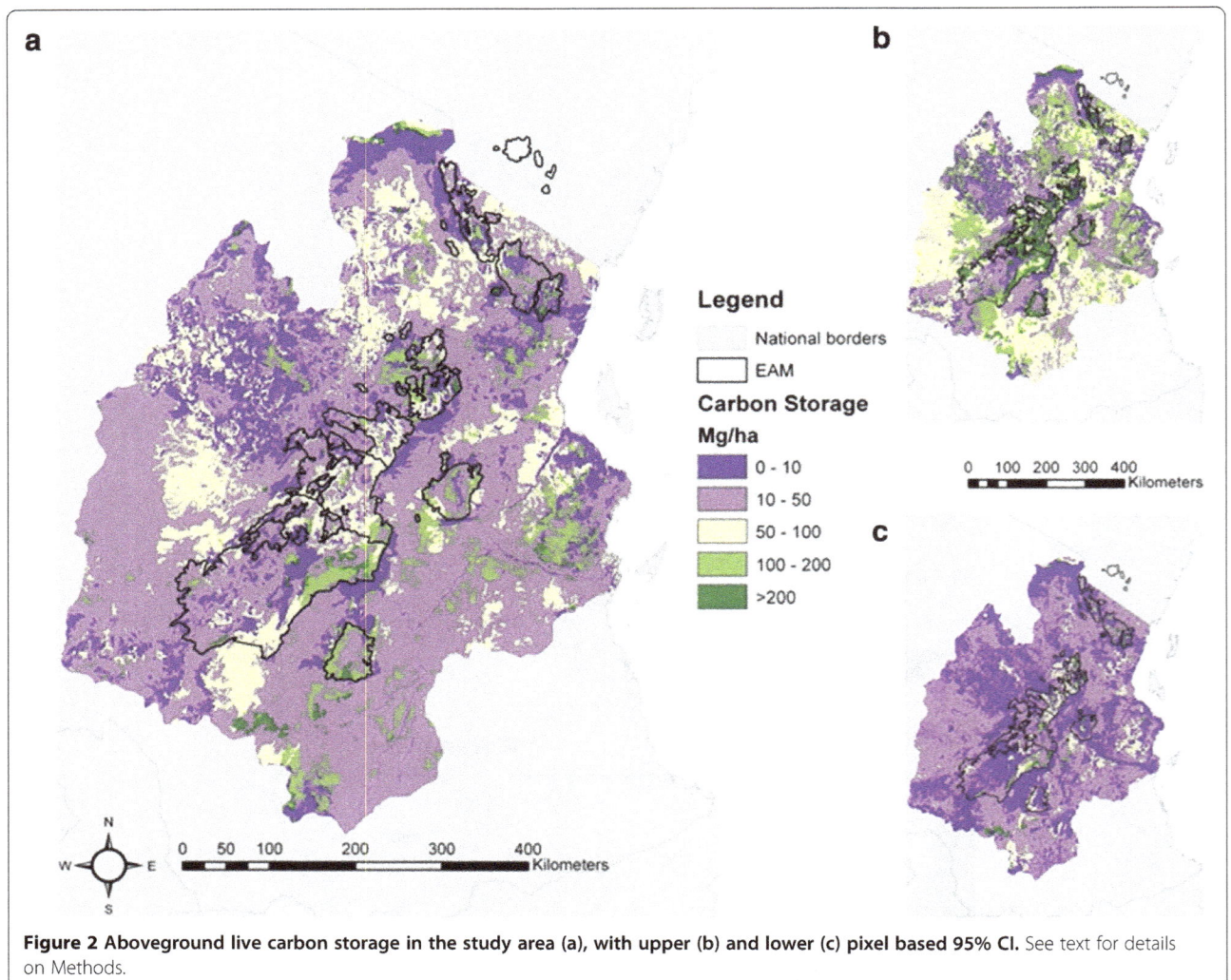

Figure 2 Aboveground live carbon storage in the study area (a), with upper (b) and lower (c) pixel based 95% CI. See text for details on Methods.

Table 1 Aboveground live carbon stored within the study area for the year 2000, estimated by this and previous studies

Study	Aboveground live carbon, Pg (95% CI range)	Methodology	Resolution (m^2)	Disturbance included?	Tanzanian on-the-ground data?
Present study* – Tier 3	1.32 (0.89-3.16)	Correlation equations derived using remotely sensed influential variables.	100	Anthropogenic variables represent human disturbance. Natural disturbance variables also included.	Yes
Willcock et al (2012)* – Original Tier 2 [42]	1.58 (1.56-1.60)	Land cover based look-up table.	100	Only where land cover categories are identified as disturbed (e.g. cropland mosaics).	Yes
Willcock et al (2012) – Harmonised Tier 2 [42]	1.64 (1.52-1.76)	Land cover based look-up table.	100	Only where land cover categories are identified as disturbed (e.g. cropland mosaics).	Yes
Baccini et al (2012) – Tier 1 [3]	2.03	Derived from MODIS and GLAS LiDAR data.	500	Partially includes disturbance through impacts on canopy heights.	Yes
Saatchi et al (2011) – Tier 1 [4]	0.83	Derived from MODIS, SRTM, QSCAT and GLAS LiDAR.	1000	Partially includes disturbance through impacts on canopy heights.	No
Hurtt et al (2006) HYDE-SAGE – Tier 1 [46]	0.63	Modelled from the Miami LU ecosystem model with cropland data from the Centre for Sustainability and the Global Environment.	~110,000	Contains simple submodels of natural plant mortality, disturbance from fire, and organic matter decomposition, as well as wood harvesting.	No
Hurtt et al (2006) HYDE – Tier 1 [46]	0.41	Modelled from the Miami LU ecosystem model.	~110,000	Contains simple submodels of natural plant mortality, disturbance from fire, and organic matter decomposition, as well as wood harvesting.	No
Baccini et al (2008) – Tier 1 [47]	0.34	Derived from MODIS and GLAS LiDAR data.	1000	Partially includes disturbance through impacts on canopy heights.	No

*This study and Willcock et al (2012) are not independent as they are derived from the same underlying data and utilise the same look-up table values.

close agreement with forward models (Tables 3 and 4; Additional file 1: Tables S1-S3).

Carbon storage (adjusted R-squared [Adj R-sq] = 0.18) is correlated positively with the natural logarithm of the population pressure with decay constant of 12.5 km (p-value < 0.001) and increased by 1 Mg ha^{-1} for every 8700 km from a road (p-value < 0.010), and every 30,000 units in the cost distance to Dar es Salaam (p-value < 0.010). Carbon storage decreased by 1 Mg ha^{-1} for every 1°C increase in mean annual monthly temperature range (p-value < 0.001), every 2.7% rise in the total available water capacity of the soil (p-value < 0.001), and every

Table 2 The mean (and 95% CI) estimates of forest characteristics investigated in this study (carbon storage, carbon sequestration, WSG, the intercept from the power law relationship and the gradient from the power law relationship) separated by land cover category

Land cover category [41]	Carbon storage (Mg ha^{-1})	Carbon sequestration (Mg ha^{-1} yr^{-1})	WSG (g cm^{-3})	The intercept from the power law relationship	The gradient from the power law relationship
Lowland Forest (<1000 m)	182 (152 to 360)	-0.91 (-7.08 to 4.29)	0.60 (0.59 to 0.60)	6.01 (2.94 to 5.17)	-0.93 (-1.04 to -0.82)
Sub-montane forest (1000-1500 m)	189 (95 to 588)	-2.02 (-11.06 to 1.29)	0.58 (0.57 to 0.58)	5.95 (3.68 to 8.23)	-1.31 (-1.48 to -1.14)
Montane Forest (1500-2000 m)	130 (62 to 702)	-2.03 (-11.85 to 1.07)	0.60 (0.59 to 0.60)	6.95 (3.51 to 10.39)	-1.57 (-1.82 to -1.32)
Upper-montane forest (>2000 m)	166 (69 to 533)	-2.08 (-10.49 to 1.23)	0.60 (0.58 to 0.60)	7.03 (4.60 to 9.45)	-1.61 (-1.93 to -1.26)
Forest mosaic	121 (55 to 485)	-1.18 (-6.69 to 2.92)	0.56 (0.56 to 0.56)	9.22 (6.98 to 11.46)	-1.90 (-1.99 to -1.81)
Closed Woodland	100 (70 to 331)	-1.24 (-7.91 to 2.63)	0.64 (06.2 to 0.65)	6.67 (4.95 to 8.60)	-1.55 (-1.85 to -1.30)
Open Woodland	51 (38 to 165)	-1.49 (-7.53 to 2.05)	0.61 (0.59 to 0.62)	6.38 (4.88 to 7.82)	-1.45 (-1.70 to -1.19)

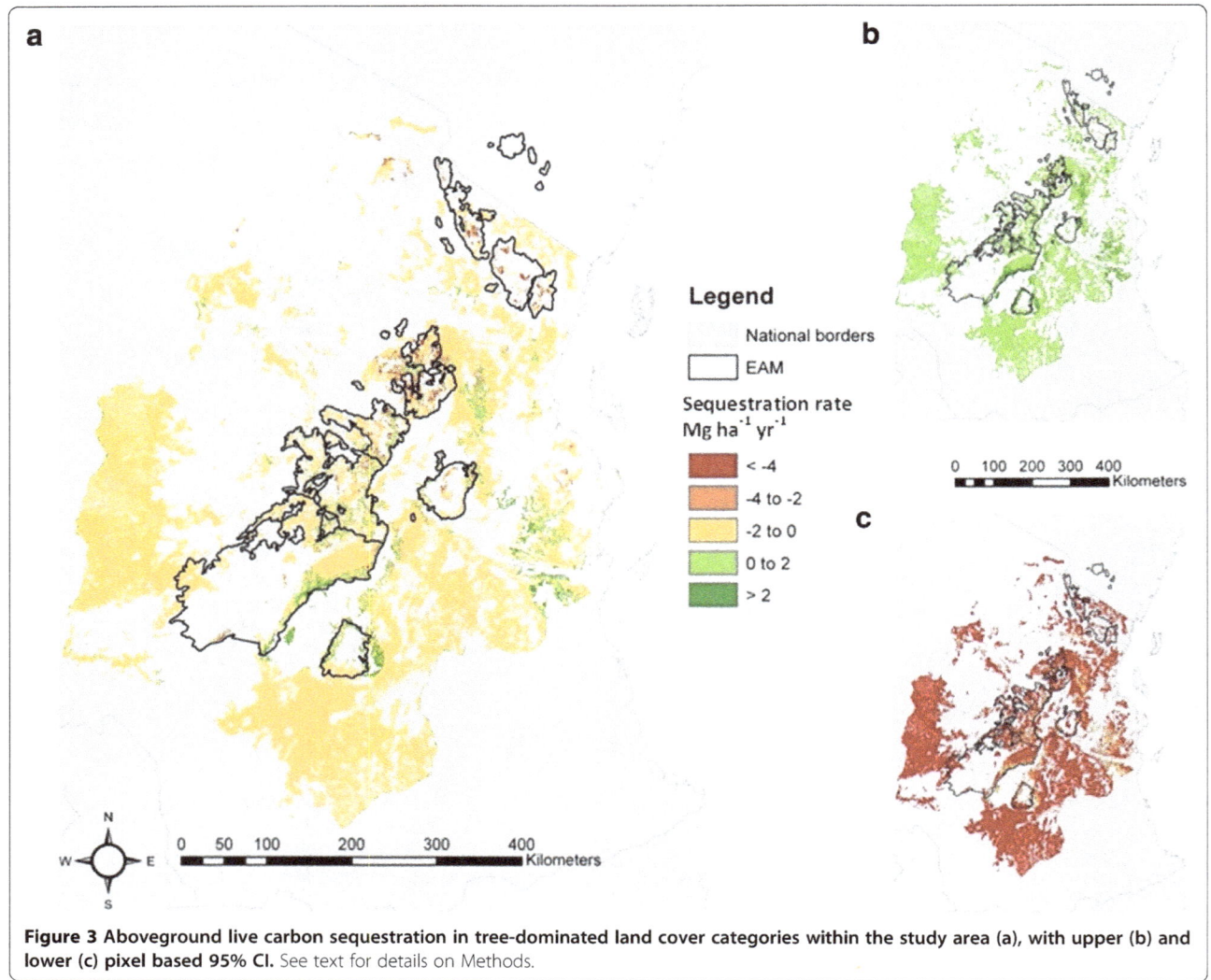

Figure 3 Aboveground live carbon sequestration in tree-dominated land cover categories within the study area (a), with upper (b) and lower (c) pixel based 95% CI. See text for details on Methods.

4.4 month increase in the mean number of dry months annually (p-value < 0.050). Carbon storage was 2.1 Mg ha^{-1} lower in areas where historical logging was present (p-value < 0.010), and 4.2 Mg ha^{-1} higher in areas under the control of local communities/governments (p-value < 0.010). Thus, carbon storage is high in areas far from the commercial capital, with a low monthly temperature range, without a dry season, that have not suffered from historical logging and are under local community/government control (Figure 4; Table 3).

The rate of carbon sequestration correlated with three principal component (PC) axes (presented in order of influence; Adj R-sq = 0.41). Carbon sequestration was negatively correlated with the soil fertility axis (PC5; p-value < 0.050), warmer temperatures and longer dry seasons (PC3; p-value < 0.050), and with increased anthropogenic disturbance (PC1; p-value < 0.010). Thus, carbon sequestration was highest in less fertile areas with

little or no drought and little anthropogenic disturbance (Table 4).

Wood specific gravity (WSG; Adj R-sq = 0.28; see Additional file 1: SI2) was most strongly affected by the annual mean burned area probability (increasing by 1 g cm^{-3} for every 0.04 increase; p-value < 0.001) and the total available water capacity of the soil (decreasing by 1 g cm^{-3} for every 82.0% increase; p-value < 0.001). Thus, WSG is higher in burnt areas with little available water (Additional file 2: Figure S1; Additional file 3: Figure S2; Additional file 1: Table S1).

The intercept of the power law relationship (an indication of potential stem density [see Additional file 1: SI3]; Adj R-sq = 0.30) was most affected by the natural logarithm of the population pressure with decay constant of 12.5 km (positive correlation; p-value < 0.001) and the mean annual monthly temperature range (increasing by 1.0 for every 1.2°C increase; p-value < 0.001). Thus, the

Table 3 The coefficients and associated p-values of the variables correlated with aboveground carbon storage using both forward and backward selection procedures

Variable (where appropriate, units are given in brackets)	Group	Forward		Backward	
		Coefficient	p-value	Coefficient	p-value
(Intercept)	n/a	-1.21E + 03	3.14E-03	-2.80E + 00	7.55E-01
Natural logarithm of the population pressure with decay constant of 12.5 km	Anthropogenic	1.06E + 00	1.06E-05	1.42E + 00	2.27E-06
Natural logarithm of the population pressure with decay constant of 16.7 km	Anthropogenic	n/a	n/a	1.42E + 00	2.27E-06
Distance to roads (km)	Anthropogenic	1.15E-04	1.09E-03	1.78E-04	1.30E-05
Historical logging – Partially logged (no logging/partially logged)	Anthropogenic	-2.10E + 00	1.09E-03	-3.83E + 00	4.97E-07
Cost distance to Dar es Salaam	Anthropogenic	3.41E-05	2.00E-03	2.58E + 00	5.46E-03
Natural logarithm of the cost distance to market towns	Anthropogenic	-6.05E-01	5.24E-02	-9.85E-01	1.89E-02
Governance – local (national/local/joint/unknown)	Anthropogenic	4.24E + 00	9.29E-03	n/a	n/a
Governance – national (national/local/joint/unknown)	Anthropogenic	-7.95E-03	9.78E-01	n/a	n/a
Governance – unknown (national/local/joint/unknown)	Anthropogenic	6.26E-01	7.10E-01	n/a	n/a
Mean annual monthly temperature range (°C)	Climatic	-9.79E-01	2.00E-16	-1.15E + 00	1.98E-13
Mean annual minimum monthly temperature (°C)	Climatic	n/a	n/a	1.09E + 00	3.07E-16
Mean annual maximum monthly temperature (°C)	Climatic	n/a	n/a	-1.15E + 00	1.98E-13
Mean number of dry months annually	Climatic	-2.28E-01	2.57E-02	-3.09E-01	5.58E-03
Total available water capacity of the soil (vol. %, -33 to -1500 kPA conforming to USDA standards)	Edaphic	-3.75E-01	1.16E-05	-8.59E-01	3.05E-05
Total nitrogen content of the soil (g kg^{-1})	Edaphic	n/a	n/a	-4.13E-01	2.50E-03
Total carbon content of the soil (g kg^{-1})	Edaphic	n/a	n/a	6.18E + 00	1.15E-03
pH of the soil (pH)	Edaphic	n/a	n/a	1.73E + 00	2.96E-02
Spatial autocorrelation term 5	Spatial	6.45E + 01	3.15E-03	6.60E + 00	1.18E-01
Spatial autocorrelation term 7	Spatial	-8.48E-01	3.57E-03	-1.71E-01	1.45E-01
Spatial autocorrelation term 4	Spatial	n/a	n/a	6.60E + 00	1.18E-01
Spatial autocorrelation term 3	Spatial	n/a	n/a	-1.71E-01	1.45E-01

density of smaller stems increases in areas with a high population pressure and large temperature fluctuations (Additional file 3: Figure S2; Additional file 4: Figure S3; Additional file 1: Table S2).

Correlations identified for the gradient of the power law relationship (an indication of the proportion of larger stems; see Additional file 1: SI3) were broadly the inverse of those identified for the intercept. The gradient of the power law relationship was most affected by the natural logarithm of the population pressure with decay constant of 20.8 km (negative correlation; p-value < 0.001) and the mean burned area probability in

Table 4 The coefficients and associated p-values of the variables correlated with aboveground carbon sequestration

Variable	Coefficient	p-value
(Intercept)	0.032	0.890
PC1	-0.112	0.006
PC3	-0.255	0.010
PC5	-0.412	0.012

the fourth quarter (decreasing by 1.0 for every 0.2 increase; p-value < 0.001). Thus, the proportion of large stems was greater in areas experiencing few disturbances from people or fire (Additional file 3: Figure S2; Additional file 5: Figure S4; Additional file 1: Table S3).

When investigating the most influential correlates of spatial differences in carbon storage and how these result from changes in either species composition affecting wood density (specific gravity) or the number of large trees present, we found that the final Tier 3 carbon storage estimates were positively correlated with both size-frequency distribution estimates (both intercept and gradient [p-values < 0.001]), and negatively correlated with WSG estimates (p- value < 0.001) and maximum height estimates (p-value < 0.001; Additional file 1: see SI4). All possible interactions were investigated and were significant (Adj R-sq = 0.35; p- values < 0.001), however, the majority of the explanatory power lay within the second order interactions (Adj R-sq = 0.33; p-values < 0.001; Additional file 1: Table S5). Broadly, WSG and the proportion of larger stems had largest influence over the carbon storage estimate. Considering only second order

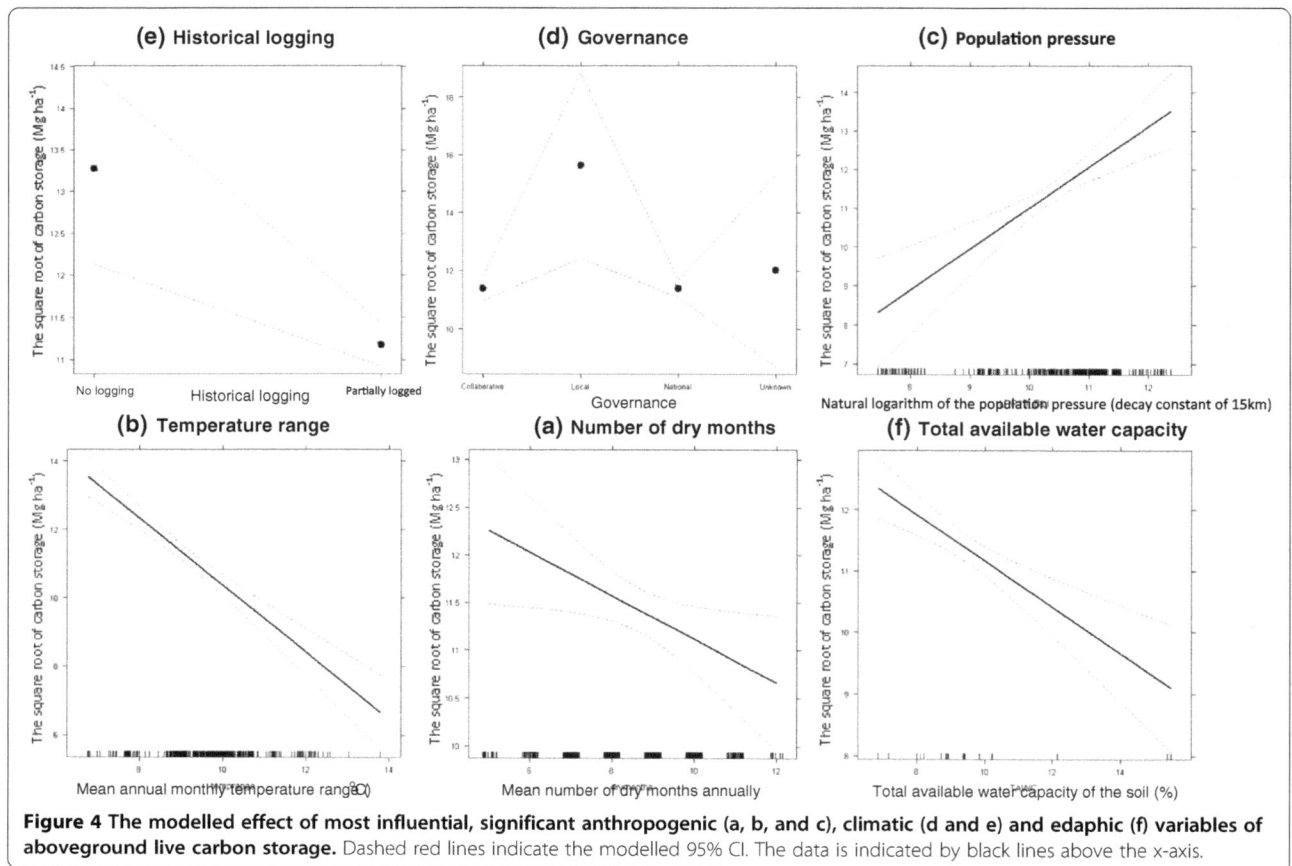

Figure 4 The modelled effect of most influential, significant anthropogenic (a, b, and c), climatic (d and e) and edaphic (f) variables of aboveground live carbon storage. Dashed red lines indicate the modelled 95% CI. The data is indicated by black lines above the x-axis.

interactions, in areas of low potential stem density, carbon storage is positively correlated with maximum canopy height (Additional file 6: Figure S5). However, the opposite correlation is observed in areas of higher stem density. Although similar interactions are observed between both size-frequency distribution estimates (gradient and intercept), the interaction between WSG and maximum canopy height is inverse, with carbon storage only showing positive correlations with maximum canopy height in areas of high WSG. Both size-frequency distribution estimates also interacted similarly with WSG, with both showing positive correlations with carbon storage in areas of low WSG, but negative correlations in areas of high WSG (Additional file 6: Figure S5). Finally carbon sequestration correlation values were positively correlated with carbon storage estimates (p-value < 0.001), indicating that areas storing the most carbon are also those that are increasing in stock at the fastest rate.

Discussion

Tier 3 correlation-based method vs. Tier 1 and 2 methods

Our estimates of 1.3 Pg C stored across the 33.9 million hectares is larger than most previous Tier 1 estimates [46-48], although below the most recently produced estimate [3] (Table 1). Underestimation of the amount of

carbon stored in the EAM region in global analyses can be a result of their poor resolution and/or application of data from other regions which may differ systematically compared to East African forests, woodlands and savannas [42]. When separated by land cover category, our locally derived carbon estimates are comparable to those presented in other local [49-52] and global studies, the latter often containing little or no data from East Africa [3,4,46,47,53]. This suggests differences between our estimates and other studies have arisen because many previous studies mapped carbon storage at lower resolution [3,4,46,47,53]. When considering homogenous landscapes, scale effects are unlikely to cause a dramatic difference in carbon estimates. However, in highly fragmented and heterogeneous landscapes, such as East Africa, the effects of scale are likely to be substantial. Forest fragments, typically of high carbon storage, may be omitted at lower resolutions, being 'replaced' by more dominant, but low carbon, land cover categories (e.g. open woodland), resulting in underestimation of carbon storage.

It must be noted that, the landscape-scale confidence intervals surrounding our Tier 3 estimates are considerably wider than those around previous estimates [3,4,42,47,53]. This result is consistent with Hill et al (2013), who also showed increasing methodological sophistication does not necessarily result in reduced uncertainty, as is often assumed [54]. Confidence intervals derived from look-up table values may show a systematic bias. The ranges provided are an artefact of the study area, the number of land cover categories and the resolution, as when summed across a large number of pixels, pixel error is mostly negated as underestimates in one part of the landscape are counterbalanced by overestimates in other parts. The 95% CI developed from correlation equations are effectively based on numerous continuous variables, containing the uncertainty relating to anthropogenic, climatic and edaphic variables, thus have many thousands of possible combinations, severely limiting the ability of the 'law of averages' to act. Hence, the 95% CI presented in this investigation may better reflect that of the actual landscape, containing more variables that make-up the complex landscape heterogeneity (i.e. improved representativeness), although this is only true for those pixels estimated using the correlation equations (86% of the EAM but only 52% of the study area). Therefore, the look-up table 95% CI presented in Willcock et al (2012), and used in this study, may underestimate uncertainty [42]. Future studies should expand the existing plot network (Figure 1), enabling the correlation equations (and improved 95% CI) to be applied to the entire study area. This process has already begun under a new WWF-REDD+ project (which focusses on better sampling the data-deficient land cover categories identified in this study [55]) and the National Forest Monitoring and Assessment (NAFORMA) project [56,57].

Links between carbon stock and influential variables

The results presented here indicate that ALC storage in tree-dominated ecosystems is correlated with anthropogenic, climatic and edaphic variables. However, in all our models there is a large amount of unexplained variation (R-squared values for our correlation models vary between 0.18 and 0.41). This is likely to be due to three main reasons (Additional file 1: SI6). Firstly, although we used the highest resolution datasets that are freely available, several of the associated variables are of relatively poor resolution across the EAM (including; wind, light and soil nutrient variables [Additional file 1: Table S6]). This is particularly important here as low resolution GIS data is unlikely to correlate well with the response variables from our plot network as many plots (with high variance [58]) may fall within a single cell [59]. Thus, our study may be biased against retaining low resolution explanatory variables in our models. Secondly, contemporary forest characteristics are the result of growth, recruitment and mortality over many years. It is difficult to obtain data on historical variables and yet these could have had a significant impact on present day carbon storage and other forest characteristics [60]. Thirdly, present day information is also lacking, for example

datasets describing physical soil properties in the study area are unavailable. Thus, future work is needed to develop additional high resolution GIS data, particularly for historic time periods.

Of the variance explained in our forward and backward models, direct anthropogenic factors are the most influential explanatory variables (as noted by the largest coefficient of explanatory variables on the response variable, in contrast to those [e.g. temperature] with smaller p-values but also smaller influence [Table 3]) and so are the focus of our remaining discussion (see Additional file 1: SI5 for discussion of climatic and edaphic variables).

Within our study area, people are clustered around high carbon areas (Figure 4). We suggest this could be due to these areas having favourable climatic conditions with more moisture for plant (and thus crop) growth. Further, the incidence of malaria is lower at high elevations [61], making these locations more habitable for human populations. Thus there is a peak in population density near the base of high-carbon montane forests [40]. Our interpretation that it is the landscape suitability driving human population density is consistent with the observation that when individual localities are followed over time, degradation at the local level caused by the population is evident [62,63]. This emphasises that our results are not proof of causation and that the drivers may be a correlate of the explanatory variables retained in our models (Additional file 1: SI6). Our results also show a decrease in carbon storage in previously logged areas and in areas nearer the commercial capital, Dar es Salaam. This confirms previous reports that areas near the capital have lower biomass due to the local demand of low grade timber by the city, as well as international demand for high grade timber via the city's port [62]; emphasising the connections between the rural and urban landscape, and how the sphere of urban influence drives change in rural ecosystems. Future investigations should use simulation modelling and direct experimentation to identify if the influential variables highlighted here can be confirmed as drivers of carbon storage and sequestration, providing a deeper understanding of the process-based relationships.

The decrease in carbon storage as a result of logging (51-77% of the ALC is retained) is of similar magnitude to other reported estimates [64]. However, the historical logging data we utilised was based on expert opinion (Additional file 1: Table S6) so, given its importance, further work developing and evaluating historical variables is needed (Additional file 1: Table S7). We observe a comparable decrease due to differing governance. Land under national control holds between 40% and 65% of the ALC stored in areas under decentralised governance. This perhaps indicates that decentralisation of management (e.g. participatory and community led forestry) is successful in our study area [37,65]. However, it is not possible to prove causation within the framework of this study. Many locally managed forests are located in the south-east of our study area within an area of naturally high carbon storage, whereas land under national control covers much larger areas, including the dry, carbon-poor east. Hence, our finding that carbon storage is higher in areas under decentralised control may be an artefact of the differing areas where this type of land management occurs. Further studies monitoring change in carbon storage over time under the two different governance regimes would enable the effect of land management to be determined.

The overall effects on carbon storage are a result of many changes in forest characteristics. Both WSG and the proportion of larger stems decrease with increasing anthropogenic disturbance, however, stem density ($\geq$ = 10 cm DBH) increases. Anthropogenic disturbance, for example logging, is often a commercial activity and results in the preferential removal of the largest, most valuable stems [62]. The more open canopy, following stem removal, would result in increased recruitment from young forest trees [66], leading to the high numbers of small stems observed. However, the opposite would be expected in woodlands and savannas, with more open canopies resulting in more grass, high fire intensity and so less recruitment [67,68]. Our results highlight how influential the negative effect of people on tropical forest carbon storage can be. This assertion is supported by data from across the tropics [69-71]. The significant impact of anthropogenic activities implies that REDD+ could, at the local scale, have significant positive impacts on carbon storage. However, careful policy designs to limit leakage of deforestation and encourage the involvement of the local population are needed to ensure REDD+ schemes achieve their carbon storage and sequestration aims [72].

Like carbon storage and its components, carbon sequestration is also correlated with anthropogenic, climatic and edaphic variables. We estimate that some localities (for example the Udzungwa Mountains National Park; Figure 4) provide a carbon sink of comparable per-area magnitude to modelled estimates in East Africa [73] and to that observed over recent decades in structurally intact African forest [7]. However, many areas of forest and woodland within the study area experience a high level of degradation and disturbance, and so are net sources. Here, we have shown that anthropogenic disturbance is a key determinant of the trend in carbon storage over time in eastern Tanzania. Important locations of high carbon losses are the Pare and Usambara mountains (Table 5), which historically have seen the highest rates of degradation and disturbance [74]. The national population of Tanzania is increasing

Table 5 Carbon stored and sequestered across the individual mountain blocks of the EAM range (the total is denoted in bold)

Eastern Arc Mountain Block [40]	Area, km²	Aboveground live carbon storage, Tg		Mean carbon sequestration, Mg ha⁻¹ yr⁻¹
		Tier 3	Willcock et al (2012) - Original Tier 2 [42]	
North Pare	510	1.93	2.38	2.60
South Pare	2,327	8.96	9.59	2.41
West Usambara	2,945	13.52	15.96	3.64
East Usambara	1,145	5.91	7.63	2.79
Nguu	1,562	9.34	12.71	1.89
Nguru	2,565	15.11	18.86	1.79
Ukaguru	3,243	13.39	20.63	1.42
Uluguru	3,057	15.92	13.91	1.35
Rubeho	7,984	36.84	40.96	1.06
Malundwe	33	0.29	0.29	1.80
Udzungwa	22,788	101.73	104.05	1.01
Mahenge	2,606	23.58	12.08	0.19
Total	**50,765**	**246.53**	**259.06**	**1.19**

[75] and this may increase the pressure on tree-dominated ecosystems which could result in the study area becoming a significant source of carbon in the future. Furthermore, the effect of increase in anthropogenic pressures could be compounded by potential decrease in carbon storage as a result of increasing temperatures [76,77] and changes in soil nutrients (see Additional file 1: SI5). However, these future effects could be complicated by increasing levels of atmospheric CO_2, varying effectiveness of legally protected areas and shifting consumption patterns.

Conclusions

Our results show that the amount of carbon stored in forests across 33.9 million ha of the Eastern Arc Mountains of Tanzania is considerable: 1.32 (0.89 to 3.16) Pg. Our estimate is significantly higher than most previous estimates. However, our more sophisticated method also has higher uncertainty, implying that other methods may substantially underestimate the uncertainty involved. Within the tree-dominated land cover categories, historical logging is the most influential direct anthropogenic factor, while the mean number of dry months is the most influential environmental factor, with an order of magnitude less impact on carbon storage. We show that WSG, size-frequency distribution variables and height variables are all important in determining carbon storage. Our estimates indicate that, between 2004 and 2008, tree-dominated communities across the study areas showed no significant change, however some areas were identified as large sinks (0.8% of the study area) and others large sources (1.4% of the study area), showing the importance of taking a landscape scale approach. The carbon maps produced and statistical relationships documented can assist policy-makers in designing policies to maintain and enhance carbon storage for climate mitigation and other ecosystem services.

Method

We collated data from 2,462 tree inventory plots within our study area (see Additional file 1: SI3), then applied a quality control and standardisation protocol. This consists of two main steps: (1) Metadata quality control; and (2) Measurement bias detection.

Firstly, all plots lacking a recorded spatial location and a fixed area were discarded (770 plots). Plots where one or more diameter at breast height (DBH) data were known to be missing were also excluded (7 plots). Furthermore, plots smaller than 0.025 ha (16 plots) were deemed to produce unreliable carbon estimates so also removed from the dataset.

Secondly, to assess possible measurement bias, i.e. not measuring over buttresses and so overestimating biomass [78], the remaining plots were grouped by the lead field researcher. Size-frequency distributions, using 10 cm size classes, were created for each of these groups. Forest size-frequency distributions are suggested to conform to the -2 power law based on metabolic scaling [79]. Although it has been argued that this rule is not globally applicable [80], many studies accept this as a theoretical maximum value for the abundance of large stems [81]. Thus, researchers with many plots above this maximum value likely measured stems around buttresses and so were removed (1 researcher, 100 Plots).

The quality control and standardisation procedure resulted in a dataset of 1,611 tree inventory plots (median 0.1 ha, mean 0.1 ha, mode 0.1 ha [43 plots with multiple censuses; median 0.1 ha, mean 0.5 ha, mode 1.0 ha]; Figure 1; see Additional file 1: SI3 for a further information) from which we calculated plot-level stand structure indices and aboveground carbon storage per unit area (see Additional file 1: SI2 for full details). We obtained the exponent and intercept of the population size-frequency distribution using the power law fit for each plot using the log-log transformation method. Whereby, for each plot, we created 10 cm bin size-frequency distributions based on DBH, and a linear model of the logarithm of frequency against the logarithm of the size class was fitted. Whilst not as accurate as the maximum likelihood estimation method, our simpler method is more stable for many of our plots, providing both the intercept and slope indicators of population structure [82].

We obtained WSG data via the phylogenetic information provided by our tree inventory plots. We used a global wood density database to extract species average WSG [83]. This procedure provided over 32,000 trees with WSG data. When this was not possible we adopted a hierarchical approach, first applying the appropriate genus average if available (~14,000 trees) before considering family average (~9,500 trees), plot average (~4,500 trees) and dataset average (~80 trees) in turn [84]. Including WSG as an additional parameter in allometric equations reduces the biomass estimation error [49,85,86].

In addition, we estimated plot biomass using moist forest tree allometry [86] based on measurements of DBH from our tree inventory plots, WSG (as described above) and height data (derived from our dataset using the best fit DBH-height equation form [Equation 5.1; see Additional file 1: SI4], if not measured in the tree inventory plots). Finally, carbon was assumed to be 50% of biomass [7].

For a smaller number of plots, multiple measurements were available over time (n = 43; mean plot size = 0.5 ha; mean measurement period = 3.9 years). We calculated changes in carbon storage rates by dividing the difference in carbon storage estimates between censuses by the number of years separating them.

For our 1,611 geo-referenced tree inventory plots, we obtained further information on variables falling into five broad categories; anthropogenic, climatic, geographic, edaphic, and pyrologic (median resolution 1.0 ha, mean resolution 22.0 ha, mode resolution 1.0 ha; Additional file 1: Table S6). Anthropogenic data, further divided into six subcategories, were obtained: (1) population pressure variables (n = 14 related variables) were obtained from Platts (2012) [87] (see Additional file 1: SI7); (2) Dar es Salaam related variables (n = 3; e.g. distance to Dar es Salaam), (3) market town related variables (n = 3; e.g. distance to market towns), and (4) infrastructure related variables (n = 2; e.g. distance to roads) were derived from available topographic maps; (5) historical logging (n = 1) from Swetnam et al (2011) [88]; and (6) governance (n = 1) from the World Database on Protected Areas [89]. Climate data were divided into three subcategories (precipitation [n = 2; maximum mean cumulative water deficit and mean number of dry months annually], temperature [n = 4; mean annual temperature, mean annual minimum monthly temperature, mean annual monthly maximum temperature, and mean annual monthly temperature range] and wind speed [n = 1]) and were derived from the Tropical Rainfall Measuring Mission [90,91], WorldClim [92,93], and United States National Aeronautics and Space Administration Surface meteorology and Solar Energy [94] datasets. Similarly, geographic data have two variables (aspect [n = 1] and incoming solar radiation [n = 1]) derived from Shuttle Radar Topography Mission [93] and National Renewable Energy

Laboratory [95,96] datasets respectively. Lastly, we extracted edaphic data (n = 6) from the International Soil Reference and Information Centre database [97,98] and fire-related variables (n = 5) derived from MODIS images [99].

We then correlated these variables with carbon storage, and following this, its components: WSG, the intercept of the power law relationship, and the gradient of the power law relationship, in each case using general linear models (see Additional file 1: SI2-5). No transformations were required to ensure a normal distribution when correlating either WSG, the intercept of the power law relationship or the gradient of the power law relationship with the individual variables. However, carbon storage estimates required a square root transformation to ensure a normal distribution within the general linear models (normality was confirmed using the Shapiro-Wilk test; p-value > 0.05). In all models, plots were weighted by the square root of their area as confidence in biomass estimation increases with the area surveyed [100,101]. Landscape scale spatial autocorrelation was accounted for by including spatial terms (latitude, longitude and the interactions between them) in the model (Additional file 1: Table S6) [102]. The numerous possible interactions were excluded from the models, as these were found to add very little explanatory power to the models, only increasing R-squared values by ~0.001 with the addition of each interaction term. All analyses were performed using R 2.12.1 [103] and mapped in ArcGIS v9.3.1 [104].

When assessing carbon sequestration (n = 43) fewer degrees of freedom were available, therefore explanatory variables need to be grouped. Therefore, we conducted a principle components (PC) analysis, obtaining five PC which explained >90% of the cumulative variance of the individual influential variables (Additional file 1: Table S4). Then, covariation of PC with carbon sequestration was assessed instead of the individual influential variables. Carbon sequestration estimates required a cube-root transformation to ensure a normal distribution within the general linear models (confirmed using the Shapiro-Wilk test; p-value > 0.05). This enabled the effect of multiple variables to be examined even with this limited dataset. PC analysis of the variables was performed on the scaled data using the prcomp package [105] within R 2.12.1 [103]. All other aspects of the model (weighting and spatial autocorrelation) were performed identically to the models for carbon storage and its components.

The most appropriate model was chosen using forward and backward stepwise selection. Forward models are more useful for inferring causal relationships [43] and so were preferentially used to infer the influential variables of carbon storage and sequestration. However,

averaging forward–backwards and backward–forwards predictions outperforms conventional selection procedures [43] and so both methods were used when estimating the spatial distributions within the study area. Akaike information criterion (AIC) was used to reduce/expand the models, with variable selection occurring when the variable reduced the mean squared error (MSE) under ten-fold cross validation [106]. Unlike model selection using R-squared, which neglects the principles of parsimony, AIC considers both model fit and complexity, resulting in better predictions and allowing inferences to be made from multiple models [107]. Model selection continued until the addition/removal of further variables able to reduce cross validation MSE no longer increased AIC, thereby producing the best-fit model with the lowest prediction error [43].

Within each category (anthropogenic, climatic, geographic, edaphic, and pyrologic), some variables were highly correlated (Additional file 1: Table S7) and this may confound the stepwise procedure as each variable does not carry enough distinct information [108]. For example, all temperature related variables (Additional file 1: Table S7) were correlated (R-squared > 0.6). However, it is unclear which correlated best with the variables of interest, e.g. carbon storage and sequestration. Many studies include mean annual temperature in biomass models [77,109], but theory suggests that it may be the temperature range driving this relationship as photosynthesis correlates with maximum temperatures, but respiration with minimum temperatures [76,110,111]. We found that, if we removed correlated variables prior to model selection, the final models were artefacts of the variables we had selected. For example, if we included mean annual temperature in the model, but not temperature range, then the significant correlations between mean annual temperature and ALC storage were found. However, these correlations were insignificant if temperature range was added to the model, with the newly added variable showing a significant effect instead. In short, the resultant models were automatically biased towards *a priori* expectations. To avoid this bias, we devised a procedure by which the influential variables included in model selection were selected by their ability to explain variation within the data of interest (e.g. carbon storage). All variables (describe above) were included in model selection. Once this had run to completion the model was assessed. The subcategory with the most correlated variables retained within the model was selected and all but the most influential, significant variable were removed. For example, if all four temperature-related variables were included in the initial model and this was the largest group of variables then this group would be selected. Then, if mean annual temperature was the most influential and significant temperature-related variable, all other temperature-related variables would be excluded in the next round of model selection. Thus, stepwise model selection was then repeated for all remaining variables. This process was repeated until no highly correlated variables remained within the model produced.

Since only landscape-scale variation was accounted for by the spatial terms already included in the model (latitude, longitude and the interactions between them; Table 1; Additional file 1: Table S6), it was necessary to investigate the effect of local-scale (<10 km^2) spatial autocorrelation [102]. To do this, the separate forward and backward models, containing no highly correlated variables (produced above), were mapped. Then, the sum of the model estimates within the maps were extracted at 1, 3, 5, 7 and 10 km^2 resolutions, and included as additional variables (representing local spatial autocorrelation terms) into the stepwise model selection process, which was re-run a final time [112]. However, in all cases, local spatial autocorrelation terms were rejected as they did not reduce cross validated MSE.

Since it was not necessary to include local spatial autocorrelation terms in the models, the preliminary maps produced above could be regarded as final spatial representations of the ten best fit models, two (forward and backward) for each of the five variables of interest (carbon storage, carbon sequestration, WSG, the intercept of the power law relationship and the gradient of the power law relationship). Each pair of maps (forward and backward) were then combined into a single, final weighted mean estimate. The ratio of the relevant cross validated MSE of the forward and backward models was used to create the weighted mean, with the model showing lowest error receiving the highest weighting [43]. Thus, we ultimately produced five maps (from ten best fit models); one each for carbon storage, carbon sequestration, WSG, the intercept of the power law relationship, and the gradient of the power law relationship. As our carbon storage estimates were derived from data representing trees with a DBH greater than or equal to 10 cm, regionally estimates of ratios from Willcock et al (2012) were used to estimate the unmeasured component of ALC storage [42], this was summed with our modelled carbon storage estimate, providing an estimate of total ALC storage.

Although the five maps produced covered the entire study area, we were concerned that extrapolating predictions beyond the range of observed predictor variables from our dataset could result in large, unquantifiable errors. Thus, we limited the models to localities where all the associate variables were within the range of that shown in our dataset, thus only interpolating within our correlation models for tree-dominated land cover categories. For any pixels outside the data range, look-up table methods were used in preference to the correlation

model estimates. Thus, for every land cover in our study area containing trees (open woodland; closed woodland; forest mosaic; lowland forest; sub-montane forest; montane forest; and upper montane forest [41]) that fell within the limits of our dataset, the estimate of carbon storage derived from the correlation equations was used. For all other land cover categories, and for those localities for which predictor variables fell outside the ranges of values used in model construction, land cover based look-up table values from Willcock et al (2012) were used to estimate ALC storage [42]. In total, look-up table values were applied to 52% of the landscape, although this was predominantly to low carbon land cover categories, with 86% of the EAM (which hold the majority of the regions tropical forest [113]) estimated using the correlation approach described above. Estimates of WSG and population structure were only made for wooded land cover categories, with estimates for areas within our dataset range being derived from the relevant correlation equations and estimates for other areas coming from land cover based look-up table values derived from the median value of our WSG and population structure data (weighted by the square root of plot size and derived via sampling with replacement 10,000 times) for each land cover category (Additional file 1: Table S8). For carbon sequestration, again, estimates were only made for wooded land cover categories for those areas inside the range of our dataset estimates derived from the correlation equations were used. However, unlike carbon storage, WSG and population structure, for areas outside the range of our dataset, a land cover based look-up table was not used as several land cover categories were poorly represented due to the small sample size available (n = 43). Instead, for pixels outside the range of the correlation-derived carbon sequestration model (16% of pixels with wooded land cover), the median value of data from our recensused plots (again weighted by the square root of plot size and derived via sampling with replacement 10,000 times) was utilised.

For every 1 ha pixel of each map derived from correlation equations, we produced 95% confidence intervals (CI). If the pixel estimate was derived from the general linear models, then the pixel 95% CI was calculated by adding and subtracting the square root of the cross validation MSE. For look-up table pixels the look up table 95% CI were used. The pixel 95% CI describes, for every pixel, the range we would expect each of our estimates to lie within. However, as we are also interested in estimating carbon storage and sequestration on a landscape scale, indications of uncertainty are also required at landscape-scale. Simply summing the pixel 95% CI to derive 95% CI of the overall landscape-scale estimates would incorrectly treat random error as a region-wide systematic bias. Thus, to derive 95% CI for landscape-scale estimates, we randomly allocated each pixel an estimate within the range dictated by its 95% pixel CI, and summed these values across the entire landscape. This process was performed 10,000 times and the median value and 95% CI (the 250th and 9,750th ranked values, which may not be equally distributed around the median) for aboveground carbon storage and sequestration in the study area were obtained.

For the final model of carbon storage estimates, we investigated how the components of carbon storage (population structure, WSG and tree height) interacted to ultimately produce the ecosystem service of carbon storage. We obtained estimates of maximum canopy height from the best fit DBH-height equation [Equation 5.1; see Additional file 1: SI4 and Additional file 7: Figure S6], and combined this spatially with our correlation model derived estimates of WSG, the intercept of the power law relationship and gradient of the power law relationship. We then correlated these against our estimates of carbon storage, allowing all possible interactions, and selected the best-fit model (via AIC) using both forwards and backwards stepwise regression.

Ethical approval for the above study was obtained from the Faculty of Environment Research Ethics Committee, in accordance with the University of Leeds research ethics policy.

Additional files

> **Additional file 1: Supporting text (including S1-7 and Tables S1-12).**
>
> **Additional file 2: Figure S1.** The spatial variation of WSG in tree-dominated land cover categories within the study area (a), with upper (b) and lower (c) pixel based 95% CI. See text for details on methods.
>
> **Additional file 3: Figure S2.** The most influential, significant influential variables on WSG (a and b), the intercept of the power law relationship (c and d), and the gradient of the power law relationship (e and f). Dashed red lines indicate 95% CI.
>
> **Additional file 4: Figure S3.** The spatial variation in the intercept of the power law relationship (a proxy measure for potential stem density) in tree dominated land cover categories within the study area (a), with upper (b) and lower (c) pixel based 95% CI. See text for details on methods.
>
> **Additional file 5: Figure S4.** The spatial variation in the gradient of the power law relationship (a proxy measure for the proportion of larger stems) in tree-dominated land cover categories within the study area (a), with upper (b) and lower (c) pixel based 95% CI. See text for details on methods.
>
> **Additional file 6: Figure S5.** The 2nd order interactions relating my carbon storage derivatives (wood specific gravity, maximum canopy height, the intercept of the power law relationship, and the gradient of the power law relationship [shown here as WSG, height, intercept, and gradient respectively]) to aboveground live carbon storage. Dashed red lines indicate 95% CI.
>
> **Additional file 7: Figure S6.** The effect of MAT on tree height for a range of DBH. The data (points) correspond to DBH ranges whereas the Gompertz model fits (solid lines) illustrate the relationship for mid-point of this range only. Dotted lines represent the 95CI of the model fits.

Abbreviations
AIC: Akaike information criteria; ALC: Aboveground live carbon; CI: Confidence interval; CV: Cross validation; CWD: Coarse woody debris; DBH: Diameter at breast height; EAM: Eastern Arc Mountains; eCEC: Effective cation exchange capacity; GIS: Geographic information systems; HYDE: History Database of the Global Environment; IPCC: Intergovernmental Panel on Climate Change; IUCN: International Union for Conservation of Nature; KITE: York Institute for Tropical Ecosystems; MAT: Mean annual temperature; MSE: Mean squared error; PC: Principal components; REDD+: Reducing Emissions from Deforestation and Forest degradation; SAGE: Centre for Sustainability and Global Environment; WSG: Wood specific gravity.

Competing interests
The authors declare that they have no competing interests.

Authors' contributions
The majority of this jointly-authored publication was led by SW. Contributions to the collaborative dataset came from PJP, AA, ND, KD, EF, JG, JH, KH, ARM, BM, PKTM, NO, EJTJ, AM, SW and RDS. The analysis was performed by SW, supervised by OLP and SLL. The manuscript was prepared by SW, with assistance from OLP, SLL, AB, PP, NDD and RM. All authors read and approved the final manuscript.

Acknowledgements
This study is part of the Valuing the Arc research programme (http://valuingthearc.org/) funded by the Leverhulme Trust (http://www.leverhulme.ac.uk/). Manuscript preparation and later analyses took place under the 'Which Ecosystem Service Models Best Capture the Needs of the Rural Poor?' project (WISER; NE/L001322/1), funded with support from the United Kingdom's Ecosystem Services for Poverty Alleviation program (ESPA; www.espa.ac.uk). ESPA receives its funding from the Department for International Development (DFID), the Economic and Social Research Council (ESRC) and the Natural Environment Research Council (NERC). SLL was funded by a Royal Society University Research Fellowship; SW additionally by the Stokenchurch Charity; OLP was supported by an Advanced Grant from the European Research Council and is a Royal Society-Wolfson Research Award holder. The funders had no role in study design, data collection and analysis, decision to publish, or preparation of the manuscript. We thank the Tanzanian Commission for Science and Technology (COSTECH), the Tanzanian Wildlife Institute (TAWIRI) and the Sokoine University of Agriculture for their support of this work, as well as all the field assistants involved. Furthermore, we would like to thank the two anonymous reviewers, whose comments and insight vastly improved the manuscript.

Author details
[1]School of Geography, University of Leeds, Leeds LS2 9JT, UK. [2]School of Biological Sciences, University of Southampton, Southampton SO17 1BJ, UK. [3]Environment Department, University of York, York YO10 5DD, UK. [4]Department of Zoology, University of Cambridge, Cambridge CB2 3EJ, UK. [5]WWF US, Washington, USA. [6]UNEP World Conservation Monitoring Centre, Cambridge CB3 0DL, UK. [7]Genetics and Conservation, Royal Botantic Garden Edinburgh, Edinburgh, UK. [8]Tanzanian Forest Conservation Group, Dar es Salaam, Tanzania. [9]Frankfurt Zoological Society, Frankfurt D-60316, Germany. [10]The Society for Environmental Exploration, London EC2A 3QP, UK. [11]STEP Program, Princeton University, Princeton 08544, USA. [12]Department of Geography, University of Florida, PO Box 117315, Gainesville, Florida, FL 32611, USA. [13]The University of Dar es Salaam, Dar es Salaam, Tanzania. [14]Centre for the Integration of Research, Conservation and Learning, Flamingo Land Ltd, Malton YO 17 6UX, UK. [15]Sokoine University of Agriculture, PO Box 3001, Morogoro, Tanzania. [16]EDGE of Existence, Conservation Programmes, Zoological Society of London, London, UK. [17]Department of Geography, Staffordshire University, Stoke-on-Trent ST4 2DF, UK. [18]Department of Bioscience, Aarhus University, Aarhus C DK-8000, Denmark. [19]Department of Geography, University College London, London WC1E 6BT, UK.

References
1. Malhi Y, Grace J: **Tropical forests and atmospheric carbon dioxide.** *Trends Ecol Evol* 2000, 15:332–337.
2. IPCC: **Land use, land-use change, and forestry.** Cambridge, UK: Cambridge University Press; 2000.

3. Baccini A, Goetz SJ, Walker WS, Laporte NT, Sun M, Sulla-Menashe D, Hackler J, Beck PSA, Dubayah R, Friedl MA, Samanta S, Houghton RA: Estimated carbon dioxide emissions from tropical deforestation improved by carbon-density maps. Nat Clim Change 2012, 2:182–185.

4. Saatchi SS, Harris NL, Brown S, Lefsky M, Mitchard ETA, Salas W, Zutta BR, Buermann W, Lewis SL, Hagen S, Petrova S, White L, Silman M, Morel A: Benchmark map of forest carbon stocks in tropical regions across three continents. Proc Natl Acad Sci 2011, 108:9899–9904.

5. Myers N, Mittermeier RA, Mittermeier CG, da Fonseca GAB, Kent J: Biodiversity hotspots for conservation priorities. Nature 2000, 403:853–858.

6. Timko JA, Waeber PO, Kozak RA: The socio-economic contribution of non-timber forest products to rural livelihoods in Sub-Saharan Africa: knowledge gaps and new directions. Int For Rev 2010, 12:284–294.

7. Lewis SL, Lopez-Gonzalez G, Sonke B, Affum-Baffoe K, Baker TR, Ojo LO, Phillips OL, Reitsma JM, White L, Comiskey JA, Djuikouo M-N, Ewango CEN, Feldpausch TR, Hamilton AC, Gloor M, Hart T, Hladik A, Lloyd J, Lovett JC, Makana J-R, Peacock J, Peh KS-H, Sheil D, Sunderland T, Swaine M, Taplin J, Taylor D, Thomas C, Votere R, Wöll H: Increasing carbon storage in intact African tropical forests. Nature 2009, 457:1003–1006.

8. Phillips OL, Malhi Y, Higuchi N, Laurance WF, Nunez PV, Vasquez RM, Laurance SG, Ferreira LV, Stern M, Brown S, Grace J: Changes in the carbon balance of tropical forests: evidence from long-term plots. Science 1998, 282:439–442.

9. Achard F, Eva HD, Stibig H-J, Mayaux P, Gallego J, Richards T, Malingreau J-P: Determination of deforestation rates of the world's humid tropical forests. Science 2002, 297:999–1002.

10. Hansen MC, Stehman SV, Potapov PV: Quantification of global gross forest cover loss. Proc Natl Acad Sci 2010, 107:8650–8655.

11. Putz FE, Zuidema PA, Pinard MA, Boot RGA, Sayer JA, Sheil D, Sist P, Elias M, Vanclay JK: Improved Tropical Forest Management for Carbon Retention. PLoS Biol 2008, 6:e166.

12. IPCC: Climate Change 2007: The Physical Science Basis. Available at http://www.ipcc.ch/pdf/assessment-report/ar4/wg1/ar4-wg1-spm.pdf [Accessed 05/01/12]. Agenda 2007, 6.

13. van der Werf GR, Morton DC, DeFries RS, Olivier JGJ, Kasibhatla PS, Jackson RB, Collatz GJ, Randerson JT: CO2 emissions from forest loss. Nat Geosci 2009, 2:737–738.

14. Dixon RK, Borwn S, Houghton RA, Solomon AM, Trexler MC, Wisniewski J: Carbon pools and fluxes of global forest ecosystems. Science 1994, 263:185–190.

15. DeFries RS, Houghton RA, Hansen MC, Field CB, Skole D, Townshend J: Carbon emissions from tropical deforestation and regrowth based on satellite observations for the 1980s and 1990s. Proc Natl Acad Sci U S A 2002, 99:14256–14261.

16. Houghton RA: TRENDS: A Compendium of Data on Global Change. Available at: http://cdiac.ornl.gov/trends/landuse/houghton/houghton. html. An update of estimated carbon emissions from land use change, typically updated every year. Oak Ridge National Laboratory, US Department of Energy; 2008.

17. Pan Y, Birdsey RA, Fang J, Houghton R, Kauppi PE, Kurz WA, Phillips OL, Shvidenko A, Lewis SL, Canadell JG, Ciais P, Jackson RB, Pacala SW, McGuire AD, Piao S, Rautiainen A, Sitch S, Hayes D: A large and persistent carbon sink in the world's forests. Science 2011, 333:988–993.

18. Hansen MC, Stehman SV, Potapov PV, Loveland TR, Townshend JRG, DeFries RS, Pittman KW, Arunarwati B, Stolle F, Steininger MK, Carroll M, DiMiceli C: Humid tropical forest clearing from 2000 to 2005 quantified by using multitemporal and multiresolution remotely sensed data. Proc Natl Acad Sci 2008, 105:9439–9444.

19. Chave J, Condit R, Muller-Landau HC, Thomas SC, Ashton PS, Bunyavejchewin S, Co LL, Dattaraja HS, Davies SJ, Esufali S: Assessing evidence for a pervasive alteration in tropical tree communities. PLoS Biol 2008, 6:e45.

20. Field CB, Behrenfeld MJ, Randerson JT, Falkowski P: Primary production of the biosphere: integrating terrestrial and oceanic components. Science 1998, 281:237–240.

21. REDD overview: REDD overview. [http://www.un-redd.org/]

22. Strassburg B, Turner RK, Fisher B, Schaeffer R, Lovett A: Reducing emissions from deforestation—The "combined incentives" mechanism and empirical simulations. Glob Environ Chang 2009, 19:265–278.

23. Kindermann G, Obersteiner M, Sohngen B, Sathaye J, Andrasko K, Rametsteiner E, Schlamadinger B, Wunder S, Beach R: Global cost estimates

24. Cattaneo A, Lubowski R, Busch J, Creed A, Strassburg B, Boltz F, Ashton R: On international equity in reducing emissions from deforestation. Environ Sci Pol 2010, 13:742–753.

25. Farris M: The sound of falling trees: integrating environmental justice principles into the climate change framework for Reducing Emissions from Deforestation and Degradation (REDD). Fordham Environ Law Rev 2010, 20:515.

26. Birdsey R, Pan Y, Houghton R: Sustainable landscapes in a world of change: tropical forests, land use and implementation of REDD+: Part II. Carbon Manag 2013, 4:567–569.

27. IPCC: Guidelines for National Greenhouse Gas Inventories. In National Greenhouse Gas Inventories Programme. Edited by Eggleston HS, Buendia L, Miwa K, Ngara T, Tanabe K. Kanagawa, Japan: Institute for Global Environmental Strategies; 2006.

28. IPCC: Good practice guidance for land use, land-use change and forestry. In IPCC National Greenhouse Gas Inventories Programme. Edited by Penman J, Gytarsky M, Hiraishi T, Krug T, Kruger D, Pipatti R, Buendia L, Miwa K, Ngara T, Tanabe K. Kanagawa, Japan: Institute for Global Environmental Strategies; 2003.

29. GOFC-GOLD A: A Sourcebook of Methods and Procedures for Monitoring and Reporting Anthropogenic Greenhouse Gas Emissions and Removals Caused by Deforestation, Gains and Losses of Carbon Stocks in Forests Remaining Forests, and Forestation. AB, Canada: GOFC-GOLD GOFC-GOLD Project Office, Natural Resources Canada Calgary; 2010.

30. Asner GP, Powell GVN, Mascaro J, Knapp DE, Clark JK, Jacobson J, Kennedy-Bowdoin T, Balaji A, Paez-Acosta G, Victoria E, Secada L, Valqui M, Hughes RF: High-resolution forest carbon stocks and emissions in the Amazon. Proc Natl Acad Sci 2010, 107:16738–16742.

31. International W: Carbon Storage in the Los Amigos Conservation Concession, Madre de Dios, Perú. Washington DC, USA: Winrock International; 2006.

32. Asner GP, Mascaro J, Clark JK, Powell G: Reply to Stoke et al.: Regarding high-resolution carbon stocks and emissions in the Amazon. Proc Natl Acad Sci 2011, 108:E13–E14.

33. Cláudia Dias A, Louro M, Arroja L, Capela I: Comparison of methods for estimating carbon in harvested wood products. Biomass Bioenergy 2009, 33:213–222.

34. Pedroni L, Dutschke M, Streck C, Porrúa ME: Creating incentives for avoiding further deforestation: the nested approach. Clim Pol 2009, 9:207–220.

35. Hardcastle P, Baird D, Harden V, Abbot PG, O'Hara P, Palmer JR, Roby A, Haüsler T, Ambia V, Branthomme A, Wilkie M, Arends E, González C: Capability and Cost Assessment of the Major Forest Nations to Measure and Monitor their Forest Carbon. Edinburg, Scotland: LTS International Ltd; 2008.

36. Romijn E, Herold M, Kooistra L, Murdiyarso D, Verchot L: Assessing capacities of non-Annex I countries for national forest monitoring in the context of REDD+. Environ Sci Pol 2012, 19–20:33–48.

37. Burgess ND, Bahane B, Clairs T, Danielsen F, Dalsgaard S, Funder M, Hagelberg N, Harrison P, Haule C, Kabalimu K, Kilahama F, Kilawe E, Lewis SL, Lovett LC, Lyatuu G, Marshall AR, Meshack C, Miles L, Milledge SAH, Munishi PKT, Nashanda E, Shirima D, Swetnam R, Willcock S, Williams A, Zahabu E: Getting ready for REDD+ in Tanzania: a case study of progress and challenges. Oryx 2010, 44:339–351.

38. Maniatis D, Malhi Y, Andr LS, Mollicone D, Barbier N, Saatchi S, Henry M, Tellier L, Schwartzenberg M, White L: Evaluating the Potential of Commercial Forest Inventory Data to Report on Forest Carbon Stock and Forest Carbon Stock Changes for REDD+ under the UNFCCC. Int J Forest Res 2011, 2011:1–13.

39. Science for a changing world. http://www.usgs.gov/default.asp.

40. Platts PJ, Burgess ND, Gereau RE, Lovett JC, Marshall AR, Mcclean CJ, Pellikka PKE, Swetnam RD, Marchant R: Delimiting tropical mountain ecoregions for conservation. Environ Conserv 2011, 38:312–324.

41. Swetnam RD, Fisher B, Mbilinyi BP, Munishi PKT, Willcock S, Ricketts T, Mwakalila S, Balmford A, Burgess ND, Marshall AR, Lewis SL: Mapping socio-economic scenarios of land cover change: a GIS method to enable ecosystem service modelling. J Environ Manag 2011, 92:563–574.

42. Willcock S, Phillips OL, Platts PJ, Balmford A, Burgess ND, Lovett JC, Ahrends A, Bayliss J, Doggart N, Doody K, Fanning E, Green JMH, Hall J, Howell KL, Marchant R, Marchant R, Marshall A, Mbilinyi B, Munishi PKT, Owen N, Swetnam RD, Topp-Jorgensen EJ, Lewis SL: Towards regional, error-bounded

landscape carbon storage estimates for data-deficient areas of the world. *PLoS ONE* 2012, **7**:e44795.

43. Platts PJ, McClean CJ, Lovett JC, Marchant R: **Predicting tree distributions in an East African biodiversity hotspot: model selection, data bias and envelope uncertainty.** *Ecol Model* 2008, **218**:121–134.

44. Stuart EA: **Matching methods for causal inference: A review and a look forward.** *Stat Sci Rev J Inst Math Stat* 2010, **25**:1.

45. Wiegand RE: **Performance of using multiple stepwise algorithms for variable selection.** *Stat Med* 2010, **29**:1647–1659.

46. Hurtt GC, Frolking S, Fearon MG, Moore B, Shevliakova E, Malyshev S, Pacala SW, Houghton RA: **The underpinnings of land-use history: three centuries of global gridded land-use transitions, wood-harvest activity, and resulting secondary lands.** *Glob Chang Biol* 2006, **12**:1208–1229.

47. Baccini A, Laporte N, Goetz SJ, Sun M, Dong H: **A first map of tropical Africa's above-ground biomass derived from satellite imagery.** *Environ Res Lett* 2008, **3**:045011.

48. Saatchi SS, Houghton RA, Dos Santos Alvalá RC, Soares JV, Yu Y: **Distribution of aboveground live biomass in the Amazon basin.** *Glob Chang Biol* 2007, **13**:816–837.

49. Marshall AR, Willcock S, Lovett JC, Balmford A, Burgess ND, Latham JE, Munishi PKT, Platts PJ, Salter R, Shirima DD, Lewis SL: **Measuring and modelling above-ground carbon storage and tree allometry along an elevation gradient.** *Biol Conserv* 2012, **154**:20–33.

50. Munishi PKT, Shear TH: **Carbon storage in afromontane rain forests of the Eastern Arc Mountains of Tanzania: their net contribution to atmospheric carbon.** *J Trop For Sci* 2004, **16**:78–93.

51. Shirima DD, Munishi PKT, Lewis SL, Burgess ND, Marshall AR, Balmford A, Swetnam RD, Zahabu EM: **Carbon storage, structure and composition of miombo woodlands in Tanzania's Eastern Arc Mountains.** *Afr J Ecol* 2011, **49**:332–342.

52. Pfeifer M, Platts PJ, Burgess ND, Swetnam RD, Willcock S, Lewis SL, Marchant R: **Land use change and carbon fluxes in East Africa quantified using earth observation data and field measurements.** *Environ Conserv* 2013, **40**:1–12. FirstView.

53. Ruesch A, Gibbs HK: **New IPCC Tier1 Global Biomass Carbon Map For the Year 2000.** Oak Ridge, Tennessee: Oak Ridge National Laboratory; 2008. Available online from the Carbon Dioxide Information Analysis Center http://cdiac.ornl.gov/ [Accessed 15/01/12].

54. Hill TC, Williams M, Bloom AA, Mitchard ETA, Ryan CM: **Are inventory based and remotely sensed above-ground biomass estimates consistent?** *PLoS ONE* 2013, **8**:e74170.

55. Burgess ND, Mwakalila S, Munishi P, Pfeifer M, Willcock S, Shirima D, Hamidu S, Bulenga GB, Rubens J, Machano H, Marchant R: **REDD herrings or REDD menace: response to Beymer-Farris and Bassett.** *Glob Environ Chang* 2013, **23**:1349–1354.

56. Kessy JF, Anderson K, Dalsgaar S: **NAFORMA Field Manual: Socioeconomic survey.** Dar es Salaam, Tanzania: Forestry and Beekeeping Division, Ministry of Natural Resources and Tourism; 2010:96.

57. Vesa L, Malimbwi RE, Tomppo E, Zahabu E, Maliondo S, Chamuya N, Nssoko E, Otieno J, Miceli G, Kaaya AK, Dalsgaar S: **NAFORMA Field Manual: Biophysical survey.** Dar es Salaam, Tanzania: Forestry and Beekeeping Division, Ministry of Natural Resources and Tourism; 2010:96.

58. Chave J, Condit R, Lao S, Caspersen JP, Foster RB, Hubbell SP: **Spatial and temporal variation of biomass in a tropical forest: results from a large census plot in Panama.** *J Ecol* 2003, **91**:240–252.

59. Baccini A, Friedl M, Woodcock C, Zhu Z: **Scaling field data to calibrate and validate moderate spatial resolution remote sensing models.** *Photogramm Eng Remote Sens* 2007, **73**:945.

60. Ramankutty N, Gibbs HK, Achard F, Defries R, Foley JA, Houghton RA: **Challenges to estimating carbon emissions from tropical deforestation.** *Glob Chang Biol* 2007, **13**:51–66.

61. Balls MJ, Bødker R, Thomas CJ, Kisinza W, Msangeni HA, Lindsay SW: **Effect of topography on the risk of malaria infection in the Usambara Mountains, Tanzania.** *Trans R Soc Trop Med Hyg* 2004, **98**:400–408.

62. Ahrends A, Burgess ND, Milledge SAH, Bulling MT, Fisher B, Smart JCR, Clarke GP, Mhoro BE, Lewis SL: **Predictable waves of sequential forest degradation and biodiversity loss spreading from an African city.** *Proc Natl Acad Sci* 2010, **107**:14556–14561.

63. Bayon G, Dennielou B, Etoubleau J, Ponzevera E, Toucanne S, Bermell S: **Intensifying weathering and land use in iron age Central Africa.** *Science* 2012, **335**:1219–1222.

64. Putz FE, Zuidema PA, Synnott T, Peña-Claros M, Pinard MA, Sheil D, Vanclay JK, Sist P, Gourlet-Fleury S, Griscom B, Palmer J, Zagt R: **Sustaining conservation values in selectively logged tropical forests: the attained and the attainable.** *Conserv Lett* 2012, **5**:296–303.

65. Topp-Jørgensen E, Poulsen MK, Lund JF, Massao JF: **Community-based monitoring of natural resource use and forest quality in Montane Forests and Miombo Woodlands of Tanzania.** *Biodivers Conserv* 2005, **14**:2653–2677.

66. Silva JNM, de Carvalho JOP, Lopes JCA, de Almeida BF, Costa DHM, de Oliveira LC, Vanclay JK, Skovsgaard JP: **Growth and yield of a tropical rain forest in the Brazilian Amazon 13 years after logging.** *Forest Ecol Manag* 1995, **71**:267–274.

67. Govender N, Trollope WSW, Van Wilgen BW: **The effect of fire season, fire frequency, rainfall and management on fire intensity in savanna vegetation in South Africa.** *J Appl Ecol* 2006, **43**:748–758.

68. Hoffmann WA: **The effects of fire and cover on seedling establishment in a Neotropical Savanna.** *J Ecol* 1996, **84**:383–393.

69. Chhatre A, Agrawal A: **Trade-offs and synergies between carbon storage and livelihood benefits from forest commons.** *Proc Natl Acad Sci* 2009, **106**:17667–17670.

70. Chhatre A, Agrawal A: **Forest commons and local enforcement.** *Proc Natl Acad Sci* 2008, **105**:13286–13291.

71. Mbwambo L, Eid T, Malimbwi RE, Zahabu E, Kajembe GC, Luoga E: **Impact of decentralised forest management on forest resource conditions in Tanzania.** In *Forests, Trees and Livelihoods*; 2012:1–17.

72. Fisher B, Lewis SL, Burgess ND, Malimbwi RE, Munishi PK, Swetnam RD, Kerry Turner R, Willcock S, Balmford A: **Implementation and opportunity costs of reducing deforestation and forest degradation in Tanzania.** *Nat Clim Chang* 2011, **1**:161–164.

73. Doherty RM, Sitch S, Smith B, Lewis SL, Thornton PK: **Implications of future climate and atmospheric CO2 content for regional biogeochemistry, biogeography and ecosystem services across East Africa.** *Glob Chang Biol* 2009, **16**:617–640.

74. Willcock S: **Long-term Changes in Land Cover and Carbon Storage in Tanzania, East Africa.** In *University of Leeds, School of Geography*; 2012.

75. NBS: **Tanzania Census 2002 - Analytical Report Volume X.** Dar es Salaam: National Bureau of Statistics, Ministry of Planning, Economy and Empowerment; 2003.

76. Clark DA, Piper SC, Keeling CD, Clark DB: **Tropical rain forest tree growth and atmospheric carbon dynamics linked to interannual temperature variation during 1984-2000.** *Proc Natl Acad Sci U S A* 2003, **100**:5852–5857.

77. Raich JW, Russell AE, Kitayama K, Parton WJ, Vitousek PM: **Temperature influences carbon accumulation in moist tropical forests.** *Ecology* 2006, **87**:76–87.

78. Phillips OL, Malhi Y, Vinceti B, Baker T, Lewis SL, Higuchi N, Laurance WF, Vargas PN, Martinez RV, Laurance S, Ferreira LV, Stern M, Brown S, Grace J: **Changes in growth of tropical forests: evaluating potential biases.** *Ecol Appl* 2002, **12**:576–587.

79. Enquist BJ, Niklas KJ: **Invariant scaling relations across tree-dominated communities.** *Nature* 2001, **410**:655–660.

80. Li H-T, Han X-G, Wu J-G: **Lack of evidence for 3/4 scaling of metabolism in terrestrial plants.** *J Integr Plant Biol* 2005, **47**:1173–1183.

81. Enquist BJ, West GB, Brown JH: **Extensions and evaluations of a general quantitative theory of forest structure and dynamics.** *Proc Natl Acad Sci* 2009, **106**:7046–7051.

82. Goldstein ML, Morris SA, Yen GG: **Problems with fitting to the power-law distribution.** *Eur Phys J B Condens Matter Complex Syst* 2004, **41**:255–258.

83. Zanne AE, Lopez-Gonzalez G, Coomes DA, Ilic J, Jansen S, Lewis SL, Miller RB, Swenson NG, Wiemann MC, Chave J: **Global wood density database.** 2009. http://hdl.handle.net/10255/dryad.235 [Accessed 5/12/2008].

84. Baker TR, Phillips OL, Malhi Y, Almeida S, Arroyo L, Di Fiore A, Erwin T, Killeen TJ, Laurance SG, Laurance WF, Lewis SL, Lloyd J, Monteagudo A, Neill DA, Patiño S, Pitman NCA, Silva JNM, Vásquez Martínez R: **Variation in wood density determines spatial patterns in Amazonian forest biomass.** *Glob Chang Biol* 2004, **10**:545–562.

85. Djomo AN, Ibrahima A, Saborowski J, Gravenhorst G: **Allometric equations for biomass estimations in Cameroon and pan moist tropical equations including biomass data from Africa.** *Forest Ecol Manag* 2010, **260**:1873–1885.

86. Chave J, Andalo C, Brown S, Cairns MA, Chambers JQ, Eamus D, Folster H, Fromard F, Higuchi N, Kira T, Lescure J-P, Nelson BW, Ogawa H, Puig H,

Riéra B, Yamakura T: **Tree allometry and improved estimation of carbon stocks and balance in tropical forests.** *Oecologia* 2005, **145**:87–99.

87. Platts PJ: **Spatial Modelling, Phytogeography and Conservation in the Eastern Arc Mountains of Tanzania and Kenya.** In *University of York, Environment Department*; 2012.

88. Swetnam RD: **Historical logging in protected areas, E. Tanzania. V2.** Cambridge, UK: Zoology Department, Cambridge University; 2011.

89. IUCN, UNEP-WCMC: **The World Database on Protected Areas (WDPA).** Cambridge, UK: UNEP- WCMC; 2010. Available at: www.protectedplanet.net [Accessed 03/05/2010)].

90. Zomer RJ, Trabucco A, Bossio DA, Verchot LV: **Climate change mitigation: A spatial analysis of global land suitability for clean development mechanism afforestation and reforestation.** *Agr Ecosyst Environ* 2008, **126**:67–80.

91. **Tropical Rainfall Measuring Mission.** http://trmm.gsfc.nasa.gov/.

92. Hijmans RJ, Cameron SE, Parra JL, Jones PG, Jarvis A: **Very high resolution interpolated climate surfaces for global land areas.** *Int J Climatol* 2005, **25**:1965–1978.

93. **Hole-filled SRTM for the globe Version 4, from the CGIAR-CSI SRTM 90m Database.** http://srtm.csi.cgiar.org.

94. **Wind Speed At 50 m Above The Surface Of The Earth.** http://eosweb.larc.nasa.gov/sse/.

95. Perez R, Ineichen P, Moore K, Kmiecik M, Chain C, George R, Vignola F: **A new operational model for satellite-derived irradiances: description and validation.** *Sol Energy* 2002, **73**:307–317.

96. **Low Resolution Solar Data.** http://www.nrel.gov/gis/.

97. Batjes NH: **SOTER-based soil parameter estimates for Southern Africa.** Volume 4. Wageningen: ISRIC - World Soil Information; 2004:27.

98. ISRIC: **SOTER and WISE-based soil property estimates for Southern Africa.** 2010. Available at http://www.isric.org/ [Accessed 17/2/2010].

99. Roy DP, Jin Y, Lewis PE, Justice CO: **Prototyping a global algorithm for systematic fire-affected area mapping using MODIS time series data.** *Remote Sens Environ* 2005, **97**:137–162.

100. Houghton RA, Lawrence KT, Hackler JL, Brown S: **The spatial distribution of forest biomass in the Brazilian Amazon: a comparison of estimates.** *Glob Chang Biol* 2001, **7**:731–746.

101. Clark DB, Clark DA: **Landscape-scale variation in forest structure and biomass in a tropical rain forest.** *Forest Ecol Manag* 2000, **137**:185–198.

102. Dormann CF, McPherson JM, Araújo MB, Bivand R, Bolliger J, Carl G, Davies RG, Hirzel A, Jetz W, Daniel Kissling W, Kühn I, Ohlemüller R, Peres-Neto PR, Reineking B, Schröder B, Schurr FM, Wilson R: **Methods to account for spatial autocorrelation in the analysis of species distributional data: a review.** *Ecography* 2007, **30**:609–628.

103. R Development Core Team: **R: A Language and Environment for Statistical Computing.** In *R Foundation for Statistical Computing.* Vienna, Austria: R Foundation for Statistical Computing; 2010.

104. ESRI: **ArcMap 9.3.1.** 1999-2009.

105. Venables WN, Ripley BD: *Modern Applied Statistics with S.* Oxford, UK: Springer; 2002.

106. Varma S, Simon R: **Bias in error estimation when using cross-validation for model selection.** *BMC Bioinforma* 2006, **7**:91.

107. Johnson JB, Omland KS: **Model selection in ecology and evolution.** *Trends Ecol Evol* 2004, **19**:101–108.

108. Chong I-G, Jun C-H: **Performance of some variable selection methods when multicollinearity is present.** *Chemom Intell Lab Syst* 2005, **78**:103–112.

109. Asner G, Flint Hughes R, Varga T, Knapp D, Kennedy-Bowdoin T: **Environmental and Biotic Controls over Aboveground Biomass Throughout a Tropical Rain Forest.** *Ecosystems* 2009, **12**:261–278.

110. Lloyd J, Farquhar GD: **Effects of rising temperatures and CO$_2$ on the physiology of tropical forest trees.** *Phil Trans Roy Soc B Biol Sci* 2008, **363**:1811–1817.

111. Graham EA, Mulkey SS, Kitajima K, Phillips NG, Wright SJ: **Cloud cover limits net CO2 uptake and growth of a rainforest tree during tropical rainy seasons.** *Proc Natl Acad Sci* 2003, **100**:572–576.

112. Maggini R, Lehmann A, Zimmermann NE, Guisan A: **Improving generalized regression analysis for the spatial prediction of forest communities.** *J Biogeogr* 2006, **33**:1729–1749.

113. Burgess ND, Butynski TM, Cordeiro NJ, Doggart NH, Fjeldsa J, Howell KM, Kilahama FB, Loader SP, Lovett JC, Mbilinyi B, Menegon M, Moyer DC, Nashanda E, Perkin A, Rovero F, Stanley WT, Stuart SN: **The biological importance of the Eastern Arc Mountains of Tanzania and Kenya.** *Biol Conserv* 2007, **134**:209–231.

Modelling forest carbon stock changes as affected by harvest and natural disturbances. I. Comparison with countries' estimates for forest management

Roberto Pilli[1]*, Giacomo Grassi[1], Werner A. Kurz[2] , Raúl Abad Viñas[1] and Nuria Hue Guerrero[1]

Abstract

Background: According to the post-2012 rules under the Kyoto protocol, developed countries that are signatories to the protocol have to estimate and report the greenhouse gas (GHG) emissions and removals from forest management (FM), with the option to exclude the emissions associated to natural disturbances, following the Intergovernmental Panel on Climate Change (IPCC) guidelines. To increase confidence in GHG estimates, the IPCC recommends performing verification activities, i.e. comparing country data with independent estimates. However, countries currently conduct relatively few verification efforts. The aim of this study is to implement a consistent methodological approach using the Carbon Budget Model (CBM) to estimate the net CO_2 emissions from FM in 26 European Union (EU) countries for the period 2000–2012, including the impacts of natural disturbances. We validated our results against a totally independent case study and then we compared the CBM results with the data reported by countries in their 2014 Greenhouse Gas Inventories (GHGIs) submitted to the United Nations Framework Convention on Climate Change (UNFCCC).

Results: The match between the CBM results and the GHGIs was good in nine countries (i.e. the average of our results is within ±25 % compared to the GHGI and the correlation between CBM and GHGI is significant at $P < 0.05$) and partially good in ten countries. When the comparison was not satisfactory, in most cases we were able to identify possible reasons for these discrepancies, including: (1) a different representation of the interannual variability, e.g. where the GHGIs used the stock-change approach; (2) different assumptions for non-biomass pools, and for CO_2 emissions from fires and harvest residues. In few cases, further analysis will be needed to identify any possible inappropriate data used by the CBM or problems in the GHGI. Finally, the frequent updates to data and methods used by countries to prepare GHGI makes the implementation of a consistent modeling methodology challenging.

Conclusions: This study indicates opportunities to use the CBM as tool to assist countries in estimating forest carbon dynamics, including the impact of natural disturbances, and to verify the country GHGIs at the EU level, consistent with the IPCC guidelines. A systematic comparison of the CBM with the GHGIs will certainly require additional efforts—including close cooperation between modelers and country experts. This approach should be seen as a necessary step in the process of continuous improvement of GHGIs, because it may help in identifying possible errors and ultimately in building confidence in the estimates reported by the countries.

Keywords: Net CO_2 emissions, Greenhouse gas inventories, European countries, Carbon Budget Model, Forest management, Harvest, Natural disturbances

*Correspondence: roberto.pilli@jrc.ec.europa.eu
[1] European Commission, Joint Research Centre, Institute for Environment and Sustainability, Via E. Fermi 2749, 21027 Ispra, VA, Italy
Full list of author information is available at the end of the article

Background

The United Nations Framework Convention on Climate Change (UNFCCC) and its Kyoto protocol (KP) recognize the role of forests in mitigating climate change. Emissions and removals from forests are included in the greenhouse gas inventories (GHGIs) submitted annually by developed countries to the UNFCCC, and typically represent by far the most important component of the "Land use, Land-use Change and Forestry" (LULUCF) sector. Inventories should follow the methodological guidance prepared by the Intergovernmental Panel on Climate Change (IPCC).

The forests in the European Union (EU, including 28 countries) cover about 165 Mha, they increased by about 4 % since 1990 and about 83 % of this area is available for wood supply [1]. According to the EU GHGI, between 1990 and 2012 the average annual sink of EU forests was about 435 Tg CO_{2eq}. year^{-1}, or about 9 % of the EU total emissions in the same period [2].

For the first commitment period of the KP (CP1, 2008–2012) the accounting of emissions and removals was mandatory for afforestation/reforestation and deforestation (AR and D, i.e. forest land-use changes since 1990) and voluntary for forest management (FM, i.e. forest existing before 1990). For the second commitment period of the KP (CP2, 2013–2020), significant revisions of accounting rules were agreed [3], as reflected in the latest IPCC guidance [4]. The major changes for the forest sector are: (1) the accounting of FM is now mandatory; (2) the FM accounting shall include the carbon (C) stock changes in the harvested wood products (HWP) pool; and (3) emissions and subsequent removals from natural disturbances may be excluded from the accounting. These changes represent new challenges for countries when developing their GHGIs.

Since the GHGIs represent the basis for assessing the effectiveness of any national climate policy, building confidence in their accuracy is of key importance for advancing the international efforts to mitigate climate change. While the GHGIs are subject to an UNFCCC expert review process, which aims to assess the adherence of GHGIs to IPCC guidance in terms of general reporting principles,[1] this expert review does not include an independent verification of the reported estimates. The verification activities should be performed by each country, as part of the process of improving the GHGI and build confidence in its reliability [5]. However, at the EU level, few countries report efforts or results of verification for the LULUCF sector [2]. In most cases, a real verification is very difficult due to the lack of truly independent and comparable data. For example, since GHGIs cover only emissions and removals from managed lands, an inherent mismatch

exists for LULUCF between GHGIs and estimates based on process studies or atmospheric methods [5]. As alternative, a largely independent comparison may be conducted between GHGIs and large-scale models (e.g. [6, 7]) that use data from National Forest Inventories (NFIs). While not a fully-independent verification, such comparisons may be very useful in building confidence in GHGI estimates and trends, improving scientific knowledge and identifying potential problems. The major challenges for this approach are to implement a model capable to reflect the latest IPCC guidance (e.g., including the HWP and natural disturbances, [4]) and to use adequate input data from the countries.

The general aim of this study is to implement a consistent methodological approach using an internationally well established forest carbon budget model to simulate for the period 2000–2012 the impacts of harvest and salvage logging, natural disturbances and land-use changes on forest CO_2 emissions and removals in all EU countries for which adequate information was available (26 countries out of 28). To this aim, the Carbon Budget Model (CBM) developed by the Canadian Forest Service [8] was used, as part of a broader effort for a comprehensive modelling framework for the forest sector [9]. The model was applied and validated at regional and national scales in Canada [10, 11] and Russia [12]. Furthermore, the CBM was successfully adapted to specific forest management conditions in Europe (e.g. uneven-aged forests, [13]), validated at regional level [14] and applied in one country case to estimate the C balance for FM [13] and AR [15].

Specific objectives of this paper are: (1) to validate the CBM against totally independent data available at the country level (for one case study) and to provide a detailed description of four representative country cases; (2) to compare FM estimates from CBM with each country's GHGI in terms of trends and levels of net CO_2 emissions for each forest C pools (living biomass, dead organic matter (DOM) and mineral soil); (3) to analyze how the main drivers affecting the living biomass (harvest and natural disturbances, including major storms and fires) affect the estimates obtained with the CBM and the GHGIs.

A companion paper [16] provides an analysis of the CBM results at the aggregate EU level, including net CO_2 emissions in the HWP pool and the impacts of forest-land use changes.

Results
Model validation

To validate our model's results with independent data sources (i.e., not used as input data by CBM), we first compared the mean annual increment and the average volume estimated by CBM (based on the equations applied by the model during the run and the values of

---

[1] i.e. Transparency, accuracy, completeness, consistency, comparability.

merchantable C stock provided for each species) with the additional, independent data reported in the Lithuanian GHGI (Fig. 1). A further comparison is made with the dead tree stems volume estimated by CBM and the values reported by NIR, based on a specific analysis until 2001 and on NFI permanent sample plots from 2002 to 2012.

The model's results can be further compared with other information for Lithuania, not fully independent of the input data used by CBM, because derived by the same

data sources (i.e., NFI). Figure 2 reports the age class evolution estimated by CBM between 1996 and 2012, compared with the original age class distribution reported by NFI 2004–2008 (attributed to 2006).

In Fig. 3 (lower panel), the net CO_2 emissions estimated by CBM (further distinguished between living biomass, DOM and soil pools) are compared with the net emissions reported by the country's GHGI (in 2014) for the land use category forest land remaining forest land (FL–FL) (Lithuania, [18]). For Lithuania, our simulation starts in 1996 when, due to the effect of insect disturbances (see the Additional file: 1 for further details), we estimated a C source, consistent with the data reported by the country, and with the mean annual volume increment reported in Fig. 1. From 1997 to 2001, the model estimates an increasing C sink, mainly due to a reduction of the regular harvest, because of the salvage of logging residues. From 2002, the C sink decreases due to the increasing harvest demand (reported in the upper panel of Fig. 3) and, after 2007, the sink again increases following the decreasing amount of harvest. Further inter-annual variations are due to the effect of storms (in 2005 and 2007, according to the information by NIR), while the effect of fires is negligible. The interannual variability in net CO_2 emissions reported in the GHGI is considerably larger than estimated in the CBM. From 2007, the forest C sink reported by the country strongly increases, from -1.9 Mt CO_2 in 2007 to -8.0 Mt CO_2 in 2008 (i.e., $+300$ %), even if the total harvest demand decreases only slightly, from about 8.1 to 8.0 million m^3 (i.e., -1.2 %). This reduction was not observed in CBM results, which report only a slightly increase in the C sink between 2007 and 2012, which is consistent with the decreasing harvest rates.

CBM results vs country GHGIs

Net CO_2 emission estimates for the period 2000–2012 as estimated using the CBM and as reported by 26 EU countries in their 2014 GHGI show a wide range of patterns (Fig. 4). Data are for the area subject to FM[2] and are reported from an atmospheric perspective, where negative values represent a sink (CO_2 removals) and positive values a source (CO_2 emissions). Results focus on CO_2 and exclude organic soils. Non-CO_2 emissions (CH_4, N_2O) from forests may be important only for specific countries, in case of drained organic soils (not included in this paper) and in case of fires, for which we report results in terms of CO_2-eq for Portugal in the Additional file: 1.

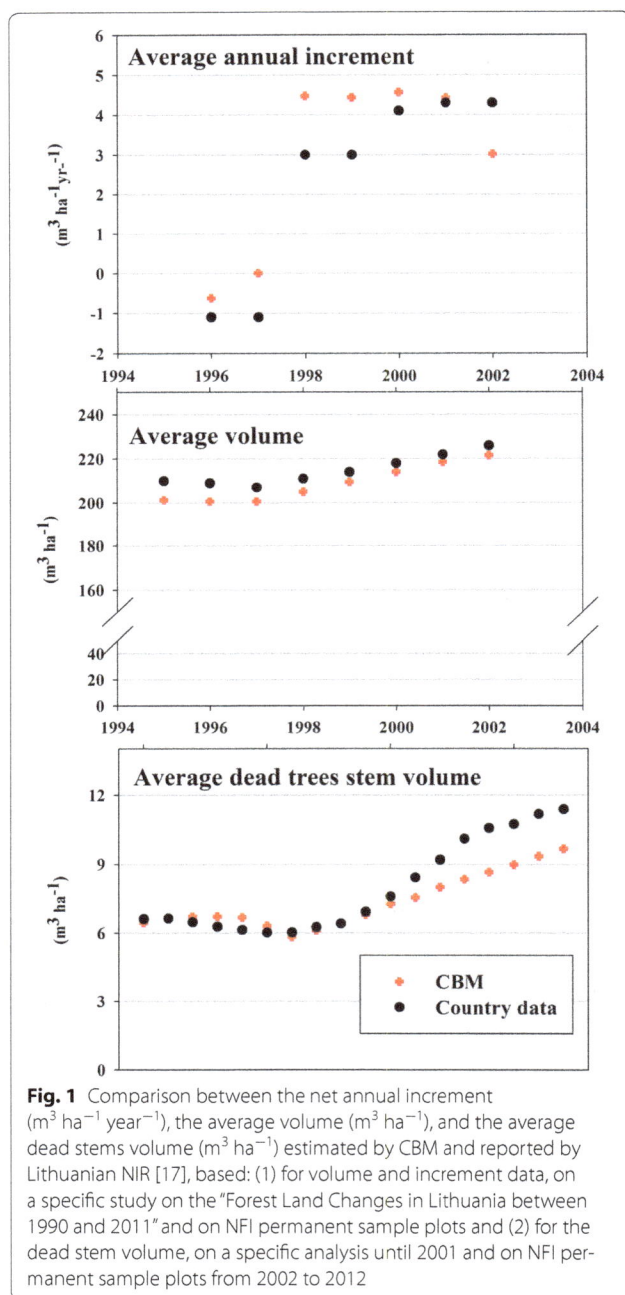

Fig. 1 Comparison between the net annual increment (m^3 ha^{-1} year^{-1}), the average volume (m^3 ha^{-1}), and the average dead stems volume (m^3 ha^{-1}) estimated by CBM and reported by Lithuanian NIR [17], based: (1) for volume and increment data, on a specific study on the "Forest Land Changes in Lithuania between 1990 and 2011" and on NFI permanent sample plots and (2) for the dead stem volume, on a specific analysis until 2001 and on NFI permanent sample plots from 2002 to 2012

---

[2] When available, FM country data from the KP-CRF tables was used for 2008-2012 (i.e., if FM had been elected during the first KP commitment period); alternatively, country data were taken from the Convention CRF tables using 'forest land remaining forest land' (FL remaining FL) as a proxy for FM.

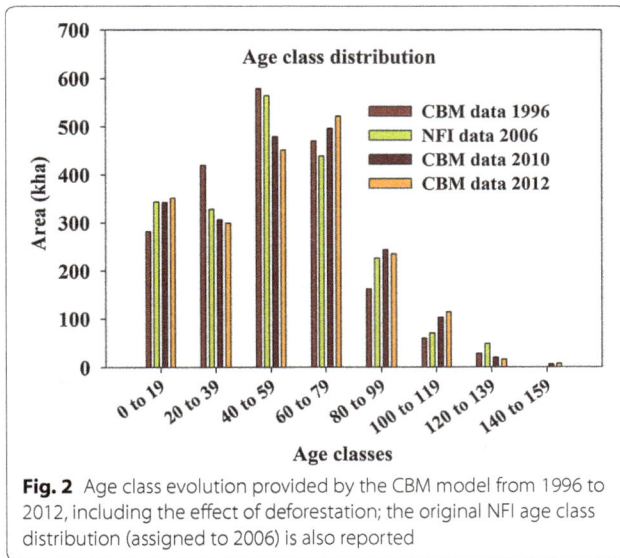

Fig. 2 Age class evolution provided by the CBM model from 1996 to 2012, including the effect of deforestation; the original NFI age class distribution (assigned to 2006) is also reported

The aggregated results at the EU level and including the harvested wood products (HWP) pool, afforestation/reforestation and deforestation, will be reported in a companion paper [16].

The results obtained from the GHGI and the CBM for these 26 countries can be assessed in terms of *level* and *trend*. For the *level*, we consider the match between CBM and each GHGI as "good" if the average net emission of CBM for the period 2000-2012 (Fig. 4) is within ± 25 %[3] compared to the GHGI. For the *trend*, Fig. 5 shows the correlation between CBM and each GHGI. In this case, we consider the match between CBM and each GHGI as "good" if the correlation is significant at $P < 0.05$.

Based on the match between CBM and GHGIs, in terms of *level* and *trend*, and on data reported in Figs. 4, 5, four different groups of countries may be distinguished:

A. Countries where CBM estimates and country data show a good match both in the *trend* and the *level*. This group includes nine countries: Croatia, Finland, Italy, Latvia, Lithuania, Portugal, Romania, Slovakia and Slovenia.
B. Countries where there is a good match in the *trend* but not in the *level*. This group includes five countries: Austria, Czech Republic, Estonia, Greece and Luxembourg,
C. Countries where there is a good match in the *level* but not in the *trend*. This group includes five countries: France, Germany, Spain, Sweden and United Kingdom.

D. Countries where the match is not good for the *level* and for the *trend*. This group includes seven countries: Belgium, Bulgaria, Denmark, Hungary, Ireland, Netherlands and Poland.

Figure 6 illustrates in more detail the results from four country cases (Discussed in "Country case studies" section.), each representative of the four groups above: Portugal (A), Austria (B), Germany (C) and Poland (D).

Discussion
Model evaluation
We implemented a consistent methodological approach to 26 EU countries, using the Carbon Budget Model to estimate the net CO_2 emissions for the period 2000–2012. To evaluate the capacity of the CBM to reproduce country data, our results can be compared with different data sources available at the country level, such as the age-class distribution reported by the NFI and the net CO_2 emissions reported by the country's GHGI. As expected, the comparison between the model results and the country GHGIs showed good agreements in level and trend for some countries and partially good for other countries. When the comparison was not satisfactory, in most cases we can identify possible reasons for these discrepancies. In many cases, however, these data are not fully independent from the NFI input data used by CBM. Where additional information is provided by independent studies (i.e., different datasets, not used by CBM), an independent validation of the model's output is possible. This is the case of Lithuania, where additional information on the living biomass increment, biomass volume and on the dead tree stem volume is available [17]. We select these parameters because increment is one of the main drivers affecting biomass growth estimated by the CBM, initial volume is the main parameter affecting biomass C stock at the beginning of the simulation and dead tree stem volume is the second major C pool with C stock changes over time, for the majority of the European countries (this is often due to the effect of natural disturbances). For Lithuania we verified that our estimates are consistent with these independent data sources. Of course, as highlighted by Vanclay and Skovsgaard [19], the effective evaluation of a forest growth model is a complex and ongoing process, that could include additional independent validations performed at the regional level [14], sensitivity analysis of the main input data, and further comparison of our estimates with other data sources, including the country-specific GHGI data (see also other comparisons reported in the Additional file:1 for additional case studies). For Lithuania, the country's GHGI reports some peaks between 2000 and 2008,

---

[3] This value is in the lower part of the range of uncertainties typically reported by EU countries for FM emissions/removals (25-50 %, EU NIR 2014).

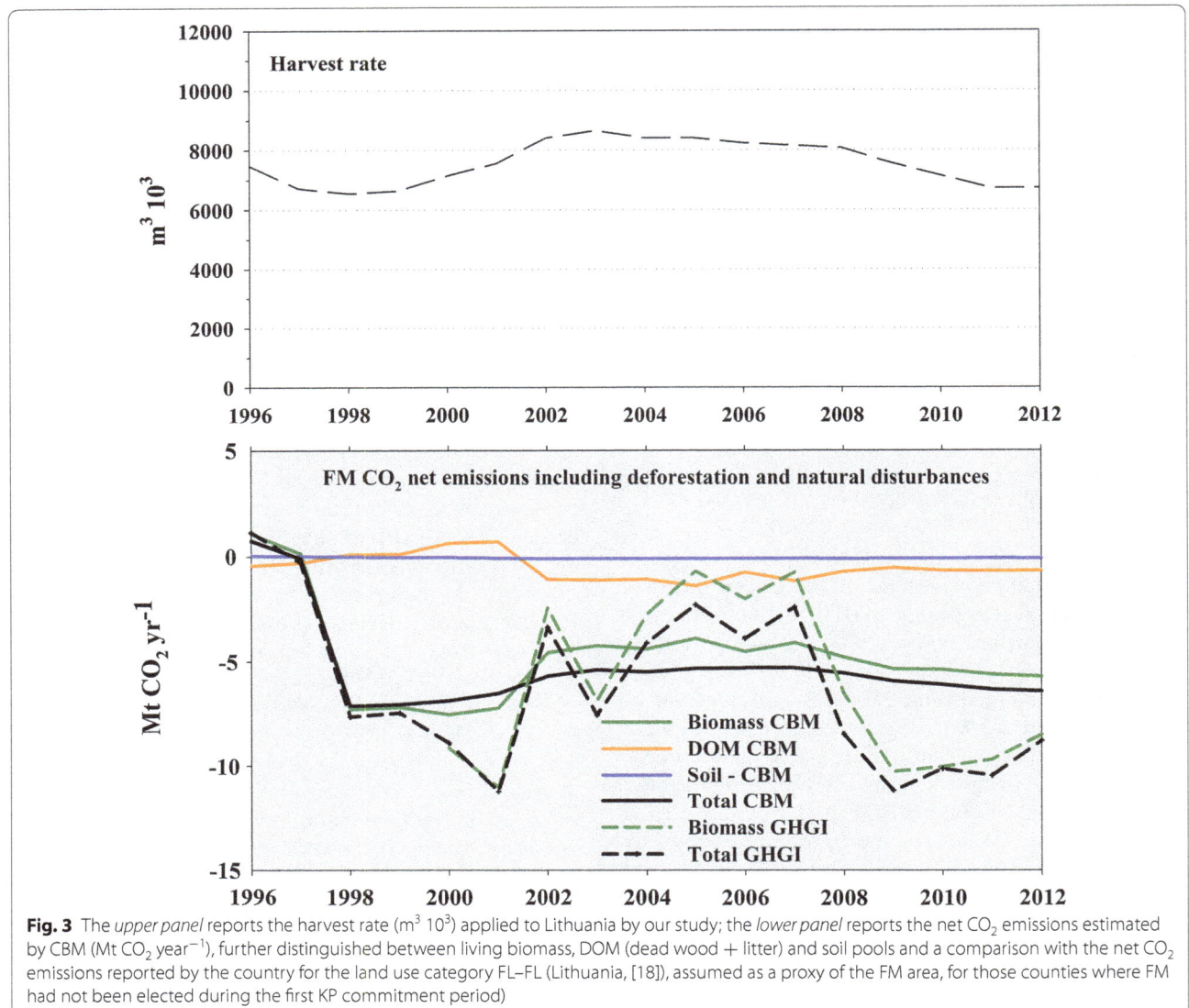

Fig. 3 The *upper panel* reports the harvest rate (m³ 10³) applied to Lithuania by our study; the *lower panel* reports the net CO_2 emissions estimated by CBM (Mt CO_2 year⁻¹), further distinguished between living biomass, DOM (dead wood + litter) and soil pools and a comparison with the net CO_2 emissions reported by the country for the land use category FL–FL (Lithuania, [18]), assumed as a proxy of the FM area, for those counties where FM had not been elected during the first KP commitment period)

not highlighted by our model (see Fig. 3, lower panel). Apart from different assumptions on the area affected by storms and on the salvage of logging residues (we considered three main disturbance events, described in details in the Additional file: 1), these differences may be even due to the interannual statistical variability associated to the stock-change approach, that can exacerbate the real variability of the C stock changes [17]. Despite this different representation of the interannual variability, the overall match between the CBM results and the Lithuania's GHGI is good, i.e. the average of our results

is within ±25 % compared to the GHGI and the correlation between CBM and GHGI is significant at P < 0.05.

Country case studies

Based on comparisons of both level and trends in CO_2 emission estimates obtained from the CBM and the country GHGIs we partitioned the 26 countries into four groups, and we discuss one representative country for each group.

For Portugal, such as for other eight countries (Group A), the CBM estimates and country data show a good

(See figure on next page.)
Fig. 4 Comparison between the net CO_2 emissions from FM reported by the countries for the period 2000–2012 (in the 2014 GHGIs, [18]) and the CBM estimates. Data are reported from an atmospheric perspective, where negative values represent a sink (CO_2 removals) and positive values a source (CO_2 emissions)

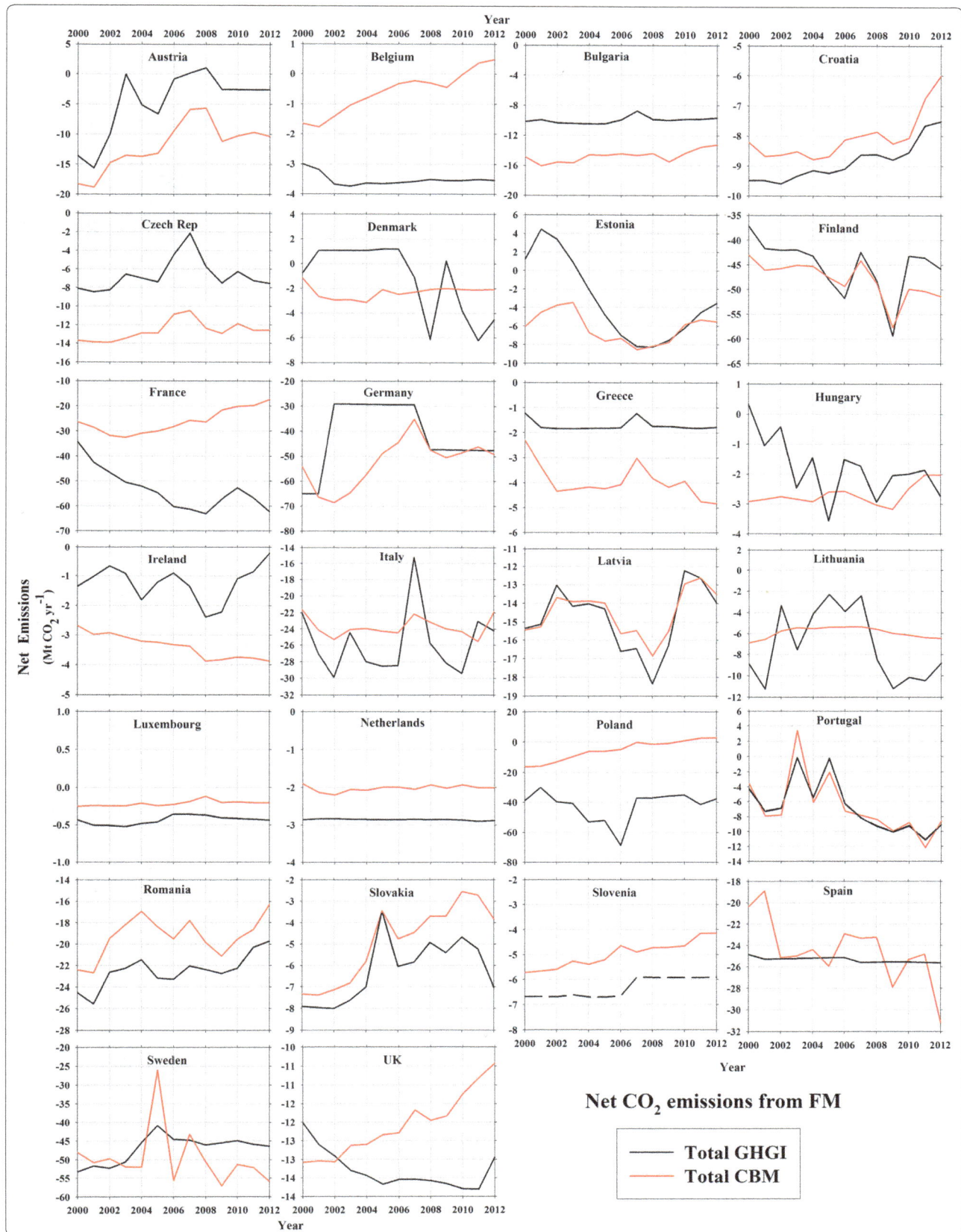

Net CO$_2$ emissions from FM

| Total GHGI |
| Total CBM |

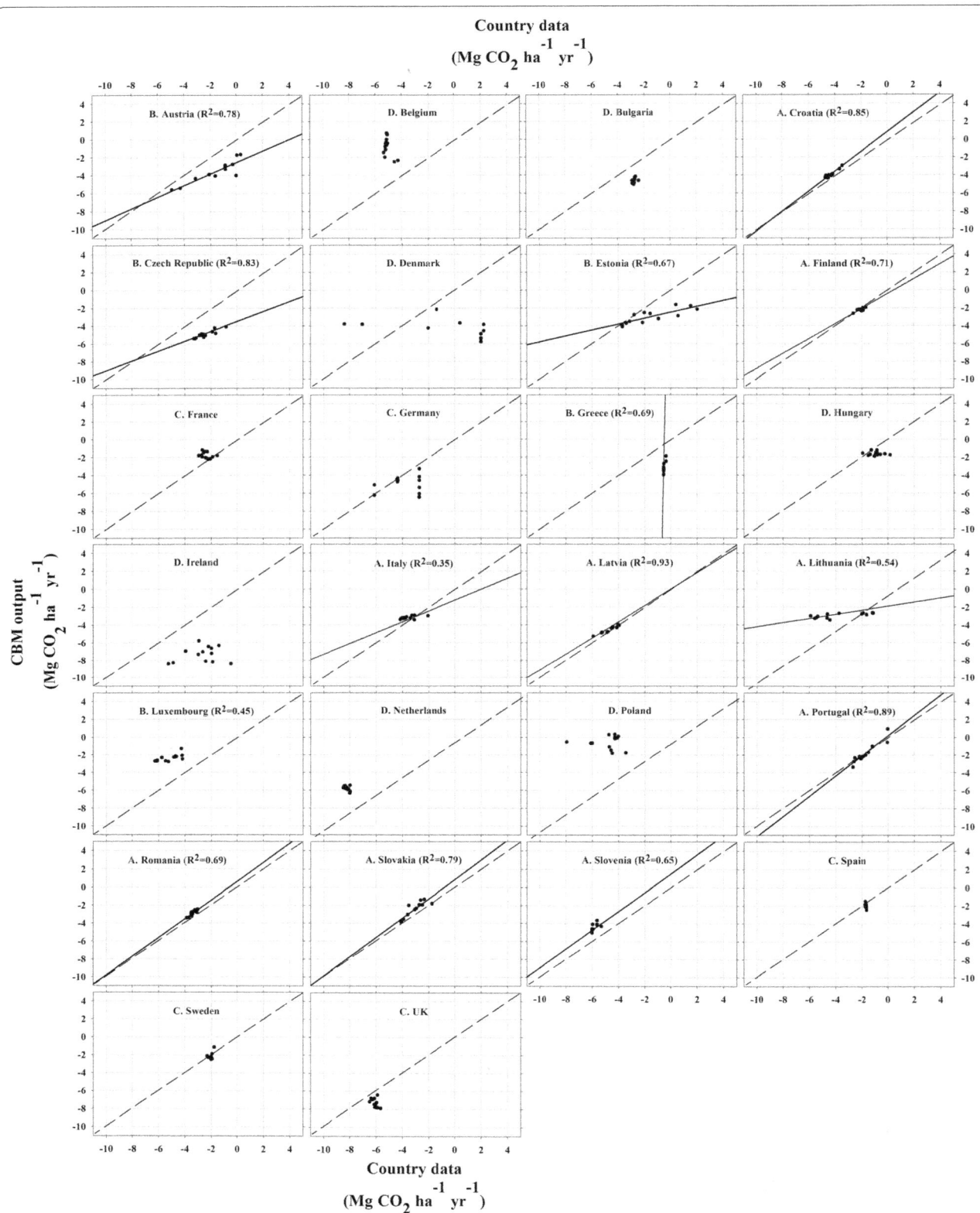

Fig. 5 Comparison between the net CO_2 emissions from FM (living biomass, DOM and mineral soil) as estimated by the CBM and reported in the countries' GHGIs. Each *point* represents one year for the period 2000–2012, as shown in Fig. 4. The *dashed line* is the 1:1 line. The *solid line* is the regression line, shown where the correlation between the CBM and GHGIs was statistically significant (P < 0.05)

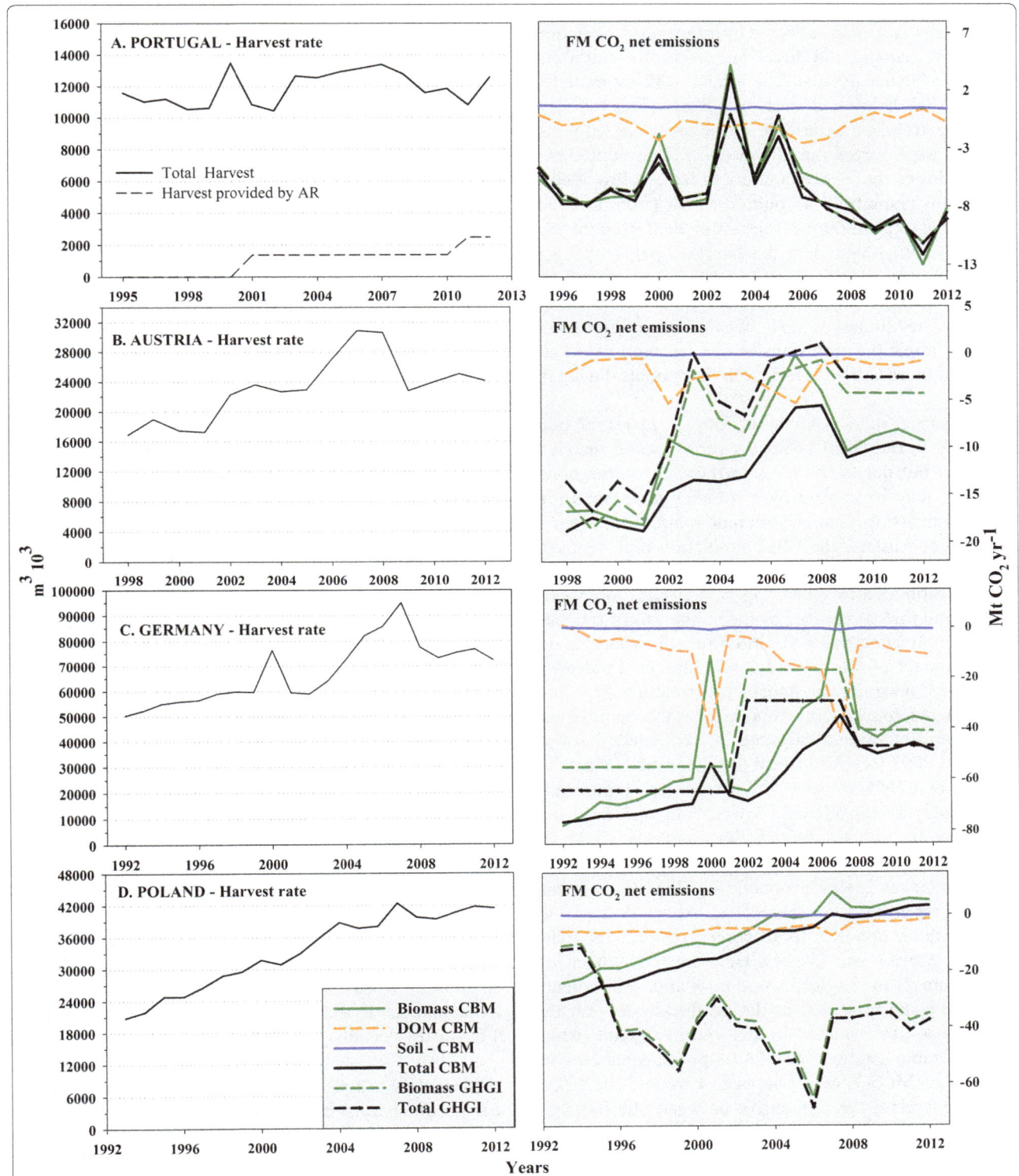

Fig. 6 Harvest rate (on the *left panels*, in $m^3\ 10^3$) and the main output provided by CBM for four representative case studies (Austria, Germany, Poland and Portugal). For each country we report the net CO_2 emissions estimated by CBM (Mt CO_2 year^{-1}, *right panel*), further distinguished between living biomass, DOM (dead wood + litter) and soil pools and a comparison with the biomass and the total net emissions reported by each country. When available, FM country data from the KP-CRF tables was used for 2008–2012 (i.e., if FM had been elected during the first KP commitment period); alternatively, country data were taken from the Convention CRF tables using 'forest land remaining forest land' (FL remaining FL) [18]. For Portugal, the amount of harvest provided by afforestation (AR) is also reported (panel **a**, *left* panel)

match both in the *trend* and the *level*. The C balance of this country is strongly affected by inter-annual variations in harvest demand and direct fire emissions (Additional file: 1 for further details). The total C sink estimated by CBM is slightly lower than the reported values but it has the same trend and the differences decrease with time (in 2011 we reported the same values). These differences may be due to the relative amount of harvest provided by Eucalyptus plantations accounted as AR (from less than 15 % of the total amount of harvest in 2002 to about 25 % in 2011, as highlighted in the harvest's panel of Fig. 6, panels A). As expected, DOM and living biomass pools showed an opposing pattern: when fires kill trees and decrease the biomass C stock, we observe an increase in DOM C pools (i.e., the transfer of C to dead wood and litter add more C than is lost from these pools during the fire).

For Austria, such as for other four countries included in Group B, the CBM estimates show a good match in the *trend* but not in the *level*. In these cases, the different level may be caused by a number of reasons (different conversion factors, different input data, etc.). In the case of Austria, the CBM simulation represents the impact of various natural disturbances. The biomass C balance estimated by CBM (Fig. 6, panels B) follows the same trend that is reported by the country until 2006 and it is strongly affected by the inter-annual variations due to the impact of storms and insect attacks. Indeed, we highlighted a significant statistical correlation ($r = 0.77$) between the total C sink reported by the country and the amount of volume damaged by bark beetle between 1998 and 2007 (see Additional file: 1 for further details). In 2003 and 2005 however, the total C sink reported by the country is considerably lower than our estimates. This may be due to different assumptions about the effect of natural disturbances in specific years. Overall, the biomass C sink estimated by CBM is consistent with the reported trend (Fig. 6). As expected, the DOM C sink has an opposite trend compared with the living biomass. Storms and insect attacks moved C from the living biomass to the dead wood pool and, subsequently salvage logging moved C to the products pool. Yet this impact was not reported by the country's data (which report a stable C source from DOM pools, equal on average to $+ 1.8$ Mt CO_2 year^{-1} between 1998 and 2012). This may also explain the differences between our estimates on the total C sink and the values estimated by country: for example, in the CBM in 2007 a strong reduction of the living biomass pools due to a storm is compensated by a corresponding increase in the DOM pools. After 2008, due to different assumptions about the average amount of harvest and about the effect of natural disturbances (country's data report a constant amount of harvest equal

to about 25 million m^3 from 2009 to 2012) our estimates are not comparable with the country because we used different harvest rates.

Germany (Group C, including five countries), represents an example where there is a good agreement in the *level* but not in the *trend*. We use it to illustrate the difference between the stock-change approach used in the GHGI and the gain-loss method used in the CBM. This methodological difference has a strong impact on the inter-annual variability of estimates as affected by harvest and natural disturbances. Overall, the total C sink estimated by CBM follows the same trend provided by the country (Fig. 6, plot C), even if the correlation is not significant at $P < 0.05$ (i.e., the threshold considered by our study). Due to the stock-change approach, the national sink estimates report three annual values, each applied to the inventory period over which observed stock changes have been annualized [20]. Compared to the reported values, our estimates show a larger inter-annual variability (in particular for the living biomass and DOM pools) due to the storms that occurred in December 1999 (assumed as 2000) and 2007. As expected, the CBM reports opposite trends in the biomass and DOM pools due to the transfer of C from living biomass to the dead wood pool. From 2008 to 2012, our estimates are fully consistent with the data reported by Germany. Further details on natural disturbances and the evolution of the age-class distribution are reported in the Additional file: 1.

For Poland, such as for other 6 countries included in Group D, the estimates differ significantly for both the *trend* and the *level*, for reasons that will require further analysis. For this country, the CBM estimates a decreasing C sink, consistent with a strong increase of the total amount of harvest reported by FAO statistics (see the left panel of Fig. 6, panel D). In contrast, Poland reports an increasing sink with increasing harvest rate. According to our estimates, the DOM pool (not reported by the country) is a C sink, because of the amount of residues left after harvest (i.e., moved from living biomass to DOM). In addition storms in 1999 and 2007 also moved C from living biomass to the dead wood pool (see Additional file: 1 for further details).

CBM results *vs.* GHGIs: impact of carbon pools coverage, harvest and natural disturbances

A first, potentially relevant factor, to be considered when comparing the CBM results with the GHGIs, is the inclusion of C pools. The CBM includes all forest C pools (living biomass, DOM and mineral soils) for all countries, but DOM and soil pools are not reported in some GHGIs. While all 26 countries report living biomass, seven do not report DOM and 14 do not report mineral soils [2]. The mineral soil is in most cases neither a large sink or source (in the CBM, and in the GHGIs). In contrast, the CBM estimates of net CO_2 emissions for DOM pools can be large when natural disturbances occur. Nevertheless, differences in the reported C pools help to explain the observed differences between the CBM and the GHGIs in only a few cases (e.g., Czech Republic, Romania, Slovakia and Slovenia). It is therefore necessary to extend the analysis to the impact of the main drivers of net CO_2 emissions, i.e. harvest and natural disturbances as indicated in both the CBM and GHGIs results.

The net CO_2 emissions from living biomass are generally correlated with the three main drivers: harvest rate, area affected by fires and area affected by storms in both the estimates from the CBM and the GHGI (Fig. 7).

The correlations shown in Fig. 7, demonstrate that for 21 out of 26 countries there is, as expected, a clear negative correlation (generally with r < −0.5) for both CBM and the countries' GHGIs, i.e. more harvest decreases the biomass sink (see for example, Croatia, Finland, Latvia, Lithuania, Portugal, Romania and Slovenia).

Within this group, in three cases (Germany, Estonia, and Slovakia) the correlation between biomass net emissions and the area affected by disturbances is negative for CBM and, surprisingly, is positive for the countries. For Estonia and Slovakia the differences may be due to different assumptions on the effect of storms on the living biomass or DOM pools (i.e., the amount of biomass moved to DOM or removed with salvage logging). For Germany, the main reason appears to be the stock-change approach applied to consecutive NFIs [20]: this approach does not capture the inter-annual variations within a measurement period caused by natural disturbances.

In other cases, despite both the CBM and the GHGIs showing a similar (negative) correlation between the biomass net CO_2 emissions and both harvest and natural disturbances, overall the match between modelled trends and the GHGI is not good (see Fig. 4). For Austria, the main difference lies in different assumptions about the mineral soil pool (which is a source in the GHGI) and partly about the DOM, since the match between the CBM and the GHGI is good for the living biomass. For France, the discrepancy between the CBM and the GHGI requires further investigation, especially with regard to possible differences about harvest assumptions and increment. For Greece, the total FM sink estimated by the CBM is higher than the values reported by the country, but different assumptions on the effect of fires (for example on the amount of biomass burned and the distribution of fires between the FM area and the unmanaged forest area) could explain some these differences. The FM sink reported by Hungary is considerably lower than our estimate and it shows a higher inter-annual variability,

for reasons that are not yet understood. For Ireland, the total sink reported by the GHGI has the opposite trend (i.e., a decreasing C sink) compared with our estimates. Ireland did not elect FM under the CP1 therefore the values reported for this country were derived from the FL remaining FL land use category and a certain amount of harvest is certainly provided by afforestation [21]; this may explain the differences observed. The sink reported by Luxembourg is considerably higher than our estimates and does not seem compatible with the harvest rate applied by CBM. The FM sink estimated by CBM for Spain is overall quite similar to the country GHGI; the main difference is that CBM shows inter-annual variability due to fires and harvest rates, while the stock-change approach implemented by Spain's GHGI masks this variability [22]. Emissions from forest fires estimated by CBM are generally lower than the CO_2 emissions reported by Spain. This is probably due to different assumptions on the amount of biomass and DOM burned. For Sweden, the differences detected on the trend may be due to the effect of storms (above all in 1999 and 2005) and an overestimate on the biomass C stock by CBM. A special case is the lack of any correlation for Italy, where the main driver of the inter-annual variability in biomass net CO_2 emissions is clearly fire (r < −0.80), as also highlighted by [13].

For 5 out of 26 countries (Belgium, Denmark, the Netherlands, Poland and UK), the correlation between biomass net CO_2 emissions and harvest rate is negative (r < 0) for CBM and, surprisingly, is positive (r > 0) for the country GHGI. In principle, this discrepancy may be explained by three reasons. First, the harvest rate applied by our study is different from the harvest reported by the country in its GHGI; even if we always tried to be consistent with the harvest reported by countries, some differences may exist due to inconsistency between different data sources (e.g. see [23]). Second, other factors (e.g. natural disturbances or rapid changes in net increment not included in our study) are a more important driver of biomass net CO_2 emissions compared to harvest; although this case does not seems very likely, it cannot be totally ruled out. Third, the estimation method used by the country in its GHGI masks the effect of harvest on the biomass carbon stock change.

For both the Netherlands and UK, a good match in both the trend and the level existed between CBM and the 2013 GHGIs, suggesting that some recent changes (in input data and/or method) were implemented for the 2014 GHGI. For Denmark, although the known most relevant storms (1999/2000 and 2005) were considered by CBM, the overall correlation between CBM and Denmark GHGI is poor (see again Fig. 4). This could potentially be explained by the method used by Denmark, where a

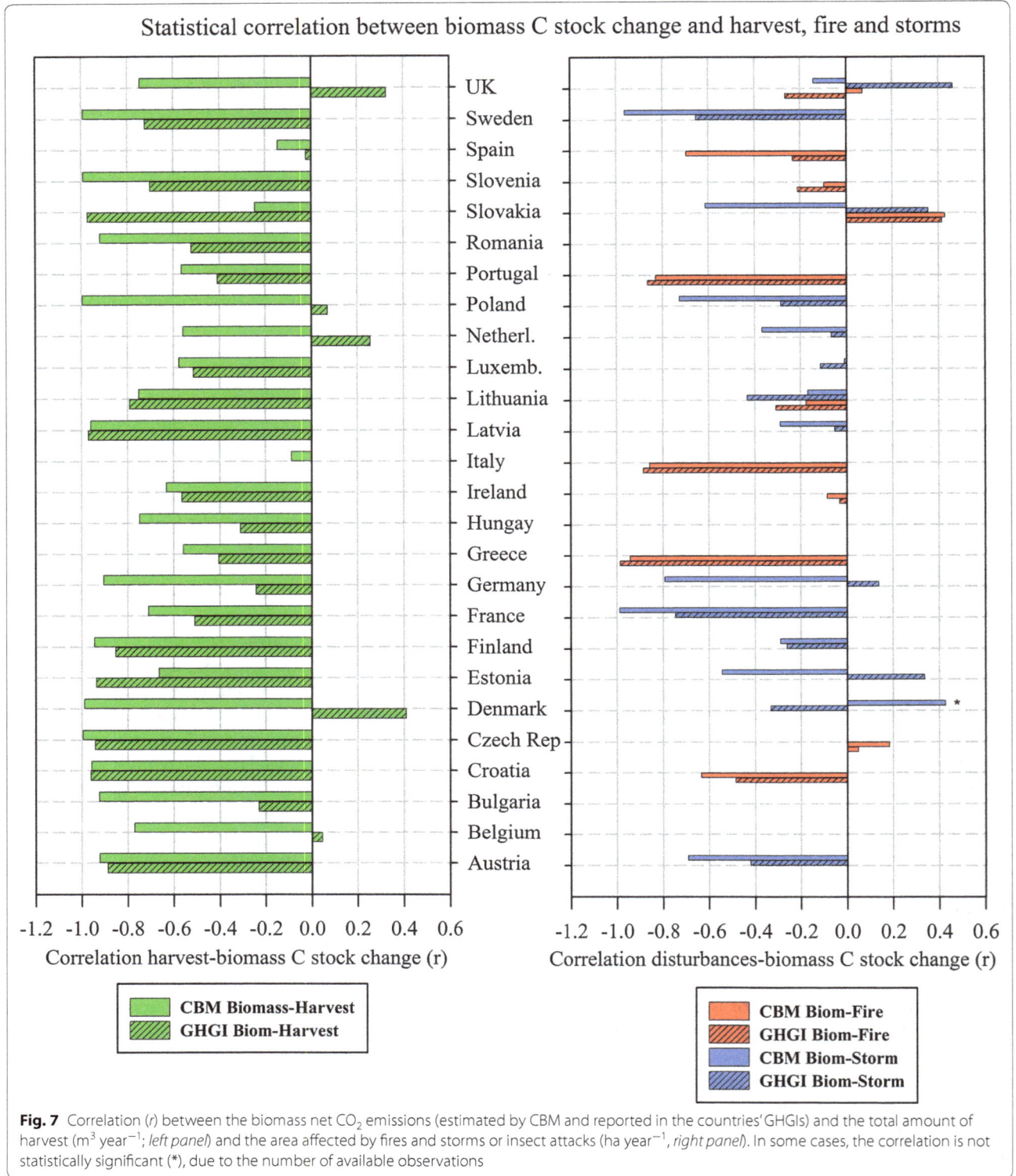

Fig. 7 Correlation (*r*) between the biomass net CO_2 emissions (estimated by CBM and reported in the countries' GHGIs) and the total amount of harvest (m^3 $year^{-1}$; *left panel*) and the area affected by fires and storms or insect attacks (ha $year^{-1}$, *right panel*). In some cases, the correlation is not statistically significant (*), due to the number of available observations

stock-change approach is implemented every year based on the information collected annually from the NFI [24]. It is possible that the interannual statistical variability of

data associated to this approach overrides all the other factors considered and exacerbates the real interannual variability of C stock changes. For Belgium and Poland further analysis is needed, to explain the observed differences between our results and the country's estimates. These may be due to the lack of data or to some incorrect assumption on the input data (i.e., the harvest).

Summary of the main differences between CBM and countries' estimates

Since the CBM and the GHGIs typically share most of the basic input data (e.g., forest area, timber volume and net increment, taken from the NFIs), we briefly discuss the level of independence of input data. Forest area will be strongly correlated between the CBM and the GHGIs, because we used whenever possible the area used by the GHGI for FM (or, for countries that did not elect FM, for FL remaining FL).

The methods used to estimate the emissions/removals per unit of area—typically the major source of uncertainty of GHGIs—differ. Eleven out of 26 countries use the *stock-change approach* in their GHGIs [5], implemented either every year (using any available new data) or at the end of each NFI cycle [25]. In these cases, the degree of independence between the CBM and GHGIs is very high because the GHGIs typically do not use net increment and harvest values (i.e. the most important drivers for the sink estimated by the CBM). Furthermore, even in the 15 countries that use the *gain-loss approach* [5]—the approach also used by the CBM—the steps needed to obtain CO_2 emission/removals are complex and introduce uncertainty, e.g. converting net increment minus disturbance losses (harvest, storm, fire) into the sink estimate. For example, the most recent data from NFI typically used in GHGIs (e.g. on net increment) are not always publicly available, and in several cases require interpretations and/or assumptions. Equations used by CBM to convert volume into C are totally independent from GHGIs. Harvest rates for the 26 countries used by the CBM are based on FAO statistics (which often require interpretation and/or adjustments, see [23], but the GHGIs may use either FAO or other national-level statistics. In summary, in most cases the methods to estimate emission/removals should be seen as largely independent between the CBM and GHGIs.

Few studies compared model results with European countries' GHGIs. The main comparison may be done with [7], where two models (EFISCEN and G4 M) were applied in 24 EU countries for the period 2000 to 2008, with the discussion focused on six countries. In comparison to that study, our analyses cover a longer period (2000–2012) and 26 countries and we include the DOM and mineral soils pool dynamics and the explicit simulation of the impact of natural disturbances. Beyond these differences—which in several cases allowed CBM to obtain a better match with GHGIs—most of the conclusions from Goen et al. [7] are valid also for our study,

e.g. (1) in several cases (i.e., for Germany), the estimation method used in the GHGIs (stock-change vs gain-loss) explains most of the differences observed, and (2) in the remaining cases, the differences seem to have country-specific reasons, like the amount of harvest used and the way harvest losses are treated.

In addition to the above, another essential aspect is the recalculations performed annually by the countries as part of the continuous process of improving their GHGIs. Our study therefore represents a "picture" in a rather dynamic process as future changes to GHGI may affect our conclusions. The frequency of the recalculations in the LULUCF sector is high: according to EU countries' GHGIs submitted between 2010 and 2014 (including the time series 1990–2008 and 1990–2012, respectively), and focusing only on FL–FL, on average every year 5–6 countries out of 26 revised emissions of the previous GHGI by 10–25 % (in terms of absolute level of emissions), and another 5–6 countries revise emissions by more than 25 %. This means that every year more than a third of the countries analyzed in this study show substantial recalculations compared to their previous GHGIs, with the biggest changes usually for the more recent years. These recalculations are due to a number of reasons (e.g. new input data, addition of pools or gases, correction of previous errors, change in methods, etc.), linked to country internal processes or to recommendations provided by the UNFCCC expert review teams. The magnitude of these recalculations is consistent with the information available on uncertainties from countries' GHGIs, which for FM in most cases fall in the range of 25–50 % [2].

Overall, given the frequency and the magnitude of the changes in GHGIs—and the associated uncertainties—for a modeler it is challenging to capture all the latest data and methods used (including possible errors) in 26 different GHGIs; an improved process to share updated information by country on an ongoing basis would certainly help. Nevertheless, the large amount of work completed by implementing the CBM in 26 countries allowed us obtain satisfactory results in most of the countries analyzed, and to understand the reasons for differences in many of the remaining cases.

Conclusions

This study implemented a consistent methodology to estimate the GHG balance in the managed forests of 26 EU countries using the CBM to estimate the historical (2000 to 2012) net CO_2 emissions from forest management (Sensu Kyoto, i.e. forest existing before 1990) as affected by harvest and natural disturbances (storms, fires and insects). In terms of number of countries, C pools and type of disturbances simulated, to our knowledge this is the most comprehensive study of its kind to date.

The comparison of CBM results with the data reported by the countries in their GHGIs shows a good match (both in the trend and in the level) in nine cases, a partially good match (either for the trend or the level) in an additional ten cases, and an un-satisfactory match in the remaining seven cases. A successful independent country-level validation of the CBM has also been performed.

Our study confirms that, in the short period (and excluding possible effects of climate change), the main factors driving the forest C sink of Europe's managed forests are the harvest rates and natural disturbances (storms for most countries). When these factors are considered in a consistent way, i.e. the gain-loss method is used in both the CBM and the GHGIs, the trends of net CO_2 emissions are very similar. Where the comparison between the CBM and the GHGIs was not fully satisfactory (for the trend and/or for the level), in most cases we provided possible explanations for the discrepancies observed, including: (1) representation of the interannual variability due to harvest and natural disturbances: while it is well simulated by the CBM, it may be masked if the country uses the stock-change approach for the GHGIs; (2) a different treatment of non-biomass pools (not reported by several countries, or reported using different assumptions compared to the CBM), or of CO_2 emissions from fires, natural mortality or other parameters (e.g. harvest residues). Beyond these explanations, some cases—e.g. where the GHGI counter intuitively reports an increasing biomass sink associated with a trend of increasing harvest rates—clearly deserve further analysis, to identify the possible cause of the discrepancy. In general, the results of the comparisons were good in those countries where the input data for the model were based on accessible recent statistics. Finally, when analyzing the discrepancies between the CBM results and the GHGIs, it should be noted that the frequent update cycle and recalculations of GHGIs can only be reflected in the model results if national statistics on harvest and disturbance rates are readily available for the model analyses.

Overall, this study documents a promising foundation for the use of the CBM both as tool to help countries in estimating the forest C dynamics (e.g., including natural disturbances) and as a potential tool to support the verification of GHGIs at the EU level using a consistent methodological approach for all countries. A systematic comparison of the CBM with the GHGIs will certainly require additional efforts—that will require close cooperation between modelers and country experts—and caution should be applied when interpreting these first results. Nevertheless, this application of consistent methods makes a useful contribution to the continuous improvement of GHGIs, because it may help in identifying possible errors, in increasing scientific understanding

and ultimately in building confidence in the estimates of emissions and removals reported by the countries by increasing consistency, transparency and completeness of the estimates.

Methods

The Carbon Budget Model (CBM-CFS3) and the main input data

The CBM is an inventory-based, yield-data driven model that simulates the stand- and landscape-level C dynamics of above- and below-ground biomass, dead organic matter (DOM: litter and dead wood) and mineral soil [8]. The model, developed by the Canadian Forest Service (the model description is available to the following URL: http://www.nrcan.gc.ca/forests/climate-change/carbon-accounting/13107), was recently applied to the Italian forests, in order to test the CBM for different European silvicultural systems, proposing a novel approach to include uneven-aged forest structures [13].

Because this work applies the same general assumptions used in the Italian case study, we provide only a short description of the model, highlighting the specific methodological assumptions related to the present study. Further details of the model can be found in [8], and its applications to European countries are found in [13–15].

The spatial framework applied by the CBM conceptually follows reporting method 1 ([4]) in which the spatial units are defined by their geographic boundaries and all forest stands are geographically referenced to a spatial unit (SPU). We considered 26 administrative units (i.e., European countries, as reported by Table 1) and 35 climatic units (CLUs, as defined by [26]) for a total of 910 SPUs. The CLU's mean annual temperatures, range from −7.5 to +17.5. Each SPU was linked to a CLU through the information provided by Corine Land Cover.

The total managed forest area of the 26 EU countries represented here covers about 138 Mha (i.e., about 82 % of the EU forest area). Two EU countries excluded from the analysis are Cyprus (no NFI data available) and Malta (very small forest area, mainly covered by shrub lands).

Within a SPU, each forest stand is characterized by age, area and seven classifiers that provide administrative and ecological information, the link to the appropriate yield curves, and parameters defining the silvicultural system such as forest composition and management type (MT), and the main use of the harvest provided by each SPU, as fuelwood or industrial roundwood. For each country, these parameters were mainly derived from NFIs. According to country-specific information, MTs may include even-aged high forests, uneven-aged high forests, coppices and specific silvicultural systems such as clearcuts (with different rotation lengths for each forest type, FT), thinnings, shelterwood systems, and partial cuttings.

Table 1 **Summary of the main parameters applied by the CBM model for each country**

Country	Original NFI year	Time step 0 (years)	CBM FM area (Mha)[a]	Harvest rate (av. 2000–2012, Mm³)	County specific biomass equations
Austria	2008	1998	3.2	22.9	X
Belgium	1999	1999	0.7	4.3	
Bulgaria	2000	2000	3.2	5.3	
Croatia	2006[b]	1996	2.0	4.6	
Czech Republic	2000	2000	2.6	17.0	X
Denmark	2004	1994	0.5	2.3	
Estonia	2000	2000	2.1	7.9	
Finland	1999	1999	21.7	55.0	
France	2008	1998	14.6	54.9	
Germany	2002	1992	10.6	74.7	X
Greece	1992[b]	1992	1.2	1.6	
Hungary	2008	1998	1.6	6.2	X
Ireland	2005	1995	0.5	2.8	
Italy	2005	1995	7.4	10.2	X
Latvia	2009	1999	3.2	15.8	X
Lithuania	2006	1996	2.0	7.7	
Luxembourg	1999	1999	0.1	0.3	
Netherlands	1997	1997	0.3	1.2	
Poland	1993	1993	8.9	37.8	
Portugal	2005	1995	3.6	12.2	X
Romania	1985	1985	6.6	17.2	X
Slovakia	2000	2000	1.9	9.0	
Slovenia	2000	2000	1.1	3.3	
Spain	2002	1992	12.6	16.8	
Sweden	2006	1996	22.6	79.5	
United Kingd.	1997	1997	2.5	9.8	
EU			137.9	480.7	8 countries

The table reports the NFI original reference year; the year since the model was applied; the FM area used by CBM at time step 0; the average harvest rate used; the countries where specific equations to convert the merchantable volume into aboveground biomass were selected. Two countries were not modeled: Cyprus (no NFI data available) and Malta (very small forest area, mainly covered by shrub lands)

[a] FM area used by CBM at time step 0. According to KP rules, FM is the area of forest in 1990, decreased by any subsequent deforestation. The FM area is taken from the official submissions made by countries to UNFCCC/Kyoto Protocol [18, 29], giving priority to data from KP-CRF tables when available (i.e., if FM had been elected during the first KP commitment period), or alternatively taking data from the Convention CRF tables (using 'forest land remaining forest land' in 1990 as a proxy for FM). To obtain FM area at time step 0, the D area reported by all countries under the Kyoto Protocol was used. Please note that CBM runs did not include forests reported as "not productive" (e.g., 0.4 Mha in Austria, 0.02 Mha in Bulgaria, 5 Mha in Sweden) and overseas territories (8.2 Mha in France)

[b] Analysis based on data from Forest Management Plans

Species-specific, stand-level equations [27] convert merchantable volume production into aboveground biomass, partitioned into merchantable stem wood, other (tops, branches, sub-merchantable size trees) and foliage components [8]. Where additional information provided by NFIs or by literature was available (see last column in Table 1), country-specific equations were selected to convert the merchantable volume into aboveground biomass [13]. If no data were available, we used the same equations selected for other countries and similar forest types (FTs, defined according to the main species). Belowground biomass is calculated using the equations provided by [28] and the annual dead wood and foliage input

is estimated as a pool-specific turnover rate (percentage) applied to the standing biomass stock.

Forest inventories typically contain no or only insufficient data on stocks in DOM and soil C pools. The model therefore uses an initialization process to estimate the size of all DOM pools at the start of the simulation and then, following IPCC guidance, links DOM dynamics to biomass dynamics. Inputs from biomass to DOM pools result from biomass litterfall and turnover as well as natural and human-caused disturbances. The DOM parameters were first calibrated in the Italian cases study (see [13], Appendix E for further details), then validated on a specific study at regional level [14] and, if necessary, further modified for specific countries, such as Finland and Sweden.

We use two sets of yield tables (YT) in these analyses [13]. Historical YTs derived from the standing volumes per age class reported by the NFI represent the impacts of growth and partial disturbances during stand development. Current YTs derived from the current annual increment reported in country NFIs represent the stand-level volume accumulation in the absence of natural disturbances and management practices.

To implement the CBM to uneven-aged FTs (when this forest structure was observed in a country), all the uneven-aged forest area was allocated to a reference age class, with the average volume equal to the volume reported by the NFI for these stands. Starting from this age class, a decreasing percentage increment was applied to the subsequent (older) age classes. We assumed that, after a certain number of years, equal to species-specific cutting cycles defined at country level, each uneven-aged stand was disturbed and moved back to the initial reference age class [13]. This approach was tested through a number of simulations in which we varied different parameters. Overall, we simulated (1) a faster (but decreasing) re-growth phase during the first period following the partial cut and (2) a decreasing growth phase during the following years.

Since this study aimed to be as comparable as possible with countries' information reported to the UNFCCC and its KP, the model was applied individually to each country and we modeled 'forest management' (FM) as the forests existing in 1990 minus any deforestation (D) since 1990. Forest area in 1990 and deforestation rates were obtained, respectively, from the 2014 GHGIs submitted by each country to the UNFCCC and to the KP [29]. The start year of the simulations (time step 0) varied between countries. FM area was reduced, during the model run, due to D between 1990 and time step 0. The D area within each country was distributed proportionally to the area of each FT. Table 1 shows the country-specific FM area at the start of model runs.

In order to provide a comparable dataset for all the EU countries, covering the period 2000–2012, when the NFI reference year was after the year 2000 (see Table 1), the original NFI age-class distribution (for even-aged forests) was rolled back by 10 years (see [13] for further details).

Harvest rate

To provide a consistent estimate of the harvest demand for all 26 EU countries, historical data on harvest were obtained from FAO statistics [30]. For some countries, the original FAOSTAT data were slightly modified to ensure consistency with other information provided by countries under the KP. The country-specific modifications applied to the original FAOSTAT data (in most cases due to different treatment of the bark fraction) are described in [23].

FAOSTAT data (modified where necessary) were further distinguished at the country level, between four compartments: Industrial Roundwood (IRW, i.e., the portion of roundwood used for the production of wood commodities) and Fuelwood (FW, i.e., wood for energy use) and between coniferous and non-coniferous (i.e., for our analysis, broadleaved) species groups [30]. For each compartment, we defined in CBM: (1) the FTs (i.e., broadleaved species for IRW and FW broadleaved species, and coniferous species for IRW and FW coniferous species), (2) the MTs (for example coppices for FW from broadleaved species) and (3) the silvicultural practices (for example thinnings for FW from coniferous species) providing the total amount of wood expected each year (the harvest target).

We assumed that the harvest rate was entirely satisfied by the FM area, considering that the possible amount of harvest provided by lands afforested or reforested (AR) since 1990 was generally negligible [15], with the exception of Portugal (see the Additional file: 1 for details).

Natural disturbances

For each country, the historical effects of storms and ice (15 countries), fires (11 countries) and insect attacks (i.e., bark beetles attacks, for 2 countries) were analysed (see Table 2 for details). We assumed that that natural disturbances occurred on the FM area, excluding possible disturbances on the afforested area.

The effect of storms was evaluated using the data reported by the FORESTORMS database [31] provided by the European Forest Institute and by specific additional information available at the country level. Depending on the available information, the effect of each event was modelled according to (1) the amount of forest biomass damaged by storm and eventually salvage logged and/or (2) the amount of area affected by the disturbance event. In the first case, we mainly modified the 'disturbance matrix' that describes the proportion of C transferred between pools and to the forest product sector or released to the atmosphere [8], in order to be consistent with the disturbance impact reported by the FORESTORMS database. In the second case, we verified that the amount of forest area affected by the disturbance event was consistent with the area reported by this database. In some cases, such as for Sweden, both these criteria were verified.

More specific information on the methodological assumptions applied to represent storms and insect attacks are reported in the Additional file: 1 for some representative case study. Since the information available on these disturbances may vary considerably by country,

Table 2 Overview of countries with natural disturbance events simulated by the CBM (*F* fire, *S* storms and ice sleets, *I* insect attacks), with information on input data used for storms (country data, National Inventory Reports, NIR or the FORESTORMS database [31] and the average annual burned area

Country	Natural disturb.	Storms, ice and insect disturbances		Fires
		Source	Vol. affected[a] (Mm³ year⁻¹)	Area burned[b] (kha year⁻¹)
Austria	S + I	Vol. based on country data	4.1	–
Belgium	–			–
Bulgaria	–			–
Croatia	F			2.3
Czech Rep.	F			0.5
Denmark	S	Vol. based on country data	0.5	–
Estonia	S	Area and vol. based on NIR	0.7	–
Finland	S	Vol. based on FORESTORMS	0.6	–
France	S	Area and vol. based on FORESTORMS	18.3	–
Germany	S	Vol. based on FORESTORMS	6.2	–
Greece	F			6.0
Hungary	–			–
Ireland	F			0.4
Italy	F			35.0
Latvia	S	Vol. based on FORESTORMS	0.7	–
Lithuania	S + F + I	Vol. based on the NIR + FORESTORMS	0.2	0.3
Luxembourg	S	Vol. based on FORESTORMS	<0.1	–
Netherlands	S	Vol. based on FORESTORMS	<0.1	–
Poland	S	Vol. based on FORESTORMS	0.4	–
Portugal	F			49.1
Romania	–			–
Slovakia	S + F	Vol. based on FORESTORMS + country data	0.8	0.6
Slovenia	S + F	Vol. based on country data	<0.1	0.1
Spain	F			35.3
Sweden	S	Vol. based on FORESTORMS + country data	7.1	–
United K.	S + F	Vol. based on FORESTORMS	<0.1	3.5
	22 countries		39.6*	134.0

[a] Average volume affected by storms, ice and insects between 2000–2012, as reported by the input data used by CBM. The interannual variations of these disturbances can vary considerably among countries (i.e., in many cases disturbances are concentrated in few big events). In some cases, further damages were considered before 2000

[b] Average area affected by fires between 2000–2012, mainly based on the data reported by National Inventory Reports*

our assumptions were adapted to the conditions in each country.

Fire disturbances were modelled according to the amount of area affected by fire, as reported by national statistics, proportionally distributed between different FTs or according to further information provided by literature (mainly, the National Inventory Reports) The disturbance matrix associated with fires was modified according to specific country-level information, to account for salvage of logging residues, commonly applied in some Mediterranean countries (i.e., Portugal). More specific information on the methodological assumptions applied to these disturbances is reported in the Additional file:1 for Portugal. As in the case of storms, our model assumptions were adapted to the specific country's conditions. When relevant (e.g., for Latvia), we also included the burning of harvest residues after a clearcut.

Model validation

For Lithuania, the information provided by CBM, based on Lithuania's NFI used as input data for the model, can be also compared and validated against some independent data, derived by specific studies[4] on living and dead tree volumes in forest land, reported by Lithuania's NIR

---

[4] "Study 1, "Forest Land Changes in Lithuania during 1990–2001" ([17], page. 349).

[17]. Further details on the methodological assumptions are reported in the Additional file: 1.

Additional file

> **Additional file 1.** Country case studies: Austria, Germany, Lithuania, Poland, Portugal

Abbreviations

AR: afforestation and reforestation; C: carbon; CBM: Carbon Budget Model; CP1: first commitment period; CP2: second commitment period; D: deforestation; DOM: dead organic matter; EU: European Union; FL: forest land; FM: forest management; FRA: forest resources assessment; FT: forest type; FW: fuelwood; GHG: greenhouse gas; GHGI: greenhouse gas inventory; HWP: harvested wood product; IPCC: Intergovernmental Panel on Climate Change; IRW: industrial roundwood; KP: Kyoto protocol; LULUCF: land use, land-use change and forestry; MT: management type; NFIs: National Forest Inventories; NIR: National Inventory Report; SPU: spatial unit; UNFCCC: United Nations Framework Convention on Climate Change; YT: yield table.

Authors' contributions

RP carried out the data analysis, in collaboration with NG. GG and WAK helped in the design of the study and the interpretation of results, and together with RP wrote the manuscript, in collaboration with RAV. All authors read and approved the final manuscript.

Author details

[1] European Commission, Joint Research Centre, Institute for Environment and Sustainability, Via E. Fermi 2749, 21027 Ispra, VA, Italy. [2] Natural Resources Canada, Canadian Forest Service, Victoria, BC V8Z 1M5, Canada.

Acknowledgements

This paper was prepared in the context of the Contract no. 31502, Administrative Arrangement 070307/2009/539525/AA/C5 between JRC and DG CLIMA. Further information was collected in the context of the AA 071201/2011/611111/CLIMA.A2. The analysis performed for each country was generally based on data public available and on additional information collected at country level, in collaboration with many colleagues and experts for each country. We especially thank Stephen Kull, Scott Morken and the Carbon Accounting Team for their indispensable technical support during this study and our colleagues, Giulia Fiorese, Viorel Blujdea and Tibor Priwitzer, who provided useful comments and suggestions.

We also thank two anonymous reviewers, who provided useful comments and suggestions to improve the manuscript.

The views expressed are purely those of the authors and may not in any circumstances be regarded as stating an official position of the European Commission or of Natural Resources Canada.

Competing interests

The authors declare that they have no competing interests.

References

1. FOREST EUROPE, UNECE, FAO. State of Europe's Forests 2015. Status and trends in sustainable forest management in Europe, 2015. URL (Access Mar 2016): http://www.foresteurope.org/docs/fullsoef2015.pdf.
2. EU NIR. Annual European Community greenhouse gas inventory 1990–2012 and inventory report 2014. Submission to the UNFCCC Secretariat. European Environment Agency, Technical report No 09/2014. URL (Access Feb 2016): https://www.google.it/search?q=Annual+European+Community+greenhouse+gas+inventory+1990%E2%80%932012+and+inventory+report+2014&ie=utf-8&oe=utf-8&gws_rd=cr&ei=fwi7VoSoJIXAOoCTroAF.
3. UNFCCC, Decision 2/CMP.7 on Land use Land use Change and Forestry, 2011. URL (Access Feb 2016): http://unfccc.int/resource/docs/2011/cmp7/eng/10a01.pdf.
4. IPCC (Intergovernmental Panel on Climate Change). Revised supplementary methods and good practice guidance arising from the Kyoto protocol. Hiraishi T, Krug T, Tanabe K, Srivastava N, Baasansuren J, Fukuda M, Troxler TG, editors. Switzerland: IPCC; 2014.
5. IPCC, Intergovernmental Panel on Climate Change. IPCC Guidelines for National Greenhouse Gas Inventories. In: Eggleston S, Buendia L, Miwa K, Ngara T, Tanabe K, editors. Agriculture, forestry and other land use, vol. 4. Japan: Hayama; 2006.
6. Böttcher H, Verkerk PJ, Mykola G, Havlik P, Grassi G. Projection of the future EU forest CO_2 sink as affected by recent bioenergy policies using two advanced forest management models. GCB Bioenergy. 2012;4(6):773–83.
7. Groen T, Verkerk PJ, Böttcher H, Grassi G, Cienciala E, Black K, Fortin M, Köthke M, Lehtonen A, Nabuurs G-J, Petrova L, Blujdea V. What causes differences between national estimates of forest management carbon emissions and removals compared to estimates of large-scale models? Environ Sci Policy. 2013;33:222–32.
8. Kurz WA, Dymond CC, White TM, Stinson G, Shaw CH, Rampley G, Smyth C, Simpson BN, Neilson E, Trofymow JA, Metsaranta J, Apps MJ. CBM-CFS3: a model of carbon-dynamics in forestry and land-use change implementing IPCC standards. Ecol Model. 2009;220:480–504.
9. Mubareka S, Jonsson R, Rinaldi F, Fiorese G, San Miguel J, Sallnas O, Baranzelli C, Pilli R, Lavalle C, Kitous A. An integrated modelling framework for the forest-Based bioeconomy. IEEE Earthzine. 2014;7(2):908802.
10. Kurz WA, Apps MJ. A 70-year retrospective analysis of carbon fluxes in the Canadian forest sector. Ecol Appl. 1999;9:526–47.
11. Stinson G, Kurz WA, Smyth CE, Neilson ET, Dymond CC, Metsaranta JM, Boisvenue C, Rampley GJ, Li Q, White TM, Blain D. An inventory-based analysis of Canada's managed forest carbon dynamics, 1990–2008. Glob Chang Biol. 2011;17:2227–44.
12. Zamolodchikov DG, Grabovsky VI, Korovin GN, Kurz WA. Assessment and projection of carbon budget in forests of Vologda Region using the Canadian model CBM-CFS (in Russian, with summary in English). Lesovedenie. 2008;6:3–14.
13. Pilli R, Grassi G, Kurz WA, Smyth CE, Bluydea V. Application of the CBM-CFS3 model to estimate Italy's forest carbon budget, 1995 to 2020. Ecol. Modell. 2013;266:144–71.
14. Pilli R, Grassi G, Cescatti A. Historical analysis and modeling of the forest carbon dynamics using the Carbon Budget Model: an example for the Trento Province (NE, Italy). Forest @. 2014;11:20–35.
15. Pilli R, Grassi G, Moris JV, Kurz WA. Assessing the carbon sink of afforestation with the Carbon Budget Model at the country level: an example for Italy. Forest. 2014;8:410–21.
16. Pilli R, Grassi G, Kurz WA, Moris JV, Viñas RA. Modelling forest carbon stock changes as affected by harvest and natural disturbances. II. EU-level analysis including land use changes, Carbon Balance and Management. 2016. **(submitted).**
17. Lithuania. Lithuania's National Inventory Report 2014. URL (last Access Mar 2015). http://unfccc.int/national_reports/annex_i_ghg_inventories/national_inventories_submissions/items/8108.php.
18. KP CRF tables, 2014. URL (last Access Mar 2015): http://unfccc.int/national_reports/annex_i_ghg_inventories/national_inventories_submissions/items/8108.php.
19. Vanclay JK, Skovsgaard JP. Evaluating forest growth models. Ecol Model. 1997;98:1–12.
20. Germany. National Inventory Report for the German Greenhouse Gas Inventory 1990–2012. Federal Environment Agency, 2014. URL (last Access Feb 2016): http://unfccc.int/national_reports/annex_i_ghg_inventories/national_inventories_submissions/items/8108.php.
21. Ireland. Ireland National Inventory Report 2014. EPA Environmental Protection Agency. URL (Access Mar 2016): http://unfccc.int/national_reports/annex_i_ghg_inventories/national_inventories_submissions/items/8108.php.
22. Spain 2014. Inventario de emisiones de gases de efecto invernadero de Espana años 1990-2012. Ministerio de Agricultura, Alimentacio y Medio Ambiete. URL (Access Mar 2016): http://unfccc.int/national_reports/

annex_i_ghg_inventories/national_inventories_submissions/items/8108. php.

23. Pilli R, Fiorese G, Grassi G. EU Mitigation Potential of harvested wood products. Carbon Balance Manage. 2015;10:6.

24. Denmark. Denmark's National Inventory Report, 2014. Aarhus University, 2014. URL (last Access Feb 2016): http://unfccc.int/national_reports/ annex_i_ghg_inventories/national_inventories_submissions/items/8108. php.

25. Blujdea V, Raul AV, Federici S, Grassi G. The EU greenhouse gas inventory for LULUCF sector: I. Overview and comparative analysis of methods used by EU member states. Carbon Manag 2016;6(5–6):247–59.

26. Pilli R. Calibrating CORINE land cover 2000 on forest inventories and climatic data: an example for Italy. Int J Appl Earth Obs. 2012;9:59–71.

27. Boudewyn P, Song X, Magnussen S, Gillis MD. Model-based, volume-to-biomass conversion for forested and vegetated land in Canada. Canadian Forest Service, Victoria, Canada, 2007 (Inf. Rep. BC-X-411). URL (Access Mar 2015): http://cfs.nrcan.gc.ca/publications/?id=27434.

28. Li Z, Kurz WA, Apps MJ, Beukema SJ. Belowground biomass dynamics in the Carbon Budget Model of the Canadian Forest Sector: recent improvements and implications for the estimation of NPP and NEP. Can J For Res. 2003;33:126–36.

29. UNFCCC CRF tables. UNFCCC Common reporting format tables, 2014. URL (last Access Feb 2016): http://unfccc.int/national_reports/annex_i_ghg_inventories/national_inventories_submissions/items/8108.php.

30. FAOSTAT. FAOSTAT data, 2013. URL (last access March 2015): http://faostat3.fao.org/home/index.html#DOWNLOAD.

31. Gardiner B, Blennow K, Carnus JM, Fleischer P, Ingemarson F, Landmann G, Lindner M, Marzano M, Nicoll B, Orazio C, Peyron JL, Reviron MP, Schelhaas MJ, Schuck A, Spielmann M, Usbeck T. Destructive storms in European forests: past and forthcoming impacts. Final report to European Commission—DG Environment, 2010. URL (Access Mar 2015): http://www.efiatlantic.efi.int/portal/databases/forestorms/.

Projecting the spatiotemporal carbon dynamics of the Greater Yellowstone Ecosystem from 2006 to 2050

Shengli Huang[1], Shuguang Liu[2*], Jinxun Liu[3], Devendra Dahal[4], Claudia Young[5], Brian Davis[4], Terry L Sohl[2], Todd J Hawbaker[6], Ben Sleeter[3] and Zhiliang Zhu[7]

Abstract

Background: Climate change and the concurrent change in wildfire events and land use comprehensively affect carbon dynamics in both spatial and temporal dimensions. The purpose of this study was to project the spatial and temporal aspects of carbon storage in the Greater Yellowstone Ecosystem (GYE) under these changes from 2006 to 2050. We selected three emission scenarios and produced simulations with the CENTURY model using three General Circulation Models (GCMs) for each scenario. We also incorporated projected land use change and fire occurrence into the carbon accounting.

Results: The three GCMs showed increases in maximum and minimum temperature, but precipitation projections varied among GCMs. Total ecosystem carbon increased steadily from 7,942 gC/m^2 in 2006 to 10,234 gC/m^2 in 2050 with an annual rate increase of 53 $gC/m^2/year$. About 56.6% and 27% of the increasing rate was attributed to total live carbon and total soil carbon, respectively. Net Primary Production (NPP) increased slightly from 260 $gC/m^2/year$ in 2006 to 310 $gC/m^2/year$ in 2050 with an annual rate increase of 1.22 $gC/m^2/year$. Forest clear-cutting and fires resulted in direct carbon removal; however, the rate was low at 2.44 $gC/m^2/year$ during 2006–2050. The area of clear-cutting and wildfires in the GYE would account for 10.87% of total forested area during 2006–2050, but the predictive simulations demonstrated different spatial distributions in national forests and national parks.

Conclusions: The GYE is a carbon sink during 2006–2050. The capability of vegetation is almost double that of soil in terms of sequestering extra carbon. Clear-cutting and wildfires in GYE will affect 10.87% of total forested area, but direct carbon removal from clear-cutting and fires is 109.6 gC/m^2, which accounts for only 1.2% of the mean ecosystem carbon level of 9,056 gC/m^2, and thus is not significant.

Keywords: Climate change; Wildfires; Land cover and land use; Carbon sequestration; Yellowstone

Background

Climate change affects ecosystem carbon dynamics through multiple pathways, including altering biogeochemical cycles (e.g., productivity), disturbance regimes (e.g., fire), and land use [1,2]. For example, climate impacts land use by influencing suitability of the landscape to support a given land use. These climate change effects are often region specific and show spatial variability [3]. The study of carbon dynamics under climate change is always challenging because of the combined influence of fire regimes, land use change, data uncertainties, and spatial heterogeneity.

Fire is a major ecosystem disturbance and more frequent and severe fires caused by warmer, drier conditions under climate change might reduce forest productivity and carbon storage [4]. In many coniferous forests, stand-replacing fires affect carbon cycling and storage over large spatial extents and long time periods [5], and increasing fire frequency with climate change has short-term and long-term effects for carbon storage [5]. For example, Westerling, et al. [6] used climate projections and examined the likely changes in occurrence, size, and spatial location of large fires (>200 hectares) in

* Correspondence: sliu@usgs.gov
[2]USGS EROS Center, 47914 252nd Street, Sioux Falls, SD 57198, USA
Full list of author information is available at the end of the article

the Greater Yellowstone Ecosystem (GYE). They found continued warming could completely transform GYE fire regimes by the mid-21st Century, with profound consequences for many species and for ecosystem processes, including carbon storage [6]. Therefore, quantifying changes in forest carbon after disturbances is essential for managing future carbon emissions, especially given the uncertainties about forest carbon storage under future climate scenarios [7].

Changes in land use, including forest harvesting, continue in many landscapes, and those changes cannot only impact climate directly through alterations in the surface-energy budget [8] but also affect carbon sequestration [4]. Land use can be projected from climate change scenarios using approaches such as IMAGE [9]. The resulting land use changes influence carbon dynamics, which is demonstrated by Karjalainen, et al. [10], who compared a management-as-usual scenario with a multifunctional management scenario and evaluated the carbon accounting using the European Forest Information Scenario Model (EFISCEN). However, forecasting future trends in land use or forest management and examining the impact on carbon remains difficult.

Many General Circulation Models (GCMs) are used to predict climate change for given emission scenarios. However, predictions vary among GCMs, which may have a major effect on carbon modeling [4]. For example, Schaphoff, et al. [11] used five GCMs from one scenario and found the increase in global Net Primary Production (NPP) ranged from 16% to 32%. Therefore, multiple GCM outputs are desired to simulate the carbon differences among different GCMs.

The spatial heterogeneity of data adds more complexity to carbon modeling under climate change for several reasons. First, the trends of climate change are non-uniform through space and time; therefore, the impact on carbon cycling shows heterogeneity [1,2]. Second, fire regimes may show geographic differences because of the spatial variation in precipitation [12]. Third, landscape patterns in forest structure and stand age need to be considered in estimates of future carbon flux across landscapes [13], which is confirmed by Smithwick, et al. [14], who quantified the carbon storage for young and mature stands and showed the variation in tree density might influence carbon flux under differing climate change scenarios. By further incorporating climate change, Smithwick, et al. [15] integrated CENTURY version 4.5 to project future carbon stocks of individual stands under different climate scenarios and fire regimes based on three GCMs, Community Climate System Model (CCSM) 3.0, Centre National de Recherches Météorologiques Circulation Model (CNRM) CM 3.0, and Geophysical Fluid Dynamics Laboratory Climate Model (GFDL CM) 2.1, forced with the A2 emissions pathway. They reaffirmed spatial

variation is critical for understanding the spatial pattern in total ecosystem carbon stocks across the landscape and ignoring the spatial variation across heterogeneous landscapes may lead to erroneous expectations on ecosystem carbon storage.

A flexible modeling approach that can incorporate sufficient interaction, contingency, and site specificity is required to examine how concurrent changes in climate, disturbance regimes, and land use influence ecosystem carbon budgets [16,17]. The GYE is a nearly intact ecosystem subject to the changes in climate, land use, and wildfire disturbance. How these concurrent changes influence the carbon dynamics at the landscape level remains unclear. The objective of this study was to use multi-source GCMs to model the spatiotemporal carbon storage in GYE associated with changes in fire, land use, and climate. We asked to what degree climate change would affect the carbon pools and fluxes in this ecosystem and whether GYE would be a carbon sink or carbon source. We hypothesized that carbon dynamics under climate change would show significant spatial variation due to the highly heterogeneous landscape in GYE. We additionally hypothesized that economy development and climate warming would result in more forest disturbance of clear-cutting and wildfires. To achieve the goal, we selected three climate change scenarios (A2, A1B, B1), then processed the CENTURY model under the General Ensemble Biogeochemical Modeling System (GEMS; [18]) using three GCMs for each scenario. We also projected the land use change and fire occurrence and incorporated their projections into the carbon accounting. By incorporating all these components, we examined the spatial and temporal carbon change of the GYE under climate change during 2006 to 2050.

Methods

Study area
GYE is one of the last remaining large, nearly intact temperate ecosystems in North America. GYE comprises 80,000-km^2 of the Rocky Mountains and encompasses two national parks, six national forests, and three wildlife refuges (Figure 1). The GYE features a continental climate of cold, snowy winters and warm, dry summers. The mean high temperatures in July are 21°C on the plateau and 24°C at mid-elevations. The mean low temperature in January is about −15°C across the region. The mean annual precipitation ranges from 600–1100 mm on the plateau to 350–650 mm at mid-elevations [19]. Lodgepole pine (*Pinus contorta*), which occupies infertile volcanic (rhyolitic) soils across the Yellowstone Plateau, and Douglas fir (*P. menziesii*), which occupies moderately fertile (non-rhyolitic, sedimentary) soils on adjacent sloping terrain, account for about two-thirds of the forested area of the GYE. Other tree species include Engelmann spruce (*Picea engelmannii*),

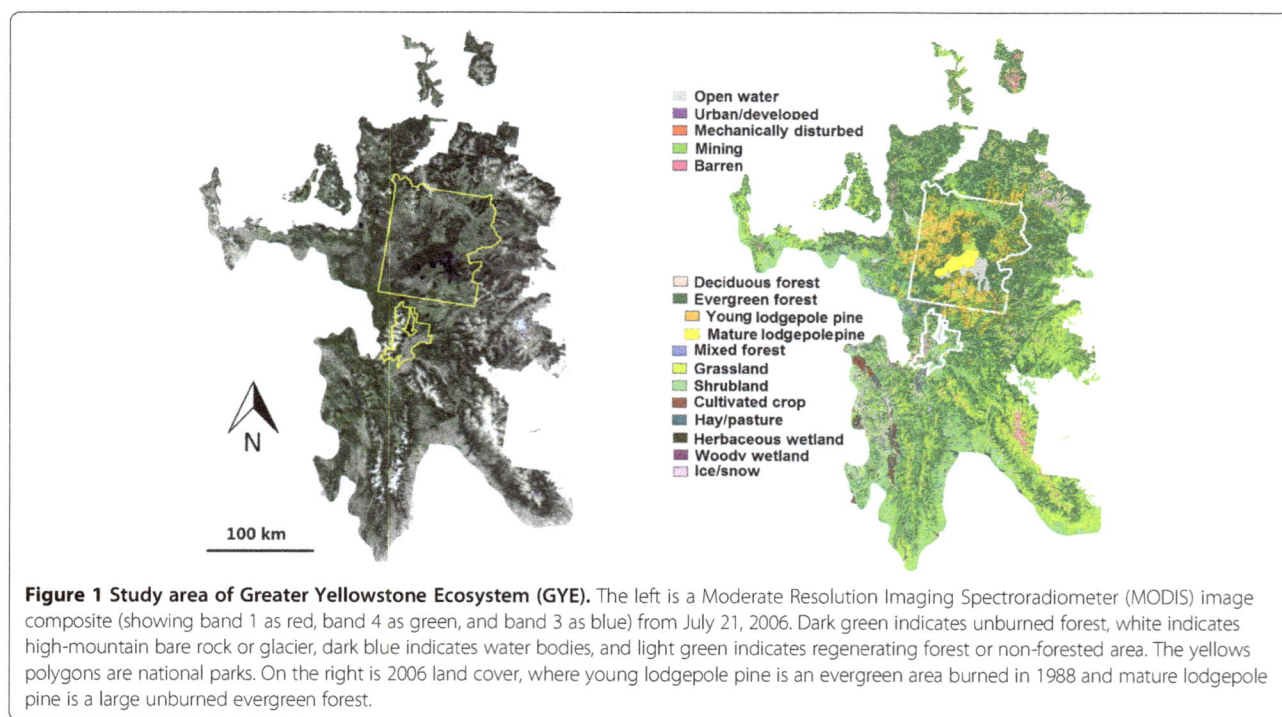

Figure 1 Study area of Greater Yellowstone Ecosystem (GYE). The left is a Moderate Resolution Imaging Spectroradiometer (MODIS) image composite (showing band 1 as red, band 4 as green, and band 3 as blue) from July 21, 2006. Dark green indicates unburned forest, white indicates high-mountain bare rock or glacier, dark blue indicates water bodies, and light green indicates regenerating forest or non-forested area. The yellows polygons are national parks. On the right is 2006 land cover, where young lodgepole pine is an evergreen area burned in 1988 and mature lodgepole pine is a large unburned evergreen forest.

subalpine fir (*Abies lasiocarpa*), and whitebark pine (*Pinus albicaulis*) on moist/high-elevation sites, or limber pine (*Pinus flexilis*) and Rocky Mountain juniper (*Juniperus scopulorum*) on dry/low-elevation sites [19].

Wildland fires are historically common in the GYE. The fire return intervals in GYE forests have been about 100–300 years for the past 10,000 years, and the intervals in the lower elevation forest-steppe vegetation is about 75–100 years [6]. An extensive fire event burned the area in 1988, which resulted in heterogeneous lodgepole pine regeneration [20]. Fine litter, branches, and foliage can be consumed and live trees can be killed during canopy fires, but the carbon lost from the pools of tree boles, downed wood, and soil is low [21]. A period of 70–100 years is required to recover the carbon losses following lodgepole pine stand-replacing fire [22].

A climate change study has indicated GYE will have elevated temperatures, reduced winter precipitation, earlier snowmelt and spring runoff, and higher potential evapotranspiration in the future [23]. Under these changes, the number of large fires has increased in the past 25 years, and this trend is expected to continue with global warming [24]. The fire rotation may decrease to <30 years with a 4.5–5.5°C warmer spring-summer temperature by mid-century [6]. Although largely undeveloped, GYE is also undergoing a transition in human demographics and economics. The population increased 58% and the area of rural lands increased 350% from 1970 to 1999 [25]. With local economic development, the communities of the GYE have undergone rapid change,

especially within the 32% of the GYE that is privately owned [25]. For example, during 1975–1995, there were increases in burned and urban areas but decreases in conifer habitats [26]. Forest harvesting in the national forests during the mid-20th Century created patchy mosaics of small, dispersed clear-cuts in some areas, but extensive portions of the GYE remain federally protected wildlands [27].

Datasets

The Intergovernmental Panel on Climate Change Special Report on Emission Scenarios (IPCC-SRES) published different scenarios exploring future emissions pathways [28]. Three IPCC-SRES scenarios (A1B, A2, and B1) were used in this study. For each scenario, we collected climate data from the Coupled General Circulation Model 3.1 (CGCM 3.1) [29], Australia's Commonwealth Scientific and Industrial Research Organization Mark 3.0 model (CSIRO–Mk3.0) [30], and the Model for Interdisciplinary Research on Climate version 3.2, medium resolution (MIROC 3.2-medres) [31]. Based on these three scenarios and three GCM models, nine data-scenario combinations were used: CGCM-A1B, CGCM-A2, CGCM-B1, CSIRO-A1B, CSIRO-A2, CSIRO-B1, MIROC-A1B, MIROC-A2, and MIROC-B1. In GYE, all climate projections show an obviously increasing trend in maximum and minimum temperatures (Figures 2 and 3), but this uniform trend is not observed for projected precipitation (Figure 4): the largest precipitation increase was from CGCM-A2 with an increasing rate of 1.9 mm/y followed by

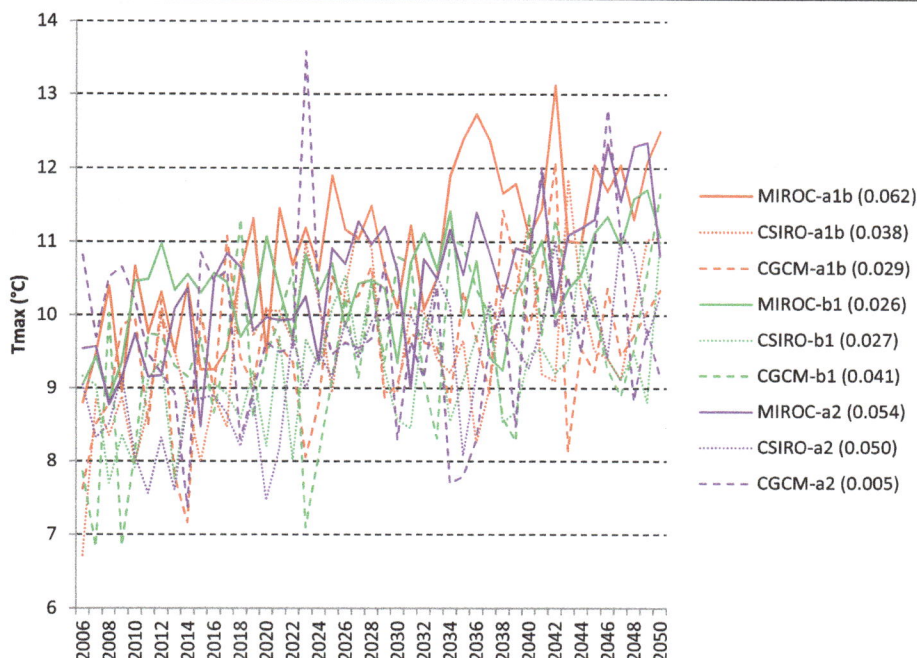

Figure 2 Projected annual average maximum temperature (T_{max}) for different data-scenario combinations. The values in the legend that are in parentheses are the slopes of linear regressions, indicating the annual increasing rates.

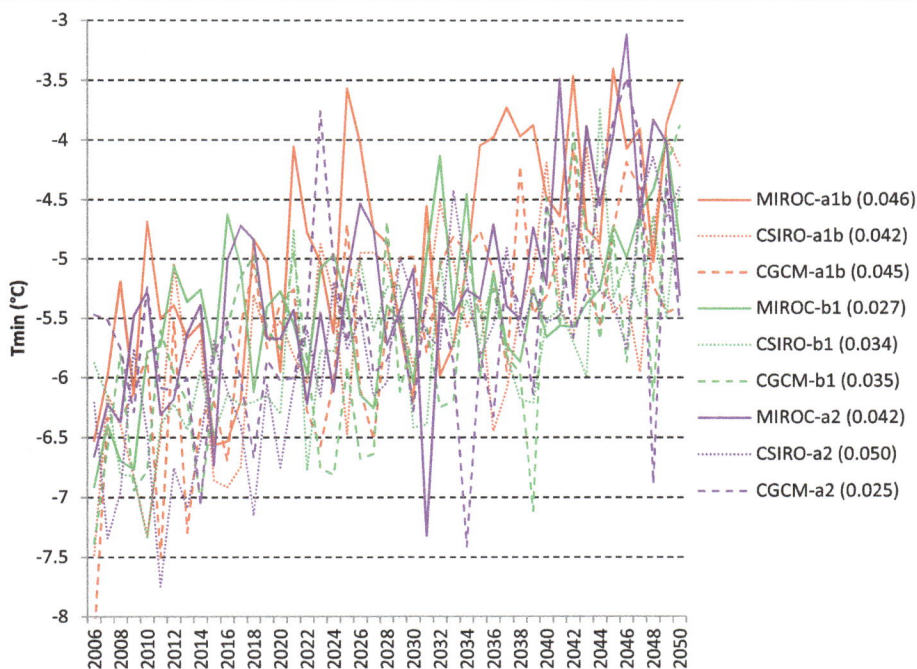

Figure 3 Projected annual average minimum temperature (T_{min}) for different data-scenario combinations. The values in the legend that are in parentheses are the slopes of linear regressions, indicating the annual increasing rates.

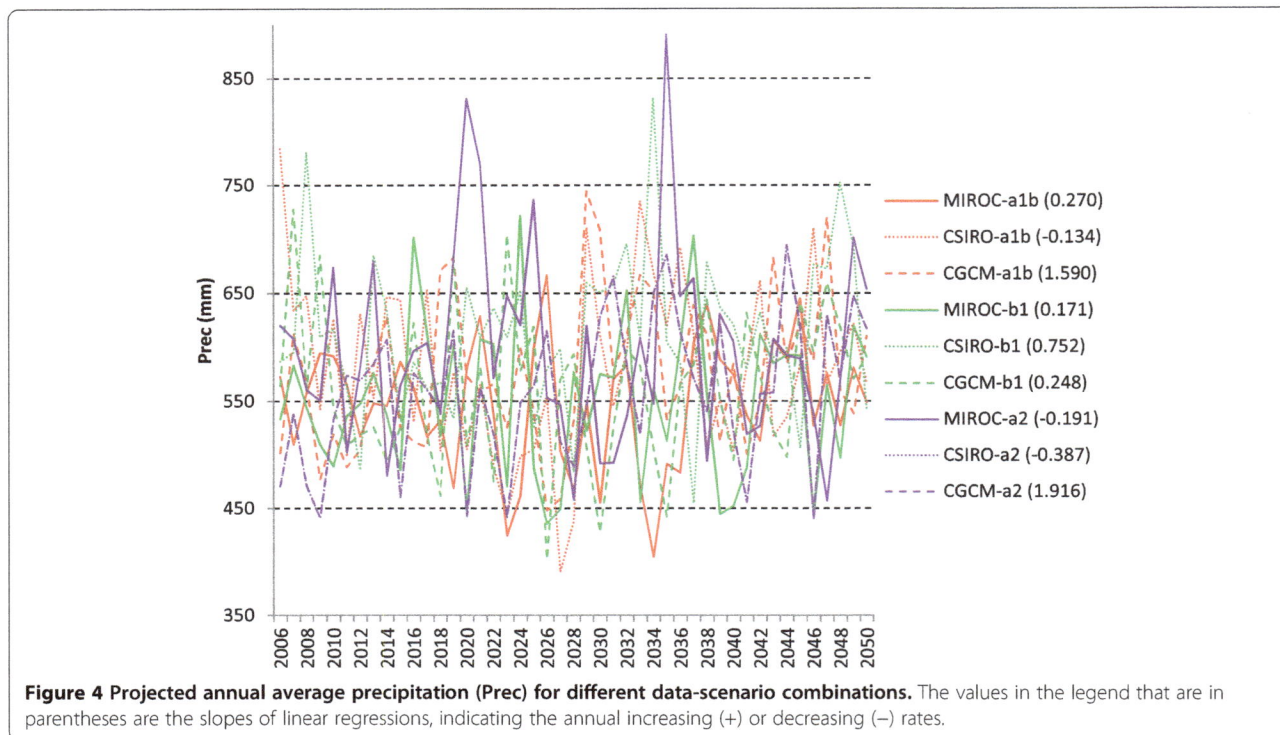

Figure 4 Projected annual average precipitation (Prec) for different data-scenario combinations. The values in the legend that are in parentheses are the slopes of linear regressions, indicating the annual increasing (+) or decreasing (−) rates.

CGCM-A1B (1.6 mm/y) and CSIRO-B1 (0.8 mm/y); the largest precipitation decrease was from CSIRO-A2 with a decreasing rate of 0.39 mm/y followed by MIROC-A2 (0.2 mm/y) and CSIRO-A1B (0.1 mm/y).

Topography (elevation, slope, and aspect) data were retrieved from the U.S. Geological Survey (USGS) National Elevation Dataset [32]. Forest data were collected from U.S. Forest Service's Forest Inventory & Analysis [33]. Soil data were mainly compiled from the Soil Survey Geographic (SSURGO) database [34], but the State Soil Geographic (STATSGO) database [35] was used where SSURGO data were not available.

Projections and modeling

We used the GEMS-CENTURY model to simulate the carbon dynamics in GYE by incorporating wildfire and land use change. The land-cover and land use projections, wildfire projections, and biogeochemical modeling are described in the following sections. The models used in this local study were validated for use at regional and national scale [36]; therefore, this study is a "bird's-eye view" of the GYE and does not take into account some specific features of this ecosystem.

Land Use and Land-Cover (LULC) projection from 2006 to 2050

IPCC-SRES storylines were designed to represent different socioeconomic development pathways, with the assumption of different driving forces such as energy sector, population growth, technological innovation, economic growth, environmental protection, and regional/global orientation [28] (Table 1). A scenario downscaling process was used to

Table 1 Assumptions about the primary driving forces affecting land-use and land-cover change

Driving forces	A1B	A2	B1
Population growth (global and United States)*	Medium. Globally, 8.7 billion by 2050, then declining; in the United States, 385 million by 2050	High. Globally, 15.1 billion by 2100; in the United States, 417 million by 2050	Medium. Globally, 8.7 billion by 2050, then declining; in the United States, 385 million by 2050
Economic growth*	Very high. U.S. per-capita income $72,531 by 2050	Medium. U.S. per-capita income $47,766 by 2050	High. U.S. per-capita income $59,880 by 2050
Regional or global orientation	Global	Regional	Global
Technological innovation	Rapid	Slow	Rapid
Energy sector	Balanced use	Adaptation to local resources	Smooth transition to renewable
Environmental protection	Active management	Local and regional focus	Protection of biodiversity

*Population and per capita income projections are from [9].

translate the coarse-scale scenario data to finer geographic scales while maintaining consistency with the original dataset and local data [37]. A global integrated assessment model (IAM) was then used to supply future projections of land use at the national scale. An accounting model was developed to refine the national-scale IAM projections and to downscale to the ecoregion where the study area is located [38]. The spatially explicit land-cover projections from 2006 to 2050 were developed using a spatially explicit LULC change model called the FOREcasting SCEnarios of land use change (FORE–SCE) model [39]. The FORE–SCE model used separate but linked "Demand" and "Spatial Allocation" components to produce spatially explicit, annual LULC maps. The "Demand" component provided aggregate-level quantities of LULC change for a region, or a "prescription" for the overall regional LULC proportions. The "Spatial Allocation" component ingested "Demand" and produced spatially explicit LULC maps using a patch-based allocation procedure [39].

It is inappropriate to assume that a simple extrapolation of historical and current trends would precisely represent the future landscape, as landscape trends change in response to socioeconomic (and climate) conditions. By using a multiple scenario approach, we could capture a range of potential future landscapes under different socioeconomic assumptions. Each of the three scenarios (see Table 1) captures different levels of clear-cutting, with the environmentally focused B1 scenario representing lower levels of clear-cutting than the economically focused A1B and A2 scenarios. In FORE-SCE, no clear-cutting was allowed to occur within National Park or Wilderness Area lands, and rural development in each of the scenarios was defined to be extremely low.

The final product was 2006–2050 annual land cover and land use maps at 250-m resolution, containing the 17 classes of Open water, Developed, Mechanical disturbed national forest, Mechanical disturbed other public forest, Mechanical disturbed private lands, Mining, Barren, Deciduous forest, Evergreen forest, Mixed forest, Grassland, Shrubland, Cultivated crop, Hay/pasture, Herbaceous wetland, Woody wetland, and Perennial snow/ice [39]. When a land was converted from one type to another (e.g., forest conversion to grassland), the carbon change was quantified based on IPCC good practice guidance [40]. Therefore, carbon removal due to forest clear-cutting could be tracked.

Wildfire projection from 2006 to 2050

Wildfire projections, driven by daily weather conditions, were generated for the study area using a spatially explicit simulation model [41]. Daily weather data were generated by temporally disaggregating the projected monthly temperature and precipitation data [42] and historical daily weather data with 1/8° spatial resolution [43]. Wind

direction and speed information was provided by the North American Regional Reanalysis [44]. The Mountain Climate Simulator (MT–CLIM) [45] was used to calculate relative humidity using the daily temperature and precipitation data. Daily live and dead fuel moistures, and wildland-fire behavior indices were then estimated using the National Fire Danger Rating System (NFDRS) algorithms [46].

Wildfire ignition locations were stochastically generated using General Linear Models (GLMs) and fit using historical weather data and fires. The spread of wildfires from individual ignition locations was simulated with the Minimum Travel Time (MTT) algorithm [47] using surface and canopy fuels [48], topography (elevation, slope, and aspect), weather (wind speed and direction), and live and dead fuel moisture data. The outputs produced by the MTT algorithm included the burned pixels as well as metrics of crown fire activity, which were used as a proxy measure of burn severity (low, medium, and high).

For each pixel burned in the simulations, the First Order Fire Effects Model (FOFEM) [49] used fuel loads along with fuel moistures to estimate the amount of forest litter and downed deadwood consumed. The consumption of duff (decaying forest litter), trees, plants, and shrubs was estimated as a function of the region, season, fuel moistures, and fuel loads. When calculating emissions with the FOFEM, 20-, 60-, and 100-percent canopy consumption was assumed for low, moderate, and high burn severity, respectively, on the basis of published literature [50]. Therefore, carbon removal by fire consumption was quantified.

Before wildfire projections were made, the ignition and spread components of the wildland-fire modeling system were calibrated with the historical Monitoring Trends in Burn Severity (MTBS) data [51]. More details of the wildfire modeling methodology can be found in Hawbaker, et al. [41,52].

Biogeochemical cycles and carbon modeling

The GEMS modeling system was used in this study. GEMS was designed to provide spatially explicit biogeochemical model simulations over large areas. It was developed to better integrate well-established ecosystem biogeochemical models and employs a Monte-Carlo–based ensemble approach to evaluate model uncertainties [18]. The underlying biogeochemical model is CENTURY 4.0 (http://www.nrel.colostate.edu/projects/century, see Metherell, et al. [53]). The details of handling data discrepancy (e.g., when comparing MODIS NPP with modeling NPP, the former could capture real-time disturbances, but the latter did not incorporate these disturbances) and data frequency (e.g., low FIA measurements probably did not capture tree mortality sufficiently) during the model initialization, calibration, and validation can be found in Liu, et al. [18].

For model initialization, soil thickness, organic carbon storage, texture (fractions of sand, silt, and clay), bulk density, and drainage were initialized from the soil database. The total soil organic carbon pool was partitioned into active (5 percent), slow (45 percent), and passive (55 percent) pools. These percentages were only used for starting initialization. The model used 10–20 years to approach soil carbon equilibrium and the values were close to SSURGO soil carbon. Forest biomass carbon pools, including the forest litter biomass, aboveground live biomass, belowground live biomass, down deadwood biomass, and standing dead biomass, were derived from the data of Forest Inventory and Analysis (FIA, http://www.fia.fs.fed.us/), forest type (evergreen, broadleaf, and mixed), and the forest age-carbon stock relation.

For model calibration, the 2001–2005 observed data were compared to modeling output and the model parameters were adjusted to minimize the difference between simulations and observations. The observed data for calibration included (1) county-based grain-yield-survey data by crop type, published by the U.S. Department of Agriculture (USDA) [33]; and (2) 250-m MODIS Net Primary Production (NPP) for forests and grasslands. During the calibration process, the potential maximum production parameter (PRDX) was adjusted to minimize the grain yield difference (i.e., modeled grain yield versus USDA county-level grain yield) and the forest NPP difference (i.e., modeled NPP versus MODIS NPP at the county level).

For model processing, CENTURY simulated NPP, photosynthetic allocation, litter fall, mortality, decomposition of plant tissues, and soil organic carbon at monthly steps from 2006 to 2050. The annual CO_2 concentration increase was included in the modeling, but CO_2 concentration remained constant in the spatial domain. Important monthly and annual carbon-related variables were output from the nine modeling combinations:

- FSYSC: Total ecosystem carbon storage
- FRSTC: Total living carbon, including both aboveground and belowground biomass
- SOMSC: Total soil carbon excluding litter and structural carbon
- CPRODA: Net carbon production (i.e., NPP)
- TCREM: Carbon removal from ecosystems by clear-cutting and fire consumption

The modeling performance was validated by comparing the simulation with the corresponding observation, which included USDA forest biomass values, aboveground biomass from the National Biomass and Carbon Dataset 2000 [54], MODIS NPP, and the USDA grain yield for 2006, 2008, and 2010.

For the nine simulation experiments, the variation of carbon outputs is expressed as the "standard deviation (V)" defined as

$$V = \sqrt{\frac{1}{N}\sum_{i=1}^{N}(x_i - \mu)^2} \qquad (1)$$

where N is the number of the data-scenario combinations of 9, x_i is each individual variable, and μ is the mean of x_i.

To compare the different effects of fires on carbon modeling, the forest area regenerating from the 1988 fires (age is 18 years in 2006) was selected as young forest and a neighboring unburned forest stand was selected as mature forest (see Figure 1).

Results

Carbon pools and fluxes for mature and young forest in 2006

For the selected mature and young forests (see Figure 1), different combinations of data and scenarios in our modeling resulted in the carbon storage and fluxes depicted in 2006 (Table 2). When averaged over the data-scenario combinations, the young and mature forest respectively had 7,874 gC/m² and 9,534 gC/m² for the total ecosystem carbon, 3,389 gC/m² and 4,310 gC/m² for total living biomass, and 301 gC/m² and 324 gC/m² for net primary production (NPP). The standard deviations in each data-scenario implied there was substantial spatial variation in carbon storage and fluxes.

Carbon change for entire GYE

We produced spatially explicit layers for variables of total ecosystem carbon, total living carbon, total soil carbon, net carbon production, and carbon removal from 2006 to 2050 at 250-m resolution. Maps for all nine modeling combinations were produced, but only the total ecosystem carbon data in 2006 and 2050 for CGCM-A2 are shown in Figure 5 as a demonstration. Figure 5 shows there are significant spatial variations in total ecosystem carbon, indicating the heterogeneity of carbon storage. In Figure 5a, the mean of total ecosystem carbon in 2006 was 7,885 gC/m² and the standard deviation was 4,688 gC/m². In Figure 5b, the mean of total ecosystem carbon in 2050 was 10,076 gC/m² and the standard deviation was 5,325 gC/m². This phenomenon of landscape heterogeneity could be observed for all map layers.

Our modeling also produced time series carbon pools and fluxes (Figure 6). In the GYE ecosystem, all nine data-scenario combinations run under the CENTURY model showed increasing trends of total ecosystem carbon, indicating incrementally more carbon will be sequestered in the GYE ecosystem. This implied GYE

Table 2 Modeling results for mature and young forest in 2006

| | cgcm | | | | | | Csiro | | | | | | miroc | | | | | | Data/scenario combination |
| | B1 | | A2 | | A1B | | B1 | | A2 | | A1B | | B1 | | A2 | | A1B | | |
	mean	std	mean	std	mean	std	mean	std	mean	std	mean	std	Mean	std	mean	std	mean	std	mean
FSYSC																			
Young	7889	2751	7805	2726	7885	2746	7898	2760	7878	2750	7874	2747	7868	2747	7853	2739	7913	2760	7874
Mature	9551	1589	9428	1554	9498	1564	9575	1600	9524	1581	9622	1633	9510	1577	9478	1561	9619	1616	9534
FRSTC																			
Young	3403	1398	3343	1378	3374	1393	3405	1399	3392	1396	3406	1389	3390	1396	3369	1389	3418	1405	3389
Mature	4322	1054	4251	1038	4274	1043	4339	1058	4300	1048	4364	1070	4300	1048	4272	1042	4364	1065	4310
CROPDA																			
Young	322	69	228	66	282	63	331	74	302	66	318	128	303	62	273	59	348	81	301
Mature	338	58	239	40	274	50	365	66	306	57	410	98	309	52	269	51	405	83	324

• Note: FSYSC refers to total ecosystem carbon storage (gC/m²); FRSTC refers to total living carbon, including both aboveground and belowground biomass (gC/m²); and CPRODA refers to net carbon production (gC/m²/year).

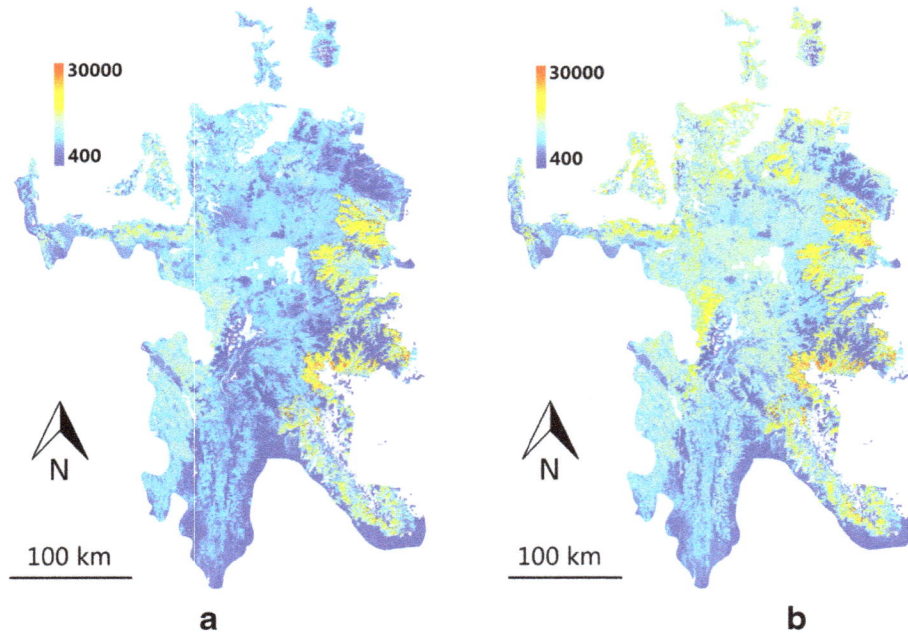

Figure 5 Total ecosystem carbon (gC/m²) in (a) 2006 and (b) 2050 resulted from CGCM-A2. Note the abrupt change in the eastern part was caused by the difference between SSURGO (right side) and STATSGO (left side). The mean difference of soil organic matter between SSURGO and STATSGO along this abrupt change line was about 6,993 gC/m².

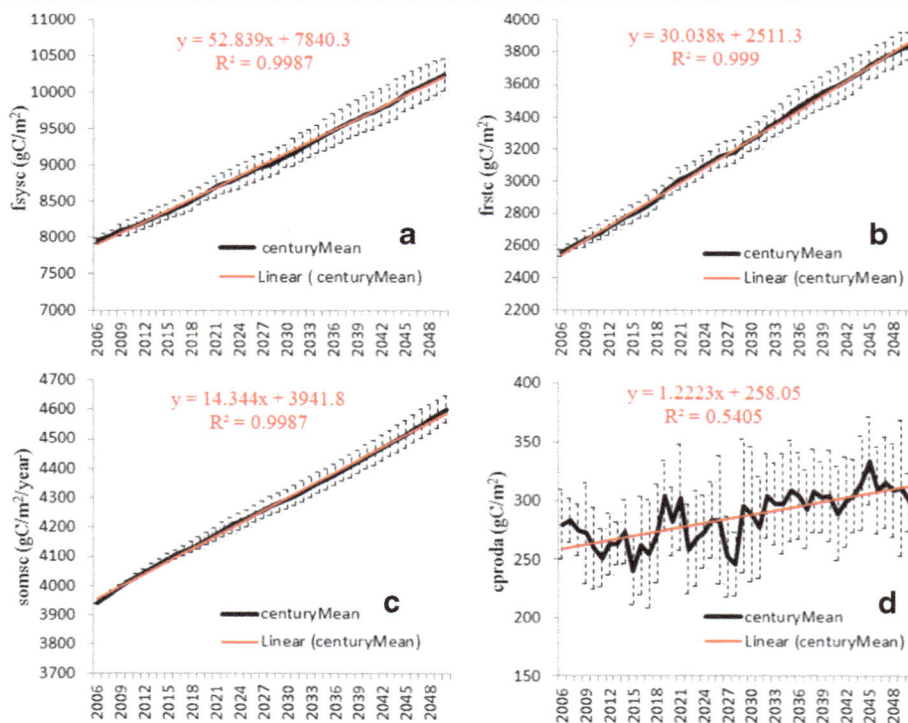

Figure 6 Modeling results of total ecosystem carbon (FSYSC) (a), live biomass carbon (FRSTC) (b), soil organic carbon (SOMSC) (c), and net primary production (CPRODA) (d) in Greater Yellowstone Ecosystem. Red lines are linear regressions, and the error bars are standard deviations calculated with equation 1.

will be a carbon sink in a future with climate change. The variance among the combinations was within 5.5%, indicating the difference was very little. Therefore, we calculated the annual mean of total ecosystem carbon with an error bar of standard deviation and performed a linear regression (Figure 6a), which shows the total ecosystem carbon increased steadily from 7,942 gC/m^2 in 2006 to approximately 10,234 gC/m^2 in 2050 (i.e., 28.9% increase), with a mean value of 9,056 gC/m^2. The extra carbon sequestration was 2,292 gC/m^2 during 2006–2050, and the average annual increasing rate, which was reflected by the slope, was 53 gC/m^2/year.

Total live carbon is an important component of total ecosystem carbon. The annual change of this carbon pool (Figure 6b) increased steadily from 2,551 gC/m^2 in 2006 to 3,833 gC/m^2 in 2050. The average annual rate increase is about 30 gC/m^2/year, which accounts for 56.6% of the annual rate increase of 53 gC/m^2/year in total ecosystem carbon. The time-series total soil carbon (Figure 6c) also steadily increased from 3,939 gC/m^2 in 2006 to 4,601 gC/m^2 in 2050. The average annual rate increase was 14.3 gC/m^2/year, which contributed 27% of the annual rate increase of 53 gC/m^2/year of total ecosystem carbon. All of these changes in total ecosystem carbon, total live carbon, and total soil carbon indicated the GYE under climate change is a carbon sink and can sequester 2,292 gC/m^2, and 83.6% of the extra carbon sequestrated during 2006–2050 is attributed to the carbon pools of live biomass and soil organic matter.

Associated with the increased total soil carbon, live biomass carbon, and total ecosystem carbon, forest productivity increased as reflected by the net carbon production (Figure 6d). Different data-scenario combinations resulted in significant differences, which could be reflected in the high standard deviations ranging from 18.3 gC/m^2/year to 57.7 gC/m^2/year during 2006–2050, indicating GCMs have great effect on NPP. However, linear regression shows net carbon production increased slightly from 260 gC/m^2/year in 2006 to 310 gC/m^2/year in 2050 with an annual rate increase of 1.22 gC/m^2/year, implying the forest productivity increased approximately 19.2% under climate change.

Forest clear-cutting, wildfires, and carbon removal

Despite the increased net carbon gain and forest productivity during 2006–2050, forest clear-cutting and fires resulted in direct carbon removal (Figure 7). The standard deviation bars show this annual carbon loss from the removal events had high variation among scenarios and the magnitude ranged from 1.4 gC/m^2/year in 2015 to 4.2 gC/m^2/year in 2039, with a mean removal of about 2.44 gC/m^2/year during 2006–2050 (Figure 7a). The linear slope of 0.02 indicates there is a very low annual rate increase of carbon removal from clear-cutting and wildfires. The total carbon removal during 2006–2050 was 109.6 gC/m^2, which accounts for only 1.2% of the mean ecosystem carbon level of 9,056 gC/m^2 over 45 years, and thus is negligible.

Clear-cutting and fires have different distribution in national parks (NP) and national forests (NF) (Figure 7b). Our modeling indicates wildfires would occur across the entire GYE. However, the clear-cutting was only distributed in NF, which can be clearly distinguished by the abrupt change along the border between NP and NF (see the

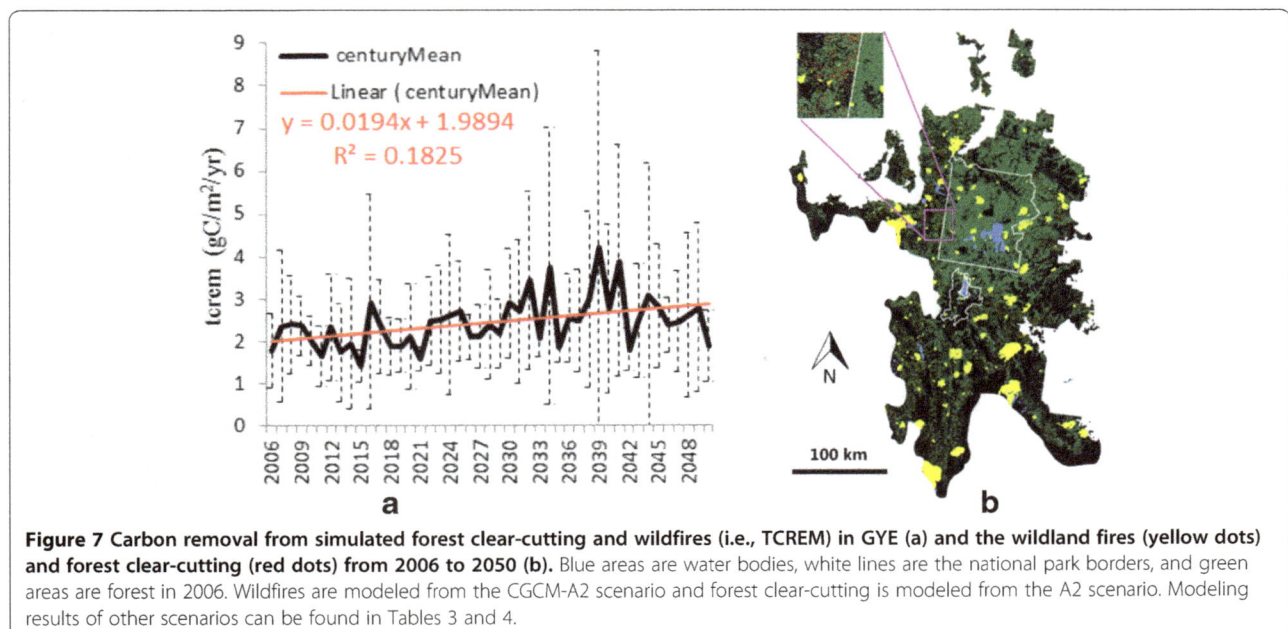

Figure 7 Carbon removal from simulated forest clear-cutting and wildfires (i.e., TCREM) in GYE (a) and the wildland fires (yellow dots) and forest clear-cutting (red dots) from 2006 to 2050 (b). Blue areas are water bodies, white lines are the national park borders, and green areas are forest in 2006. Wildfires are modeled from the CGCM-A2 scenario and forest clear-cutting is modeled from the A2 scenario. Modeling results of other scenarios can be found in Tables 3 and 4.

magnified portion in Figure 7b). The clear-cutting ratio in NF ranged from 4.87% under the B1 scenario to 6.34% under the A1B scenario with an average ratio of 5.64% (Table 3). In addition, Table 4 shows the burned forest ratio in NP ranged from 3.19% under the CSIRO-B1 scenario to 12.25% under CSIRO-A1B with an average ratio of 6.08%. Table 4 also shows 5.72%–7.49% (mean is 6.55%) of the forested area in NF would be burned under future climate change. With clear-cutting and wildfires combined, 6.11% of the forested area in NP and 12.19% of the forested area in NF would be affected. Together, the area of clear-cutting and wildfires in GYE would account for 10.87% of total forested area during 2006–2050.

It should be emphasized that within national parks, fires were modeled, but clear-cutting was not allowed. Any increased forest disturbance in national park land would be solely due to a more active fire regime. For forest disturbance related to clear-cutting outside of national park lands and national forest lands, the use of multiple scenarios allows us to examine multiple socioeconomic pathways affecting forest management in the region. We believe it is more insightful to provide a range of potential futures than to model what will "probably" happen. The B1 scenario maintained levels of clear-cutting similar to 2005. The other two scenarios projected increases due to the socioeconomic assumptions within those scenarios.

Discussion

GCM outputs are the basis for carbon projection under climate change. Our precipitation change ranged from –0.387 mm/y to 1.916 mm/y (Figure 4). In the study of Smithwick, et al. [14], the precipitation was predicted to increase 21 mm [from Hadley (HAD) source] to 32 mm [from Canadian Climate Center (CCC) source] during 1994–2100 (i.e., 0.198 mm/y from HAD and 0.302 mm/y from CCC). This change rate was higher than found in this study. In Smithwick, et al. [14], the average annual maximum temperatures were expected to increase 2.8°C (HAD) to 4.3°C (CCC) (i.e., 0.026°C/y and 0.041°C/y), which falls within or is greater than the upper limit of our projection data (Figure 2). However, the average annual minimum temperatures used by Smithwick, et al. [14] were expected to increase 4.7°C (HAD) to 9.1°C (CCC) (i.e., 0.044°C/y and 0.086°C/y), which falls

Table 3 The area of forest cutting within national forests under climate scenarios

	National forest		
	B1	A2	A1B
Clear-cutting area (km²)	1288	1510	1677
Clear-cutting ratio (%)*	4.87	5.71	6.34

*In 2006, forested area within NP is 7,321 km² and forested area within NF is 26,457 km².

within our projection data (Figure 3). The difference may be attributed to at least two reasons. First, precipitation prediction is more difficult than temperature prediction, which may lead to greater variance among GCMs. Second, the time scale of their projection was 1994–2100, which was longer than our time scale of 2006–2050. The variation among different GCMs indicated a more robust data source to be used for future carbon modeling.

Wildfires result in forest age mosaics that affect carbon storage. Our NPP values for 18-year-old forest stands and mature forest were 301 gC/m² and 324 gC/m², respectively; these values are comparable to results from previous studies. Litton, et al. [55] examined how aboveground NPP (ANPP) and belowground NPP (BNPP) varied with fire-initiated differences in tree density and stand age in lodgepole pine stands in Yellowstone National Park. They found the annual ANPPs were 59, 122, 156, and 218 gC/m² and the annual BNPPs were 68, 237, 306, and 382 gC/m² for low-, moderate-, and high-density young stands (13-year-old) and mature stands, respectively. Their values indicate the annual NPP, which is the addition of ANPP and BNPP, was 316 gC/m² for 13-year-old forest stands and 600 gC/m² for mature forest stands. Our NPP for 18-year-old forest stands was 301 gC/m², which was lower than the measured NPP value of 316 gC/m² by Litton, et al. [55] but higher than the modeled 2004 NPP of 245–253 gC/m² for early and middle successional lodgepole pine stands by Crabtree, et al. [56]. Our NPP for mature forest was 324 gC/m², which was 46% lower than the value of 600 gC/m² by Litton, et al. [55] but 28 ~ 35% higher than the 2004 NPP value of 240–253 gC/m² for middle and late successional lodgepole pine stands reported by Crabtree, et al. [56]. In general, both our young and mature forest NPP were lower than the measured NPP of Litton, et al. [55]. One possible reason is that these stands were located in the areas where STATSGO soil data were used (see Figure 5). The nutrition supplies, which affect the plant production, were lower in the coarser STATSGO database than in SSURGO.

In our study, the total living carbon of young forest in 2006 (regenerating from 1988 fires) was 3,389 gC/m². Kashian, et al. [57] measured carbon pools for 77 lodgepole pine stands in and around Yellowstone National Park (YNP) along a 300-year chronosequence. They showed the live vegetation carbon can be modeled from stand age using a Michaelis-Menton function. According to this function, the forest burned in 1988 (with an age of 18 in 2006) is 3,417 gC/m². Our result of 3,389 gC/m² agreed with the prediction of 3,417 gC/m² very well. However, the mature forest (with an age of 100–300) predicted by the expression of Kashian, et al. [57] is 7,775–9,560 gC/m², which is higher than our result of 4,310 gC/m². Nevertheless, Kashian, et al. [57] also showed their total living biomass carbon had

Table 4 The area of forest fires within national parks and national forests under climate change scenarios during 2006–2050

	National park									National forest								
	cgcm			csiro			miroc			cgcm			csiro			miroc		
	B1	A2	A1B	B1	A2	A1B	B1	A2	A1B	B1	A2	A1B	B1	A2	A1B	B1	A2	A1B
Burned forest (km^2)	344	322	441	233	275	897	706	510	280	1837	1607	1664	1515	1821	1981	1810	1689	1675
Burned forest (%)*	4.70	4.40	6.02	3.19	3.75	12.25	9.65	6.97	3.82	6.94	6.07	6.29	5.72	6.88	7.49	6.84	6.38	6.33

*In 2006, forested area within NP is 7,321 km^2 and forested area within NF is 26,457 km^2.

significant variation among mature forest stands, and our value of 4,310 gC/m^2 still fell within their range.

Our results showed the young forest in 2006 had a total ecosystem carbon of 7,874 gC/m^2 and the mature forests have a total ecosystem carbon of 9,534 gC/m^2. Kashian, et al. [57] found the total ecosystem carbon increased with stand age following a Michaelis-Menton function. Using this function, the 1988 burned forest (with an age of 18 in 2006) would have a mean total ecosystem carbon of 10,328 gC/m^2, which is 31% higher than the value determined for our study. Nevertheless, they also showed the measured total ecosystem carbon for young forest ranged from 6,000 to 15,600 gC/m^2. In our study, as depicted in Figure 5, the soil carbon represented by STATSGO, which was used in our young forest area, was 6,993 gC/m^2 lower than that of SSURGO, which represents closer to actual conditions. If the bias of 6,993 gC/m^2 was simply added, our result would be 14,867 gC/m^2, which falls within the range of 6,000 to 15,600 gC/m^2 of Kashian, et al. [57]. Similarly, using their function, the mature forest (assuming an age of 100–300 years) would have a total ecosystem carbon of 14,815 to 16,745 gC/m^2, which is higher than our value of 9,534 gC/m^2. When the bias of 6,993 gC/m^2 was simply applied, our total ecosystem carbon for mature forest was 16,527 gC/m^2, which was close to the 14,815–16,745 gC/m^2 predicted by Kashian, et al. [57]. The value of 16,527 gC/m^2 was also close to the 17,079 gC/m^2 measured by Litton, et al. [55] and 17,900 gC/m^2 modeled by Smithwick, et al. [15] for mature lodgepole pine stands.

Whether an ecosystem is a carbon sink or source is important for climate mitigation. We found GYE will be a carbon sink under climate change; this finding agrees with Smithwick, et al. [14], who also found the increasing carbon sequestration and suggested the potential for an increase in net carbon storage in GYE lodgepole pine forests under projected future climates. We found the forest productivity increased approximately 19.2% under climate change. This finding coincides with Melillo, et al. [58], who found that temperate ecosystem net primary productivity increased under climate change due to the effect of elevated temperature in enhancing the mineralization of nitrogen in the soils. The elevated temperature was also observed in our study, as shown in Figures 2 and 3. Our

general increasing trend in forest productivity is consistent with Smithwick, et al. [14], who revealed the same trend, but our magnitude of 19.2% is lower than their estimates of 25% (from HAD) and 36% (from CCC). The reasons for the difference came from the different climate data sources, but this was also caused by the different time scale (our 2006–2050 versus their 1994–2100): for many climate data sets, temperature and precipitation change much more after 2050 than before 2050.

GCMs have uncertainties and can influence carbon modeling. We found the variance of the total ecosystem carbon among the data-scenario combinations was within 5.5%, indicating the difference was very little. This finding is in contrast to Morales, et al. [59], who showed the choice of the GCM strongly influenced carbon balance in Europe. However, this finding agrees with Smithwick, et al. [15], who showed total ecosystem carbon stocks in GYE varied little (<10 percent) among future climate scenarios for a given A2 emission and fire-event pathway. Clearly, geography is attributed to this variation, indicating different landscapes have different carbon sensitivity to climate change.

The changing climate, wildfires, land use, and biogeochemical processes (e.g., forest productivity) comprehensively alter the carbon sequestration in a spatially heterogeneous manner. Understanding how concurrent changes in climate, disturbance regimes, and land use affect carbon storage in a spatially explicit manner and the accompanying uncertainties are critical but also challenging. Our method combined the concurrent changes in climate, fire, and land use for carbon modeling from 2006 to 2050 at 250 m resolution. We could also quantify the uncertainties from GCMs by processing the same CENTURY model with data from different sources. The result can help GYE stakeholders manage carbon sequestration as an important ecosystem service, and the methodology developed in this study can be applied to other regions to reveal spatiotemporal carbon dynamics under climate change. However, there are several potential areas of improvement from our current approach.

First, bark beetles (*Curculionidae: Scolytinae*) are a major native disturbance agent in most temperate coniferous forests. Since 1999, a warming climate in the Northern Rockies has coincided with beetle eruptions, which have exceeded historical records of the previous 125 years [60,61]. The outbreak of bark beetles can influence more land area than wildfires and result in a change in structure, function, and composition of forest ecosystems [19], and the impact could convert the forest from a small net carbon sink to a large net carbon source both during and immediately after the outbreak [62,63]. Beetle outbreaks are occurring throughout the entire distribution of the GYE [64]. Climate change has contributed to the unprecedented extent and severity of this outbreak [62]. Given the current mortality caused by bark beetles and projections for the future, this disturbance would shift the balance toward reduced photosynthesis capability and greater forest floor and soil carbon accumulation due to overstory tree mortality and subsequent coarse woody debris formation. If this disturbance is considered, GYE may even shift to a carbon source. Therefore, future carbon modeling needs improvement in considering bark beetle infestation.

Second, Romme, et al. [23] projected the probable effects on several representative species and community types in the GYE and found the extent of alpine vegetation in the ecosystem decreased in all scenarios. Bartlein, et al. [65] also projected the biotic response to future climate changes in Yellowstone and found the range of high-elevation species decreases and some species become regionally extirpated. The species redistribution affects ecosystem carbon sequestration, but given the time span modeled of 44 years (2006 to 2050), broad-scale change in vegetation communities may have been unlikely. For a longer time span, such as 100 years, vegetation shifting may be necessarily taken into account.

Third, after fire, net carbon loss to the atmosphere can persist for over a century [66], and fire intervals in coniferous forest are often more than 100 years [5]. This indicates that understanding the carbon cycle of a full fire cycle requires a time scale beyond 100 years. The period of our study was from 2006 to 2050, which is much shorter than this time scale.

Finally, the modeling system used in this study was originally developed for carbon analysis at the national scale. When the national-scale system was applied to the local study such as GYE, some issues can arise and there is a potential to reduce the uncertainty. For example, the models were not fine-tuned specifically for GYE, and the previous fire events such as the big fire of 1988 were not adequately represented in model simulation. Furthermore, both STATSGO and SSURGO soil databases were used in our modeling. Because the STATSGO database has less detailed information than the SSURGO database, the soil organic matter and nutrient supplies differ, which affects the simulation of ecological processes such as net primary production and soil organic decomposition. It is necessary to refine the soil data such as

Yellowstone soil database [67] to improve the modeling results.

Conclusion

The GYE is a temperate ecosystem that, based on the assumptions behind climate projections, will likely be subject to an elevated temperature, but change in precipitation in future decades varies among GCMs. With the changing climate, wildfires, land use, and the processes regulating the carbon cycle will be changed simultaneously. The concurrent change will lead to increasing total ecosystem carbon from 7,942 gC/m^2 in 2006 to approximately 10,234 gC/m^2 in 2050 with an annual rate increase of 53 gC/m^2/year. This finding indicates climate change can enhance the carbon sink characteristics of the GYE.

With an elevating temperature, the NPP will increase approximately 19.2%. Total live biomass carbon will increase from 2,551 gC/m^2 in 2006 to 3,833 gC/m^2 in 2050 with an annual rate increase of 30 gC/m^2/year. Soil organic matter will increase from 3,939 gC/m^2 in 2006 to 4,601 gC/m^2 in 2050 with an annual rate increase of 14.3 gC/m^2/year. These two carbon pools explained 56.6% and 27% of the extra carbon sequestration, respectively. This finding indicates the capability of vegetation is almost double that of soil in potential for sequestering extra carbon.

Clear-cutting and fires have different distributions in national parks (NP) and national forests (NF), with fires occurring in both NF and NP but clear-cutting mainly occurring in NF. In NF, 5.64% of the forest will be cut and 6.55% will be burned during 2006–2050. In NP, 6.08% of the forest will be burned during 2006–2050. Together, clear-cutting and wildfires in GYE will affect 10.87% of total forested area during 2006–2050. This finding indicates clear-cutting and wildfires under climate change may have great effect on this ecosystem, although the direct carbon removal by these events is insignificant.

Competing interests
The authors declare that they have no competing interests.

Authors' contributions
All authors have made substantial contributions to the analysis and interpretation of data, have been involved in drafting the manuscript and revising it critically for important intellectual content, and have given final approval of the version to be published.

Authors' information
SH is now at USDA Forest Service, Region 5, Remote Sensing Lab 3237 Peacekeeper Way, Suite 201, McClellan, CA 95652, shenglihuang@fs.fed.us.

Acknowledgements
This work was supported by the U.S. Geological Survey LandCarbon Program. The authors greatly thank Mr. Thomas Adamson for revising the English and Drs. Zhengxi Tan and Shuang Li for their help. Any use of trade, product, or firm names is for descriptive purposes only and does not imply endorsement by the U.S. Government.

Author details

[1]ASRC Federal InuTeq, Contractor to the U.S. Geological Survey (USGS) Earth Resources Observation and Science (EROS) Center, 47914 252nd Street, Sioux Falls, SD 57198, USA. [2]USGS EROS Center, 47914 252nd Street, Sioux Falls, SD 57198, USA. [3]Contractor to USGS Western Geographic Science Center, 345 Middlefield Rd, Menlo Park, CA 94025, USA. [4]Stinger Ghaffarian Technologies (SGT), Inc., Contractor to the USGS EROS Center, Sioux Falls, SD 57198, USA. [5]Innovate!, Inc. Contractor to the USGS EROS Center, Sioux Falls, SD 57198, USA. [6]U.S. Geological Survey, Denver, CO, USA. [7]U.S. Geological Survey, Reston, VA, USA.

References

1. Xia J, Chen J, Piao S, Ciais P, Luo Y, Wan S. Terrestrial carbon cycle affected by non-uniform climate warming. Nat Geosci. 2014;7(3):173–80.

2. Luo Y. Terrestrial Carbon-Cycle Feedback to Climate Warming. Annu Rev Ecol Evol Syst. 2007;38:683–712.

3. Scheller RM, Mladenoff DJ. A spatially interactive simulation of climate change, harvesting, wind, and tree species migration and projected changes to forest composition and biomass in northern Wisconsin. USA Glob Chang Biol. 2005;11(2):307–21.

4. Medlyn BE, Duursma RA, Zeppel MJB. Forest productivity under climate change: a checklist for evaluating model studies. Wiley Interdiscip Rev Clim Chang. 2011;2(3):332–55.

5. Kashian DM, Romme WH, Tinker DB, Turner MG, Ryan MG. Carbon Storage on Landscapes with Stand-replacing Fires. Bioscience. 2006;56(7):598–606.

6. Westerling AL, Turner MG, Smithwick EAH, Romme WH, Ryan MG. Continued warming could transform Greater Yellowstone fire regimes by mid-21st century. Proc Natl Acad Sci. 2011;108(32):13165–70.

7. McKinley DC, Ryan MG, Birdsey RA, Giardina CP, Harmon ME, Heath LS, et al. A synthesis of current knowledge on forests and carbon storage in the United States. Ecol Appl. 2011;21(6):1902–24.

8. Pielke RA, Marland G, Betts RA, Chase TN, Eastman JL, Niles JO, et al. The influence of land-use change and landscape dynamics on the climate system: relevance to climate-change policy beyond the radiative effect of greenhouse gases. Phil Roy Soc Lond A: Math Phys Eng Sci. 2002;360(1797):1705–19.

9. Strengers B, Leemans R, Eickhout B, de Vries B, Bouwman L. The land-use projections and resulting emissions in the IPCC SRES scenarios scenarios as simulated by the IMAGE 2.2 model. GeoJournal. 2004;61(4):381–93.

10. Karjalainen T, Pussinen A, Liski J, Nabuurs G-J, Erhard M, Eggers T, et al. An approach towards an estimate of the impact of forest management and climate change on the European forest sector carbon budget: Germany as a case study. For Ecol Manage. 2002;162(1):87–103.

11. Schaphoff S, Lucht W, Gerten D, Sitch S, Cramer W, Prentice IC. Terrestrial biosphere carbon storage under alternative climate projections. Clim Change. 2006;74:97–122.

12. Schoennagel T, Veblen TT, Romme WH, Sibold JS, Cook ER. Enso and pdo variability affect drought-induced fire occurrence in Rocky Mountain subalpine forests. Ecol Appl. 2005;15(6):2000–14.

13. Euskirchen ES, Chen J, Li H, Gustafson EJ, Crow TR. Modeling landscape net ecosystem productivity (LandNEP) under alternate management regimes. Ecol Model. 2002;154:75–91.

14. Smithwick EAH, Ryan MG, Kashian DM, Romme WH, Tinker DB, Turner MG. Modeling the Effects of Fire and Climate Change on Carbon and Nitrogen Storage in Lodgepole Pine (Pinus contorta) Stands. Glob Chang Biol. 2009;15:535–48.

15. Smithwick EAH, Westerling AL, Turner MG, Romme WH, Ryan MG. Vulnerability of landscape carbon fluxes to future climate and fire in the Greater Yellowstone Ecosystem. In: Andersen C, editor. Questioning Greater Yellowstone's Future: Climate, Land Use, and Invasive Species; Proceedings of the 10th Biennial Scientific Conference on the Greater Yellowstone Ecosystem; October 11–13. Yellowstone National Park: Yellowstone Center for Resources; 2011. p. 9.

16. Loudermilk EL, Scheller RM, Weisberg PJ, Yang J, Dilts TE, Karam SL, et al. Carbon dynamics in the future forest: the importance of long-term successional legacy and climate–fire interactions. Glob Chang Biol. 2013;19(11):3502–15.

17. Scheller R, Van Tuyl S, Clark K, Hom J, La Puma I. Carbon Sequestration in the New Jersey Pine Barrens Under Different Scenarios of Fire Management. Ecosystems. 2011;14(6):987–1004.

18. Liu S, Liu J, Young CJ, Werner JM, Wu Y, Li Z, et al. Baseline carbon storage, carbon sequestration, and greenhouse-gas fluxes in terrestrial ecosystems of the Western United States. In: Zhu Z, Reed BC, editors. Baseline and projected future carbon storage and greenhouse-gas fluxes in ecosystems of the Western United States. Professional Paper 1797: U.S. Department of the Interior, U.S. Geological Survey. 2012: 45–63.

19. Donato DC, Harvey BJ, Romme WH, Simard M, Turner MG. Bark beetle effects on fuel profiles across a range of stand structures in Douglas-fir forests of Greater Yellowstone. Ecol Appl. 2013;23(1):3–20.

20. Potter C, Li S, Huang S, Crabtree RL. Analysis of sapling density regeneration in Yellowstone National Park with hyperspectral remote sensing data. Remote Sens Environ. 2012;121:61–8.

21. Campbell J, Donato D, Azuma D, Law B. Pyrogenic carbon emission from a large wildfire in Oregon, United States. J Geophys Res Biogeosci. 2007;112(G4):G04014.

22. Schoennagel T, Veblen TT, Romme WH. The Interaction of Fire, Fuels, and Climate across Rocky Mountain Forests. Bioscience. 2004;54(7):661–76.

23. Romme WH, Turner MG. Implications of Global Climate Change for Biogeographic Patterns in the Greater Yellowstone Ecosystem. Conserv Biol. 1991;5(3):373–86.

24. Westerling AL, Hidalgo HG, Cayan DR, Swetnam TW. Warming and earlier spring increase western US forest wildfire activity. Science. 2006;313(5789):940–3.

25. Gude PH, Hansen AJ, Rasker R, Maxwell B. Rates and drivers of rural residential development in the Greater Yellowstone. Landsc Urban Plan. 2006;77(1–2):131–51.

26. Parmenter AW, Hansen A, Kennedy RE, Cohen W, Langner U, Lawrence R, et al. Land use and land cover change in the Greater Yellowstone Ecosystem: 1975–1995. Ecol Appl. 2003;13(3):687–703.

27. Turner M, Donato D, Romme W. Consequences of spatial heterogeneity for ecosystem services in changing forest landscapes: priorities for future research. Landsc Ecol. 2013;28(6):1081–97.

28. Nakicenovic, N., Alcamo, J., Davis, G., De Vries, B., Fenhann, J., Gaffin, S., et al. Special report on emissions scenarios, working group III, Intergovernmental Panel on Climate Change [IPCC]. Cambridge University Press. 2000: ISBN 0, 521(80493), 0.

29. Joyce LA, Price DT, McKenney DW, Siltanen RM, Papadopol P, Lawrence K, et al. High resolution interpolation of climate scenarios for the conterminous USA and Alaska derived from general circulation model simulations. Fort Collins: CO General Technical Report RMRS–GTR–263; 2011.

30. Gordon HB, Rotstayn LD, McGregor JL, Dix MR, Kowalczyk EA, O'Farrell SP, et al. The CSIRO Mk3 Climate System Model. In: CSIRO Atmospheric Research Technical Paper No. 60, Commonwealth Scientific and Industrial Research Organisation Atmospheric Research, Aspendale, Victoria, Australia. 120: 130 pp, http://www.cmar.csiro.au/e-print/open/gordon_2002a.pdf.

31. Hasumi H, Emori S. K-1 Coupled GCM (MIROC) Description. In: vol. K-1 Technical Report: Center for Climate System Research. Tokyo, Japan: University of Tokyo; 2004.

32. U.S. Geological Survey. National Elevation Dataset. 2012, available online at http://viewer.nationalmap.gov/viewer/. Accessed [12/17/2012].

33. USDA Forest Service. Forest inventory and analysis national program: U.S. Department of Agriculture, Forest Service database. 2012.

34. U.S. Department of Agriculture NRCS. Soil Survey Geographic (SSURGO) Database. U.S. Department of Agriculture ARS: Natural Resources Conservations Service database; 2009. available online at http://websoilsurvey.sc.egov.usda.gov/App/WebSoilSurvey.aspx. Accessed [12/14/2012].

35. Soil Survey Staff. Natural Resources Conservation Service. In. United States Department of Agriculture. Web Soil Survey. available online at http://websoilsurvey.nrcs.usda.gov/. Accessed [12/17/2012].

36. Zhu Z, Reed B (eds). Baseline and projected future carbon storage and greenhouse-gas fluxes in ecosystems of the Western United States. U.S. Department of the Interior, U.S. Geological Survey; Professional Paper 1797. 2012.

37. van Vuuren DP, Lucas PL, Hilderink H. Downscaling drivers of global environmental change: Enabling use of global SRES scenarios at the national and grid levels. Glob Environ Chang. 2007;17(1):114–30.

38. Sleeter BM, Sohl TL, Wilson TS, Sleeter RR, Soulard CE, Bouchard MA, et al. Projected land-use and land-cover change in the Western United States. In: Zhu Z, Reed BC Editors. Baseline and projected future carbon storage

and greenhouse-gas fluxes in ecosystems of the Western United States. Professional Paper 1797: U.S. Department of the Interior, U.S. Geological Survey. 2012: 65–86.

39. Sohl TL, Sleeter BM, Zhu Z, Sayler KL, Bennett S, Bouchard M, et al. A land-use and land-cover modeling strategy to support a national assessment of carbon stocks and fluxes. Appl Geogr. 2012;34:111–24.

40. Penman, J, Gytarsky, M, Hiraishi, T, Krug, T, Kruger, D, Pipatti, R, et al. Good practice guidance for land use, land-use change and forestry. Institute for Global Environmental Strategies. 2003.

41. Hawbaker T, Zhu Z. Projected future wildland fires and emissions for the Western United States. In: Zhu Z, Reed B., editors. Baseline and projected future carbon storage and greenhouse-gas fluxes in ecosystems of the Western United States. U.S. Geological Survey Professional Paper 1797. 2012: 192. available at http://pubs.usgs.gov/pp/1797/.

42. Maurer EP, Brekke L, Pruitt T, Duffy PB. Fine-resolution climate projections enhance regional climate change impact studies. EOS Trans Am Geophys Union. 2007;88:504.

43. Maurer EP, Wood WA, Adam JC, Lettenmaier DP, Nijssen B. A long-term hydrologically based dataset of land surface fluxes and states for the conterminous United States. J Climate. 2002;15:3237–51.

44. Mesinger F, DiMego G, Kalnay E, Mitchell K, Shafran PC, Ebisuzaki W, et al. North American regional reanalysis. Bull Am Meteorol Soc. 2006;87(3):343. –+.

45. Glassy JM, Running SW. Validating Diurnal Climatology Logic of the MT-CLIM Model Across a Climatic Gradient in Oregon. Ecol Appl. 1994;4(2):248–57.

46. Burgan RE. 1988 revisions to the 1978 national fire-danger rating system, Southeastern Forest Experiment Station, Research Paper SE-273. 1988.

47. Finney MA. Fire growth using minimum travel time methods. Can J Forest Res. 2002;32(8):1420–4.

48. Rollins MG. LANDFIRE: a nationally consistent vegetation, wildland fire, and fuel assessment. Int J Wildland Fire. 2009;18(3):235–49.

49. Reinhardt E, Keane RE. FOFEM: The first-order fire effects model adapts to the 21st century. In: Fire Science Brief. Joint Science Program; 2009. available at http://digitalcommons.unl.edu/cgi/viewcontent.cgi?article=1143&context=jfspbriefs.

50. Spracklen DV, Mickley LJ, Logan JA, Hudman RC, Yevich R, Flannigan MD, et al. Impacts of climate change from 2000 to 2050 on wildfire activity and carbonaceous aerosol concentrations in the western United States. J Geophys Res: Atmos. 2009;114(D20), D20301.

51. Eidenshink J, Schwind B, Brewer K, Zhu ZL, Quayle B, Howard S. Project for monitoring trends in burn severity. Fire Ecology. 2007;3(1):3–20.

52. Hawbaker T, Zhu Z. Baseline wildland fires and emissions for the Western United States. In: Zhu Z, Reed B editors.Baseline and projected future carbon storage and greenhouse-gas fluxes in ecosystems of the Western United States. U.S. Geological Survey Professional Paper 1797. 2012: 192. available at http://pubs.usgs.gov/pp/1797/.

53. Metherell AK, Harding LA, Cole CV, Parton WJ. CENTURY soil organic matter model environment, technical documentation, Agroecosystem Version 4.0, Great Plains System Research Unit Technical Report 4. 1993.

54. Kellndorfer J, Walker W, Pierce L, Dobson C, Fites JA, Hunsaker C, et al. Vegetation height estimation from Shuttle Radar Topography Mission and National Elevation Datasets. Remote Sens Environ. 2004;93(3):339–58.

55. Litton CM, Ryan MG, Knight DH. Effects of tree density and stand age on carbon allocation patterns in postfire lodgepole pine. Ecol Appl. 2004;14:460–75.

56. Crabtree R, Potter C, Mullen R, Sheldon J, Huang S, Harmsen J, et al. A modeling and spatio-temporal analysis framework for monitoring environmental change using NPP as an ecosystem indicator. Remote Sens Environ. 2009;113(7):1486–96.

57. Kashian DM, Romme WH, Tinker DB, Turner MG, Ryan MG. Postfire changes in forest carbon storage over a 300-year chronosequence of Pinus contorta-dominated forests. Ecol Monogr. 2013;83(1):49–66.

58. Melillo JM, McGuire AD, Kicklighter DW, Moore B, Vorosmarty CJ, Schloss AL. Global climate change and terrestrial net primary production. Nature. 1993;363(6426):234–40.

59. Morales P, Hickler T, Rowell DP, Smith B, Sykes MT. Changes in European ecosystem productivity and carbon balance driven by regional climate model output. Glob Chang Biol. 2007;13(1):108–22.

60. Raffa KF, Aukema BH, Bentz BJ, Carroll AL, Hicke JA, Turner MG, et al. Cross-scale Drivers of Natural Disturbances Prone to Anthropogenic Amplification: The Dynamics of Bark Beetle Eruptions. Bioscience. 2008;58(6):501–17.

61. Hatala JA, Crabtree RL, Halligan KQ, Moorcroft PR. Landscape-scale patterns of forest pest and pathogen damage in the Greater Yellowstone Ecosystem. Remote Sens Environ. 2010;114(2):375–84.

62. Kurz WA, Dymond CC, Stinson G, Rampley GJ, Neilson ET, Carroll AL, et al. Mountain pine beetle and forest carbon feedback to climate change. Nature. 2008;452(7190):987–90.

63. Reed DE, Ewers BE, Pendall E. Impact of mountain pine beetle induced mortality on forest carbon and water fluxes. Environ Res Lett. 2014;9(10):105004.

64. Logan JA, Macfarlane WW, Willcox L. Whitebark pine vulnerability to climate-driven mountain pine beetle disturbance in the Greater Yellowstone Ecosystem. Ecol Appl. 2010;20(4):895–902.

65. Bartlein PJ, Whitlock C, Shafer SL. Future Climate in the Yellowstone National Park Region and Its Potential Impact on Vegetation. Conserv Biol. 1997;11(3):782–92.

66. Crutzen PJ, Goldammer JG. Fire in the environment: the ecological, atmospheric and climatic importance of vegetation fires. report of the Dahlem workshop, Berlin, 15–20 March, 1992. In: Fire in the environment: the ecological, atmospheric and climatic importance of vegetation fires Report of the Dahlem workshop, Berlin, 15–20 March, 1992. John Wiley & Sons; 1993.

67. Rodman A, Shovic H, Thoma D. Soils of Yellowstone National Park. Yellowstone National Park, Wyoming: Yellowstone Center for Resources., YCR-NRSR-96-2; 1996.

4

Spatial distribution of temporal dynamics in anthropogenic fires in miombo savanna woodlands of Tanzania

Beatrice Tarimo[1,2*], Øystein B Dick[3], Terje Gobakken[1] and Ørjan Totland[1]

Abstract

Background: Anthropogenic uses of fire play a key role in regulating fire regimes in African savannas. These fires contribute the highest proportion of the globally burned area, substantial biomass burning emissions and threaten maintenance and enhancement of carbon stocks. An understanding of fire regimes at local scales is required for the estimation and prediction of the contribution of these fires to the global carbon cycle and for fire management. We assessed the spatio-temporal distribution of fires in miombo woodlands of Tanzania, utilizing the MODIS active fire product and Landsat satellite images for the past ~40 years.

Results: Our results show that up to 50.6% of the woodland area is affected by fire each year. An early and a late dry season peak in wetter and drier miombo, respectively, characterize the annual fire season. Wetter miombo areas have higher fire activity within a shorter annual fire season and have shorter return intervals. The fire regime is characterized by small-sized fires, with a higher ratio of small than large burned areas in the frequency-size distribution ($\beta = 2.16 \pm 0.04$). Large-sized fires are rare, and occur more frequently in drier than in wetter miombo. Both fire prevalence and burned extents have decreased in the past decade. At a large scale, more than half of the woodland area has less than 2 years of fire return intervals, which prevent the occurrence of large intense fires.

Conclusion: The sizes of fires, season of burning and spatial extent of occurrence are generally consistent across time, at the scale of the current analysis. Where traditional use of fire is restricted, a reassessment of fire management strategies may be required, if sustainability of tree cover is a priority. In such cases, there is a need to combine traditional and contemporary fire management practices.

Keywords: Burned area, Carbon stocks, Fire history, Frequency-size distribution, Landsat, Miombo woodland, MODIS, Surface fires

Background

Anthropogenic fires are historically an integral component of African savannas. They strongly influence the composition, structure and distribution of mesic savannas in particular, where tree cover is not constrained by climatic conditions [1–4]. Fire regimes in African savannas, including the frequency and season of burning, are mainly human regulated [5]. The variability of fire regimes in African savannas is more dependent on human drivers than on climate and thus human drivers may regulate the future of savanna fire regimes under changing climate conditions [6–9]. Human activities associated with fire ignitions and fragmentation of the landscape play a key role in determining the occurrence of fire and resulting spatial extents of burned areas [10–14]. Tropical savannas, predominantly in Africa, contributes the highest proportion of the global burned area [15], and their contribution to biomass burning emissions is substantial [16, 17]. The role of these fires as a management tool or as a threat to woody cover, and in the global carbon cycle, vary within savannas and is dependent on the fire regime. Efforts to change fire regimes in favor of

*Correspondence: beatrice.tarimo@nmbu.no
[1] Department of Ecology and Natural Resource Management, Norwegian University of Life Sciences, P.O. Box 5003, 1432 Ås, Norway
Full list of author information is available at the end of the article

management priorities, such as carbon sequestration, are being challenged in the light of traditional fire regimes that are more suited for the sustainability of savannas [10, 18–21]. Viable fire management plans aiming at maintenance of stored carbon requires an understanding of historical fire regimes at local scales, which is generally lacking for many parts of African savannas. This understanding is required for precise estimates of the contribution of savanna fires in the global carbon dynamics. Characterization of current fire regimes at local scales is required in order to set references against which assessment of changes in burning practices and their contribution to the carbon cycle will be made [22]. This is of particular importance in fire-adapted ecosystems, such as miombo woodlands, that also support a wide range of human subsistence activities.

Fire is regarded essential to the structure and stability of miombo woodlands [23, 24]. Intense fires suppress tree biomass when their frequency is higher than the rate of tree regeneration and growth [23, 25, 26]. Frequent and intense fires threatens the maintenance of stored carbon stocks, and consequently undermines the potential benefits of activities that comprise the reducing emissions from deforestation and forest degradation (REDD+) policy instrument [27]. Fire contributes to long-term degradation that, although significant, has proven difficult to quantify and monitor, and thus receive less attention in REDD+ negotiations compared to deforestation [28–30]. In addition, they impede the enhancement of carbon stocks for REDD+ payments and sustainability of tree cover at large. Tree recruitment and succession are constrained by recurrent fires [23, 31], which instead facilitates grass encroachment and colonization that may fuel more intense and frequent fires [32, 33]. Exclusion of fire on the other hand facilitates tree dominance of the ecosystem [34], which limits the growth of light demanding grasses and consequently fuel loading. The timing of burning further regulates fire effects, such that late dry season fires have adverse effects on both vegetation and soils, whereas prescribed early dry season fires may be a beneficial management tool [23, 26]. Fire management is of crucial importance for successful forest management [23]. However, it is impaired by the limited understanding on which controlled burning treatments are beneficial for respective components of woodland savannas, coupled with the socio-economic dependency from their surroundings, which play a major role in shaping fire regimes. Characterization of the long-term fire regime will contribute to the ongoing efforts to quantify carbon stocks and fluxes for the purposes of monitoring and verification in the context of REDD+ policy framework and for better fire management practices in general.

A key challenge to both the estimation of carbon fluxes from fires and fire management efforts in African savannas is lack of complete and consistent fire records. In Tanzania, the vast majority of fire events stem from anthropogenic ignitions for different purposes, including farm preparation, pasture management, hunting, honey harvesting, charcoal production, arsons, and for security around settlements and roads [35]. Fire records are limited to a few isolated areas that implement fire management plans. In the absence of long term systematic ground fire records, satellite data forms a unique source of the recent fire history [e.g. 36, 37]. Since tropical savanna fires are fueled mainly by grasses and litter, they sweep the ground surface and leave tree crowns and soil sub-surface unaffected. The resulting burned scars persist for a few weeks only [38–41]. Therefore, frequent observations are required to capture most of the area burned in the course of a fire season. Monthly composites of observation of fire events may be representative of the spatial and temporal distribution of African savanna fires [42–44]. Although the use of different satellite systems provides multiple acquisitions every month, data availability is constrained by cloud cover and other limitations.

Datasets on active fires and burned areas derived from along track scanning radiometer (ATSR), SPOT-VEGETATION and moderate resolution imaging spectroradiometer (MODIS), among other satellite sensors, are available in the public domain. They provide fire patterns at a coarse spatial resolution and at very short temporal coverages. However, comparisons of burned areas derived from coarse resolution (1 km) with those derived from finer resolutions (e.g. 30 m) satellites, show that the majority (up to 90%) of small burned areas characteristic of fragmented fires in tropical savannas, are not detected by coarse resolution burned area products [41, 42, 44, 45]. The low detectability of small-burned areas by coarse spatial resolution products limit the efficacy of these products at smaller spatial scales when detailed information is required. There is thus a need to quantify spatial and temporal fire patterns and resulting burned extents at finer resolutions than those available in the public domain.

The availability of Landsat satellite images in the public domain provides the opportunity to extract burned area records since the early 1970s. Methods are being developed for (semi)automatic burned area mapping at finer spatial resolution e.g. [46, 47], which facilitate frequent and complete mapping at local and regional scales. However, few studies have employed the utility of these methods in African savannas. Thus, burned area records are still missing despite the availability of satellite images. We aim at assessing the fire history during the past ~40 years and respective spatial patterns from satellite based data.

Burned areas are mapped by fuzzy classification using spectral indices that include infrared wavelengths, since they are more sensitive to fire induced changes than other spectral combinations [48–50]. We discuss the derived fire return intervals, seasonality and burned extents in Tanzanian miombo relative to those from other African savannas, and the observed frequency-size statistics relative to those reported from other ecosystems. We highlight the consistency in the fire regime across spatial and temporal scales and point out priority areas requiring further analyses and reassessment of management practices.

Results
Validation of detected burned areas
Table 1 summarizes classification accuracy analysis of detected burned areas. Omissions of burned pixels are mainly in the partially burned areas, which are not included in Table 1.

Based on an independent validation, the overall performance of the fuzzy classification when including partially burned areas was 57%, which is not as good as that of the completely burned areas (Table 1). It should however be noted that the definition of fuzzy membership scores to distinguish burned from partially burned areas (see "Validation of detected burned areas" in the "Methods" section) on one hand, and the subjective element of the result of the visual interpretation on the other hand, might have had an impact on the quantification of the performance of the fuzzy classification.

Spatial and temporal patterns of burned areas
Burned patch sizes
For each particular year, the majority (up to the third quartile) of burned patches are less than five hectares in size (Fig. 1). The annual median of burned patch size ranged from 0.8 to 1.4 ha. Small burned patches are more common in wet miombo than in dry miombo areas, with annual median ranging from 0.8 to 1.4 ha and 0.7 to 1.8 ha, respectively. Relatively few and very occasional big fires may reach sizes of up to ~60,000 ha. These account for a large proportion of the total area burned but they tend to decrease in frequency during the 1972–2011

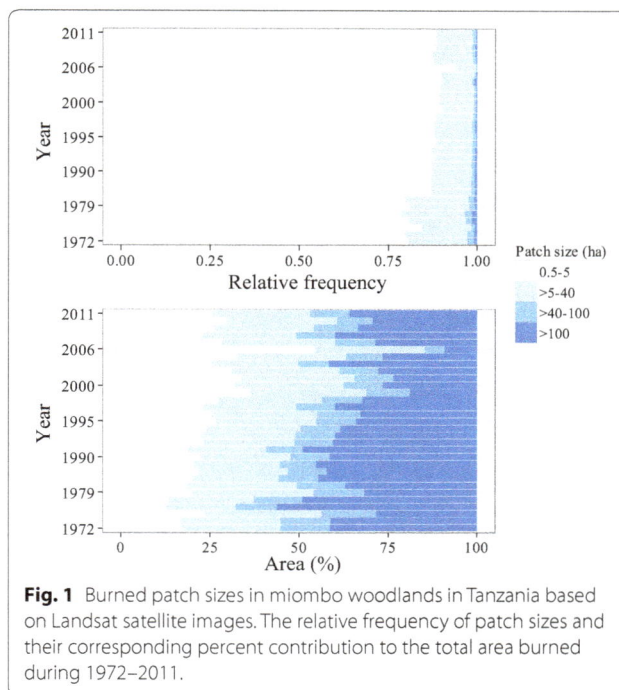

Fig. 1 Burned patch sizes in miombo woodlands in Tanzania based on Landsat satellite images. The relative frequency of patch sizes and their corresponding percent contribution to the total area burned during 1972–2011.

period, relative to other size classes. Overall, small-burned patches, which are a common occurrence over spatial and temporal scales, account for more of the total area burned than large burned patches.

Frequency-size distribution of burned patches
Frequency-size statistics of burned areas suggest a fire regime dominated by small-sized fires with scaling, $\beta = 2.16 \pm 0.04$, with $r^2 = 0.99$ for the whole woodland during 1972–2011. Wet miombo has a slightly smaller scaling, $\beta = 2.13 \pm 0.03$, with $r^2 = 0.99$ relative to dry miombo where scaling, $\beta = 2.15 \pm 0.04$, with $r^2 = 0.99$. Given the high number of annual burned patches, it was deemed relevant to analyze the annual frequency-size distributions. Annual analyses resulted in scaling ranging from $\beta = 1.89 \pm 0.04$ to $\beta = 2.53 \pm 0.15$, with $r^2 > 0.98$ for the whole miombo woodland. Similarly, wet miombo has a slightly smaller scaling than dry miombo from annual analyses, ranging from $\beta = 1.71 \pm 0.17$ to $\beta = 2.50 \pm 0.19$, with $r^2 > 0.95$ and from $\beta = 1.82 \pm 0.05$ to $\beta = 2.57 \pm 0.43$, with $r^2 > 0.94$, respectively.

Burned extents
Figure 2 presents patterns of burned areas detected from Landsat images, summarized at a 5 × 5 km grid. At this scale, fire incidences appear to be consistently within the same spatial extents. Temporal differences in the extent burned per window show an irregular spatial trend. Annually, up to 13.7% and 12.6% of the total area with available

Table 1 Omission and commission errors of burned pixels

Class	Samples		Errors	
	Correctly classified	Incorrectly classified	Omission (%)	Commission (%)
Burned	1,022	365	26.3	0.7
Not burned	7,826	7	0.1	4.5

Kappa coefficient = 0.82.

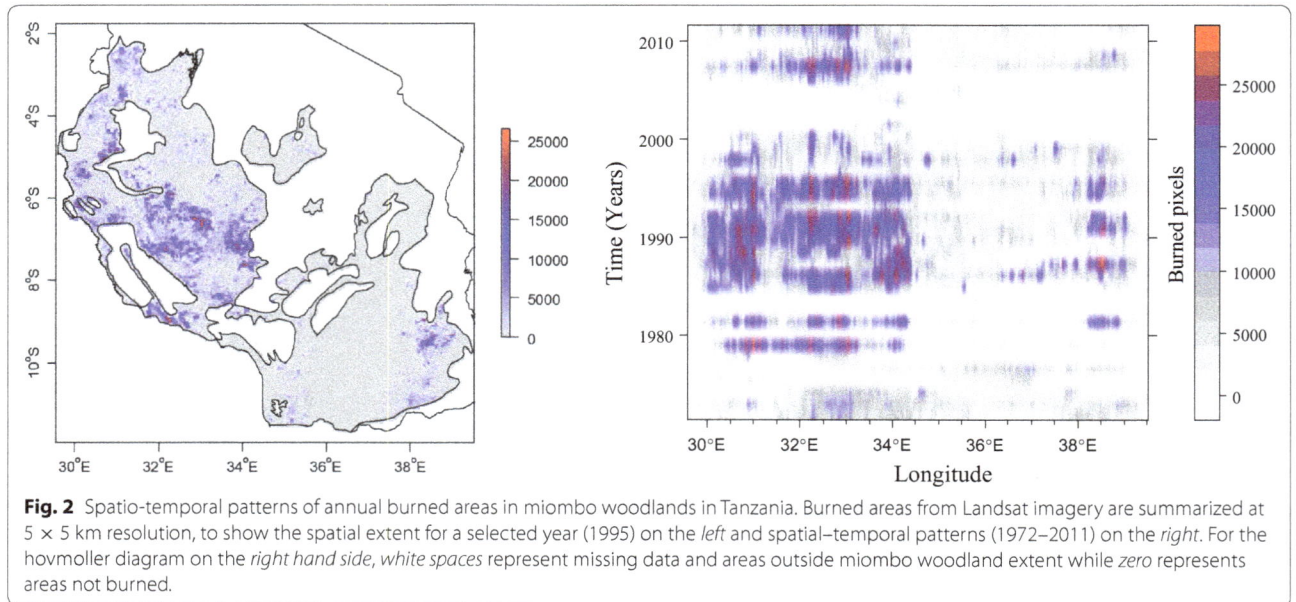

Fig. 2 Spatio-temporal patterns of annual burned areas in miombo woodlands in Tanzania. Burned areas from Landsat imagery are summarized at 5 × 5 km resolution, to show the spatial extent for a selected year (1995) on the *left* and spatial–temporal patterns (1972–2011) on the *right*. For the hovmoller diagram on the *right hand side*, *white spaces* represent missing data and areas outside miombo woodland extent while *zero* represents areas not burned.

imagery was detected as burned in wet and dry miombo, respectively. When combined with partially burned areas, up to 65.8% and 42.1% of wet and dry miombo, respectively, was detected as burned annually. For the whole miombo woodland in Tanzania up to 11.3% is burned annually, while when combined with partially burned areas, up to 50.6% of the woodland is affected by fire annually. Table 2 provides a decadal summary of the contribution of wet and dry miombo areas to the total area burned

for the whole woodland. In this table, comparisons are more reliable between dry and wet miombo for the same duration than between durations due to differences in the number of years with available data for each location.

Spatial and temporal patterns of active fires
Early and late dry season burning
A west-to-east transition of fire events from early to late burning is observed in Fig. 3. The sudden drop of

Table 2 Burned area characteristics in (a) dry and (b) wet miombo areas in Tanzania during 1972–2011

Duration	P (ha)	A (%)	M [Dry] (%)	RI	Scaling (β)	A_P
(a) Dry miombo						
1972–1979	930–53,300	3.6–12.6	5.2 [2.9]–11.3 [9.8]	1.6	1.82–2.11	NA
1980–1989	5.4–27,270	3.7–6.5	4.5 [2.2]–8.6 [3.6]	1.8	1.97–2.57	14.3–26.9
1990–1999	1,649–64,650	0.6–8.0	0.8 [0.4]–10.0 [5.1]	2.4	2.00–2.15	10.0–34.1
2000–2011	676.4–21,090	0.7–6.1	1.0 [0.5]–6.6 [4.3]	2.8/3.0[a]	1.93–2.29	4.3–23.6

Duration	P (ha)	A (%)	M [Wet] (%)	RI	Scaling (β)	A_P
(b) Wet miombo						
1972–1979	1,199–29,050	6.7–11.0	7.6 [2.2]–5.2 [2.3]	1.4	1.83–2.12	NA
1980–1989	8.2–15,760	5.9–11.2	4.5 [2.3]–8.6 [5.0]	1.6	2.04–2.50	16.3–31.3
1990–1999	1,166–36,530	1.6–13.7	0.8 [0.4]–10.0 [4.9]	1.4	2.02–2.14	13.3–52.1
2000–2011	9.1–15,160	0.2–7.7	0.8 [0.1]–6.6 [2.3]	2.0/2.1[a]	1.71–2.26	1.2–25.3

Values in square brackets represents the percent contribution of dry/wet miombo to M.

P the range of largest annual burned patches detected, may have aggregated with time during the fire season, A the range of the total area burned in dry/wet miombo as a percentage of the dry/wet miombo area with data, A_P the total area partially burned in dry/wet miombo as percentage of the total dry/wet miombo area with data, M the total area burned in miombo at the same time when A was recorded, as percentage of miombo area with data, RI Fire return interval observed for every 2,500 ha from Landsat satellite images.

[a] Based on MODIS detected fires for every 314 ha for the period 2001–2013.

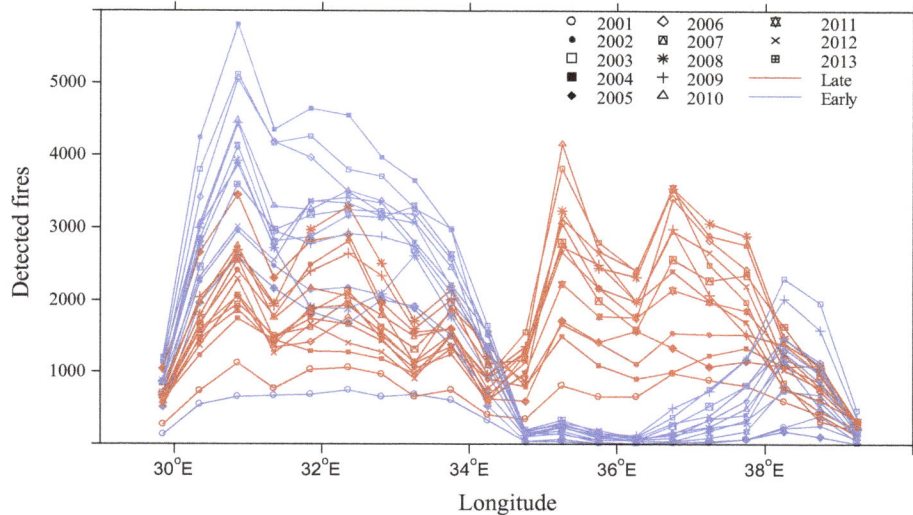

Fig. 3 Spatial distribution of MODIS detected fires in miombo woodlands in Tanzania. July was defined to mark the end of early dry season burning for the entire woodland; local variations are likely to occur.

incidences at ~35.75°E is partly explained by the extent of miombo woodland areas (see Fig. 4) and it marks a distinction between an early dry season burning dominated west to a late dry season burning dominated east.

Based on the number of detected fires each month the fire season peaks during the first part of the dry season in July (Fig. 5). To investigate the effect of early dry season burning on late dry season fires, fire radiative power (FRP) values of late dry season fires were compared for those fires which were either close (within 1 km; i.e. approximately within the same fire pixel) or far (>1 km) from early dry season fires during the same fire season. There is no significant reduction, at 95% confidence level, of FRP values in the late dry season fires, which were close to early dry season burned areas than those far from them.

Fire activity
The combined characteristics of detected active fires provide a composite estimate of fire activity given in Fig. 6. Fire activity is consistently high in the western part of the woodland with the exception of areas along its northeastern border. An increasing systematic westward reduction in fire activity is observed along this border during 2001–2013 (Fig. 6). This reduction is associated with the expansion of croplands when interpreted in the context of the GLC-Share land cover types [51]. On the other hand, there is a shift from high to low fire activity between years on the central, south and eastern parts of the woodland.

Fire return interval
The mean fire return interval for a circular area of ~314 ha, centered at the location of MODIS detected fires was 2.7 years (range 1–13 years) between 2001 and 2013. When the analysis was performed for every 2,500 ha during 1972–2011, based on burned areas detected from Landsat images, the interval was reduced to 2.1 years.

Discussion
Historical fire regimes are best reconstructed from long-term consistent ground records, charcoal deposition in soils or fire scars on trees with annual growth rings [52–55]. In the absence of these, fire history in miombo woodlands of Tanzania was documented from Landsat satellite images and MODIS detected active fires for the past ~40 years. Both fire prevalence and burned extents have recently decreased (Table 2). This decrease is likely an outcome of a number of contributing factors, including a reduction in miombo woodland coverage through e.g. conversion of the woodland into permanent cultivated fields and fire management practices in some parts of the woodland. Burned areas and detected active fire events are consistently within the same spatial coverage (Figs. 2, 6), at the scale of the current analysis. The lack of an independent burned area perimeter for validation restricted our analysis to burned pixels with the highest confidence, which underestimate the total area burned. When thoroughly validated, an analysis including partially burned areas might increase fire activity and shorten fire return intervals in some parts of the woodland. This however, will not affect the general patterns presented in this study.

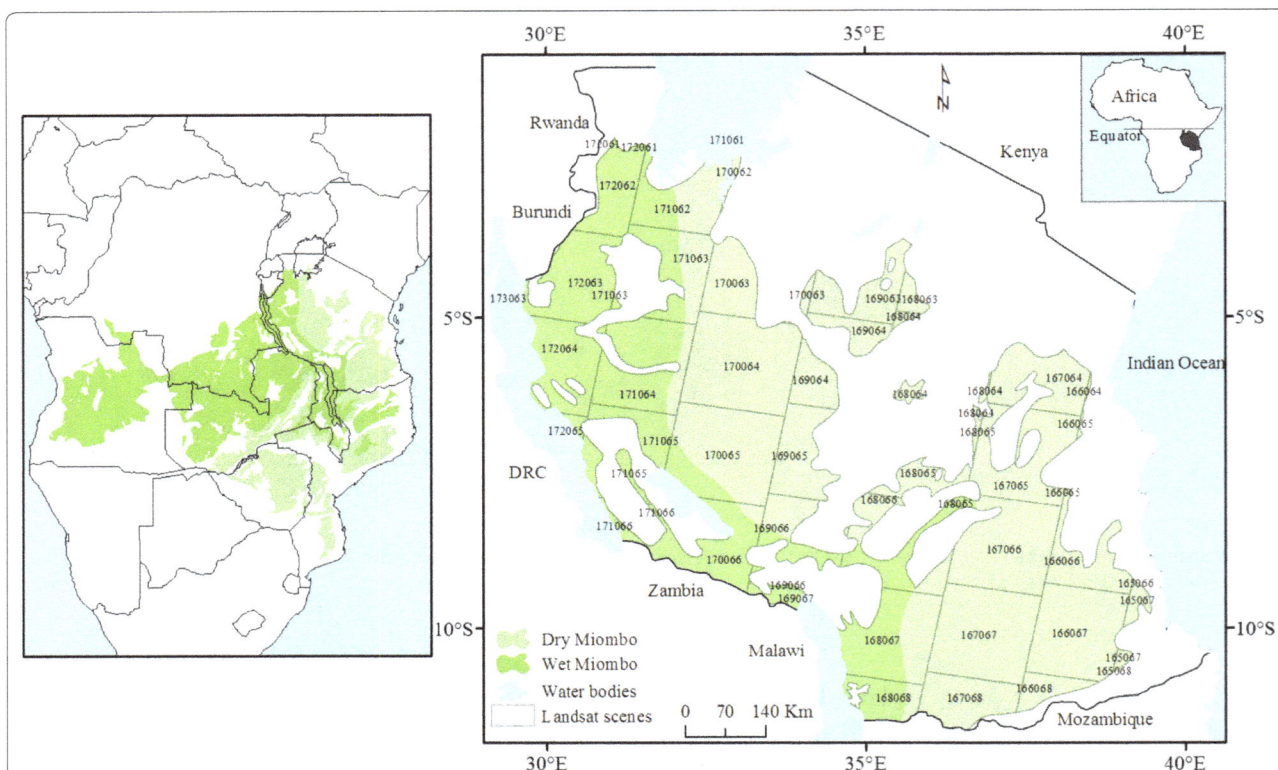

Fig. 4 Distribution and classification of miombo woodlands in Tanzania. The map is based on White's vegetation map of Africa [74]. *Numbers* show the identification (path and row numbers) and extents of Landsat TM/ETM+ scenes.

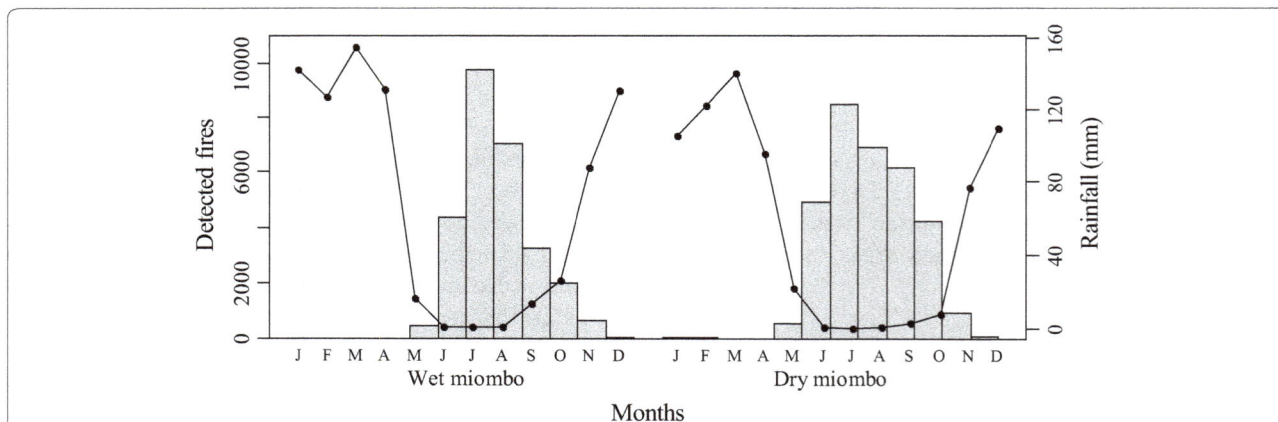

Fig. 5 The fire season in wet and dry miombo areas in Tanzania. *Bars* show mean monthly-detected fires for the period 2001–2013 based on MODIS active fires dataset (see "Data sets and preprocessing" section), and *lines* mean monthly rainfall climatology for 1983–2012 period. Rainfall data is sourced from TARCAT [96].

Fire prevalence

About 46% of the woodland area had a mean fire return interval of <2 years for every 314 ha during 2001–2013 period. Field observations in a dry miombo site have shown a mean fire return interval of 1.6 years in Zambia [23], while a return interval of 3 years on a regional scale was observed based on satellite data [24]. In a global study, the fire return interval for African savannas and grasslands has been reduced, from 4.8 years early in 1900 to 3.6 years towards 2000 [15]. Fire return interval

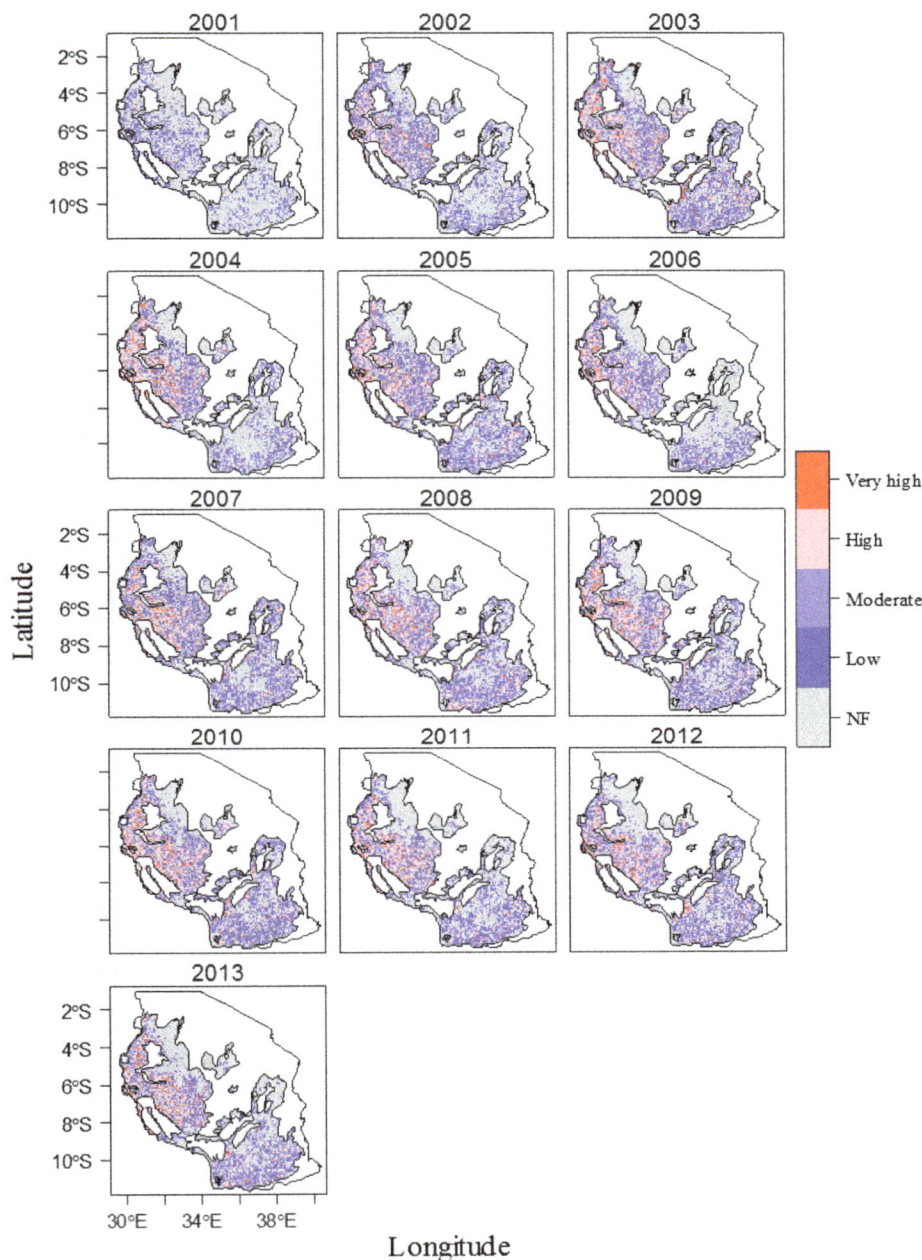

Fig. 6 Fire activity in miombo woodlands in Tanzania. Fire activity based on density, proximity and fire radiative power of MODIS detected fire events, and annual length of the fire season, at a 5 × 5 km resolution. *NF* represents areas with no fire activity.

and burning seasonality have selective effects on different components of the woodland, as they influence the intensity of fires and extents burned. Results from a study that combined field observations and modelling, from miombo sites in Zimbabwe and Mozambique, show that at least 2 years are required between successive low intensity burns to allow tree establishment and development [25]. Although spatial variation is expected within the scale at which fire activity and return intervals are estimated in the present study, our results indicate that almost half of the woodland is ignited at a return interval that threaten the longer-term sustainability of the tree cover. However, the resulting burned patterns (see "Burned extents" section) indicates an offset, to some extent, of the effect of recurrent fires. About 74% of the woodland area had a mean return interval of <2 years for every 2,500 ha between 1972 and 2011. Therefore, frequent fires have been part of some portions of the woodland for the past ~40 years, when satellite data is available. It is important to note that there is a seasonal and inter-annual variation of burned patches within each of the 2,500 ha and thus the return interval varies at smaller scales. Shorter fire return intervals are observed in wet miombo (Table 2), which are the same areas where annual fire activity is consistently high (Fig. 6). Within dry miombo, shorter return intervals persist in western- as compared to eastern- dry miombo areas. The mean fire return interval was 2.5 and 3.8 years for western- and eastern- dry miombo areas, respectively, for the period 2001–2013. On wider spatial and temporal extents, western- and eastern- dry miombo areas have 1.8 and 2.9 years of fire return intervals, respectively, for the period 1972–2011. Western parts of the woodland, including both dry and wet miombo areas, have higher fire prevalence than eastern parts of the woodland, which consist mainly of dry miombo areas. The higher fire prevalence is mainly a result of the interacting effects of rainfall patterns that influence fuel availability, and ignition sources. In southern African savannas, shorter return intervals occurred in higher rainfall areas but interacted with soil properties and herbivory, over a time period encompassing fire suppression, natural fires and controlled fires [55]. Similarly, in western African savannas, higher fire prevalence occurred in relation to increasing rainfall but interacted with both vegetation type and choices by herders and farmers to burn at different times during the fire season [11]. Rainfall influences productivity of grasses that make up the fuel load, but the fire prevalence is ultimately dependent on human influences on ignitions, fire season and extents burned [6, 7, 11, 56].

The west to east dominance of early and late dry season burning, respectively (Fig. 3), might be explained by differences in the length of the dry season. Parts of the western side of the study area receives light rains during September, from north and extending southward. These light rains continues through the main rainy season, thus reducing the length of the fire season. Central and eastern parts have a unimodal rainfall pattern and thus remain relatively dry until the beginning of another season in November/December, facilitating conditions favorable for late dry season fires.

The observed reduction in fire activity from north towards west (Fig. 6) is associated with expansion of croplands. The expansion of croplands in this area is likely a response to growing mining activities and respectively settlements in the Geita and Kahama districts, north of the study area. Expansion of croplands has had similar effect in northern hemisphere African savannas, where decreasing annual burned area occurred with increasing croplands [57]. As with fire activity, croplands had smaller burned extents when compared to vegetated cover types in the GLC-Share Database [51] for the extent of the study area. Based on GLC-Share cover types, our results show that up to 1.6% of the croplands are burned annually compared to 4.3% of grasslands, 2.6% of tree covered areas, 3.9% of shrubs covered areas and 10% of herbaceous vegetation, aquatic or regularly flooded. The datasets used to compile GLC-Share database for Tanzania is from 2001. Therefore, the values presented above are within a decadal range; from 1995 to 2005.

Burned extents
Burned patch sizes
Small-burned patches, less than five hectares in size, are the most prevalent across spatial and temporal scales. Smaller burned patches are a common occurrence across tropical savannas and are mainly associated with traditional fire management practices [18, 21, 58, 59]. Burned patches with similar sizes were associated with farm preparation in a neighboring Mozambican savanna [60]. Similarly, small fires within African savannas have been associated with agricultural activities and fragmentation of the landscape as a result of high population densities [14, 61]. These fires, burning small patches at a time progressively during the dry season, are generally a desirable management tool and are less damaging to savanna woodlands [18, 62, 63], unless they escape to burn unintended areas. Most of the cases of escaped fires are associated with clearing of new farms as opposed to burning agricultural residues in established agricultural areas. As discussed in the "Fire prevalence" section, our results show that burned extents were smaller in croplands than in other cover types. We could not quantify the effect of other sources of fire on burned patch sizes. However, Butz [21] has observed an increase is large accidental fires and a decrease in small fragmented fires, within a pastoral community in the savannas of northeastern Tanzania. In this area, a decline in nomadic pastoralism has occurred with a trend towards sedentarization and diversified livelihoods [64, 65]. Butz identified changes in rainfall patterns, population growth and fire suppression policies as the drivers of the change in the fire regime. Similar drivers of change in fire regimes persist in western African savannas [8]. In general, competition over land areas increases with a growing population, leading to changes in socioeconomic practices [64] and increasing land fragmentation. Consequently, fire regimes including the frequency, season and sizes of burned areas vary with localized adaptation to these changes in the context of the landscape pattern, and may be influenced by public policies and rainfall patterns [8, 14, 61, 66]. Fires with a higher threat are those ignited within woodland areas where tenure accessibility

and private uses are restricted. This threat is associated with increasing homogeneity and buildup of fuels with decreasing human activities [61]. In such cases, the prevailing weather regulates the spread of a fire when ignited [7], as opposed to the human control that fragments the landscape with small fires. In a recent analysis of MODIS burned area product (at 500 m resolution) for the whole of Tanzania between 2000 and 2011, up to 77% of the annual burned area in the country was detected on gazetted land [67]. Although the causes of these fires were not evaluated, it is less likely that they all stem from control burning for fire management purposes. Similarly, protected areas in the southern hemisphere African savannas had relatively larger burned areas than outside protected areas [13]. In the Llanos savannas of Columbia, relatively larger burned areas associated with hunting were observed in a national park compared to indigenous reserves and ranches [68]. In western African savannas, higher densities of fire events occurred in protected areas of Burkina Faso and lower populated areas of Mali [12, 69]. At a smaller scale, the highest tree mortality associated with fire in central Zambia occurred within an encroached part of a national park [70]. This highlights priority fire causes and affected areas that need further detailed analyses and probably a reassessment of management practices. Combining traditional and contemporary fire management practices may achieve reduction in burned extents and consequently biomass burning emissions [71].

We observed a larger scaling of frequency-size distribution, indicating a higher ratio of small relative to large fires, in this study as compared to other ecosystems, e.g. in the United States and Spain [53, 72]. In a global study of fire size distribution, Hantson et al. [14] have also observed a dominance of small fires in our study area, with a similar range of the scaling parameter, β (see "Frequency-size distribution of burned patches" section, Table 2). Small fires are recurrent in both dry and wet miombo areas but dry miombo areas experiences both smaller and larger fires than wet miombo areas (Table 2). The difference in size classes of burned patches in wet and dry miombo contributes to the slightly smaller scaling of frequency-size distribution in wet miombo, which implies large burned patches contribute slightly more to the total area burned in wet than in dry miombo. These large burned patches are partly a result of aggregation of smaller fires during the fire season. Generally, large fires are rare but small fires accumulate to cover extended areas in the course of a fire season each year. Archibald et al. [13] found similar contribution of small fires to the annual burned area in southern hemisphere Africa savannas. Small recurrent fires reduce the risk of occasional large fires, which have recently occurred in areas where fire suppression strategies are enforced.

Partially burned areas

Partially burned areas were defined to include intermixed pixels groups that are burned, partially burned and those with a diminishing char signature. They cover relatively wider extents than completely burned areas each year, ranging between 3.2 and 40.6% and 0.8–11.3%, respectively, for the whole woodland. Table 2 provides a decadal summary for wet and dry miombo areas. Between 9 and 14% of Tanzania's area was detected as burned annually during 2000–2011 from a lower (500 m) resolution burned area product [67]. Of this burned area, 69% occurred in the woodland. Miombo woodland areas covers approximately 90% of the forested areas in Tanzania, implying that much of the burned areas in the country are not detected at the lower resolution. Lower detection rates are possibly higher in the mixed burned–unburned pixels. Similar to completely burned areas, western parts of the woodland have relatively larger extents of partially burned areas, predominantly in wet miombo, than eastern parts of the woodland. Rigorous validation was not performed for partially burned areas, thus they were not further analyzed. However, they provide crucial information for understanding vegetation dynamics, which requires the season and severity of fires at specific areas.

Conclusions

We have documented the recent fire regime, for the past ~40 years, of the miombo woodland areas of Tanzania at spatial and temporal resolutions that have not been recorded before, to the best of our knowledge. The observed fire patterns for the past 40 years show that the majority of fire events occur in the western parts of miombo woodlands, consisting of wet miombo and western dry miombo areas. Fire events on the western parts of the woodland occur mainly during the first part of the dry season. Thus, an early dry season fire peak characterizes the west while a late dry season fire peak characterizes the east. Almost half of the woodland area has fire return intervals of <2 years. Return intervals are shorter in wet than in dry miombo areas. Short return intervals limit fuel loading and therefore prevents large intense fires. Human activities play a major role in shaping fire regimes. Mainly small sized fires characterize the regime across spatial and temporal scales. Occasional large fires are more frequently detected in dry than in wet miombo areas. Management strategies need to address spatially specific needs of wet and dry miombo areas, in the light of their fire regimes and socio-economic context.

Methods
Study area
The study area is miombo woodlands in Tanzania (Fig. 4). Miombo woodlands are disturbance driven moist

savannas that are shaped by natural and anthropogenic disturbances, to a larger extent, than by nutrient and water availability [1, 73]. They occur on nutrient poor soils and generally experience a warm-to-hot climate with a dry cold season [24]. The average annual rainfall ranges between 600 and 1,500 mm and falls during 5–6 months [23], followed by an extended dry period. Wet miombo areas, which receives more than 1,000 mm of average annual rainfall, are distinguished from dry miombo areas receiving less than 1,000 mm of average annual rainfall [74]. The woodland is characterized by wooded canopy species and an understory consisting of shrubs and light demanding grass species [24]. The annual production of these flammable, 0.5–2 m tall, grasses every rain season followed by accumulation of litter from the deciduous trees, makes miombo woodland highly susceptible to annual fires.

The fire season extends from the onset of the dry season to its end (Fig. 5), although isolated burning events may occur throughout the year at different localities. During the fire season, individual fires burn small patches at a time with the exception of very occasional big fires in areas where fuel load is accumulated and continuous. Towards the end of the fire season, a mosaic of burned and unburned patches occur.

Data sets and preprocessing

Landsat Level 1 Terrain (L1T) corrected product satellite images and MODIS collection 5 Level 2 MOD14/MYD14 active fire product form the major data source for this study. We derive fire patterns from the two datasets independently and compare results. Similar patterns will indicate that the datasets are representative of the fire patterns in the study area. This is important because although Landsat provides a finer spatial resolution its temporal resolution (16 days), and further limitation by cloud cover, may limit detection of savanna fires. On the other hand, MODIS detected active fires provide a more complete coverage at high temporal resolution but its temporal coverage is relatively short (since 2000) compared to that of Landsat (since 1972). In addition, MODIS can detect small active fires that are not captured by coarse resolution burned area products [45]. Thus, combining the two datasets benefits from their complementary availability, spatial and temporal characteristics [41].

All available Landsat images were downloaded from the USGS Global Visualization Viewer [75], to cover the study extent (Fig. 4) and for the period 1972–2011. Availability was constrained by image quality, predominantly percentage cloud cover within the study extent. Thus, a complete spatio-temporal dataset was difficult to achieve. For each year, processing and analysis was performed for areas where at least one image was available during the fire season. A total of 1,835 scenes, among them 234 MSS, 1,284 TM and 317 ETM+ SLC-On, were processed. Landsat TM imagery was preferred over MSS imagery for the period when both were available. Each image was converted to at-surface reflectance using the Dark Object Subtraction (DOS) method [76–78].

MODIS active fire data for the whole country were downloaded from the Fire Information for Resource Management System (FIRMS) [79], for the period between November 2000 and December 2013. MOD14/MYD14 provides, among others, coordinates of detected fires (the center of fire pixels at 1 km resolution), their acquisition date and time and respective FRP. Fire locations within miombo were retrieved and categorized as early dry season burning (January–July) or late dry season burning (August–December). Isolated fire events during the wet season were included in respective dry season burning based on the month of their detection. July was chosen to mark the end of early dry season burning for the entire woodland area, consistent with prescribed early burning between May and July in some parts of Tanzania [80–82]. This distinction was made to capture patterns of fire during the dry season, since the timing of burning influences the intensity and spread of a fire and thus its effects, such that fire management through prescribed burning is recommended during early dry season [23].

Spatial and temporal patterns of burned areas and active fires

Burned areas were detected by means of fuzzy classification of spectral indices derived from Landsat satellite images while spatial patterns of active fires were analyzed based on MODIS dataset. We limited our analysis to prevalence, burned extents and spatial patterns of active fires and burned patches. The general processing flow is summarized in Fig. 7. Analyses were performed in GRASS GIS [83] and R version 3.1.0 [84].

Training and testing of the fuzzy classification

A total of 523,092 pixels were sampled by visual interpretation from representative scenes for training and testing purposes. Burned areas were identified based on color composites with SWIR, NIR and VIS bands in RGB display, an approach that has been employed to extract image based training, and testing samples [46, 47, 85]. Active fires captured by Landsat satellite images and the analyst's field experience were utilized in line with the color composites. These formed the basis for selection of spectral indices.

Spectral Indices used for fuzzy classification

Spectral indices commonly used for burned area mapping were identified based on literature review. Eleven

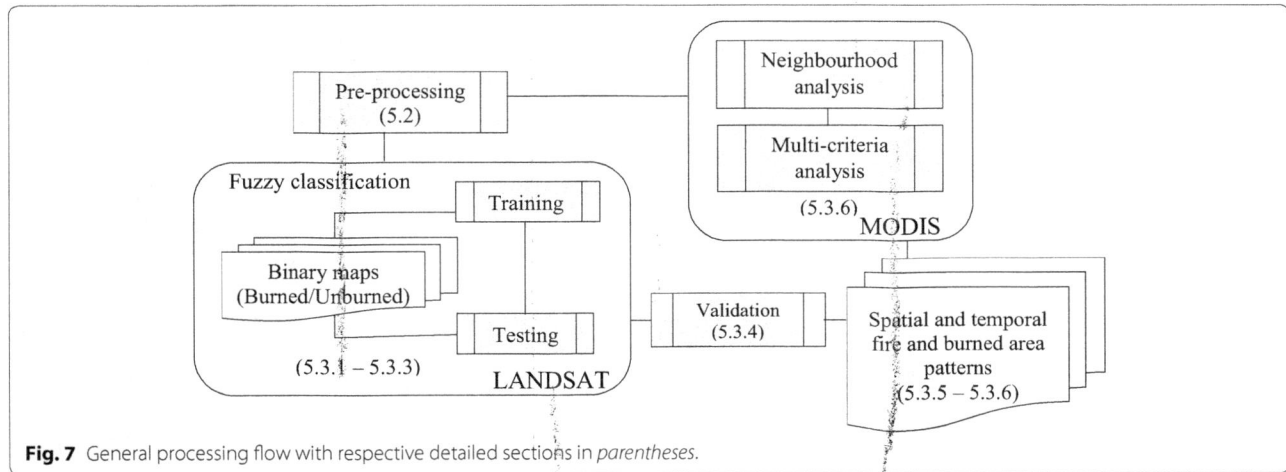

Fig. 7 General processing flow with respective detailed sections in *parentheses*.

indices with a potential for discriminating burned areas in sparsely vegetated areas were selected and tested. These included BAI [49], BAIM [86], $BAIM_L$ [46], GEMI [87], MIRBI [50], NBR_S [46], NBR_L [88], NBR_2 [47], NDVI, SARVI [89] and SAVI [90]. The range of values from individual scenes and the differences among scenes, for which burned pixels were well separated from other cover types, was determined for each index to form the base for fuzzy sets definition (see "Fuzzy membership rules" section). Penalized logistic regression was then employed to analyze the discrimination performance of burned from unburned pixels for each spectral index. An analysis combining sampled pixels from all scenes was also done to investigate how well results at scene level could be generalized.

Fuzzy membership rules

Fuzzy discrimination employs membership rules that are defined in terms of fuzzy sets [91], whose elements differentiate definite members from definite non-members and those with some level of uncertainty as to whether they are members or not. Fuzzy classification was experimented for each index individually and for different combinations of indices. Indices and combinations thereof were selected (Table 3) for fuzzy set definition based on how well they distinguished burned from unburned areas. Selected indices conformed to regression results (see "Spectral Indices used for fuzzy classification" section).

Validation of detected burned areas

Due to the lack of an independent burned area perimeter for validation, completely burned areas were distinguished from partially burned and unburned areas, based on their membership scores, for the purpose of

Table 3 Spectral indices used in fuzzy classification

Indices	Use
$BAIM_L$ and MIRBI	Detect burned areas at different post-fire conditions
$BAIM_L$ and threshold	Mask bare soil, water[a], topographic and cloud shadows
NBR_L and threshold	Distinguish active fires from other features
BAI	Detect burned areas on MSS imagery

$BAI = 1/(\rho_{c2} - \rho_2)^2 + (\rho_{c4} - \rho_4)^2$; $\rho_{c2} = 0.1$, $\rho_{c4} = 0.06$.
$BAIM_L = 1/(\rho_{c4} - \rho_4)^2 + (\rho_{c7} - \rho_7)^2$; $\rho_{c4} = 0.05$, $\rho_{c7} = 0.2$.
$MIRBI = 10 \times \rho_7 - 9.8 \times \rho_5 + 2$.
$NBR = (\rho_4 - \rho_7)/(\rho_4 + \rho_7)$.

ρ_2 = Band 2 of MSS on Landsat 4-5 and Band 5 of MSS on Landsat 1-3.

ρ_4 = Band 4 of MSS on Landsat 4-5, Band 7 of MSS on Landsat 1-3 and Band 4 of TM/ETM+.

ρ_7 = Band 7 of TM/ETM+ .

[a] Permanent water bodies were manually masked out from fuzzy classification results.

restricting further analysis to definite burned areas. Partially burned areas consisted of intermixed pixel groups of burned, partially burned and those with a diminishing char signature (Fig. 8). These areas were not included in subsequent analyses but they indicate the spatial and temporal extents of areas affected by fire each year. Validation of completely burned areas, which are referred to as burned areas, was performed based on visual interpretation of randomly selected samples, from another set of representative scenes different from those used for training and testing fuzzy classification. To validate the performance of the fuzzy classification when including also the partially burned areas, we employed visual analysis and unsupervised clustering. This approach, combining visual interpretation and unsupervised clustering, is suitable for discriminating burned areas in African savannas [42].

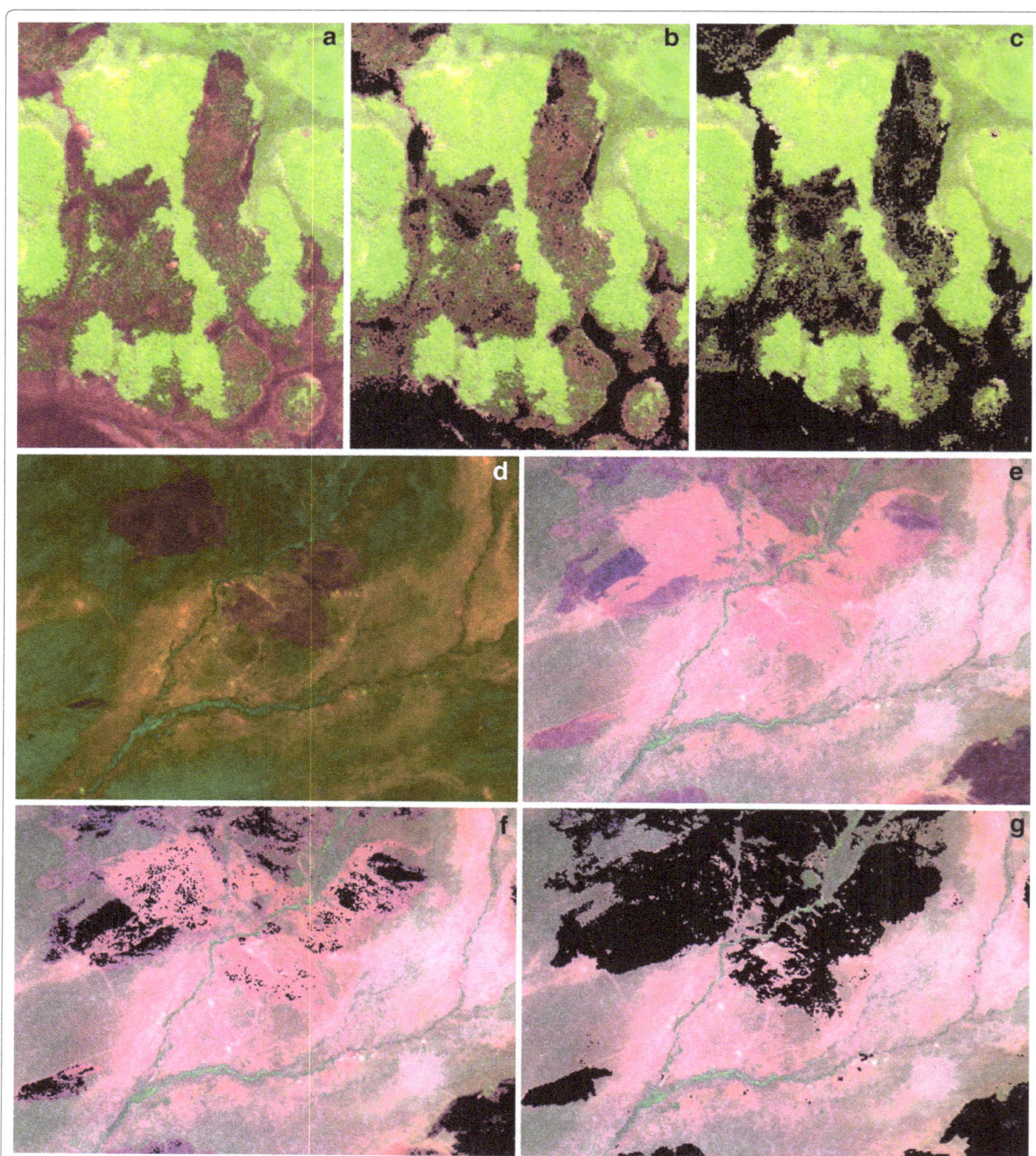

Fig. 8 Illustration of areas defined as partially burned. The *top panel* shows an area with mixed burned and unburned pixels (**a**). The *middle panel* shows burned patches at the beginning of the fire season (**d**) and the same area later (**e**) during the fire season. Burned areas (**b**) and (**f**), consist of contiguous groups of burned pixels with a definite fire scar. A mix of burned and unburned pixels and those with a diminishing fire scar were defined as partially burned, shown in (**c**) and (**g**) when combined with burned areas. Detection of burned areas with a diminishing fire scar is desirable when an image from an earlier date during the fire season is not available.

We adapted the approach described in [47] where three 1,000 × 1,000 pixels image subsets were visually interpreted to delineate burned/partially burned area perimeters. An independent image analyst examined these visually interpreted burned areas with support of false color composites (bands 432 and 741 as RGB) in combination with clustering of the bands 741 data subset, utilizing ERDAS Imagine 2014. The results were then used to validate the combined burned and partially burned area.

Burned patch sizes and spatio-temporal variation in burned extents

The sizes of burned patches were calculated based on contiguous burned pixels at scene level, while burned extents from annual mosaics after accounting for multiple detections between acquisitions. The fire return interval based on detected burned areas was determined by overlaying a 5 × 5 km grid on annual burned area maps, thus a return interval for every 2,500 ha. Grid cells containing burned patches >0.5 ha were considered affected by fire in respective years and provided a crude estimate of fire return interval for each cell. We use this estimate for consistency across the spatial and temporal extents with different Landsat data availability and for comparison with MODIS data (see "Occurrence and spatial patterns of active fires" section). The 0.5 ha threshold was selected based on reported burned patch sizes from anthropogenic fire sources in neighboring Mozambican savannas [60]. Frequency-size distributions of burned patches >0.1 and >0.4 ha for TM/ETM+ and MSS imagery, respectively, were examined for the period 1972–2011. Frequency densities of patch size classes were analyzed in log–log space, where the slope coefficient, β, provided the scaling of burned patch sizes i.e. the ratio of the number of large to small fires [53].

Occurrence and spatial patterns of active fires

The spatial association of detected fires was examined by Ripley K function for inhomogeneous spatial patterns [92]. Annual fire activity was derived at a 5 × 5 km resolution as a composite measure of active fire characteristics, including density and proximity of fires, annual duration of the fire season and range of FRP values. These combined characteristics of active fires provides a classification of the fire activity that is related to the fire regime [93]. The FRP values, for instance, are associated with the type of vegetation burned [94] and the density of fires is a good predictor of burned areas [13]. The grid resolution was based on both an optimal choice for spatial aggregation when comparing datasets with different resolution as applied in [44, 95] and for practical handling purposes. Fire return interval based on detected active fires was examined by a neighborhood analysis within a 1-km distance from each detected fire.

Thus, a return interval for an area of ~314 ha, which is the area of a circle of 1 km radius, centered at the location of detected active fires. This distance was selected to reflect the ground size of detected fire pixels. Results for each year provided fire return interval given locations of detected fires for that year and their average provided mean return interval for all years.

Abbreviations
BAI: Burned Area Index; BAIM: MODIS Burned Area Index; FIRMS: Fire Information for Resource Management System; FRP: Fire Radiative Power; GEMI: Global Environment Monitoring Index; MIRBI: Mid-Infrared Burn Index; MODIS: Moderate Resolution Imaging Spectroradiometer; NBR: Normalized Burn Ratio; NDVI: Normalized Difference Vegetation Index; SARVI: Soil and Atmospherically Resistant Vegetation Index; SAVI: Soil Adjusted Vegetation Index.

Authors' contributions
BT planned and implemented the study and prepared the manuscript. ØBD performed an independent visual analysis. TG planned some of the methods and prepared the manuscript. ØT prepared the manuscript. All authors contributed in revising the manuscript. All authors read and approved the final manuscript.

Author details
[1] Department of Ecology and Natural Resource Management, Norwegian University of Life Sciences, P.O. Box 5003, 1432 Ås, Norway. [2] Department of Geoinformatics, School of Geospatial Sciences and Technology, Ardhi University, P.O. Box 35176, Dar es Salaam, Tanzania. [3] Department of Mathematical Sciences and Technology, Norwegian University of Life Sciences, P.O. Box 5003, 1432 Ås, Norway.

Acknowledgements
We are very grateful to the Climate Change Impacts, Adaptation and Mitigation (CCIAM) Program in Tanzania for funding this study. We thank Professor Fred Midtgaard and Professor Seif Madoffe for initiating this study and for valuable comments on the manuscript. We also thank the anonymous reviewers for the very pertinent comments that greatly improved a previous version of the manuscript.

Compliance with ethical guidelines

Competing interests
The authors declare that they have no competing interests.

References
1. Sankaran M, Hanan NP, Scholes RJ, Ratnam J, Augustine DJ, Cade BS et al (2005) Determinants of woody cover in African savannas. Nature 438(7069):846–849
2. Staver AC, Archibald S, Levin S (2011) Tree cover in sub-Saharan Africa: rainfall and fire constrain forest and savanna as alternative stable states. Ecology 92(5):1063–1072
3. Lehmann CE, Anderson TM, Sankaran M, Higgins SI, Archibald S, Hoffmann WA et al (2014) Savanna vegetation–fire–climate relationships differ among continents. Science 343(6170):548–552
4. Bond W, Woodward F, Midgley G (2005) The global distribution of ecosystems in a world without fire. New Phytol 165(2):525–538
5. Archibald S, Staver AC, Levin SA (2012) Evolution of human-driven fire regimes in Africa. Proc Natl Acad Sci 109(3):847–852
6. Van Der Werf GR, Randerson JT, Giglio L, Gobron N, Dolman A (2008) Climate controls on the variability of fires in the tropics and subtropics. Glob Biogeochem Cycles 22:GB3028

7. Archibald S, Nickless A, Govender N, Scholes R, Lehsten V (2010) Climate and the inter-annual variability of fire in southern Africa: a meta-analysis using long-term field data and satellite-derived burnt area data. Glob Ecol Biogeogr 19(6):794–809

8. Laris P (2013) Integrating land change science and savanna fire models in West Africa. Land 2(4):609–636

9. Archibald S, Lehmann CE, Gómez-Dans JL, Bradstock RA (2013) Defining pyromes and global syndromes of fire regimes. Proc Natl Acad Sci 110(16):6442–6447

10. Bowman DM, Balch J, Artaxo P, Bond WJ, Cochrane MA, D'Antonio CM et al (2011) The human dimension of fire regimes on Earth. J Biogeogr 38(12):2223–2236

11. Mbow C, Nielsen TT, Rasmussen K (2000) Savanna fires in east-central Senegal: distribution patterns, resource management and perceptions. Human Ecol 28(4):561–583

12. Laris P, Caillault S, Dadashi S, Jo A (2015) The human ecology and geography of burning in an unstable savanna environment. J Ethnobiol 35(1):111–139

13. Archibald S, Scholes R, Roy D, Roberts G, Boschetti L (2010) Southern African fire regimes as revealed by remote sensing. Int J Wildland Fire 19(7):861–878

14. Hantson S, Pueyo S, Chuvieco E (2015) Global fire size distribution is driven by human impact and climate. Glob Ecol Biogeogr 24(1):77–86

15. Mouillot F, Field CB (2005) Fire history and the global carbon budget: a 1 × 1 fire history reconstruction for the 20th century. Glob Change Biol 11(3):398–420

16. Andreae MO, Atlas E, Cachier H, Cofer WR III, Harris GW, Helas G et al (1996) Trace gas and aerosol emissions from savanna fires. Biomass Burn Glob Change 1:278–295

17. van der Werf GR, Randerson JT, Giglio L, Collatz G, Mu M, Kasibhatla PS et al (2010) Global fire emissions and the contribution of deforestation, savanna, forest, agricultural, and peat fires (1997–2009). Atmos Chem Phys 10(23):11707–11735. doi:10.5194/acp-10-11707-2010

18. Bird RB, Codding BF, Kauhanen PG, Bird DW (2012) Aboriginal hunting buffers climate-driven fire-size variability in Australia's spinifex grasslands. Proc Natl Acad Sci 109(26):10287–10292

19. Moussa K, Bassett T, Nkem J (eds) (2011) Changing fire regimes in the Cote d'Ivoire savanna: implications for greenhouse emissions and carbon sequestration. In: Sustainable Forest Management in Africa: some solutions to natural forest management problems in Africa. Proceedings of the sustainable forest management in Africa Symposium. Stellenbosch, 3–7 November 2008, 2011, Stellenbosch University, Stellenbosch, South Africa

20. Laris P, Wardell DA (2006) Good, bad or 'necessary evil'? Reinterpreting the colonial burning experiments in the savanna landscapes of West Africa. Geogr J 172(4):271–290

21. Butz RJ (2009) Traditional fire management: historical fire regimes and land use change in pastoral East Africa. Int J Wildland Fire 18(4):442–450

22. GOFC-GOLD (2014) A sourcebook of methods and procedures for monitoring and reporting anthropogenic greenhouse gas emissions and removals associated with deforestation, gains and losses of carbon stocks in forests remaining forests, and forestation. GOFC-GOLD Report version COP20-1, GOFC-GOLD Land Cover Project Office, Wageningen University, The Netherlands

23. Chidumayo EN (1997) Miombo ecology and management: an introduction. Intermediate Technology Publications Ltd (ITP), London

24. Frost P (1996) The ecology of miombo woodlands. In: Campbell BM (ed) The miombo in transition: woodlands and welfare in Africa. CIFOR, Bogor, pp 11–57

25. Ryan CM, Williams M (2011) How does fire intensity and frequency affect miombo woodland tree populations and biomass? Ecol Appl 21(1):48–60

26. Trapnell C (1959) Ecological results of woodland burning experiments in northern Rhodesia. J Ecol 47(1):129–168

27. UNFCCC (2011) Report of the conference of the parties on its sixteenth session, held in Cancun from 29 November to 10 December 2010, Adendum, Part Two: action taken by the conference of the parties at its sexteenth session, FCCC/CP/2010/7/Add.1. United Nations Framework Convention on Climate Change, Bonn, Germany

28. Ahrends A, Burgess ND, Milledge SA, Bulling MT, Fisher B, Smart JC et al (2010) Predictable waves of sequential forest degradation and biodiversity loss spreading from an African city. Proc Natl Acad Sci 107(33):14556–14561

29. Barlow J, Parry L, Gardner TA, Ferreira J, Aragão LE, Carmenta R et al (2012) The critical importance of considering fire in REDD+ programs. Biol Conserv 154:1–8

30. Ryan CM, Hill T, Woollen E, Ghee C, Mitchard E, Cassells G et al (2012) Quantifying small-scale deforestation and forest degradation in African woodlands using radar imagery. Glob Change Biol 18(1):243–257

31. De Michele C, Accatino F, Vezzoli R, Scholes R (2011) Savanna domain in the herbivores-fire parameter space exploiting a tree–grass–soil water dynamic model. J Theor Biol 289:74–82

32. Accatino F, De Michele C, Vezzoli R, Donzelli D, Scholes RJ (2010) Tree–grass co-existence in savanna: interactions of rain and fire. J Theor Biol 267(2):235–242

33. Bowman DM, MacDermott HJ, Nichols SC, Murphy BP (2014) A grass–fire cycle eliminates an obligate-seeding tree in a tropical savanna. Ecol Evol 4(21):4185–4194

34. Bond WJ, Keeley JE (2005) Fire as a global 'herbivore': the ecology and evolution of flammable ecosystems. Trends Ecol Evol 20(7):387–394

35. Kikula IS (1986) The influence of fire on the composition of Miombo woodland of SW Tanzania. Oikos 46(3):317–324

36. Giglio L, Randerson J, Van der Werf G, Kasibhatla P, Collatz G, Morton D et al (2010) Assessing variability and long-term trends in burned area by merging multiple satellite fire products. Biogeosciences 7(3):1171–1186

37. Russell-Smith J, Ryan PG, Durieu R (1997) A LANDSAT MSS-derived fire history of Kakadu National Park, monsoonal northern Australial, 1980–94: seasonal extent, frequency and patchiness. J Appl Ecol 34(3):748–766

38. Trigg S, Flasse S (2000) Characterizing the spectral-temporal response of burned savannah using in situ spectroradiometry and infrared thermometry. Int J Remote Sens 21(16):3161–3168

39. Chuvieco E, Opazo S, Sione W, Valle HD, Anaya J, Bella CD et al (2008) Global burned-land estimation in Latin America using MODIS composite data. Ecol Appl 18(1):64–79

40. Armenteras D, Romero M, Galindo G (2005) Vegetation fire in the savannas of the Llanos Orientales of Colombia. World Resour Rev 17(4):531–543

41. Boschetti L, Roy DP, Justice CO, Humber ML (2015) MODIS–Landsat fusion for large area 30 m burned area mapping. Remote Sens Environ 161:27–42

42. Laris PS (2005) Spatiotemporal problems with detecting and mapping mosaic fire regimes with coarse-resolution satellite data in savanna environments. Remote Sens Environ 99(4):412–424

43. Giglio L, Randerson JT, van der Werf GR (2013) Analysis of daily, monthly, and annual burned area using the fourth-generation global fire emissions database (GFED4). J Geophys Res Biogeosci 118(1):317–328. doi:10.1002/jgrg.20042

44. Roy DP, Boschetti L (2009) Southern Africa validation of the MODIS, L3JRC, and GlobCarbon burned-area products. IEEE Trans Geosci Remote Sens 47(4):1032–1044

45. Randerson J, Chen Y, Werf G, Rogers B, Morton D (2012) Global burned area and biomass burning emissions from small fires. J Geophys Res Biogeosci (2005–2012) 117:G04012

46. Bastarrika A, Chuvieco E, Martín MP (2011) Mapping burned areas from Landsat TM/ETM+ data with a two-phase algorithm: balancing omission and commission errors. Remote Sens Environ 115(4):1003–1012

47. Stroppiana D, Bordogna G, Carrara P, Boschetti M, Boschetti L, Brivio P (2012) A method for extracting burned areas from Landsat TM/ETM+ images by soft aggregation of multiple Spectral Indices and a region growing algorithm. ISPRS J Photogramm Remote Sens 69:88–102

48. Barbosa PM, Grégoire J-M, Pereira JMC (1999) An algorithm for extracting burned areas from time series of AVHRR GAC data applied at a continental scale. Remote Sens Environ 69(3):253–263

49. Chuvieco E, Martin MP, Palacios A (2002) Assessment of different spectral indices in the red-near-infrared spectral domain for burned land discrimination. Int J Remote Sens 23(23):5103–5110

50. Trigg S, Flasse S (2001) An evaluation of different bi-spectral spaces for discriminating burned shrub-savannah. Int J Remote Sens 22(13):2641–2647. doi:10.1080/01431160110053185

51. Latham J, Cumani R, Rosati I, Bloise M (2014) FAO global land cover (GLC-SHARE) Beta-Release 1.0 Database, Division LaW

52. Kasin I, Blanck Y, Storaunet KO, Rolstad J, Ohlson M (2013) The charcoal record in peat and mineral soil across a boreal landscape and possible linkages to climate change and recent fire history. Holocene 23(7):1052–1065

53. Malamud BD, Millington JD, Perry GL (2005) Characterizing wildfire regimes in the United States. Proc Natl Acad Sci USA 102(13):4694–4699

54. Scott AC (2000) The Pre-Quaternary history of fire. Palaeogeogr Palaeoclimatol Palaeoecol 164(1):281–329

55. Van Wilgen B, Biggs H, O'regan S, Mare N (2000) Fire history of the savanna ecosystems in the Kruger National Park, South Africa, between 1941 and 1996. S Afr J Sci 96(4):167–178

56. Le Page Y, Oom D, Silva J, Jönsson P, Pereira J (2010) Seasonality of vegetation fires as modified by human action: observing the deviation from eco-climatic fire regimes. Glob Ecol Biogeogr 19(4):575–588

57. Andela N, van der Werf GR (2014) Recent trends in African fires driven by cropland expansion and El Nino to La Nina transition. Nature Climate Change 4(9):791–795

58. Laris P (2002) Burning the seasonal mosaic: preventative burning strategies in the wooded savanna of southern Mali. Human Ecol 30(2):155–186

59. Mistry J, Berardi A, Andrade V, Krahô T, Krahô P, Leonardos O (2005) Indigenous fire management in the cerrado of Brazil: the case of the Krahô of Tocantíns. Human Ecol 33(3):365–386

60. Shaffer LJ (2010) Indigenous fire use to manage savanna landscapes in Southern Mozambique. Fire Ecol 6(2):43–59. doi:10.4996/fireecology.0602043

61. Archibald S, Roy DP, Wilgen V, Brian W, Scholes RJ (2009) What limits fire? An examination of drivers of burnt area in Southern Africa. Glob Change Biol 15(3):613–630

62. Laris P (2011) Humanizing savanna biogeography: linking human practices with ecological patterns in a frequently burned savanna of southern Mali. Ann Assoc Am Geogr 101(5):1067–1088

63. Brockett B, Biggs H, Van Wilgen B (2001) A patch mosaic burning system for conservation areas in southern African savannas. Int J Wildland Fire 10(2):169–183

64. McCabe JT, Leslie PW, DeLuca L (2010) Adopting cultivation to remain pastoralists: the diversification of Maasai livelihoods in northern Tanzania. Human Ecol 38(3):321–334. doi:10.1007/s10745-010-9312-8

65. Nkedianye D, de Leeuw J, Ogutu JO, Said MY, Saidimu TL, Kifugo SC et al (2011) Mobility and livestock mortality in communally used pastoral areas: the impact of the 2005–2006 drought on livestock mortality in Maasailand. Pastoralism 1(1):1–17. doi:10.1186/2041-7136-1-17

66. Hudak AT, Fairbanks DH, Brockett BH (2004) Trends in fire patterns in a southern African savanna under alternative land use practices. Agric Ecosyst Environ 101(2):307–325

67. FAO (2013) A fire baseline for Tanzania. Sustainable forest management in a changing climate. FAO-Finland Forestry Programme. Dar es Salaam, Tanzania

68. Romero-Ruiz M, Etter A, Sarmiento A, Tansey K (2010) Spatial and temporal variability of fires in relation to ecosystems, land tenure and rainfall in savannas of northern South America. Glob Change Biol 16(7):2013–2023

69. Caillault S, Ballouche A, Delahaye D (2014) Where are the 'bad fires' in West African savannas? Rethinking burning management through a space–time analysis in Burkina Faso. Geogr J. doi:10.1111/geoj.12074

70. Chidumayo E (2002) Changes in miombo woodland structure under different land tenure and use systems in central Zambia. J Biogeogr 29(12):1619–1626

71. Russell-Smith J, Cook GD, Cooke PM, Edwards AC, Lendrum M, Meyer C et al (2013) Managing fire regimes in north Australian savannas: applying Aboriginal approaches to contemporary global problems. Front Ecol Environ 11(s1):e55–e63

72. Moreno M, Malamud B, Chuvieco E (2011) Wildfire frequency–area statistics in Spain. Procedia Environ Sci 7:182–187

73. Bond W, Midgley G, Woodward F (2003) What controls South African vegetation-climate or fire? S Afr J Bot 69(1):79–91

74. White F (1983) The vegetation of Africa: a descriptive memoir to accompany the UNESCO/AETFAT/UNSO vegetation map of Africa by F White. Natural Resources Research Report XX, UNESCO, Paris, France

75. USGS Global Visualization Viewer. http://glovis.usgs.gov/. Accessed 13 April 2015

76. Chavez PS (1996) Image-based atmospheric corrections-revisited and improved. Photogramm Eng Remote Sens 62(9):1025–1036

77. Song C, Woodcock CE, Seto KC, Lenney MP, Macomber SA (2001) Classification and change detection using Landsat TM data: when and how to correct atmospheric effects? Remote Sens Environ 75(2):230–244

78. Tizado EJ (2013) i.landsat.toar: calculates top-of-atmosphere radiance or reflectance and temperature for Landsat MSS/TM/ETM+/OLI. In: GRASS Development Team (ed) Geographic Resources Analysis Support System (GRASS 7) user's manual: open source geospatial foundation project. http://grass.osgeo.org

79. Fire Information for Resource Management System. https://earthdata.nasa.gov/data/near-real-time-data/firms. Accessed 13 April 2015

80. Nssoko E (2004) Community-based fire management in the Miombo woodlands: a case study from Bukombe District, Shinyanga, Tanzania. Aridlands No 55, May/June 2004

81. Luoga E, Witkowski E, Balkwill K (2005) Land cover and use changes in relation to the institutional framework and tenure of land and resources in eastern Tanzania miombo woodlands. Environ Dev Sustain 7(1):71–93

82. Hassan SN, Rija AA (2011) Fire history and management as determinant of patch selection by foraging herbivores in western Serengeti, Tanzania. Int J Biodivers Sci Ecosyst Serv Manage 7(2):122–133

83. GRASS Development Team (2012) Geographic Resources Analysis Support System (GRASS) Software. Open Source Geospatial Foundation Project. http://grass.osgeo.org

84. R Core Team (2014) R: a language and environment for statistical computing. R Foundation for Statistical Computing, Vienna, Austria. http://www.R-project.org/

85. Koutsias N, Karteris M (2000) Burned area mapping using logistic regression modeling of a single post-fire Landsat-5 Thematic Mapper image. Int J Remote Sens 21(4):673–687

86. Martín M, Gómez I, Chuvieco E (eds) (2005) Performance of a burned-area index (BAIM) for mapping Mediterranean burned scars from MODIS data. In: Proceedings of the 5th international workshop on remote sensing and GIS applications to forest fire management: fire effects assessment. Universidad de Zaragoza, GOFC GOLD, EARSeL, Paris

87. Pinty B, Verstraete M (1992) GEMI: a non-linear index to monitor global vegetation from satellites. Vegetatio 101(1):15–20

88. Key CH, Benson NC (1999) Measuring and remote sensing of burn severity: the CBI and NBR. Poster Abstract. In: Neuenschwander LF, Ryan KC (eds) Proceedings joint fire science conference and workshop, vol II, Boise, ID, 15–17 June 1999. University of Idaho and International Association of Wildland Fire, p 284

89. Huete A, Liu H, Batchily K, Van Leeuwen W (1997) A comparison of vegetation indices over a global set of TM images for EOS-MODIS. Remote Sens Environ 59(3):440–451

90. Huete AR (1988) A soil-adjusted vegetation index (SAVI). Remote Sens Environ 25(3):295–309

91. Jasiewicz J (2011) A new GRASS GIS fuzzy inference system for massive data analysis. Comput Geosci 37(9):1525–1531

92. Baddeley AJ, Møller J, Waagepetersen R (2000) Non-and semi-parametric estimation of interaction in inhomogeneous point patterns. Stat Neerl 54(3):329–350

93. Chuvieco E, Giglio L, Justice C (2008) Global characterization of fire activity: toward defining fire regimes from Earth observation data. Glob Change Biol 14(7):1488–1502

94. Giglio L, Csiszar I, Justice CO (2006) Global distribution and seasonality of active fires as observed with the terra and aqua moderate resolution imaging spectroradiometer (MODIS) sensors. J Geophys Res Biogeosci (2005–2012) 111:G02016

95. Eva H, Lambin EF (1998) Remote sensing of biomass burning in tropical regions: sampling issues and multisensor approach. Remote Sens Environ 64(3):292–315

96. Maidment RI, Grimes D, Allan RP, Tarnavsky E, Stringer M, Hewison T et al (2014) The 30 year TAMSAT African rainfall climatology and time series (TARCAT) data set. J Geophys Res Atmos 119(18):10619–10644. doi:10.1002/2014JD021927

5

A framework for estimating forest disturbance intensity from successive remotely sensed biomass maps: moving beyond average biomass loss estimates

T. C. Hill[1,2,3*], C. M. Ryan[2] and M. Williams[2,3]

Abstract

Background: The success of satellites in mapping deforestation has been invaluable for improving our understanding of the impacts and nature of land cover change and carbon balance. However, current satellite approaches struggle to quantify the intensity of forest disturbance, i.e. whether the average rate of biomass loss for a region arises from heavy disturbance focused in a few locations, or the less severe disturbance of a wider area. The ability to distinguish between these, very different, disturbance regimes remains critical for forest managers and ecologists.

Results: We put forward a framework for describing all intensities of forest disturbance, from deforestation, to widespread low intensity disturbance. By grouping satellite observations into ensembles with a common disturbance regime, the framework is able to mitigate the impacts of poor signal-to-noise ratio that limits current satellite observations. Using an observation system simulation experiment we demonstrate that the framework can be applied to provide estimates of the mean biomass loss rate, as well as distinguish the intensity of the disturbance. The approach is robust despite the large random and systematic errors typical of biomass maps derived from radar. The best accuracies are achieved with ensembles of $\geq$1600 pixels ($\geq$1 km^2 with 25 by 25 m pixels).

Summary: The framework we describe provides a novel way to describe and quantify the intensity of forest disturbance, which could help to provide information on the causes of both natural and anthropogenic forest loss—such information is vital for effective forest and climate policy formulation.

Keywords: Deforestation, Forest, Degradation, Disturbance, Intensity, Biomass, REDD, Satellite, Remote sensing, Carbon

Background

Tropical deforestation has been estimated to occur at a rate of 13 million ha per year [1], with an associated net loss of forest biomass of 1.3 ± 0.7 Pg C year^{-1} [2]. Remote sensing has been successful at mapping global deforestation [3, 4]. However, deforestation presents a simplified view of forest disturbance that ignores the many graduations of lower intensity, but often widespread, forest disturbance and degradation [5]. Forest disturbance refers

to the mechanisms which limit biomass by causing its destruction [6]. The impact of forest disturbance is highly variable, leading to the total or partial loss of biomass through a diverse range of natural (e.g. disease, droughts, fires, herbivory and windstorms) and/or anthropogenic processes (e.g. urbanisation, agriculture, selective logging, and fires). Remotely sensed information on the spatial extent and intensity of forest disturbance would be extremely useful for managers and ecologists in attempts to develop a mechanistic understanding of forest degradation and the processes of forest disturbance [7, 8].

Unfortunately, whilst satellites have the coverage to provide global information on forest disturbance, very

*Correspondence: t.c.hill@exeter.ac.uk
[1] Department of Earth and Environmental Science, University of St Andrews, Irvine Building, North Street, St Andrews, UK
Full list of author information is available at the end of the article

few studies have attempted to do so [9]. The focus of remote sensing has remained on mapping deforestation (e.g. [3]) and arises from a number of technological limitations which include, amongst other factors, observation precision and the relative "directness" of the observation [10–12]. Ideally remote sensing measurements would have a direct dependence on the quantity being estimated; however current optical satellite derived estimates of forest disturbance are not able to achieve this ideal. Instead, optical satellite measurements of forest disturbance rely on observing physical properties that are expected to correlate with disturbance (e.g. changes in leaf area that can be detected by satellites due to a change in the absorption in the chlorophyll spectral bands). When these correlations change, indirect measures cannot be expected to provide a robust estimate of disturbance.

As Synthetic aperture radar (SAR) is sensitive to the structural properties of forests it can be thought of as a more direct measure of forest biomass than the passive optical alternatives [8, 13, 14]. SAR provides one of the only viable data sources currently available for global monitoring of forest disturbance at moderate spatial resolutions. However, the low signal to noise ratio of SAR leads to poor precision and large uncertainty in the estimated biomass of individual pixels [13] and this equates to high levels of uncertainty when detecting biomass change at the pixel level. Filtering can be used to improve the signal to noise ratio of SAR, but this improvement comes at the expense of a reduction in the effective spatial resolution of the biomass change estimates. Therefore remote sensing can provide regional estimates of biomass change or fine scale maps of deforestation, but it cannot yet be said to truly determine the intensity of forest degradation across scales [13].

In this study we set out a novel framework for quantifying the intensity of forest disturbance from successive remotely sensed biomass maps. This approach can describe both focused high intensity forest loss and low intensity, widespread degradation. We provide an example methodology to exploit this framework using SAR data with realistic observation errors. We explore the strengths and limitations of this approach using an observation system simulation experiment.

Results and discussion
A framework for quantifying the intensity of forest disturbance

The framework that we propose describes forest disturbance for an area in which all satellite observation pixels are assumed to experience the same disturbance regime. That is, whilst each pixel might or might not be disturbed, each pixel within each area has the same probability of being disturbed and the same relative loss of biomass when disturbed. We consider an ensemble of n remote sensing pixels, which are observed on two successive dates ($t = 1$ and $t = 2$). This approach builds on an earlier framework set out in Williams [15] that used the biomass distribution at a single point in time. The first advantage of this new approach is that it allows ensemble statistics to be calculated, negating the limitations of poor signal to noise of individual biomass pixels by estimating ensemble's mean fractional loss of biomass per year (E_M). The second advantage of the framework is that it permits E_M to be split into two factors for the ensemble: the probability of disturbance per year for each ensemble pixel member (E_P) and the fractional loss of biomass per disturbance for each disturbed ensemble pixel member (E_I), Eq. 1.

$$E_M \approx E_P E_I \tag{1}$$

Both E_M and its factors E_P and E_I can take values ranging from 0–1. Where, for example, an $E_P = 0.05$ would imply a 1 in 20 chance of each pixel being disturbed in a year. An $E_I = 0.2$ implies that, if disturbed, a pixel will lose 20 % of its biomass. Combining these example factors would lead to the expectation of $E_M = 0.01$, or a 1 % reduction in mean biomass for the ensemble. The ensemble size can be picked to balance the competing requirements of high precision on estimates of E_M, E_P and E_I and meeting the assumption of a continuous disturbance regime.

The inclusion of the factors E_P and E_I allows a flexible description of biomass loss, without the arbitrary distinction between deforestation and lower intensity forest disturbance. Forest disturbance (other than deforestation) can be represented by the parameter space $0 < E_M < 1$, $0 < E_I < 1$ and $0 < E_P \leq 1$ (Fig. 1). There are also several special cases: the total deforestation of the ensemble area ($E_M = 1$, $E_I = 1$ and $E_P = 1$); partial deforestation within the area ($0 < E_M < 1$, $E_I = 1$ and $0 < E_P < 1$); and no disturbance ($E_M = 0$).

Describing forest disturbance using this framework has a number of advantages over traditional descriptors as it avoids the need for an arbitrary threshold for a forest cover loss used in other studies, e.g. [16]. In turn, this allows for a more nuanced description of forest disturbance than is possible with categorical land cover classes, or measures of forest cover.

Using the framework to estimate biomass loss and disturbance intensity

Using the disturbance framework, our observation system simulation experiments (OSSE) show it is possible to robustly estimate the mean biomass loss for the ensemble (E_M) and also the disturbance regime, as described by

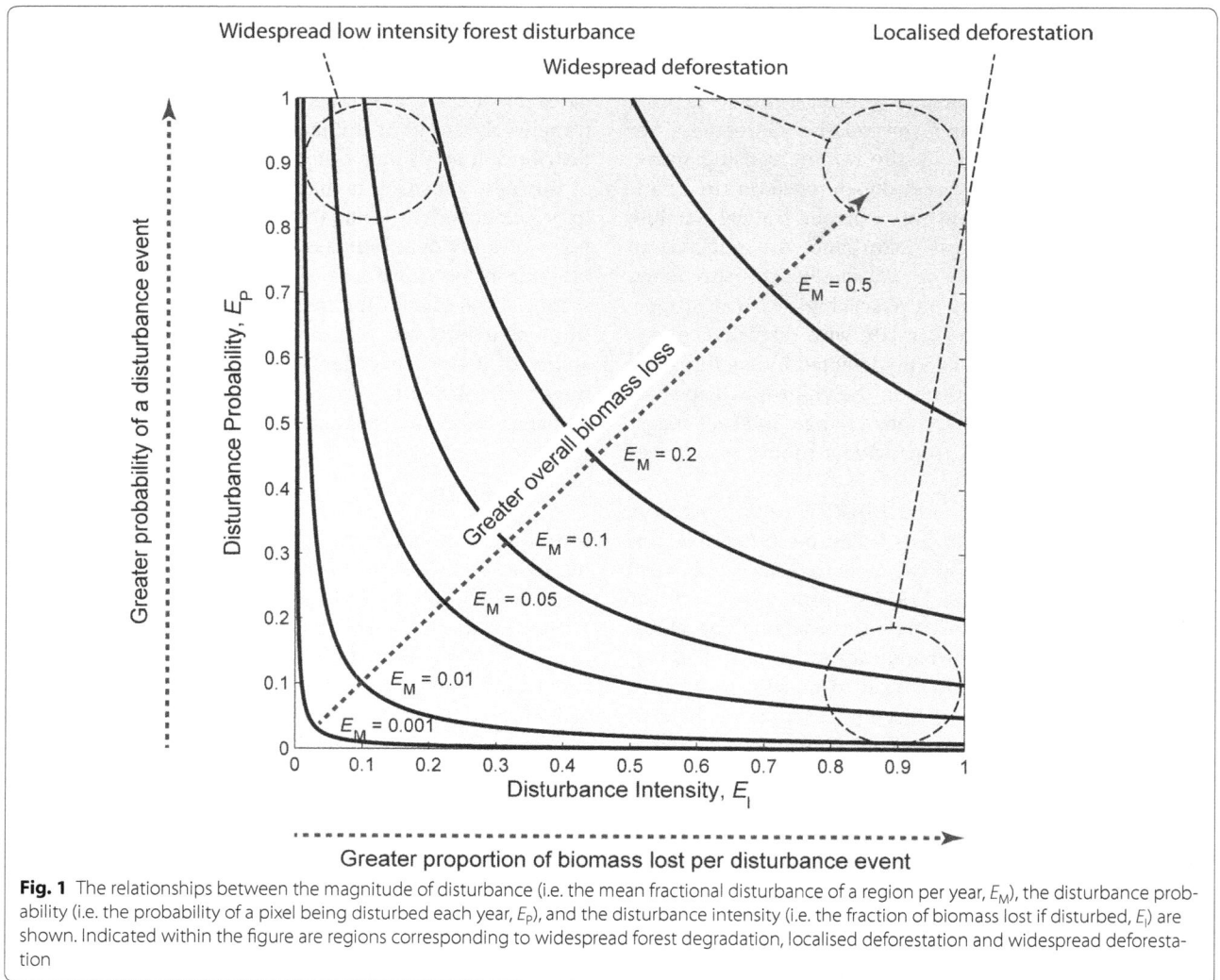

Fig. 1 The relationships between the magnitude of disturbance (i.e. the mean fractional disturbance of a region per year, E_M), the disturbance probability (i.e. the probability of a pixel being disturbed each year, E_P), and the disturbance intensity (i.e. the fraction of biomass lost if disturbed, E_I) are shown. Indicated within the figure are regions corresponding to widespread forest degradation, localised deforestation and widespread deforestation

E_P and E_I (Fig. 2; Table 1). E_M is better constrained than either E_I or E_P, reflecting the direct impact of E_M on the mean of the biomass distribution versus the more variable impacts of the E_I or E_P on the second and third order moments: standard deviation and skew. With larger mean disturbances the constraints on E_I or E_P improve, presumably due to the larger number of pixels effected and/or greater impact on the observed biomass.

How the framework is affected by observation bias

It is highly likely that biomass maps used with the framework will not only have random noise, but also systematic observation errors [13], therefore we test the framework's sensitivity to bias. The inclusion of a realistic observation bias of ±160 gC m^{-2} (±1.6 tC ha^{-1}) [13] had the largest impacts on the estimates of E_M, which showed a bias of 0.004 (0.4 %). However E_I and E_P were less affected

(Fig. 3; Table 1) and it was still possible to distinguish high and low intensity disturbance regimes. The bias in the estimate for E_M is consistent with change in the mean biomass that is implied by a bias of 160 gC m^{-2}, and would therefore apply to any other approach to estimating biomass loss.

What is the optimal ensemble size?

The accuracy of predictions using the framework are best when considering an ensemble size of $n \geq 1600$, whilst for $n \leq 400$, the precision of our estimates drops rapidly (Fig. 4; Table 1). Therefore, for the 25 m pixels typical of current SAR biomass estimates, the recommended ensemble size covers an area of at least 1 km^2 or 100 ha. The implication is that it must be reasonable to assume that the disturbance regime is constant across areas of at least 100 ha. We expect this minimum area to be

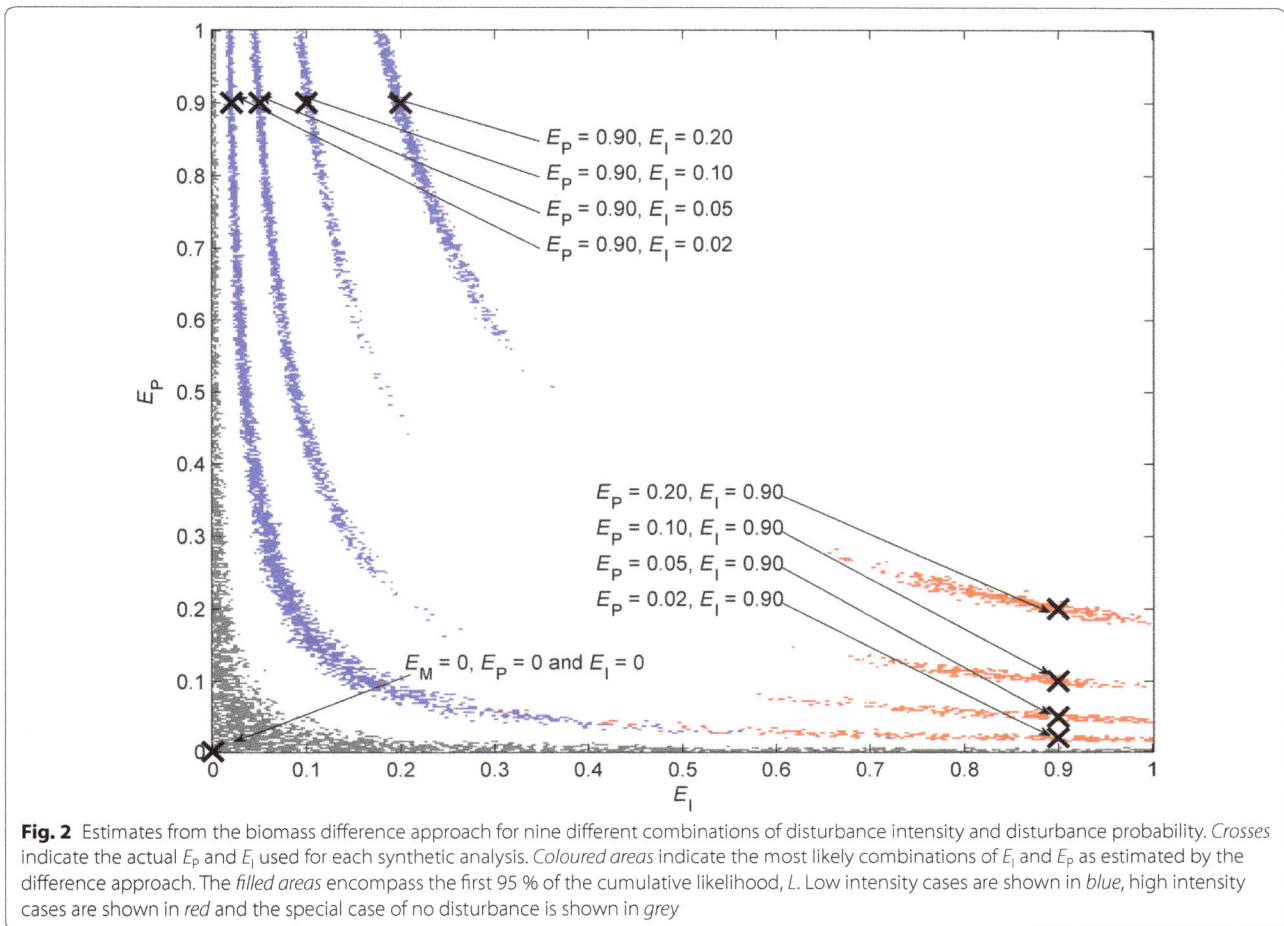

Fig. 2 Estimates from the biomass difference approach for nine different combinations of disturbance intensity and disturbance probability. *Crosses* indicate the actual E_P and E_I used for each synthetic analysis. *Coloured areas* indicate the most likely combinations of E_I and E_P as estimated by the difference approach. The *filled areas* encompass the first 95 % of the cumulative likelihood, L. Low intensity cases are shown in *blue*, high intensity cases are shown in *red* and the special case of no disturbance is shown in *grey*

robust, at least for ALOS PALSAR data, but the sensitivity to ensemble size should checked in each new study. It is worth noting that there is no requirement for these areas to be rectangular, or even contiguous. The grainy texture evident in the estimates from the analyses is due to the simulation of stochastic disturbances (Figs. 2, 3, 4). This graining is reduced with increasing ensemble size, but can be further reduced, at significant computational expense, by averaging repeat runs of the maximum likelihood estimation.

Alternative formulations of the framework

We formulated the framework in terms of a fractional change in biomass for both E_M and E_I. This is not the only possible formulation, nor is it necessarily the optimal for all situations; however, it is a mathematically simple approach to maintaining a positive (i.e. plausible) biomass for each pixel. The implicit assumptions of the formulation are: 1) that all pixels are equally likely to be disturbed, and 2) that when disturbed a fixed fraction is lost, irrespective of the starting biomass. It is possible to imagine scenarios that would not be well described

by our scheme where (say) selective logging only targets the largest trees (i.e. E_P is high for pixels above a threshold biomass and zero for all others). These scenarios do not contradict our first assumption of common disturbance regime for the ensemble and it should be possible to restate the parameters used in the framework to accommodate a particular set of assumptions about the disturbance regime. However the inclusion of more complicated mathematical representations and, specifically, more parameters, can be expected to increase the challenge of estimating the parameters of any new framework formulation and decrease its general applicability.

Limitations to the framework

Finally there are two notable caveats: Firstly, the success of the approach is strongly tied to the ability to characterise the random error of biomass estimates. The design of the analysis mitigates some of the errors as biomass errors that are consistent between the two biomass maps will be removed by this differencing. Whilst Ryan et al. (2012) report normal errors in the biomass domain, other studies assume errors will actually be normal in

Table 1 The maximum (97.5 % CI), minimum (2.5 % CI) and 95 % CI range (maximum–minimum) for estimates of E_M, E_I and E_P are shown for each of the test

Figs.	Description	Ensemble size (n)	Actual E_I	Actual E_P	Actual E_M	Est. E_I (max/min/ range)	Est. E_P (max/min/ range)	Est. E_M (max/min/ range)
2	High intensity	1600	0.9	0.02	0.018	1.000/0.178/0.823	0.100/0.015/0.085	0.023/0.012/0.011
2	High intensity	1600	0.9	0.05	0.045	1.000/0.580/0.420	0.083/0.043/0.040	0.053/0.040/0.013
2	High intensity	1600	0.9	0.1	0.090	0.993/0.618/0.375	0.148/0.093/0.055	0.097/0.084/0.014
2	High intensity	1600	0.9	0.2	0.180	0.998/0.655/0.343	0.283/0.180/0.103	0.195/0.168/0.027
2	Low intensity	1600	0.02	0.9	0.018	0.720/0.013/0.708	1.000/0.020/0.980	0.023/0.011/0.012
2	Low intensity	1600	0.05	0.9	0.045	0.265/0.040/0.225	1.000/0.163/0.838	0.050/0.038/0.012
2	Low intensity	1600	0.1	0.9	0.090	0.208/0.088/0.120	1.000/0.443/0.558	0.100/0.086/0.014
2	Low intensity	1600	0.2	0.9	0.180	0.363/0.173/0.190	1.000/0.508/0.493	0.188/0.168/0.020
2	No disturbance	1600	0	0	0.000	1.000/0.000/1.000	1.000/0.000/1.000	0.006/0.000/0.006
3	High intensity: Bias = 160 gC m^{-2}	1600	0.9	0.05	0.045	1.000/0.653/0.348	0.060/0.038/0.023	0.043/0.035/0.008
3	High intensity: Bias = 0 gC m^{-2}	1600	0.9	0.05	0.045	1.000/0.468/0.533	0.098/0.038/0.060	0.049/0.035/0.014
3	High intensity: Bias = −160 gC m^{-2}	1600	0.9	0.05	0.045	1.000/0.518/0.483	0.108/0.050/0.058	0.060/0.049/0.011
3	Low intensity: Bias = 160 gC m^{-2}	1600	0.05	0.9	0.045	0.188/0.035/0.153	1.000/0.215/0.785	0.044/0.032/0.012
3	Low intensity: Bias = 0 gC m^{-2}	1600	0.05	0.9	0.045	0.325/0.040/0.285	1.000/0.148/0.853	0.052/0.039/0.012
3	Low intensity: Bias = −1600 gC m^{-2}	1600	0.05	0.9	0.045	0.283/0.048/0.235	1.000/0.173/0.828	0.059/0.046/0.012
4	Area = 2000 by 2000 m	6400	0.05	0.9	0.045	0.128/0.045/0.083	1.000/0.350/0.650	0.048/0.042/0.006
4	Area = 1000 by 1000 m	1600	0.05	0.9	0.045	0.298/0.043/0.255	1.000/0.150/0.850	0.053/0.041/0.013
4	Area = 500 by 500 m	400	0.05	0.9	0.045	0.848/0.040/0.808	1.000/0.045/0.955	0.061/0.033/0.028
4	Area = 250 by 250 m	100	0.05	0.9	0.045	0.980/0.023/0.958	1.000/0.028/0.973	0.074/0.019/0.055

The 95 % CI range is expected encompass the actual E_M, E_I and E_P used to generate the synthetic observations

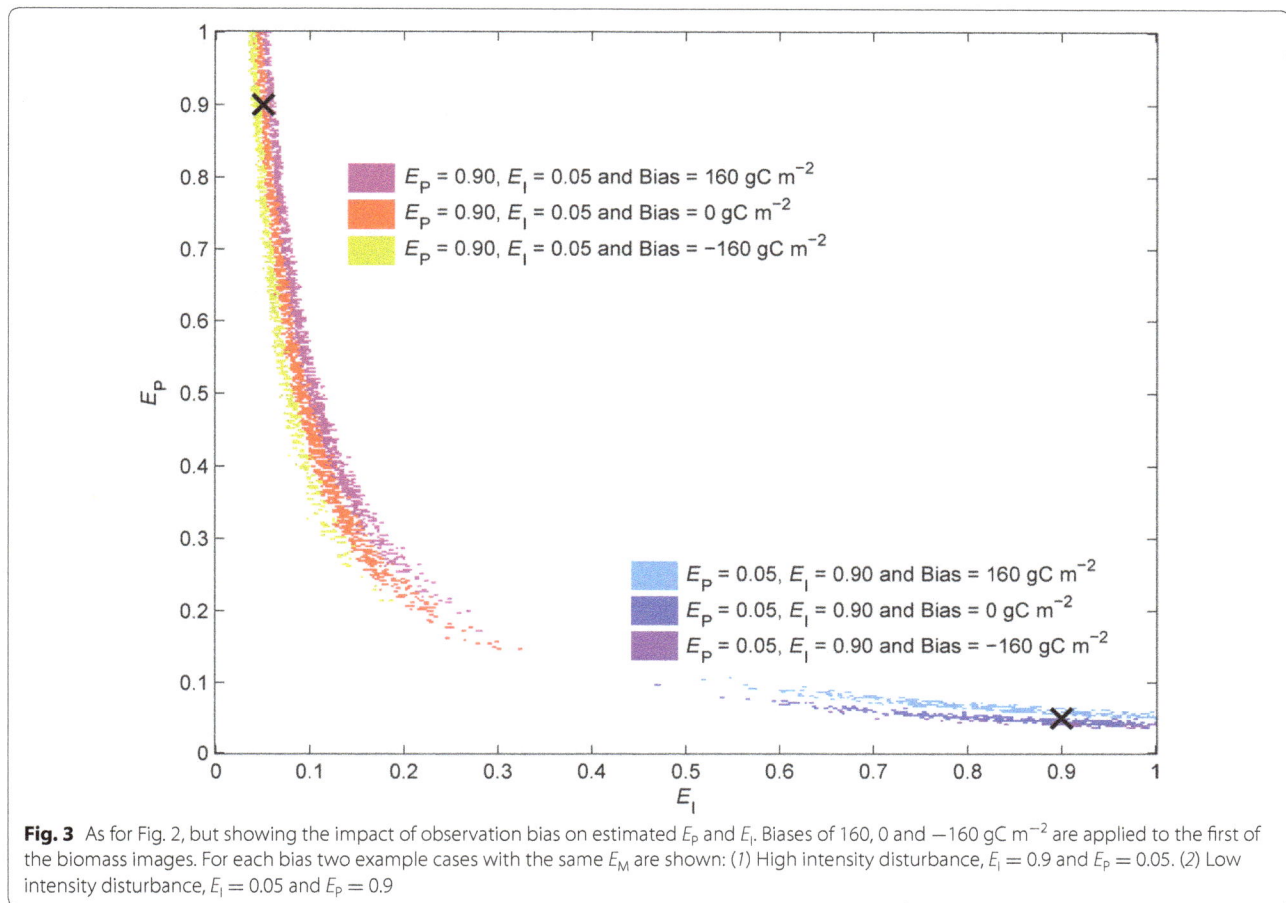

Fig. 3 As for Fig. 2, but showing the impact of observation bias on estimated E_P and E_I. Biases of 160, 0 and -160 gC m^{-2} are applied to the first of the biomass images. For each bias two example cases with the same E_M are shown: (1) High intensity disturbance, $E_I = 0.9$ and $E_P = 0.05$. (2) Low intensity disturbance, $E_I = 0.05$ and $E_P = 0.9$

the dB (i.e. $\log_{10}$) domain [17] which would result in log-normal, asymmetric errors on biomass estimates. We therefore reran the synthetic experiments with 0.5 dB error and achieved similar results. The second caveat is that the biomass differencing approach assumes the carbon model (A-DALEC) is unbiased. This assumption is, by definition, valid in a observation system simulation experiment (OSSE). However in practice the productivity of the ecosystem is not known perfectly, and so estimates of E_M can be expected to be biased. However the assumption is not as crude as it at first might seem: ignoring the production term (as is implicitly done by most deforestation algorithms) makes the assumption that the forest otherwise in steady state. This issue is likely to be more severe in field sites where less information is available on which to provide independent estimate of the rates of aboveground biomass accumulation, e.g. [18].

Conclusions

From a management and policy perspective it is important to be able to distinguish between the different intensities of biomass loss as they may be associated with different disturbances mechanisms (e.g. low intensity disturbances are likely driven by a need for timber or fuel, and high intensity disturbances driven by a need for agricultural land) [19]. However, current estimates of deforestation and biomass loss are not adequate for estimating lower intensity forest disturbance. Theoretically, high resolution biomass loss estimates could provide fine-scale estimates of forest degradation, but the current precision is not adequate and there is no immediate prospect of this changing, partly due to the speckle and other noise in SAR imagery [20]. The framework that we have described is a pragmatic representation of forest degradation that uses ensemble statistics to mitigate the poor precision of current SAR biomass estimates [13, 17, 20]. SAR is expected to remain a key technique for global mapping of forest biomass, and the framework we propose is compatible with the upcoming BIOMASS and new L-band satellites [14]. Using an OSSE we have shown that is possible to robustly estimate the parameters of the framework to describe forest disturbance, provided that two successive biomass maps, separated by at least one year, are available. We suggest that using

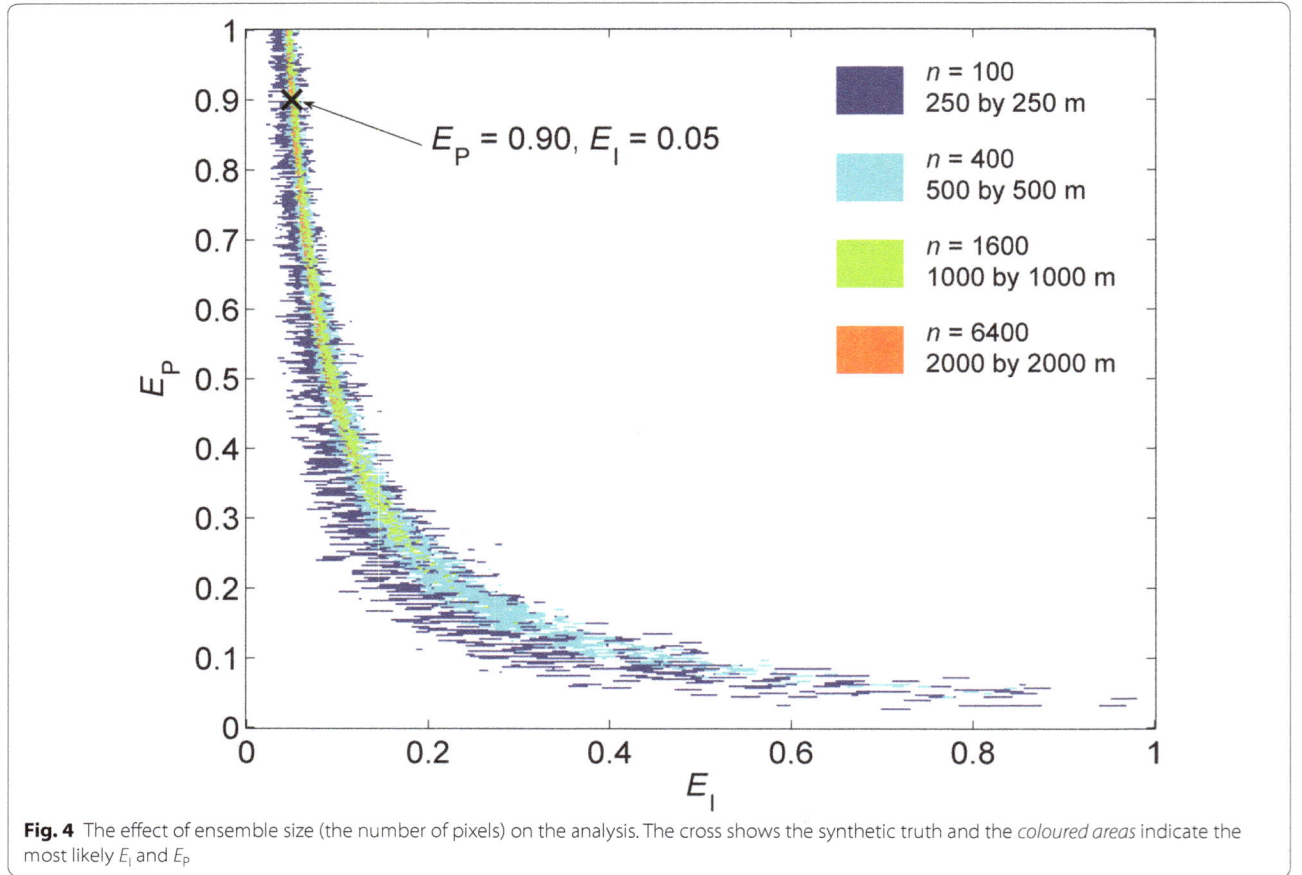

Fig. 4 The effect of ensemble size (the number of pixels) on the analysis. The cross shows the synthetic truth and the *coloured areas* indicate the most likely E_I and E_P

a similar framework will allow remote sensing studies to provide more relevant constraints on estimates of land-use change.

Methods

Estimating the framework parameters

A number of approaches could be taken to estimating E_M, E_I and E_P, we chose to use maximum likelihood estimation. We identify the combinations of E_I and E_P that allow a simulations to most closely match the observed changes in the biomass distribution from time $t = 1$ to time $t = 2$, for an ensemble of n pixels.

The observed biomass (O_i^t) of the ith pixel at time t is sum of the actual biomass (B_i^t) and the measurement noise (N_i^t), Eq. 2. The observed biomass at time $t + 1$, follows the same logic (Eq. 3). The actual biomasses are related via the growth G_i^t and disturbance D_i^t of each pixel in the ensemble (Eq. 4).

$$O_i^t = B_i^t + N_i^t \quad \text{for} \ i = 1, 2, 3, \ldots, n \quad (2)$$

$$O_i^{t+1} = B_i^{t+1} + N_i^{t+1} \quad \text{for} \ i = 1, 2, 3, \ldots, n \quad (3)$$

$$B_i^{t+1} = B_i^t + G_i^t - D_i^t \quad \text{for} \ i = 1, 2, 3, \ldots, n \quad (4)$$

Similarly we are able to simulate the observed biomass at time t (Os_i^t) based on the simulation biomass (Bs_i^t) added to the simulated noise (Ns_i^t), Eq. 5. The observed biomass at time t (Os_i^{t+1}) is calculated from the simulation biomass (Bs_i^t) through the sum of the simulated growth (Gs_i^t), the simulated disturbance (Ds_i^t), and the simulated noise (Ns_i^{t+1}), Eq. 6.

$$Os_i^t = Bs_i^t + Ns_i^t \quad \text{for} \ i = 1, 2, 3, \ldots, n \quad (5)$$

$$Os_i^{t+1} = Bs_i^t + Gs_i^t - Ds_i^t + Ns_i^{t+1} \quad \text{for} \ i = 1, 2, 3, \ldots, n \quad (6)$$

where Ds_i^t is related to the disturbance parameters in Eq. 7.

$$Ds_i^t = \begin{cases} 0 & \text{if} \ U([0, 1]) \geq E_p \\ E_I Bs_i^t & otherwise \end{cases} \quad \text{for} \ i = 1, 2, 3, \ldots, n \quad (7)$$

We model Gs_i^t using the A-DALEC model, a biogeochemical model of carbon cycling in forests, which runs at an annual time step (see Additional file 1 A for full details) [15, 21]. Ns_i^t and Ns_i^{t+1} are based on observed pixel uncertainty in SAR biomass estimates [13]. We cannot base Bs_i^t directly on the observed biomass O_i^t, as the observation noise N_i^t cannot be determined (and thus removed) for each pixel, and so the A-DALEC model is used to spin-up an estimate of Bs_i^t, see Additional file 1 section B for full details. The equations above assume that $t = 1$ and $t = 2$ are separated by 1 year. When the separation is more than 1 year, as in the case of our OSSE, the base Bs_i^t should have the annual growth and disturbance applied once for each year of separation.

95 % confidence intervals for the factors E_P and E_I are found using a maximum likelihood estimation routine, see Additional file 1 C for full details. This routine finds the most likely combinations of E_P and E_I based on minimising the differences between the observed and simulated ensemble distributions; $\Delta O = O^{t+1} - O^t$ and $\Delta Os = Os^{t+1} - Os^t$.

Testing the framework parameter estimates

To allow the technique to be assessed against a known 'truth' we use a set of observation system simulation experiment (OSSE). In the OSSE we simulate two successive (noisy) biomass observations which have observation noise and a definable disturbance regime applied. The OSSE observations are derived by spinning up the A-DALEC model to match ALOS observations within the Gorongosa and Nhamatanda districts of Sofala province in central Mozambique. The area is dominated by dry miombo woodland, the dominant woodland type in Southern Africa, see Additional file 1 B. The first OSSE observation, $t = 1$, is taken at the end of the 1st year, the second observation, $t = 2$, is taken at the end of the 5th year, i.e. separated by 4 years. Separate spin-ups are performed for the ΔO and ΔS simulations. Three tests are performed; the first to assess the ability of the approach to identify E_M, E_I and E_P, the second to assess the impact of observation bias and the third to determine impact of ensemble size.

Determining the accuracy of the framework parameter estimates: test 1

A set of nine analyses were used to test the ability of our approach to identify E_M, E_I and E_P. Each analysis was performed for the same size ensemble of $n = 1600$ pixels, which equates to an area of 1000 by 1000 m given the 25 by 25 m pixel size of the ALOS biomass maps [13]. We perform OSSEs with four levels of mean disturbance $E_M = 0, 0.018, 0.045, 0.090$ and 0.180. These disturbance fractions equate to annual biomass loss rates of 0, 1.8, 4.5, 9 and 18 %. For each nonzero E_M we simulate a high intensity ($E_I \gg E_P$) and low intensity ($E_I \ll E_P$) disturbance scenario. In the high intensity cases only a few pixels are disturbed, but when disturbed 90 % of the biomass is lost (i.e. $E_I = 0.9$). In the low intensity cases, 90 % of the

pixels are disturbed (i.e. $E_P = 0.9$), but each disturbance results in a small loss of biomass at the pixel level. The various combinations of factors are shown in Table 1.

Determining the impact of observation bias on the framework parameter estimates: test 2

To simulate the impact of observation bias we perturb the synthetic experiments using the bias observed in the actual biomass maps [13]. Only the first biomass observation (i.e. at time $t = 1$) is biased by $-160, 0$, and 160 gC m^{-2}. The impact of the bias is assessed using both a low intensity case (i.e. $E_I = 0.05$ and $E_P = 0.9$), and a high intensity case (i.e. $E_I = 0.9$ and $E_P = 0.05$). In both cases the OSSE has an $E_M = 0.045$ and an ensemble size of $n = 1600$ pixels.

Determining the impact of ensemble size on the framework parameter estimates: test 3

The impact of varying the ensemble size is tested for $n = 6400, 1600, 400$, and 100. These ensembles equate to areas of 2000 by 2000 m (6400 pixels), 1000 by 1000 m (1600 pixels), 500 by 500 m (400 pixels), and 250 by 250 m (100 pixels). The areas are centred on the same location. We analyses these areas for a single low intensity disturbance regime $E_M = 0.045$, $E_I = 0.05$ and $E_P = 0.9$.

Additional file

Additional file 1. A framework for estimating forest disturbance intensity from remotely sensed biomass maps: moving beyond average biomass loss estimates.

Authors' contributions
The study was designed by all authors. TH performed the analysis and produced the figures and tables and wrote the manuscript. All co-authors contributed substantially to the interpretation of results and revision of the manuscript. All authors read and approved the final manuscript.

Author details
[1] Department of Earth and Environmental Science, University of St Andrews, Irvine Building, North Street, St Andrews, UK. [2] School of GeoSciences, The University of Edinburgh, Edinburgh, UK. [3] The NERC National Centre for Earth Observation, St Andrews, UK.

Acknowledgements
This work was funded by ESA (Contract No 4000102042/10/NL/CT), the NERC National Centre for Earth Observation, the NERC CarbonFusion project (NE/D000874/1), the Mpingo Conservation and Development Initiative and the EU Framework 7 I-REDD + project. ESA and JAXA provided the ALOS PALSAR imagery through C1P.7493 and the 4th Research Agreement for the Advanced Land Observing Satellite-2. The authors would like to thank Klaus Scipal and Shaun Quegan for advice during the study. Finally we would like to thank the three anonymous reviewers for their helpful comments.

Competing interests
The authors declare that they have no competing interests.

References

1. FAO. Global forest resources assessment 2010: main report. FAO forestry paper. Rome: Food and agriculture organization of the United Nations; 2010. p. 2010.

2. Pan Y, Birdsey RA, Fang J, Houghton R, Kauppi PE, Kurz WA, et al. A large and persistent carbon sink in the world's forests. Science. 2011;333(6045):988–93. doi:10.1126/science.1201609.

3. Achard F, Beuchle R, Mayaux P, Stibig H-J, Bodart C, Brink A, et al. Determination of tropical deforestation rates and related carbon losses from 1990 to 2010. Glob Chang Biol. 2014;20(8):2540–54. doi:10.1111/gcb.12605.

4. Hansen MC, Stehman SV, Potapov PV. Quantification of global gross forest cover loss. Proc Natl Acad Sci USA. 2010;107(19):8650–5. doi:10.1073/pnas.0912668107.

5. Mertz O, Müller D, Sikor T, Hett C, Heinimann A, Castella J-C, et al. The forgotten D: challenges of addressing forest degradation in complex mosaic landscapes under REDD+. Geografisk Tidsskrift-Danish J Geogr. 2012;112(1):63–76. doi:10.1080/00167223.2012.709678.

6. Grime JP. Evidence for existence of three primary strategies in plants and its relevance to ecological and evolutionary theory. Am Nat. 1977;111(982):1169–94. doi:10.1086/283244.

7. Agarwal DK, Silander JA Jr, Gelfand AE, Dewar RE, Mickelson JG Jr. Tropical deforestation in Madagascar: analysis using hierarchical, spatially explicit, Bayesian regression models. Ecol Model. 2005;185(1):105–31. doi:10.1016/j.ecolmodel.2004.11.023.

8. Ryan CM, Berry NJ, Joshi N. Quantifying the causes of deforestation and degradation and creating transparent REDD + baselines: a method and case study from central Mozambique. Appl Geogr. 2014;53:45–54. doi:10.1016/j.apgeog.2014.05.014.

9. Herold M, Roman-Cuesta RM, Mollicone D, Hirata Y, Van Laake P, Asner GP, et al. Options for monitoring and estimating historical carbon emissions from forest degradation in the context of REDD+. Carbon Balance Manag. 2011;6(1):13. doi:10.1186/1750-0680-6-13.

10. Hill TC, Williams M, Bloom AA, Mitchard ETA, Ryan CM. Are inventory based and remotely sensed above-ground biomass estimates consistent? PLoS One. 2013;8(9):e74170. doi:10.1371/journal.pone.0074170.

11. Woodhouse IH, Mitchard ETA, Brolly M, Maniatis D, Ryan CM. Radar backscatter is not a 'direct measure' of forest biomass. Nature Clim Chang. 2012;2(8):556–7.

12. Saatchi S, Ulander L, Williams M, Quegan S, LeToan T, Shugart H, et al. Forest biomass and the science of inventory from space. Nature Clim Chang. 2012;2(12):826–7.

13. Ryan CM, Hill TC, Woollen E, Ghee C, Mitchard E, Cassells G, et al. Quantifying small-scale deforestation and forest degradation in African woodlands using radar imagery. Glob Chang Biol. 2012;18:243–57. doi:10.1111/j.1365-2486.2011.02551.x.

14. Le Toan T, Quegan S, Davidson MWJ, Balzter H, Paillou P, Papathanassiou K, et al. The BIOMASS mission: mapping global forest biomass to better understand the terrestrial carbon cycle. Remote Sens Environ. 2011;115(11):2850–60. doi:10.1016/j.rse.2011.03.020.

15. Williams M, Hill TC, Ryan CM. Using biomass distributions to determine probability and intensity of tropical forest disturbance. Plant Ecol Divers. 2013;6(1):1–13. doi: 10.1080/17550874.2012.692404.

16. Harris NL, Brown S, Hagen SC, Saatchi SS, Petrova S, Salas W, et al. Baseline map of carbon emissions from deforestation in tropical regions. Science. 2012;336(6088):1573–6. doi:10.1126/science.1217962.

17. Mitchard ETA, Saatchi SS, Woodhouse IH, Nangendo G, Ribeiro NS, Williams M et al. Using satellite radar backscatter to predict above-ground woody biomass: a consistent relationship across four different African landscapes. Geophys Res Lett. 2009;36. doi: 10.1029/2009GL040692.

18. Williams M, Ryan CM, Rees RM, Sarnbane E, Femando J, Grace J. Carbon sequestration and biodiversity of re-growing miombo woodlands in Mozambique. For Ecol Manage. 2008;254(2):145–55. doi:10.1016/j.foreco.2007.07.033.

19. Frolking S, Palace MW, Clark DB, Chambers JQ, Shugart HH, Hurtt GC. Forest disturbance and recovery: a general review in the context of space-borne remote sensing of impacts on aboveground biomass and canopy structure. J Geophys Res Biogeosci. 2009;114. doi: 10.1029/2008jg000911.

20. Williams M, Hill TC, Ryan CM, Peylin P. Final report BIOMASS level 2 product to flux study. ESA contract No 4000102042/10/NL/CT. Nordwijk: The European Space Research and Technology Centre; 2011.

21. Williams M, Schwarz PA, Law BE, Irvine J, Kurpius MR. An improved analysis of forest carbon dynamics using data assimilation. Glob Chang Biol. 2005;11(1):89–105.

6

These are the days of lasers in the jungle

Joseph Mascaro[1*], Gregory P Asner[2], Stuart Davies[3], Alex Dehgan[4] and Sassan Saatchi[5]

Abstract

For tropical forest carbon to be commoditized, a consistent, globally verifiable system for reporting and monitoring carbon stocks and emissions must be achieved. We call for a global airborne LiDAR campaign that will measure the 3-D structure of each hectare of forested (and formerly forested) land in the tropics. We believe such a database could be assembled for only 5% of funding already pledged to offset tropical forest carbon emissions.

In a precious 152 minutes on the Lunar surface, Neil Armstrong and Buzz Aldrin raced to collect rocks, dust and photographs that are an enduring fascination in museums and laboratories—testament to the awesome effort expended to touch them. They left behind a discarded landing assembly, sprawling trails of footprints, and the flag of the United States. But they also left behind a science experiment about the size of a coffee table. This hunk of honeycombed metal may be the most important scientific legacy of the Apollo Program. It's called a laser retroreflector, and it revolutionized our understanding of the Moon [1].

Laser ranging technology is straightforward: it works like radar or sonar but with lasers. Point a laser at a target and collect the reflected light bouncing back; with a stop watch fit for Einstein, you can measure the time delay between the laser shot and the return bounce to determine the distance to your target.

Since the experiment began, we've learned that the Moon's orbit is widening by 3.8 centimeters per year, slowing the rotation of the Earth by about 2.3 seconds per century. These changes give us the leap second, and in a few million years will make February 29th obsolete. The revelation has also dramatically influenced our thinking about the formation of the Earth-Moon system some 4.5 billion years ago.

Just as the Moon's history was disrobed by laser ranging 50 years ago, Earth's tropical forests are giving up their secrets to the light. Airborne light detection and ranging—called LiDAR—has over the last ten years become a key tool that ecologists use to understand physical variation in tropical forests across space and time [2,3]. Like an MRI of the human brain, LiDAR probes the intricate three-dimensional architecture of the forest canopy, unveiling carbon that forests keep out of the atmosphere, and also the mounting threats to that carbon storehouse: drought, fire, clandestine logging and brash gold-mining operations [4]. Even the quintessential natural disturbance of the sun-filled light gap—long thought to enhance the incredibly high species diversity of tropical forests—has been deconstructed by laser technology [5].

Laser ranging in tropical forests is such a game-changing technology that science results can scarcely get through peer-review before they are dwarfed by still larger-scale studies. In a decade, laser power on commercial-grade LiDARs has skyrocketed and costs have plummeted. These improvements in LiDAR technology allow airplanes to fly faster, higher and farther, covering more forest area in a single day than every ground-based survey that has ever been collected in the history of tropical ecology. To estimate the amount of carbon stored in a 50-hectare tropical forest monitoring plot on the ground—the largest field plot in the world—takes a team of 12 people about eight months: a slog of rain and mud and snakes with tape measures and data log books [6]. Today's airborne LiDARs can get you to within about 10% [7,8] of the same carbon estimate in *eight seconds*.

It is this staggering contrast in scale between LiDAR and fieldwork that led us here: Before this decade is out, we could directly assess the carbon stock of every single square hectare of tropical forest on Earth. We could do it just as well as if we were standing there in the flesh with tape measures in hand. And we could do it for far less than what we have already spent to offset carbon emissions from forests.

* Correspondence: jmascaro@usaid.gov
[1]American Association for the Advancement of Science, 1300 New York Ave NW, Washington, DC 20001, USA
Full list of author information is available at the end of the article

"A loose affiliation of millionaires and billionaires"

The United Nations Framework Convention on Climate Change is the international venue in which tropical forest carbon accounting programs and practices have been negotiated, the most visible of these being REDD+ (Reduced Emissions from Deforestation and Degradation). When first imagined, the precursor of REDD+ was straightforward: landowners and governments would draw revenue from a global carbon market if they could verifiably reduce their forest-sector carbon emissions to the atmosphere. The policy has at times seemed like a real possibility, but has recently faced major setbacks.

At present, more than a dozen countries are participating in preliminary work to develop REDD+ implementation strategies. These efforts involve the building of infrastructure and expertise, including consulting and contracting with various academic groups and non-government organizations to estimate carbon stocks and emissions. Pledged REDD+ funding now exceeds $5 billion [9]. These funds come partly from donors but primarily from governments.

Although the UNFCCC has outlined general guidelines for implementing a monitoring, reporting and verification program, most REDD+ projects remain in the demonstration phase, and thus the funds are being spent in myriad ways. Because of the urgency in developing the program to meet scattershot international deadlines and twisting geopolitical expectations—current efforts to assess tropical forest carbon are highly inconsistent.

But for REDD+ to function, a consistent, globally verifiable system for reporting and monitoring carbon stocks and emissions in tropical forests must be achieved in the near future. Without such an accounting system, the amount of avoided carbon emissions from deforestation and degradation achieved would not be quantifiable and thus the program would not be tractable.

The time has come for a brute-force effort to directly assess the carbon stock for all of the world's tropical forests by 2020. Airborne LiDAR is uniquely suited for this role because it can be collected, standardized, reported and verified in a simple manner by both a landholder and any third party.

We call for a global airborne LiDAR campaign that will measure the 3-D structure of each hectare of the Earth's non-barren land surface between the Tropic of Capricorn and Tropic of Cancer. Our targets are approximately 1,500,000,000 individual hectares of the richest and fastest disappearing habitats in the world, and the human-used lands that have replaced them.

"Staccato signals of constant information"

It is easy in principle, though logistically nightmarish, to measure carbon in tropical forests. A strict constructionist would cut, dry and weigh the biomass of the world's forests. But this is a self-defeating enterprise. As a result, it is likely that no one has measured carbon over a single hectare of tropical forest, even with the most detailed field surveys. For a century ecologists and foresters have relied on allometric estimation in lieu of carbon measurements to translate field surveys of tree diameters, heights and wood densities into whole-forest carbon estimates. Given a volume with known dimensions and density, one would estimate its mass in a similar fashion.

As the new kid on the block, LiDAR has been tacked onto the back end—initially thought of as kind of large-scale helper to field surveys. Carbon estimates from the field have been treated as something inherently closer to the real thing than measurements made by LiDAR—ground "Truth" with a capital "T". This is perhaps understandable historically, but vis-à-vis actual carbon, there is no such thing as ground truth: both field and LiDAR efforts rely on allometry to convert measurements into carbon estimates [10]. *Prior* to using these measurements for carbon estimation, they exist as standardized, spatially explicit, archivable and verifiable data—the needed substrate for a REDD-type accounting program.

Due to the constancy of the underlying measurements, both field and LiDAR data could provide the needed information if they covered every hectare on Earth. But, in the case of field surveys, this is impossible. The surveys that do exist measure a tiny amount of actual forest, and so what might be verified is widely spaced. And to avoid fraud and protect landowners, many governments keep their plot locations secret. Satellite LiDAR data remain sparse, providing only extrapolated, coarse-resolution carbon estimates with very high uncertainties [11], and there is no prospect of wall-to-wall coverage in the near future [12]. By 2020, airborne LiDAR could give us a direct measurement of 3-D forest structure for every hectare in the tropics: a standardized database from which to build a carbon economy.

"Don't cry, baby, don't cry"

The 1993 film *Jurassic Park* is filled with brilliant insights into science, from Malcolm's cautionary tale on the wisdom of cloning to Sattler's diatribe about human control over nature. The best of these is professor Alan Grant's reaction to a sonar return of a still-entombed velociraptor skeleton. The lowly grad student showing him the return quips: "a few more years and we won't have to dig anymore". He replies, dismayed: "Where's the fun in that?"

Ecologists often believe that embracing LiDAR technology will somehow mean the end of fieldwork. It won't. LiDAR has initially enhanced the role of fieldwork: even though many plots were established before carbon assessment became a key objective in tropical forest science, these plots are a critically important resource for understanding how LiDAR 3-D measurements can be used to estimate carbon stocks. Very soon, plot-based harvests will allow remote sensing data to relate directly to measurements of carbon stocks in closed canopy forests—something that has recently been accomplished in a tropical savanna [13]. And beyond carbon applications, LiDAR data are revolutionizing our understanding of basic tropical forest ecology, from environmental controls to succession e.g., [5,14,15].

The other reason to smile? The price tag. Our ambitious plan can be accomplished for far less than what we have already spent on avoided deforestation. Aircraft leasing, data collection and processing costs for 30 days of flying can reasonably be limited to $500,000 [16]. Using this monthly sampling unit, collecting at an average of 100,000 hectares per day, a fleet of ten aircraft could do the job in four years at US$250 million, or just 5% of pledged REDD + funding.

At that rate, we can lose a year and a half to torrential rain and still get the job done.

Acknowledgements
Comments by two anonymous reviewers substantially improved this manuscript.

Author details
[1]American Association for the Advancement of Science, 1300 New York Ave NW, Washington, DC 20001, USA. [2]Department of Global Ecology, Carnegie Institution for Science, 260 Panama St., Stanford, CA 94305, USA. [3]Forest Global Earth Observatory, Center for Tropical Forest Science, Smithsonian Tropical Research Institute, PO Box 37012, Washington, DC 20013, USA. [4]Conservation X Labs, 2380 Champlain St. NW #203, Washington, DC 20009, USA. [5]Jet Propulsion Laboratory, California Institute of Technology, 4800 Oak Grove Drive, Pasadena, CA 91109, USA.

References
1. Dickey JO, Bender PL, Faller JE, Newhall XX, Ricklefs RL, Ries JG, Shelus PJ, Veillet C, Whipple AL, Wiant JR, Williams JG, Yoder CF: **Lunar laser ranging: a continuing legacy of the Apollo program.** *Science* 1994, **265**:482–490.
2. Asner GP, Powell GVN, Mascaro J, Knapp DE, Clark JK, Jacobson J, Kennedy-Bowdoin T, Balaji A, Paez-Acosta G, Victoria E, Secada L, Valqui M, Hughes RF: **High-resolution forest carbon stocks and emissions in the Amazon.** *Proc Natl Acad Sci U S A* 2010, **107**(38):16738–16742.
3. Goetz S, Dubayah R: **Advances in remote sensing technology and implications for measuring and monitoring forest carbon stocks and change.** *Carbon Manag* 2011, **2**(3):231–244.
4. Asner GP, Llactayo W, Tupayachi R, Ráez Luna E: **Elevated rates of gold mining in the Amazon revealed through high-resolution monitoring.** *Proc Natl Acad Sci U S A* 2013, **111**(15):5604–5609.
5. Kellner JR, Asner GP: **Convergent structural responses of tropical forests to diverse disturbance regimes.** *Ecol Lett* 2009, **12**(9):887–897.
6. Hubbell SP, Foster RB, O'Brien ST, Harms KE, Condit R, Wechsler B, Wright SJ, Loo de Lao S: **Light-gap disturbances, recruitment limitation, and tree diversity in a Neotropical forest.** *Science* 1999, **283**(5401):554–557.
7. Mascaro J, Detto M, Asner GP, Muller-Landau HC: **Evaluating uncertainty in mapping forest carbon with airborne LiDAR.** *Remote Sens Environ* 2011, **115**(12):3770–3774.
8. Zolkos SG, Goetz SJ, Dubayah R: **A meta-analysis of terrestrial aboveground biomass estimation using lidar remote sensing.** *Remote Sens Environ* 2013, **128**:289–298.
9. *Voluntary REDD+ Database.* http://reddplusdatabase.org/; Accessed 5/11/2014.
10. Clark DB, Kellner JR: **Tropical forest biomass estimation and the fallacy of misplaced concreteness.** *J Veg Sci* 2012, **23**(6):1191–1196.
11. Mitchard ETA, Saatchi SS, Baccini A, Asner GP, Goetz SJ, Harris NL, Brown S: **Uncertainty in the spatial distribution of tropical forest biomass: a comparison of pan-tropical maps.** *Carb Bal Manag* 2013, **8**:1–13.
12. Goetz S: **The lost promise of DESDynI.** *Remote Sens Environ* 2011, **115**(11):2751.
13. Colgan MS, Asner GP, Swemmer T: **Harvesting tree biomass at the stand level to assess the accuracy of field and airborne biomass estimation in savannas.** *Ecol Appl* 2013, **23**(5):1170–1184.
14. Detto M, Muller-Landau HC, Mascaro J, Asner GP: **Hydrological networks and associated topographic variation as templates for the spatial organization of tropical forest vegetation.** *PLoSONE* 2013, e76296.
15. Mascaro J, Asner GP, Dent DH, DeWalt SJ, Denslow JS: **Scale-dependence of aboveground carbon accumulation in secondary forests of Panama: a test of the intermediate peak hypothesis.** *Forest Ecol Manage* 2012, **276**:62–70.
16. Asner GP, Knapp DE, Martin RE, Tupayachi R, Anderson CB, Mascaro J, Sinca F, Chadwick KD, Sousan S, Higgins M, Farfan W, Silman MR, Llactayo WA, Neyra AF: *The Carbon Geography of Perú.* Berkeley, CA: Minuteman Press; 2014:64. ISBN: 978-0-9913870-7-6 (English edition) and ISBN: 978-0-9913870-6-9 (Spanish edition).

7

Advancing reference emission levels in subnational and national REDD+ initiatives: a CLASlite approach

Florian Reimer[1*], Gregory P Asner[2] and Shijo Joseph[3]

Abstract

Conservation and monitoring of tropical forests requires accurate information on their extent and change dynamics. Cloud cover, sensor errors and technical barriers associated with satellite remote sensing data continue to prevent many national and sub-national REDD+ initiatives from developing their reference deforestation and forest degradation emission levels. Here we present a framework for large-scale historical forest cover change analysis using free multispectral satellite imagery in an extremely cloudy tropical forest region. The CLASlite approach provided highly automated mapping of tropical forest cover, deforestation and degradation from Landsat satellite imagery. Critically, the fractional cover of forest photosynthetic vegetation, non-photosynthetic vegetation, and bare substrates calculated by CLASlite provided scene-invariant quantities for forest cover, allowing for systematic mosaicking of incomplete satellite data coverage. A synthesized satellite-based data set of forest cover was thereby created, reducing image incompleteness caused by clouds, shadows or sensor errors. This approach can readily be implemented by single operators with highly constrained budgets. We test this framework on tropical forests of the Colombian Pacific Coast (Chocó) – one of the cloudiest regions on Earth, with successful comparison to the Colombian government's deforestation map and a global deforestation map.

Keywords: Chocó; Colombia; Deforestation; Forest cover; Forest degradation; REDD+; Reference emissions

Background

Reducing emissions from deforestation and forest degradation, and enhancing the carbon stocks (REDD+), remains a key strategy for mitigating climate change. Unlike many previous conservation efforts, REDD+ is constructed on the principles of additionality against a baseline or reference emission level (REL), with no displacement of emissions to neighboring areas (leakage). It is noted here that the United Nations Framework Convention on Climate Change (UNFCCC) and the World Bank's Forest Carbon Partnership Facility (FCPF) use the term REL, while the Verified Carbon Standard (VCS) applies the term baseline. They are synonymous, as long as they are reported as greenhouse gas emissions in units of tons equivalent to CO_2 (tCO_2e). REDD+ intends to follow a hierarchical nested approach where project, subnational, and national initiatives contribute to the reduction in emission from deforestation and degradation. A consistent system that works across scales is therefore important for operationalizing REDD+, ensuring no displacement in the emission, and also to avoid potential double counting issues. The UNFCCC in its Warsaw Framework for REDD+ [1] specifies that such national forest monitoring systems "should provide data and information that are transparent, consistent over time, and are suitable for measuring, reporting and verifying anthropogenic forest-related emissions by sources and removals by sinks, forest carbon stocks, and forest carbon stock and forest area change".

The role of remote sensing in measuring and monitoring forest area, and assessing its structural and functional attributes, has been well documented [2-4]. However, the REDD+ projects are often located in the humid tropics where a number of prevalent atmospheric and ground conditions, such as cloud cover, haze and uneven topography, often disrupt a satellite sensor's ability to provide high quality observations of the land surface [5]. Moreover,

* Correspondence: freimer@uni-bonn.de
[1]Center for Development Research (ZEF), Group Börner, Rheinische Friedrich-Wilhelm University, Walter-Flex-Str. 3, 53113 Bonn, Germany
Full list of author information is available at the end of the article

spatial infrastructure, data access and technological expertise are key determinants of remote sensing capacity in countries within the tropics. This data limitation problem has been heavily reported and continues to be discussed [4-10]. Although operational monitoring of deforestation is reasonably possible with medium resolution remote sensing data such as Landsat, as evidenced from the Brazilian government's program [11], establishing such a scheme at global scale is still underway. Progress has been made by the Global Forest Change program [12], and recent initiatives such as mapping of annual deforestation rate using Landsat data in the Congo basin [13], Sumatra [14], Colombia [15] and Peru [16] are also notable in this context. More importantly, such a system must make use of the satellite image pixel-based time series data compositing to minimize cloud, haze and other atmospheric artefacts that severely limit Landsat and other medium-resolution optical satellite data.

Here we report on the performance of a forest cover and deforestation mapping tool developed by Asner et al [17] for the operational monitoring of REDD+ landscapes in order to advance the readiness activities in carbon accounting frameworks. CLASlite is intended for a non-expert user to quickly assess the regional distribution of forest cover, deforestation and degradation. This makes it particularly appropriate for the establishment of subnational to national reference levels in tropical regions with reduced satellite image quality and technical resources. CLASlite is intended to help the REDD+ community achieve rapid and reliable estimates of forest cover and deforestation. Here we test the efficacy of CLASlite in the context of new developments with sub-national and jurisdictional REDD+ initiatives. We also report on the performance of some new CLASlite modules such as the reduced masking option and deforestation artifact remover, and we elaborate on their effects on REDD+ reference levels and give recommendations for good practice. Furthermore we compare our mapping results with those of the Colombian national Institute of Hydrology, Metrology and Environmental Studies (IDEAM) [15] and of the new global maps recently made available by Hansen et al. [12].

Study area

The study region includes the Colombian municipality of Ancandi and the northern portion of Unguía, in the Department of Chocó. The southern border is formed by the common area coverage of Landsat image scene path-row 10 / 54 (Figure 1), with an area of about 1,900 km^2. The approach presented here can be scaled up to regions of varying number of scenes and sub-national dimensions.

Several municipalities form the northernmost portion of land of the Department of Chocó. They are bordered

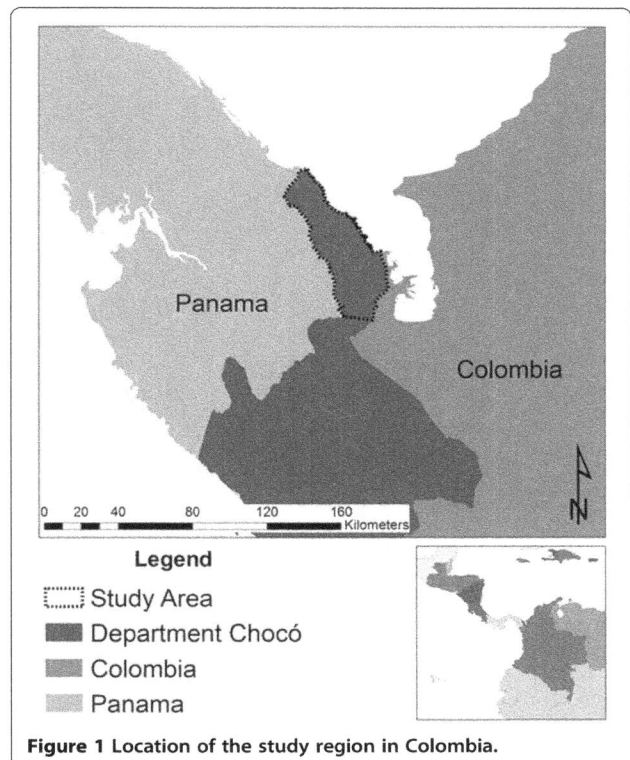

Figure 1 Location of the study region in Colombia.

to the west by Panama's Darién province and to the east by the Urabá Gulf of the Caribbean Sea, where the Rio Atrato forms a distinct delta known as "Bocas del Atrato" within the municipality of Turbo in neighboring department of Antioquia. Apart from logging, small-scale agriculture, fishery and cattle ranching, land-use in the study area includes illicit crops (*Erythroxylum* spp.) and activities around trafficking and contraband. Forests in the study region are exclusively humid Neotropical evergreen broadleaf in lowland, sub-montane, and pre-montane elevation ranges (1-1,400 m a.s.l.). Its peninsular geography at the Isthmus of Panama between the Caribbean Sea and the Pacific Ocean results in consistently heavy cloud cover. IDEAM estimates mean precipitation of 2,500-3,000 mm yr^{-1} in the region [18]. However, the Chocó harbors areas with > 9,000 mm of rainfall annually and is well known to be one of the rainiest regions on Earth. Studies suggest that the southern Department of Cauca (San Miguel de Micay) potentially has the highest recorded rainfall on Earth with an annual mean of 12,892 mm from 1971-2000 [18]. As a result, the Pacific coast of Colombia presents an extremely challenging case for optical remote sensing of forest cover and change. This challenge makes it an excellent laboratory to test new remote sensing approaches, and comparisons between monitoring systems can give us important information on the effects of monitoring design on mapped deforestation and therefore REDD+ reference emission levels [19].

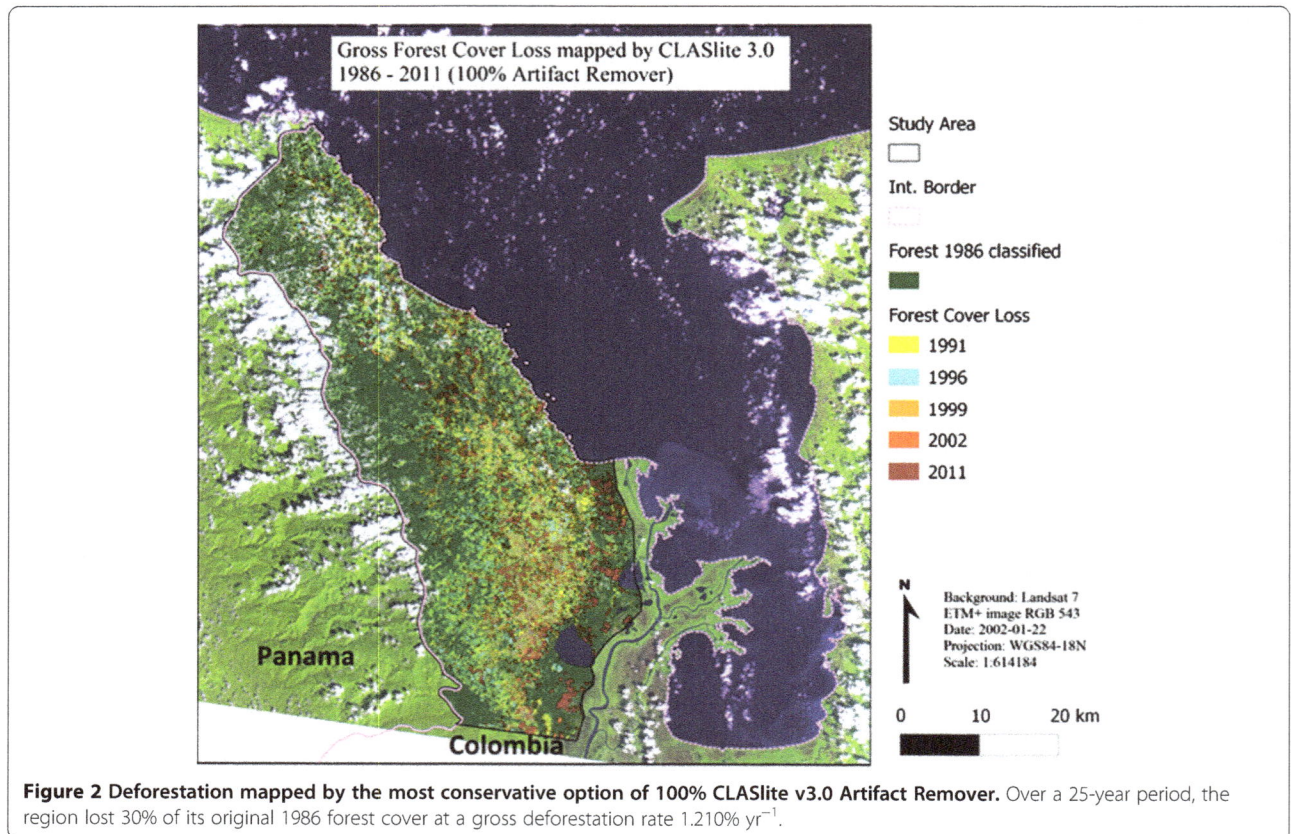

Figure 2 **Deforestation mapped by the most conservative option of 100% CLASlite v3.0 Artifact Remover.** Over a 25-year period, the region lost 30% of its original 1986 forest cover at a gross deforestation rate 1.210% yr^{-1}.

Results and discussion
Observed deforestation from mosaicked fractional covers
The multi-temporal analysis covering 25 years (1986 – 2011) detected 30,681.3 ha of forest cover loss, which represents 26.31% of 1986 cover, or a deforestation rate of 1291 ha yr^{-1} (Figure 2). Such a long-term average rate is already applicable for a REDD+ REL, and can be extrapolated to the forest cover remaining at project start date [20-24] (Table 1).

Rates of forest loss represent a convenient way of reporting deforestation in a globally comparable way [12]. However, there is little agreement among land-use change modelers as to whether projected deforestation rates should be predicted from average observed rates of forest cover loss or average observed rates of forest area lost. Using any rate of loss for prediction introduces problems, as the rate of loss not only depends equally on accurate mapping of the current area or cover, but also of accurate mapping of the original forest area or cover. In addition, rate of loss introduces a semantic problem: Researchers may quantify deforestation over large spatial and temporal extents, however, the process actually forms in local decision makers' minds in terms of absolute areas, with no concept of the regional or national rate of forest loss. Simply speaking, farmers know how much land they need to clear for the expansion of a

given activity, but they usually have little concept of how this area scales relative to the regional rate of loss. This may lead them to believe, if they have cleared 10,000 ha every year over the last 10 years, this quantity is the same as that converted regionally in the business as usual scenario. The land-holders do not know if those 10,000 ha cleared represent 1%, 0.1% or 0.01% of annual loss. REDD+ RELs predict emissions from forest carbon loss based on emission factors per activity type in Greenhouse Gas (GHG) emissions in tons equivalent to CO_2 per ha (tCO_2e ha^{-1}). Therefore estimating emissions through predicted values of absolute forest area loss also in hectares per year is more straightforward and transparent than using a rate of loss in percent re-applied to remaining forest area. This is particularly true if the predicted future quantity of loss is not a stable average, but a function of the historic trend in quantities of loss [25].

Deforestation artifact remover test
CLASlite 3.0 offers the user an option to apply a Deforestation Artifact Remover (DFAR) ranging in value from 0-100%. Unaltered forest change outputs may include unwanted artifacts (false positives) caused by the influence of clouds, unmasked cloud edges, cloud shadows, topography, and water boundaries. For Landsat imagery, the user can define desired settings for artifact removal

Table 1 Observed forest cover loss in study area by CLASlite v3.0 mosaic approach

Time A_1	Time A_2	Forest cover A_1 [ha]	Forest cover A_2 [ha]	Def_{period} [ha]	Deforestation rate yr^{-1} [%]
1986.049	1991.219	116,623.17	108,208.53	8,414.64	−1.45
1991.219	1996.520	113,263.83	109,904.49	3,359.34	−0.57
1996.520	1999.542	108,906.39	104,548.95	4,357.44	−1.35
1999.542	2002.498	106,949.25	104,992.11	1,957.14	−0.62
2002.498	2011.194	104,992.11	92,399.40	12,592.71	−1.47
1986.049	**2011.194**	**116,623.17**	**85,941.90**	**30,681.27**	**−1.21**

in the deforestation image. In the standard operating mode, most artifacts are eliminated prior to analysis by the CLASlite pixel exclusion algorithm. With the user-selected DFAR value of 100%, CLASlite eliminates all pixels it recognizes as potential false positives. In contrast, at 0%, CLASlite does not eliminate any of these potential false positives. In order to assess the impact of this tool the time-series of mosaicked fraction cover images was run with the parameter set at DFAR values equal to 0%, 50%, and 100%. The accumulated deforested area over the monitoring period with a DFAR value of 0% was 137.6% of the area under 100% (Figure 3).

As a general rule, all measurements, assumptions and models used in carbon projects like REDD+ should be "emission reduction conservative" [26]. This means, if a choice of methods and parameters is to be made between equally justifiable approaches, the preferred option should result in more conservative estimates of GHG emission reductions attributed to a climate change mitigation action in the end. Historical forest cover change monitoring is an essential element of the reference emission level of REDD+. Net GHG emission reductions creditable to a climate change mitigation intervention are calculated as a difference between REL and observed emissions from

forest cover change (Measure, Report, Verify – MRV) in years of project operation [1,20-24,26]. This means that options should be selected to report conservative historical forest cover change to avoid risks of inflating the baseline emissions by measurement decisions.

Therefore, for a final report of historical deforestation, we choose the most conservative DFAR 100%. We also recommend this as the default for CLASlite 3.0 monitoring efforts with the aim to generate data for a REDD+ REL. Deviation by users from this conservative approach should be justified. This study does not analyze the forest degradation output of CLASlite 3.0. Should CLASlite 3.0 also be applied to generate data for REDD+ REL credit, it seems prudent to apply the same conservative option of 100% to the "Degradation Artifact Remover".

Mosaicked vs. single scene time-steps

Our study also included a comparison of deforestation monitoring using a mosaic of multiple scenes versus single-scene per time-step input approach to CLASlite 3.0. We sought to determine if the new, mosaicked approach significantly increased the change area monitored in the observation periods and if the detection under both approaches is valid. To this end, the individual

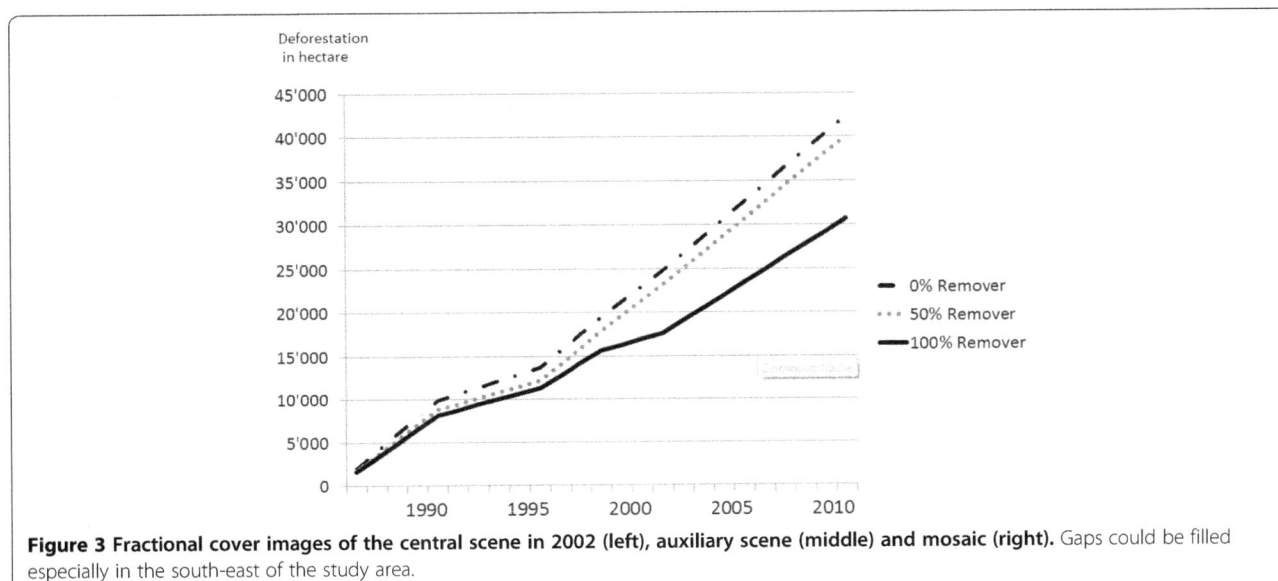

Figure 3 Fractional cover images of the central scene in 2002 (left), auxiliary scene (middle) and mosaic (right). Gaps could be filled especially in the south-east of the study area.

fractional cover maps of the central scenes were analyzed in a multi-temporal forest cover change analysis with the same default thresholds for the CLASlite 3.0 decision tree [27].

The single-scene approach, using only the central scenes with the same time steps as described earlier, thus following a conventional CLASlite processing, managed detected 45.59% (13,988.3 ha) of the deforestation mapped from 1986-2011 by the mosaicked fraction cover scenes (30,681.3 ha). In line with the principle of conservativeness [26], the most conservative option, 100% for DFAR was used. In all time steps, the mosaicked scene approach produced more forest cover change than the single scene approach – which is likely due to the addition of artifact-free pixels integrated from additional scenes to the observation area.

The mosaicked scene approach not only added 54.41% to the absolute forest cover change of the single scene approach, but also showed good spatial overlap with the single scene change results. Spatial overlap per time-step ranged from 37.05% (1991-1996) to 94.36% (1996-1999), averaging 62.43% over the entire observation period. This still leaves a substantial portion of single scene change results that are not picked up by the mosaicked approach.

We also analyzed how additional detected areas of forest change in the filled mosaic pixels might influence the change results. The additional forest change from the mosaicked approach, located in areas previously without data in the single-scene approach, varied from 7.52% (1986-1991) to 74.42% (2002-2011). On average, additional pixels in the mosaic approach resulted in a 39.11% in forest cover change over the single scene approach.

A detailed visual inspection of both forest cover outputs with the original Landsat imagery indicated a high probability of the mapped deforestation to be valid in both approaches. Therefore we conclude mosaicking fractional cover images, can aid in the assessment of forest cover in a greater proportion of the pixels allowing better detection and quantification of deforestation in environments of very low image quality due to persistent cloud cover or even sensor failures (Landsat 7 ETM+ SLC-off).

Comparison with previous work

For the study region, two independent deforestation datasets were available, one from IDEAM [15], the other from the University of Maryland, UMD [12]. A comparison of deforestation output from different remote sensing approaches can help to quantify the impact of monitoring approaches on estimated RELs [19]. REDD+ emission reductions are only useful if they are achieved relative a realistic REL, and quantification of historical deforestation is a central element to REL construction. In absence of historical, spatially extensive ground data, which are almost never available, it is not possible to verify or falsify the three datasets we compare here [26,28,29]. Instead, we compare the predicted range of REL values calculated from the deforestation results over the years 2000-2012 from the three different approaches, and we compare the predicted range of reference emission levels.

The three deforestation datasets covered different observation periods, possibly complicating the interpretation of the results. The study presented considered the longest period of 25 years from 1986-2011. The IDEAM dataset [15] covered 1990-2012, while UMD only covered [12] 2000-2012. To look at long-term temporal dynamics, we compare the three datasets in the 1990-2010 period in terms of accumulated deforestation per hectare (Figure 4). In the case of UMD [12], we incorporate the deforestation output of the CLASlite analysis of 2000 (lower compared to IDEAM) to allow for graphical comparison (Table 2).

Overall, IDEAM exhibited the lowest temporal variability and mid-levels of accumulated loss. Our results using CLASlite demonstrated moderate temporal variability and the highest accumulated loss, while the UMD method had the greatest temporal variability due to the annual temporal resolution and the lowest levels of accumulated loss. When the resulting imagery was visually inspected, the UMD method generally under-estimated deforestation losses compared to losses resulting from the CLASlite or IDEAM approaches. For IDEAM [15], the historical average rate of forest area loss was 1019.2 ha yr^{-1} between 2000 and 2010. Using CLASlite v3.0 with 100% artifact remover, the rate was 1288.9 ha yr^{-1}, and for UMD the calculated loss was 603.2 ha yr^{-1}. None of the three datasets indicated a clear increasing or decreasing trend in annual forest loss, so no regression was warranted [20].

We quantified the impact of the three deforestation datasets on the REL establishment. For simplicity we applied an average emission facto of 500 tCO$_2$e ha^{-1} for each hectare of deforestation representing a single land-use change from a uniform forest to agriculture. Earlier we referred to the principle of conservativeness for GHG emission reduction quantification to justify our choice for the set of strict CLASlite v3.0 parameters that usually result in the lowest deforestation values. One might consider choosing the UMD approach for its extremely low historical average forest loss rate compared to the CLASlite and IDEAM output. However, such a decision would ignore two critically important factors:

a) The CLASlite mosaic-approach and IDEAM approach show a great similarity. Although the two datasets switch in terms of accumulated deforestation mapped in 2005, their overall

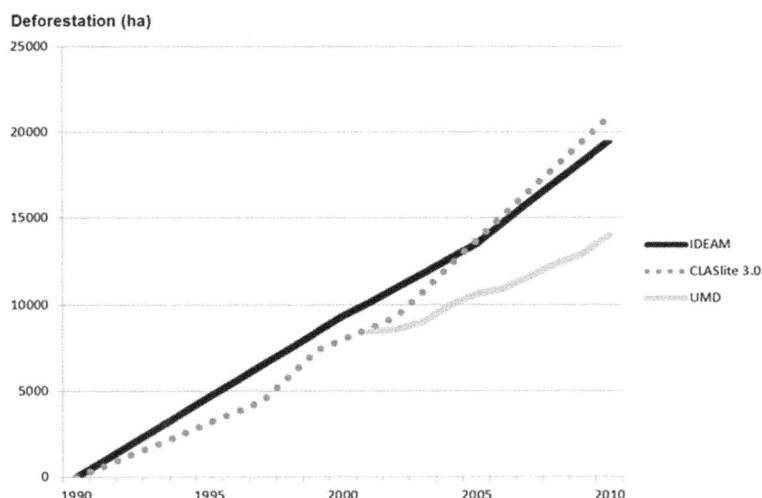

Figure 4 Accumulated deforestation in hectares, analyzing the same fractional cover mosaics with different parameters for the "Deforestation Artifact Remover" of CLASlite 3.0.

accumulated outputs are, after 20 years, different by only 6.23%. Both approaches indicate an accumulated deforestation of about 20,000 ha in the period of 1990-2010 (20,890 ha CLASlite vs 19,506 ha IDEAM).

b) The UMD dataset detects less than 50% of the output mapped 2000-2010 by the CLASlite mosaic-approach, and just 56% of the IDEAM mapping for the same period. Remembering that we are dealing with a region with chronic cloud and shadow cover, it is possible that the global-scale automated UMD technique is not able to detect much of the deforestation in the region. Our visual inspection of the satellite data strongly supports the hypothesis that, in areas where IDEAM and CLASlite map deforestation, UMD misses true change visible in the raw satellite imagery.

The UMD approach uses a Landsat composite image from the greenest pixel calculation provided by Google Earth Engine [30]. We reviewed the Landsat greenest pixel product of [30] for the study region, and evaluated its use as a basis for the image mosaic approach. We have found that using the greenest NDVI or other metrics for image or pixel-scale mosaicking such as with the Google product severely reduces the amount of apparent forest cover lost over time, and in some regions it vastly over-estimates forest recovery. Moreover, the automated approach may include blurry and noisy data in the Google composites. Manual selection results in a much more consistent and reliable set of image inputs for use in deforestation and forest degradation detection and monitoring over time.

The UMD dataset is a very interesting experiment in automated global landcover mapping and change detection. It sparked a lively scientific discussion where it drew much acclaim and some criticism [31], including responses from the authors [32]. It provides an understanding of the general trajectory of forest cover change in any given country without differentiating between natural forest and plantation cover changes, while the later might include tree energy crops such as oil palm. This lack of ability to differentiate between natural old growth forests and plantation is true for all three approaches compared here [12,15,17]. The results can be viewed as indications of relative changes in forest dynamics (e.g. comparing the deforestation rate of an earlier period to a later period), but it the use to actually map absolute rates of old growth forest cover change for REDD+ purposes should be considered with caution and local interpretation.

All automatic land cover classification products can be validated for recent time steps by ground truth data in the form of confusion matrices [28], but ground truth data was not available to this study. For land cover & land use change products referring to periods several years in the past, map validation remains challenging. In a few instances, historic high resolution imagery such as aerial photography can be applied, but generally periods further in the past coincide with a lack of high

Table 2 Annual reference emission level calculated from three different deforestation datasets

	Emission factor [tCO$_2$e/ha]	Average loss 2000-2011 [ha]	Annual REL [tCO$_2$e]
IDEAM	500	1,019.23	509,615
CLASlite	500	1,288.95	644,473
UMD	500	603.24	301,619

resolution imagery and systematically sampled ground truth data.

Outlook for subnational RELs

REDD+ continues to be a concept in active development, and it has substantially evolved from a vague idea of "payments for forest carbon sequestration and storage" to real tests, for example in the Noell-Kempf Project [9] or the Kariba REDD+ Project [33]. REDD+ is taking a prominent role in international climate change mitigation negotiations [1,22], and continues evolve through the development of a variety standards certifying REDD+ projects (CCBS, VCS, PlanVivo, ACR, CAR [34,35]). The latest progress in this field was the publication of the VCS Jurisdictional and Nested REDD+ Requirements [21], the Warsaw UNFCCC decisions on REDD+ [1], and the FCPF Carbon Fund Methodological Framework [24]. On the other side, the first embryonic developments of compliance carbon trading schemes accepting and actively supporting international REDD+ offsets [34] are taking shape with the integration process of the system of payments for ecosystem services in the Brazilian state of Acre and the California compliance carbon offset and trading scheme. The link between the two could potentially be a verification of Acre's GHG emission reductions by [21], and an acceptance of this approach into the California system.

From a REDD+ perspective, it should be noted that applying a REL built on forest carbon emissions has profound implications for whether a performance-based conservation program is adequately compensated. A REL built on an inaccurately low deforestation rate poses significant risks. For example, if due to underreporting, the REL only captures 50% of the average historic loss, a comparison with the REL built on underreporting would show no emission reduction against a true 50% decrease today. This would reduce or eliminate any real emission reduction based on REDD+, thus preventing financial resources from being allocated to a successful emissions reduction activity. The efficient allocation of financial resources to the most cost-effective climate change mitigation actions is a key rationale for REDD+, MRV and performance based payments [1,3,9,22,24,26,28,34].

The role of NASA's Landsat mission continues to stand out as primary data source for tropical historic land cover and land use change analysis, although the variety of sources is ever expanding and barriers to satellite data access are decreasing over time. Without entering an in-depth analysis on reasons for Landsat's dominance, certain factors likely play a role: free access, easy catalog search, long-time continuity and a broad body of scientific research supporting the use of Landsat. However, the failures of Landsat 7 ETM+ SLC instrument 2003 and of Landsat 5 TM late 2011 have left data gaps in many areas until the new Landsat 8 became operational in June 2013.

Approaches such as the framework presented in this study, to be implemented by single platform users or sophisticated cloud-computing approaches [16], can help to bridge these gaps, reduce uncertainties in historic observations and standardize the comparability of results by the application of the CLASlite modules for high quality image pre-processing and classification. This new approach can advance subnational REDD+ baselines and reference emission levels, considered crucial for verifiable emission reductions and forest carbon finance efforts.

Conclusions

The proposed approach extends the traditional application of CLASlite for forest cover change monitoring and adds to CLASlite's automation, speed, lowered technical barriers and mapping accuracy a new way to address incomplete imagery as a result of from clouds, shadows and sensor-failures. While this study only used automatic cloud and shadow masking of CLASlite, users can also easily decide on inclusion or exclusion of regions of interest, by manually drawn polygons (e.g. to mask smoke from fires distorting fractional cover values).

The proposed framework could support extended spatial and temporal observation coverage of a given region and monitoring period, where incomplete images and sensor failures limit spatial coverage and number of time steps. This may, in turn, contribute to increased validity, reduced workload and cost, and an inclusion of more time steps per monitoring period to capture forest cover change dynamics more aptly. This can also increase the fit of currently proposed regression equations to predict the quantity of future deforestation [20]. The mosaicked fractional cover scene approach can be a useful extension of observable area in time-series with the aim of detecting deforestation with CLASlite 3.0 when a user in a tropical region without cloud free season does not have access to more sophisticated approaches that select the greenest pixel per location [16]. The presented approach also supports experienced CLASlite users with limited resources who wish to expand their observable area in their time steps instantly with techniques already at their disposal.

Fractional cover values per pixel are products of a normalizing, yet iterative process that takes reflectance properties into account to find a best fit from spectral libraries. So, if not obscured by atmospheric phenomena missed by the cloud & shadow masking module (e.g. haze) or partial sensor failures (SLC-off pixel with values in some but not all 6 spectral bands), each pixel has the same validity whether coming from a July or March scene of the year of interest. This is mainly true for seasonal tropical forests, which are a common focus of CLASlite and our case study region.

A limitation to the proposed approach remains the subjective selection of a central scene and its auxiliary scenes to construct a meta-time step. This leaves the possibility that from the central scene not all sub-optimal pixels are masked and therefore are not replaced by mosaicking with the auxiliary scenes even if those would have clear-sky information at this location. Thus, information distorted by atmospheric phenomena (e.g. haze) potentially not accurately corrected in the modules of CLASlite would enter the final mosaic. Such sub-optimal pixels are avoided by more sophisticated approaches that select the greenest pixel per location using NDVI from all imagery taken of the year to generate an annual composite of maximum difference in photosynthetic vegetation (PV) between forests and non-forest land covers [27].

Calculating deforestation rates, we correct for the fact that central scenes are not the same acquisition date per year, by calculating an annual deforestation rate normalized for the number of days between time steps. We justify our recommendation to use absolute values of predicted area of deforestation and degradation for REDD+ REL development instead of rates.

As noted in earlier reviews of CLASlite deforestation outputs [19] and in the User Guide itself [27], the results of CLASlite – whether from a single-scene or multi-scene approach – should be carefully inspected and interpreted. Applying the principle of conservativeness for REDD+ projects – a manual subtraction of perceived false positive change results is always allowable, e.g. by vectorization of results and editing.

Methods

Landsat data availability

Freely available Landsat imagery from the United States Geological Survey was characterized by high cloud cover throughout the study region. Between 2002 and 2009, Landsat 5 Thematic Mapper (TM) imagery also had a time gap, leaving us with only Landsat 7 Enhanced Thematic Mapper (ETM+) data containing the Scan Line Corrector (SLC) error that occurred in 2003, rendering each image missing pixel data in stripes across the outer ~30% of each image. Mosaicking of multispectral imagery from different acquisition dates often brings changed radiance properties, however, various approaches have been developed and applied [36-38]. This leaves the user with the option to classify incomplete imagery separately, assess map accuracy separately, and later mosaic land-cover classification products. Such efforts with incomplete imagery require much more work, and are ultimately severely limited by a scarcity of valid ground truth data for classification training and for map validation. This incomplete imagery problem that plagues many large-scale multi-temporal monitoring efforts of forest cover change can be remedied by using CLASlite's unique ability to calculate fractional cover values invariant from radiance value properties through standardized atmospheric correction and

the iterative fitting of reflectance values to spectral libraries of typical fractional covers. Given the limited Landsat data availability for our study region, we selected a monitoring period from 1986 to 2011 – a 25 year period covered by six time steps (1986, 1991, 1996, 1999, 2002, and 2011), and therefore five sub-periods with deforestation (net forest change) analysis.

As the main advantage of the presented approach is a low barrier improvement to forest cover monitoring in highly clouded areas, it should be noted that other technologies, such as InSAR (Interferometric synthetic aperture radar) also hold great value for observation of forest cover in highly cloudy regions. Most prominently, the PALSAR (Phased Array type L-band Synthetic Aperture Radar) sensor on the Japanese ALOS (Advanced Land Observation Satellite) has been repeatedly applied for this purpose [39,40], including the Pacific Coast of Colombia (Niels Wielaard, personal communication). Please also see section 2.9.3 of [26] for discussion of the technology in the context of REDD+. There are some qualifications to be made about InSAR based deforestation monitoring: Though not insurmountable, the processing of InSAR imagery requires an even more specialized expert-knowledge than the application of multi-spectral imagery in a semi-automated process such as CLASlite v3.0 already demands. Several factors such as terrain relief particularly pronounced in the Chocó department require careful correction in order to avoid distortions to land cover & land use change results. For the analysis of historic deforestation, the continuity of imagery time-series and free data availability is an important aspect, where NASA´s Landsat mission remains unmatched. Recognizing the great potential and contribution of InSAR applictions to tropical forest monitoring, we focus in this study on presenting a low barrier approach for users with limited resources.

Image calibration in CLASlite

CLASlite contains an automated set of algorithms that converts Landsat and other satellite images from raw digital number (DN) recordings to final maps of forest cover and forest change (both deforestation and forest disturbance) [17]. CLASlite's approach includes four major automated steps, and several minor yet important "bad-image" data masking steps [27]. First, the raw DN images are converted to top-of-atmosphere radiance images using sensor offset and gain values provided by the satellite data source (e.g. USGS for Landsat). Then the radiance images are converted to apparent surface reflectance using a combination of atmospheric correction with the 6S model [41] and, if needed, haze correction [18]. Within this step CLASlite 3.0 offers the a standard set of masking parameters, which use optical and thermal channels from Landsat to remove conservatively mask or remove portions of the image that contain atmospheric phenomena like haze, clouds, cloud edges and shadows, and topographic shade [17]. The user can select the "Reduced Masking" option, in order to decrease the area masked by altering the sensitivity of the optical and thermal channels to cloud and cloud shadow spectral signatures. In this study we found that the masking process widely avoided problems of deforestation over-detection, but also conservatively masked many pixels with apparently valid DN values for which a valid fractional cover calculation seemed reasonable.

For forest cover change, the original forest cover (1986 in our case) is important in order to map change from it. To maximize our chances of picking up valid forest cover change, we utilized the "Reduced Masking" option for the original forest cover of 1986, but the standard masking for the later time-steps of change detection – thus increasing our valid original cover, but mapping change still conservatively.

Generating fraction covers by AutoMCU

Image calibration is followed by the most important step in CLASlite: the Automated Monte Carlo Unmixing (AutoMCU) algorithm [42,43], which is applied on the image, providing the fractional cover of photosynthetic vegetation (PV), non-photosynthetic vegetation (NPV), and bare substrate (S) on a scale from 0-100% cover in every image pixel. Critical to this study, the AutoMCU employs spectral endmember libraries for PV, NPV and S that are derived from thousands of hyperspectral measurements made using field, airborne and spaceborne imaging spectrometers [17]. Because the PV, NPV and S spectral libraries already incorporate enormous variation in reflectance properties of land covers, including under widely varying atmospheric conditions, the probabilistic approach usually leads to a very stable result in each pixel, even if the data come from different sensors or times of

the year (as long as the data are not heavily cloud or atmospherically contaminated). This provides leverage for compositing different spatial subsets of AutoMCU output to allow for mosaicking the clearest pixels throughout a region otherwise heavily contaminated by clouds over time.

The final step of CLASlite takes the outputs from the reflectance and AutoMCU (PV, NPV, S) steps, and applies a series of decision trees to estimate forest cover on single-date imagery and forest change on multi-temporal images [17]. The decision trees have been steadily expanded and improved to allow for multiple tropical forest types, from lowland to montane forests. These decision trees are mostly empirically derived from validation studies in the field, and by input from the CLASlite user community [27]. In total, 17 Landsat 5 TM & Landsat 7 ETM+ scenes were obtained and processed using the CLASlite approach.

Mosaicking fractional covers

Mosaicking the individual fractional cover of CLASlite follows a minimal invasive approach to not distort DN values. The image identified as the best central scene of the time-step is used as top image, the other auxiliary scenes below. DN value -1 which represents masked areas in the fractional cover is used as "see through value" in order to allow fill-in from the auxiliary scenes. No color-matching is applied, so each valid fractional cover value per pixel entering the final mosaicked image is the same as coming out of CLASlite's original processing. The results of increasing areas filled with multispectral reflection information is shown in Figure 5 below – base image left, auxiliary fill image middle and mosaicked image on the right.

To define which images were candidates for mosaicking of their fractional cover images, rules were established to facilitate multi-temporal analysis. At each time step we identified a central scene of best quality. Additionally up

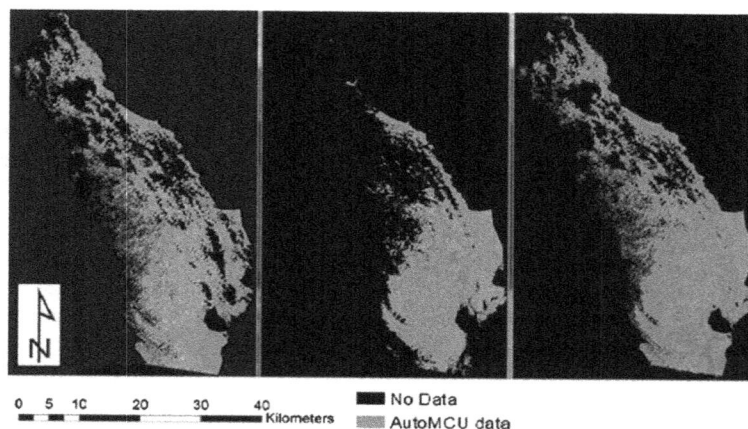

Figure 5 Comparison of accumulated deforestation output of the IDEAM, CLASlite v3.0 and UMD methogologies.

to three other auxiliary scenes were selected within the temporal range of +/- 12 month around the central scene. Because forest types in the project area are evergreen, we assumed minimal seasonal effects on reflectance due to phenology. In addition, the spectral libraries within the AutoMCU sub-module of CLASlite allow for some degree of phenological variation in the spectra, with minimal effects on the fractional cover estimates.

Estimation of deforestation rate

Under Decision 11/CP.7, the UNFCCC defined deforestation as: "the direct, human induced conversion of forested land to non-forested land." This requires the application of a threshold between forested and non-forested land. We apply the forest definition reported by the Colombian Designated National Authority (DNA) to the UNFCCC CDM Executive Board:

Minimum forest area: 1 hectare
Tree crown cover value: 30%
Tree height (or in situ potential to reach it): 5 meters

Deforestation rate is an important parameter to express deforested area comparable between all locations and scales. Puyravaud [44] suggested a standardized approach to calculate deforestation rates which has hence been widely applied, e.g. as baseline approach in [20], which can be applied to develop REDD+ RELs [21-23,25]. To calculate an annual deforestation rate, it is necessary to adjust for the fact that satellite scenes per time step may not fall in the same month. A simple calculation using only years would look like:

Deforestation rate $yr^{-1} = \{[1/(\text{Year } A_2 - \text{Year } A_1)] \times \log (A_2/A_1)\} \times 100$, where:

A_1 = Forest Area at beginning of time step
A_2 = Forest Area at end of time step
Year A_1 = Year of beginning of time step
Year A_2 = Year of end of time step

This, however, could lead to distorted results as it would assume that a forest is always observed in the same month of each year, and thus the number of months between beginning and end of a time step implicitly being (Year A_2 − Year A_1) × 12. As is often the case, this is not the situation in our study, as the months of acquisition of our central scene varied from January to July.

Fortunately it is simple to include dates of image acquisition into Puyravaud's [44] equation, adding the day count per year as a digit number. The exact day count is divided by 365.25 thus giving values ranging from 0.000 to 0.999. For example 17th of July 1999 is translated to 1999.542. This way the difference between time A_2 and time A_1 accounts for the acquisition dates of the central image per time.

Deforestation rate $yr^{-1} = \{[1/(\text{time } A_2 - \text{time } A_1)] \times \log (A_2/A_1)\} \times 100$

where:
A_1 = Forest Area at beginning of time step
A_2 = Forest Area at end of time step
time A_1 = Year and day count as digit number of beginning of time step
time A_2 = Year and day count as digit number of end of time step

This topic becomes relevant, if for development of a REDD+ REL to project future deforestation from historic trends, rates of loss per year are used and not averages of absolute observed deforested area.

Abbreviations

ACR: American Carbon AutoMCU: Automated Monte Carlo Unmixing algorithm; CAR: Climate Action Reserve CCBS: The Climate, Community & Biodiversity Alliance; CDM: Clean Development Mechanism; CLASlite: Carnegie Landsat Anlysis System lite; DN: Digital Number; ETM+: Enhanced Thematic Mapper; EU ETS: European Union Emissions Trading Scheme; FAO: Food and Agricultural Organisation of the United Nations; FRA: Global Forest Resources Assessments; FCPF: Forest Carbon Partnership Facility of the World Bank; GHG: Green House Gas; GIS: Geographical Information System; IDEAM: The National Institute of Hydrology, Metrology and Environmental Studies or Instituto de Hidrología, Meteorología y Estudios Ambientales de Colombia; INPE: INSTITUTO NACIONAL DE PESQUISAS ESPACIAIS – Brazilian National Institute of Space Research; MRV: Measuring, Reporting & Verification; NPV: Non-photosynthetic Vegetation; NDVI: Normalized Difference Vegetation Index; PlanVivo: Forest carbon community certification standard operated by the Plan Vivo foundation, a registered Scottish charity; PRODES: Project Deforestation (Projeto Desmatamento in port.); PV: Photosynthetic Vegetation; REDD+: Reducing Emissions from Deforestation and Degradation and carbon stock enhancement; REL: Reference Emission Level; S: Bare Substrate; SLC: Scan Line Corrector; TM: Thematic Mapper; UMD: University of Maryland; UNFCCC: United Nations Framework Convention on Climate Change; USGS: United States Geological Survey; VCS: Verified Carbon Standard; OLI: Operational Land Imager.

Competing interests
The authors declare that they have no competing interests.

Authors' contributions
FR carried out the remote sensing analysis, designed the comparison approach and drafted the manuscript. GA has lead the development of the CLASlite software & approach, has given inputs on the remote sensing & comparison approach, provided literature indications and revised the manuscript. JS has given inputs to National & Sub-National REDD+ REL, provided literature indications and revised the manuscript. All authors read and approved the final manuscript.

Acknowledgments
We thank R. Martin and Patrick Brooke-Bailey for editing of the manuscript. S.J. is supported by the CGIAR Research Program on Forests Trees and Agroforestry. CLASlite is made possible through the support of the Gordon and Betty Moore Foundation, John D. and Catherine T. MacArthur Foundation, and the endowment of the Carnegie Institution for Science. We thank D.V. Giraldo Tirado for his contribution to access of Colombian datasets.

Author details
[1]Center for Development Research (ZEF), Group Börner, Rheinische Friedrich-Wilhelm University, Walter-Flex-Str. 3, 53113 Bonn, Germany. [2]Department of Global Ecology, Carnegie Institution for Science, 260 Panama Street, Stanford 94305CA, USA. [3]Forest and Environment Program, Center for International Forestry Research, Jalan CIFOR, Bogor 16115, Indonesia.

References

1. UNFCCC. Modalities for national forest monitoring systems. Decision 11/CP.19 2013:31-32. Online: http://unfccc.int/resource/docs/2013/cop19/eng/10a01.pdf

2. Asner GP. Painting the world REDD: addressing scientific barriers to monitoring emissions from tropical forests. Environ Res Lett. 2011:6-021002.

3. Estrada M, Joseph S. Baselines and monitoring in local REDD+ projects. In: Angelsen A, Brockhaus M, Sunderlin WD, Verchot LV, editors. Analyzing REDD+. Bogor: CIFOR; 2012. p. 247-60.

4. Joseph S, Herold M, Sunderlin WD, Verchot LV. REDD+ readiness: early insights on monitoring, reporting and verification systems of project developers. Environ Res Lett. 2013;8:34-8.

5. Joseph S, Herold M, Sunderlin W, Verchot L. Challenges in operationalizing remote sensing in climate change mitigation projects in developing countries. Geoscience and Remote Sensing Symposium (IGARSS). Melbourne, VIC: IEEE International; 2013. p. 2752-5.

6. Joseph S, Murthy MSR, Thomas AP. The progress on remote sensing technology in identifying tropical forest degradation: a synthesis of the present knowledge and future perspectives. Environmental Earth Sciences. 2011;64:731-41.

7. Dutschke M. Key issues in REDD+ verification. Bogor, Indonesia: Center for International Forestry Research; 2013.

8. Hardcastle PD, Baird D. Capability and cost assessment of the major forest nations to measure and monitor their forest carbon. Report prepared for the Office of Climate Change. Penicuick: Office of Climate Change; 2008.

9. Wunder S. The economics of deforestation: the example of Ecuador. London: MacMillan and St. Martin Press in association with St. Anthony's College; 2000.

10. De Sy V, Herold M, Achard F, Asner GP, Held A, Kellndorfer J, et al. Synergies of multiple remote sensing data sources for REDD+ monitoring. Curr Opin Environ Sustain. 2012;4-6:696-706.

11. INPE. Metodologia para o Calculo da Taxa Anual de Desmatamento na Amazonia Legal. Sao José dos Campos, 2006. Online: http://www.obt.inpe.br/prodes/metodologia.pdf

12. Hansen MC, Potapov PV, Moore R, Hancher M, Turubanova SA, Tyukavina A, et al. High-Resolution Global Maps of 21st-Century Forest Cover Change. Science. 2013;342:850.

13. Potapov PV, Turubanova SA, Hansen MC, Adusei B, Broich M, Alstatt A, et al. Quantifying forest cover loss in Democratic Republic of the Congo, 2000-2010, with Landsat ETM+ data. Remote Sens Environ. 2012;122:106-16.

14. Margono BA, Turubanova S, Zhuravleva I, Potapov P, Tyukavina A, Baccini A, et al. Mapping and monitoring deforestation and forest degradation in Sumatra (Indonesia) using Landsat time series data sets from 1990 to 2010. Env Research Letters. 2012;7-3:034010.

15. IDEA. Protocolo de Procesamiento digital de imagines para la Cuantificación de la Deforestación en Colombia nivel nacional escala gruesa y fina. 2011. Online: https://documentacion.ideam.gov.co/openbiblio/Bvirtual/022106/022106.htm

16. MINAM – Ministerio del Ambiente de Perú. Memoria Técnica de la Cuantificación de los cambios de la Cobertura de Bosque a No Bosque por Deforestación en el ámbito de la Amazonía Peruana. Periodo 2009-2010-2011. 2012. Online: http://geoservidor.minam.gob.pe/geoservidor/deforestacion.aspx

17. Asner GP, Knapp DE, Balaji A, Paez-Acosta GP. Automated mapping of tropical deforestation and forest degradation: CLASlite. J Appl Remote Sens. 2009;3 (033543):1-24.

18. IDEAM. Atlas Climatico de Colombia. 2005. Online: https://documentacion.ideam.gov.co/openbiblio/Bvirtual/019711/019711.htm

19. Reimer F, Börner J, Wunder S. Monitoring Deforestation for REDD. An overview of options for the Juma Sustainable Development Reserve Project. CIFOR Technical Brief, Rio de Janeiro, 2012. Online: http://fas-amazonas.org/versao/2012/wordpress/wp-content/uploads/2009/10/ACTUALIZED-08-06-CIFOR-UP.pdf

20. VCS. Methodology for Avoided Unplanned Deforestation – VM0015 v1.1. 2012a. Online: http://www.v-c-s.org/methodologies/VM0015

21. VCS. 2012b. Jurisdictional and Nested REDD+ (JNR) Requirements v3.0. Online: http://www.v-c-s.org/sites/v-c-s.org/files/Jurisdictional%20and%20Nested%20REDD%2B%20Requirements%2C%20v3.1.pdf

22. REDD Research and Development Center. REDD-plus COOKBOOK - How to Measure and Monitor Forest Carbon. Tsukuba, Japan: Forestry and Forest Products Research Institute. 2012.

23. UNFCC. Decision 12/CP.17 on guidance on systems for providing information on how safeguards are addressed and respected and modalities relating to forest reference emission levels and forest reference levels as referred to in decision 1/ CP.16. Decision 12/CP.17 2011. Online: http://unfccc.int/files/meetings/durban_nov_2011/decisions/application/pdf/cop17safeguards.pdf

24. FCPF Carbon Fund. Methodological Framework. 2013. Online: http://www.forestcarbonpartnership.org/sites/fcp/files/2013/Dec2013/FCPF%20Carbon%20Fund%20Meth%20Framework%20-%20Final%20December%2020%202013%20posted%20Dec%2023rd.pdf

25. Soares-Filho B, Nepstad DC, Curran LM, Cerqueira GC, Garcia RA, Ramos CA, et al. Modelling conservation in the Amazon basin. Nature. 2006;440:520-3.

26. GOFC-GOLD. A sourcebook of methods and procedures for monitoring and reporting anthropogenic greenhouse gas emissions and removals associated with deforestation, gains and losses of carbon stocks in forests remaining forests, and forestation. GOFC-GOLD Report version COP18 (GOFC-GOLD Land Cover Project Office, Wageningen University, The Netherlands). 2012. Online: http://www.gofcgold.wur.nl/redd/sourcebook/GOFC-GOLD_Sourcebook.pdf

27. Carnegie Institution. CLASlite 3.0 User Guide. 2013. Online: http://claslite.stanford.edu/en/support/links.attachment/63/download

28. Congalton RG. A Review of Assessing the Accuracy of Classifications of Remotely Sensed Data. Remote Sens Environ. 1991;37:35-46.

29. Story M, Congalton RG. Accuracy assessment: A user's perspective. Photogramm Eng Remote Sens. 1986;52:397-9.

30. Google Inc. Deprecated Landsat 5 Annual Greenest-Pixel TOA Reflectance Composite. 2012 Online: https://earthengine.google.org/#detail/LANDSAT%2FL5_L1T_ANNUAL_GREENEST_TOA

31. Tropek R, Sedláček O, Beck J, Keil P, Musilová Z, Símová I, et al. Comment on "High-resolution global maps of 21st-century forest cover change". Science. 2014;344:981-d.

32. Hansen MC, Potapov PV, Margono B, Turubanova SA, Tyukavina A. Response to Comment on "High-resolution global maps of 21st-century forest cover change". Science. 2014;344:981-e.

33. Carbon Green Africa. Project Design Document of the Kariba REDD+ Project. 2012 Online: https://vcsprojectdatabase2.apx.com/myModule/Interactive.asp?Tab=Projects&a=2&i=902&lat=-16.8184067184111&lon=28.7615526227228

34. REDD OFFSET WORKING GROUP, THE. California, Acre and Chiapas. Partnering to Reduce Emissions from Tropical Deforestation. 2013 Online: http://greentechleadership.org/documents/2013/07/row-final-recommendations-2.pdf

35. Merger E, Dutschke M, Verchot L. Options for REDD+ Voluntary Certification to Ensure Net GHG Benefits, Poverty Alleviation, Sustainable Management of Forests and Biodiversity Conservation. Forests. 2011;2:550-77.

36. Pringle MJ, Schmidt M, Muir JS. Geostatistical interpolation of SLC-off Landsat ETM+ images. ISPRS J Photogramm Remote Sens. 2009;64(6):654-64.

37. Maxwell SK, Schmidt GL, Storey JC. A multi-scale segmentation approach to filling gaps in Landsat ETM+ SLC-off images. Int J Remote Sensing. 2007;28-23:5339-56.

38. Zhang C, Li W, Travis D. Gaps-fill of SLC-off Landsat ETM+ satellite image using a geostatistical approach. Int J Remote Sensing. 2007;28-22:5103-22.

39. Avtar R, Takeuchi W, Sawada H. Full polarimetric PALSAR-based land cover monitoring in Cambodia for implementation of REDD policies. Int J of Dig Earth. 2011;10:1-21.

40. Gaveau DLA, Sloan S, Molidena E, Yaen H, Sheil D, Abram NK, et al. Four Decades of Forest Persistence, Clearance and Logging on Borneo. PLoS ONE. 2014;9-7:1-11.

41. Kotchenovy SY, Vermote EF, Raffaella M, Klemm jr FJ. Validation of a vector version of the 6S radiative transfer code for atmospheric correction of satellite data. 2006 Online: http://6s.ltdri.org/6S_code2_thiner_stuff/Kotchenova_et_al_2006.pdf

42. Carlotto MJ. Reducing the effects of space-varying, wavelength-dependent scattering in multispectral imagery. Int J Remote Sens. 1999;20:3333-44. Online: http://carlotto.us/MJCtechPubs/pdfs/1999ijrsDeHaze.pdf.

43. Asner GP, Heidebrecht KB. Spectral unmixing of vegetation, soil and dry carbon in arid regions: Comparing multi-spectral and hyperspectral observations. Int J Remote Sensing. 2002;23-3:939-58.

44. Puyravaud JP. Standardizing the calculation of the annual rate of deforestation. For Ecol Manage. 2003;177:593-6.

8

Time-series maps of aboveground carbon stocks in the forests of central Sumatra

Rajesh Bahadur Thapa*, Takeshi Motohka, Manabu Watanabe and Masanobu Shimada

Abstract

Background: Efforts to reduce emissions from deforestation and forest degradation in tropical Asia require accurate high-resolution mapping of forest carbon stocks and predictions of their likely future variation. Here we combine radar and LiDAR with field measurements to create a high-resolution aboveground forest carbon stock (AFCS) map and use spatial modeling to present probable future AFCS changes for the Riau province of central Sumatra.

Results: Our map provides spatially explicit estimates of the AFCS with an accuracy of ±23.5 Mg C ha^{-1}. According to this map, the natural forests in the province currently store 265 million Mg C, with a density of 72 Mg C ha^{-1}, as aboveground biomass. Using a spatially explicit modeling technique we derived time-series AFCS maps up to the year 2030 under three forest policy scenarios: business as usual, conservation, and concession. The spatial patterns of AFCS and their trends under different scenarios vary on a local scale, and some areas are highlighted that are at eminent risk of carbon emission. Based on the business as usual scenario, the current AFCS could decrease by 75 %, which may lead to the release of 747 million Mg CO_2. The other two scenarios, conservation and concession, suggest the risk reductions by 11 and 59 %, respectively.

Conclusion: The time-series AFCS maps provide spatially explicit scenarios of changes in AFCS. These data may aid in planning Reducing Emissions from Deforestation and forest Degradation in developing countries projects in the study area, and stimulate the development of AFCS maps for other regions of tropical Asia.

Keywords: Forest carbon, Deforestation, PALSAR, AFCS, Aboveground biomass, REDD+, Riau

Background

Across the world, existing tropical forest landscapes are undergoing rapid deforestation due to natural disasters, as well as the increasing demand for agricultural land, wood products, energy, and developmental projects. Currently, global forest areas account for 3.85 billion ha, or 26 % of the Earth's land surface [1], but this area is decreasing at around 13 million ha per year [2]. As deforestation continues, the Earth becomes more susceptible to potentially negative impacts on ecosystems and the overall climate system due to the associated effects on carbon balance, biodiversity, soil, water regulation, and weather patterns. Currently, emissions caused by deforestation worldwide are considered to be very high, and are likely to continue in this way for the coming decades.

Tropical forest regions in particular are major potential sources of carbon emissions [3–7]. Reducing Emissions from Deforestation and forest Degradation in developing countries (REDD+) [8] is one of the key global initiatives that aims to conserve forests and reduce carbon emissions. The goal of REDD+ is to connect investors to forest users and offer an economic portfolio for the retention of forest carbon and the avoidance of deforestation, while also slowing the drivers of land use change. As a result, the initiative contributes indirectly to biodiversity conservation by helping to reduce habitat loss and ensure the continuation of normal ecosystem services; hence, it is considered a sustainable option for the maintenance of forests. Meaningful implementation of REDD+ requires accurate, high-resolution, spatially explicit maps of forested areas and forest carbon stocks, as well as predictions of their change in the future. Therefore, efforts to improve the methods for mapping forest extents and

*Correspondence: rajesh.thapa@jaxa.jp; thaparb@gmail.com
Earth Observation Research Center, Japan Aerospace Exploration Agency (JAXA), 2-1-1 Sengen, Tsukuba, Ibaraki 305-8505, Japan

forest-related carbon stocks, as well as identifying their changes, have been advancing in many parts of the world, including tropical Asia [5, 9–13]. Remote sensing and spatial modeling techniques offer a practical means to monitor and examine changes in forest cover, analyze the implications of forest policies, predict spatial patterns of forest cover in the future, and relate these patterns to carbon stock densities [9, 14–16].

Accurate mapping of aboveground forest carbon stocks (AFCS) using spaceborne satellite data is still very challenging due to the requirement of a large amount of in situ data for forest carbon estimation model calibration and validation. Although traditional plot-based field measurements of AFCS have proven most accurate, they are costly and difficult to implement for large areas with dense tropical forests. Studies [13, 14, 17] have shown that light detection and ranging (LiDAR) techniques allow the accurate measurements of geographically referenced vertical forest structures, including canopy height, volume, and biomass. Using LiDAR data, an allometric model for AFCS can be developed with a relatively small number of field measurements [13, 17]. Modeling results can be used to extend the field data, providing spatially extensive and detailed forest attribute data, that can be used to calibrate AFCS predictive models build around a wide variety of spaceborne data, including synthetic aperture radar (SAR) and optical imageries covering larger areas [14, 18].

The integration of both airborne and spaceborne remote sensing techniques has offered the opportunity to more precisely map forest cover and related carbon stocks over wider areas at suitable spatiotemporal scales [6, 13, 14, 19]. However, the potential application of optical spaceborne remote sensing data in Asian tropical forest regions is limited, due to the frequent appearance of clouds and haze, as well as the insensitivity of sensing systems to the variability of biomass with a multi-layer canopy in highly dense forests. In contrast, spaceborne SAR is not limited by these factors as it penetrates clouds to image the Earth's surface regardless of weather conditions or solar illumination. Among the available spaceborne SAR systems, the Advanced Land Observation Satellite (ALOS) Phase Arrayed L-band SAR (PALSAR) operating at a wavelength of 23.6 cm is very sensitive to forest structure, yielding valuable information for the mapping of forest cover [20–22] and AFCS measurements [23, 24]. However, studies have shown that saturation remains a dominant issue when directly estimating AFCS using SAR data in high biomass areas [25–29]. A consideration of multi-temporal SAR data with multiple polarizations and the use of rule-based algorithms can help to mitigate the saturation problem and improve AFCS estimations [14, 24].

In addition to remote sensing techniques, spatial modeling is required in order to visualize and quantify the future variations of AFCS [3, 9, 15]. Future trends are reliant on the past processes of deforestation, and represent a consolidation of the relationships between time, space, and driving factors. A logically developed spatial model incorporates these relationships and extrapolates the likelihoods of various forest spatial patterns into the future [15, 16]. Such models offer a means of examining the implications of different forest policies on AFCS, allowing appropriate measures to control deforestation and retain AFCS to be formulated. In this study, our aim is to create a baseline AFCS map of a tropical forest in Asia and to estimate its future AFCS patterns under different forest policy frameworks. The Riau Province in Indonesia (Fig. 1) was chosen as the study site, due to its high carbon emissions as a result of deforestation compared with other provinces in the country [30]. Currently, 5.54 million people live in the province, with an annual growth rate of 3.6 % [16, 31]. Pulp and paper, oil-palm, rubber, and petroleum products are the main sources of income, while the forest landscape has also become a major provider of land in recent years. An ever increasing population and demanding economic activities have increased deforestation and forest degradation, which is ultimately threatening the forest carbon stocks, peat drainage, and biodiversity in the province.

Field measurements, LiDAR data, time-series PALSAR data, and a rule-based algorithm were used together to create a baseline AFCS map with high spatial resolution. The spatial model developed by Thapa et al. [16] was applied to visualize and assess the implications of different forest policies on future AFCS.

Results and discussion

Table 1 summarizes the field measurement data collected over 87 plots, divided according to the various forest types in the study area. The forests are diverse and exhibit high variability in AFCS of different regions. Around 47 % of the measurement plots are within natural forests, including peat swamps (21.8), dry moist forest (10.3), regrowth (5.7), and mangrove areas (9.2). The remaining plots are within plantation forests including rubber (11.5), acacia (10.3), oil palm (23.0), and coconut (8.0) forests. Such a range of forest types is host to AFCS of 1.18–334.10 Mg C ha^{-1}. Among the plots, oil palm was found to have the lowest AFCS, while the natural dry moist forest areas had the highest. The regrowth forest areas had AFCS of 84.58–164.49 Mg C ha^{-1}, ranging between that of dry moist forests and peat swamps. Among the plantation forests, rubber plantations had both the highest mean and the maximum AFCS. Across the full range of values, the oil palm plantations had the

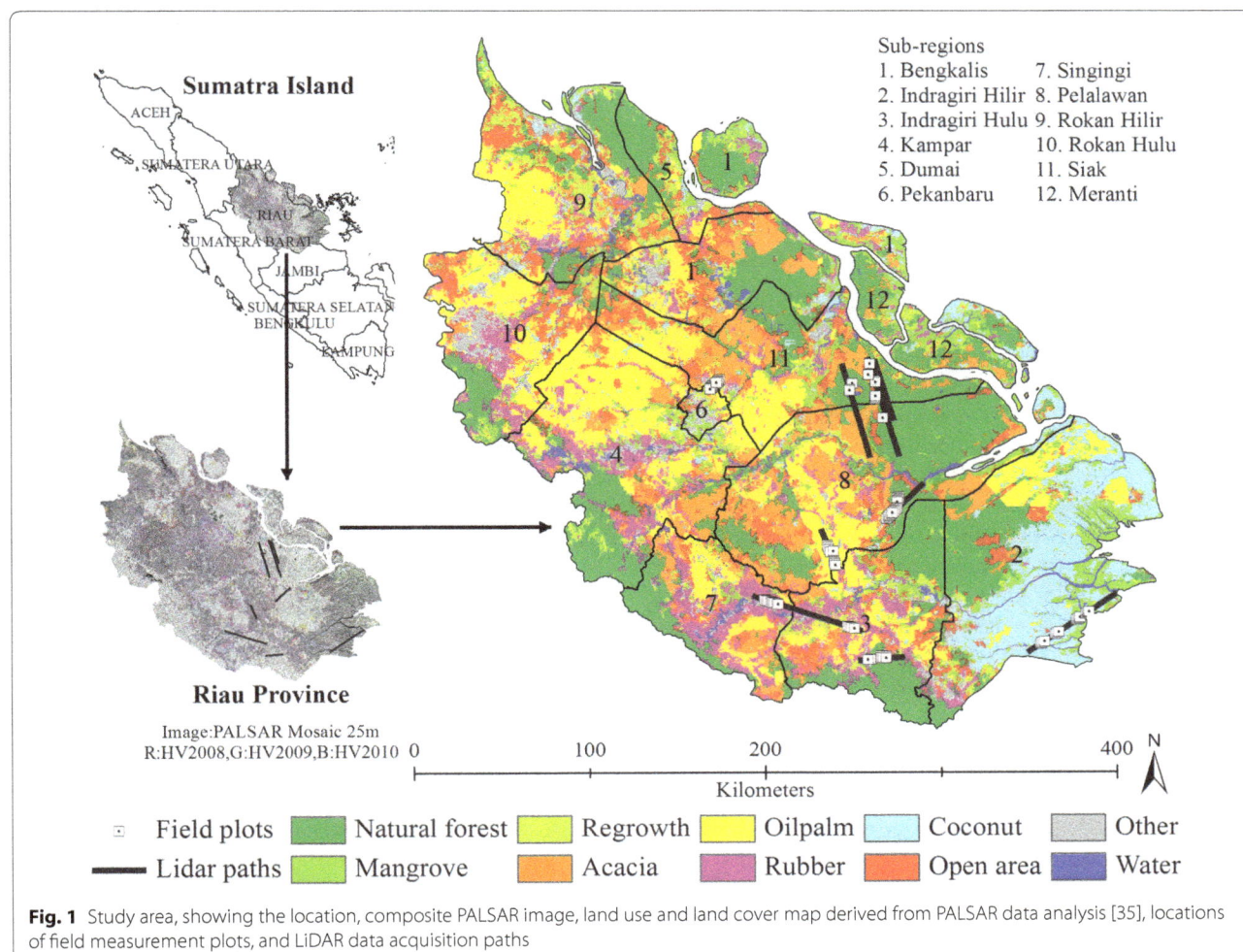

Fig. 1 Study area, showing the location, composite PALSAR image, land use and land cover map derived from PALSAR data analysis [35], locations of field measurement plots, and LiDAR data acquisition paths

Table 1 Summary of field measurement plots and AFCS estimates by forest types

Forest types	Field plots in %	AFCS (Mg C ha^{-1})	
		Mean	Range
Natural forests			
Peat swamp	21.8	114.06	69.70–173.62
Dry moist	10.3	198.10	106.69–334.10
Regrowth	5.7	125.15	84.58–164.49
Mangrove	9.2	26.55	9.02–42.99
Plantation forests			
Rubber	11.5	38.82	19.10–58.88
Acacia	10.3	32.39	21.04–46.38
Oil palm	23.0	8.27	1.18–20.94
Coconut	8.0	16.00	6.87–27.60

AFCS is 47 % of the field measured aboveground biomass

lowest carbon stock of all the forest types in the study region.

The maximum likelihood algorithm (MLA) mapping procedure was used to create four maps using the PALSAR gamma-naught image from 2010, three AFCS maps to which the Lee, Frost, and median filters had been applied, respectively, with a 3 × 3 window, and an AFCS map without any filters for comparison. Validation was performed with three individual statistical measures to assess the filtering effects. The median filter provided the best AFCS map, with a RMSE and bias of 27.59 and −0.83 Mg C ha^{-1} and an index of agreement (D) of 0.74. The other two filters, Lee and Frost, gave RMSEs with biases of 28.03 and −1.28, and 27.47 and −3.67 Mg C ha^{-1}, and D values of 0.71 and 0.74, respectively. The map created without the application of any filters had an

RMSE and bias of 30.09 and -2.71 Mg C ha^{-1}, respectively, and a D of 0.59. The D values are similar in both the Frost and the median filters, although the RMSE indicates a slightly better performance with the Frost filtered map, the bias is comparatively high. Overall, these results suggest that the median filter provides a good AFCS mapping product.

Four additional AFCS maps were created using the 2010 dataset and four increasing sizes of the filtering window with the median filter. The mapping results gradually improved as the window size increased from 3×3 to 9×9. Compared with the 3×3 window size, the map at the 9×9 window size had a reduced RMSE of 25.54 Mg C ha^{-1} with a minimal bias of 0.06 Mg C ha^{-1} and showed high similarity between the predicted and observed AFCS, as indicated by the index of agreement (D = 0.814). However, a decrease in performance was observed with the application of the 11×11 window size, with a RMSE and bias of 25.83 and 0.31 Mg C ha^{-1} and a D of 0.812, respectively. This indicated that investigation into larger window sizes is unnecessary; the observed degradation is likely due to overgeneralization of the image data at larger window sizes. As such, a 9×9 median filter was applied to PALSAR data from 2009, which was used together with the 2010 data

while running the MLA. Inclusion of the 2009 data set improved the AFCS mapping by reducing the mapping uncertainty, and the RMSE and bias dropped to 23.49 and 1.13 Mg C ha^{-1}. The value of D also improved to 0.843, indicating a very high similarity between the predicted and observed AFCS values (Fig. 2). Consideration of multi-temporal mosaic data sets can provide better estimates, indicating a normalization of climatic conditions and increasing the potential of replicating the modeling results in other similar tropical regions. It should be noted that after removing the bias in the RMSE, the error was reduced to 23.47 Mg C ha^{-1}.

The level of error in this map may have resulted from several unquantifiable factors, including differences in field measurement processes, inaccuracies in the field and LiDAR allometric equations, the slope correction method for the PALSAR mosaic data, and the time difference between field and remote sensing measurements [24]. However, the mapping error produced herein is low, and to date remains unmatched in similar studies of tropical forest regions, such as Kalimantan [27], the Amazon [17], and other areas [19]. Owing to the low mapping error and the high level of similarity between predicted and observed AFCS in such a diverse tropical forest, this map was used as a baseline for a spatial model that

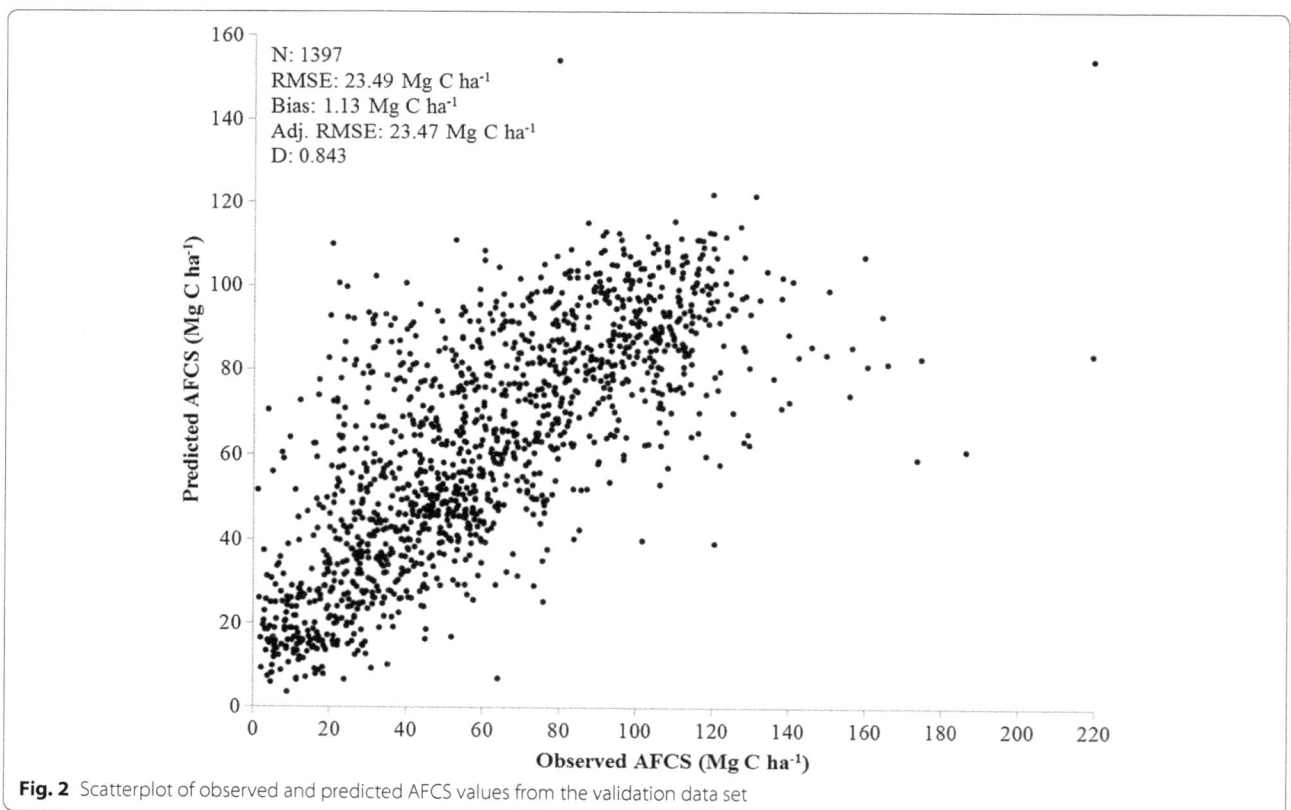

N: 1397
RMSE: 23.49 Mg C ha^{-1}
Bias: 1.13 Mg C ha^{-1}
Adj. RMSE: 23.47 Mg C ha^{-1}
D: 0.843

Fig. 2 Scatterplot of observed and predicted AFCS values from the validation data set

calculates future changes in the forest carbon footprint of the study area.

Figure 3 illustrates the baseline AFCS map for the Riau province, in which interesting spatial patterns, with AFCS ranging from <1 to 334 Mg C ha^{-1}, can be observed. The spatial variation of AFCS across this map indicates that the majority of areas have carbon densities between 100 and 200 Mg C ha^{-1}. High carbon density areas are mostly found within the northern and south-eastern parts of the province, as well as along its western margin. The central area has a generally low carbon density from north to south. These patterns result from the relative distributions of natural forests and plantation forests. The low carbon density areas are mostly covered by plantation forests, agricultural land, and urban forests, while the higher density areas are correlated with existing natural forests, including peat swamps, dry moist forests, and regrowth. The slightly higher AFCS values on the islands in the central-eastern region and on the southeastern margin of the province represent mangrove forests.

The results of this study suggest that the AFCS of the natural forest areas in Riau province, which cover around 3.68 million ha, was 265.57 million Mg at the time of measurement (Table 2). The AFCS density of the natural forest is on average 71.99 Mg C ha^{-1} across the province. Among the different districts, Indragiri Hilir in the southeastern part of the province has the highest AFCS density, at 77.99 Mg C ha^{-1}, whereas Pekanbaru in the central region has the lowest density, at 43.63 Mg C ha^{-1}. The Pekanbaru district is dominated by urban areas and contains fewer forested areas than the other districts, and therefore has a very little AFCS of 0.33 million Mg. In contrast, the Pelalawan district has the highest AFCS in the province, at 48.79 million Mg, which amounts to 18.37 % of the total AFCS stored in natural forests across the study area. However, the AFCS density in this district is slightly less than the provincial average. In terms of overall AFCS density, the Indragiri Hulu and Meranti districts have a similar quality of forest, as they both have a density of 75 Mg C ha^{-1}. Interestingly, the Bengkalis

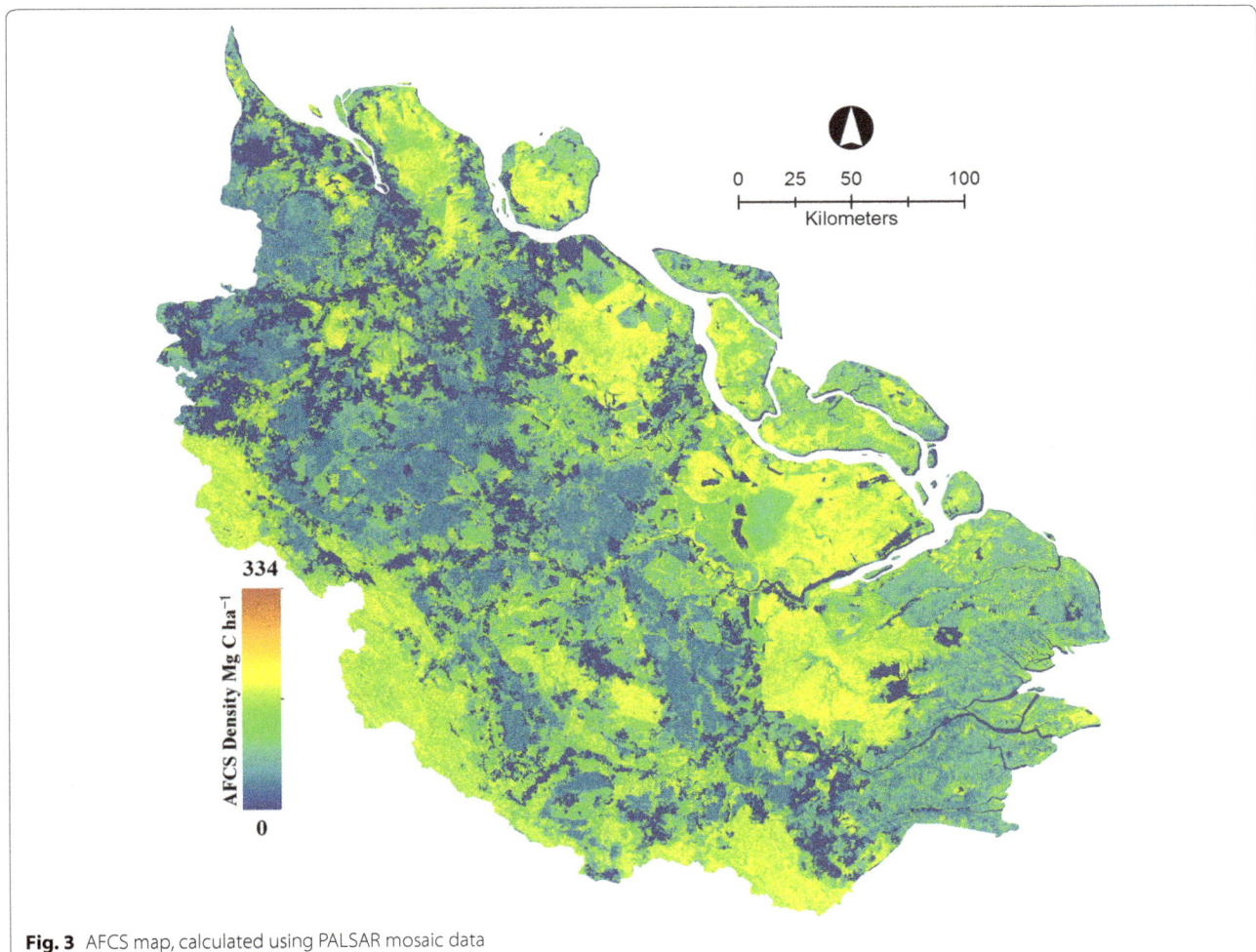

Fig. 3 AFCS map, calculated using PALSAR mosaic data

Table 2 Quantity of AFCSs extracted for natural forest areas and their distribution by district

Districts	Forest area in ha	AFCSs		
		In million Mg C	% distribution	Density in Mg C ha^{-1}
Bengkalis	370,636	26.74	10.07	72.15
Indragiri Hilir	288,914	22.53	8.49	77.99
Indragiri Hulu	443,704	33.31	12.54	75.06
Kampar	433,434	30.50	11.48	70.37
Dumai	101,173	7.22	2.72	71.36
Pekanbaru	7526	0.33	0.12	43.63
Singingi	307,621	20.93	7.88	68.04
Pelalawan	679,426	48.79	18.37	71.81
Rokan Hilir	276,571	18.57	6.99	67.13
Rokan Hulu	217,973	15.64	5.89	71.74
Siak	370,296	26.63	10.03	71.92
Meranti	191,670	14.39	5.42	75.07
Total	3,688,944	265.57	100.00	71.99

Natural forest area is calculated using the Thapa et al. [16] map for 2010

and Siak districts contain a similar amount of AFCS, each with around 10 % of the total carbon stocks, but the AFCS in the Bengkalis districts is slightly higher. Additionally, although the total AFCS of the Dumai district is lower, its density is higher than that in the Singingi district. The Indragiri Hulu district has the second largest AFCS overall, at 33.31 million Mg, and has a density similar to that of the Meranti district. Finally, the Kampar, Rokan Hulu, and Rokan Hilir districts within the northwestern part of the province contain considerable forest carbon stocks, but their densities are slightly lower than the provincial average.

We overlaid the AFCS baseline map derived in this study with the observed and simulated scenario-wide forest maps of Thapa et al. [16]. Figure 4 contains information on the distributions of conservation areas, concession areas, as well as illustrating the AFCS map for natural forest cover in 2010, and scenario-wide simulated AFCS maps for the years 2015, 2020, 2025, and 2030. The changes in spatial patterns of AFCS are clear when comparing the two policy scenarios, the Government-Forest Conservation (G-FC) and Government-Concession for Plantation and Logging (G-CPL) policies, with the Business as Usual (BAU) approach. If the past deforestation processes continue without any policy implementation, as shown by the BAU policy scenario, then the AFCS will be persistently released from most of the forested areas. The AFCS removal will likely occur in the most ecologically delicate areas, including the peat swamps and conservation areas in the northeast, and the dry-forested areas in the southwest of the study area. This indicates that the current land use

change-inducing activities pose an extremely serious threat to AFCS, and immediate measures are required to ensure sustainability and forest protection. The spatial trends in AFCS observed under the G-CF policy scenario are somewhat comparable to those of the BAU scenario, with the exception of the forest conservation zones. The AFCS hosting areas remain fairly large under the G-CF scenario, due to the impact of the policy on forest protection. For instance, the forests in the designated conservation areas of 2010 remain intact for the forthcoming decades, retaining their AFCS. However, deforestation pressures will still affect the regions outside the conservation areas, rapidly releasing AFCS from the districts in the northern part of the study area through 2015 and 2020. Under the G-CPL scenario, the geographic distribution of AFCS across the province was better retained compared with that of the other scenarios. AFCS removal will likely occur only in the concession lands for plantations and selective logging. If the G-CPL policy is implemented without modification, then entire districts will retain a considerable amounts of their AFCS, even by the end of 2030.

Figure 5 illustrates quantifications of the predicted AFCS changes under the three policy scenarios by district for every 5 years until 2030. These charts represent important information by showing the spatial variations in future forest carbon reserves. The AFCS curves for each of the different scenarios show the dynamic variability of carbon stocks over time. In all districts and at all times, the BAU line resides at the bottom of the chart, indicating the greatest loss to AFCS. Thus, if the previous tendencies continue, the AFCS in the Pekanbaru

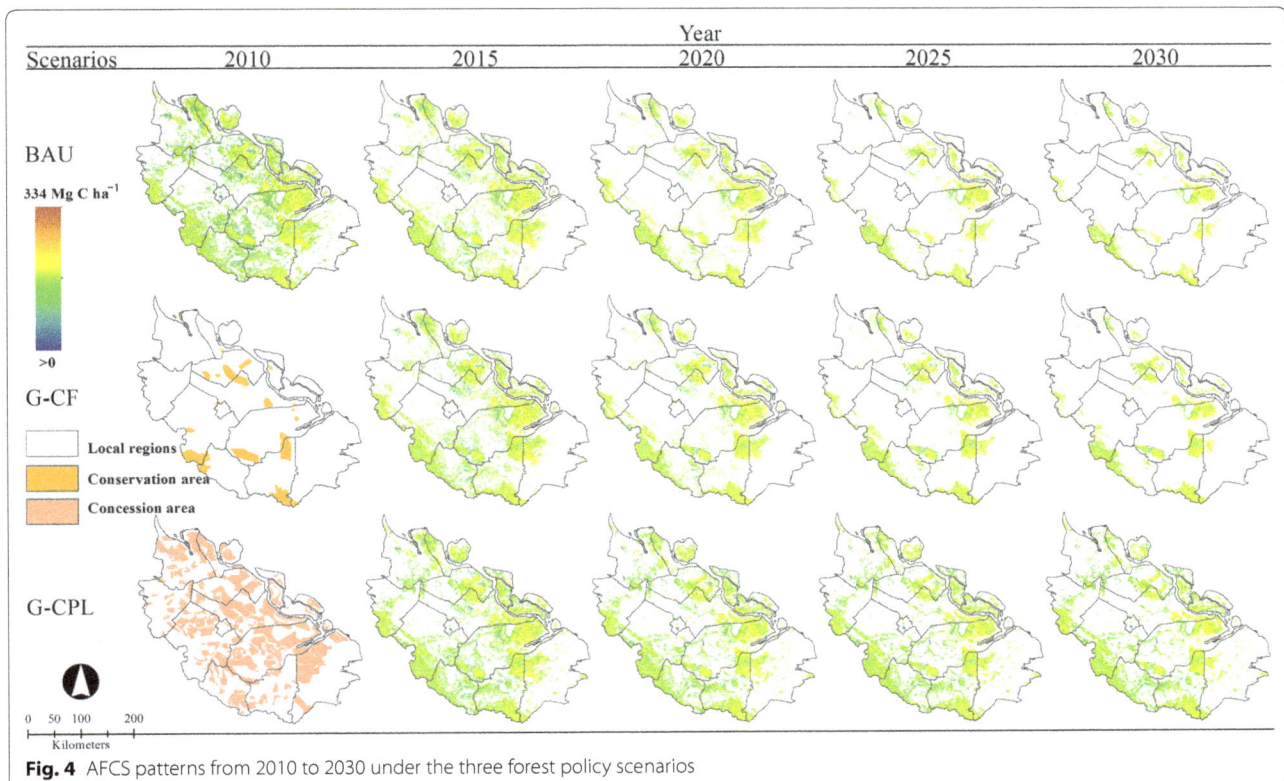

Fig. 4 AFCS patterns from 2010 to 2030 under the three forest policy scenarios

district will be almost entirely lost by 2030. In addition, the Dumai, Meranti, Rokan Hulu, and Rokan Hilir districts are likely to face severe damage to their AFCS balances over the next two decades. These districts will also face similar problems under the conservation scenario. The AFCS emissions in the Pelalawan district under the BAU expected to be high, due to the large number of reserves in this region compared with the other districts. The most rapid declines in AFCS up to 2015 under BAU are expected in the Siak, Indragiri Hilir, and Rokan Hilir districts. Over the study period, the changes in projected AFCS between the G-CF and BAU scenarios gradually widen in the Kampar, Pelalawan, Siak, and Indragiri Hulu districts. This indicates that the G-CF scenario somehow retains the stability of forest carbon reserves to a greater degree than the BAU scenario. In comparison, the G-CPL policy scenario estimates the retention of relatively high AFCS in the Meranti, Singini, Indragiri Hulu, and Kampar districts, even by 2030. Remarkably, the estimated AFCS emissions in Indragiri Hulu, Kampar, Rokan Hulu, and Singingi districts under the G-CPL policy scenario are very low over the study period. In contrast, rapidly declining levels of AFCS are detected in the Siak, Dumai, and Indragiri Hilir districts under the G-CPL policy scenario. The remaining districts, including Indragiri Hulu, Rokan Hulu, Meranti, and Singingi, will contain higher

AFCS by 2030 in the G-CPL scenario than under the other scenarios.

Figure 6 presents the expected forest carbon emissions from the province as a whole under each scenario in five-year intervals up to 2030. The effects of the different scenarios on the projected carbon emission differ with the passing time. If the existing trend continues, as evidenced by the BAU, around 747.61 million Mg CO_2, representing 75 % of the current AFCS, will likely be in the atmosphere by 2030. The trend indicates that the emissions will be greater in earlier years, meaning that two thirds of the forest carbon will be emitted into the air over the next 10 years, potentially resulting in globally adverse environmental consequences. In comparison, the trend of the G-CF scenario suggests some measure of success in the form of reduction in emissions of 20 million Mg CO_2 by 2015, although these double by 2020, and reach approximately 84.3 million Mg CO_2, a reduction of 11.27 % as compared to the BAU in 2030. In contrast, the emissions under the G-CPL scenario appear to differ remarkably from those under the G-CF and the BAU scenarios. The implementation of this policy will gradually slow the emission of forest carbon stocks by controlling the deforestation to a greater degree than the other scenarios. Thus, under the G-CPL scenario, the estimated carbon emissions will be around 305 million Mg CO_2 in

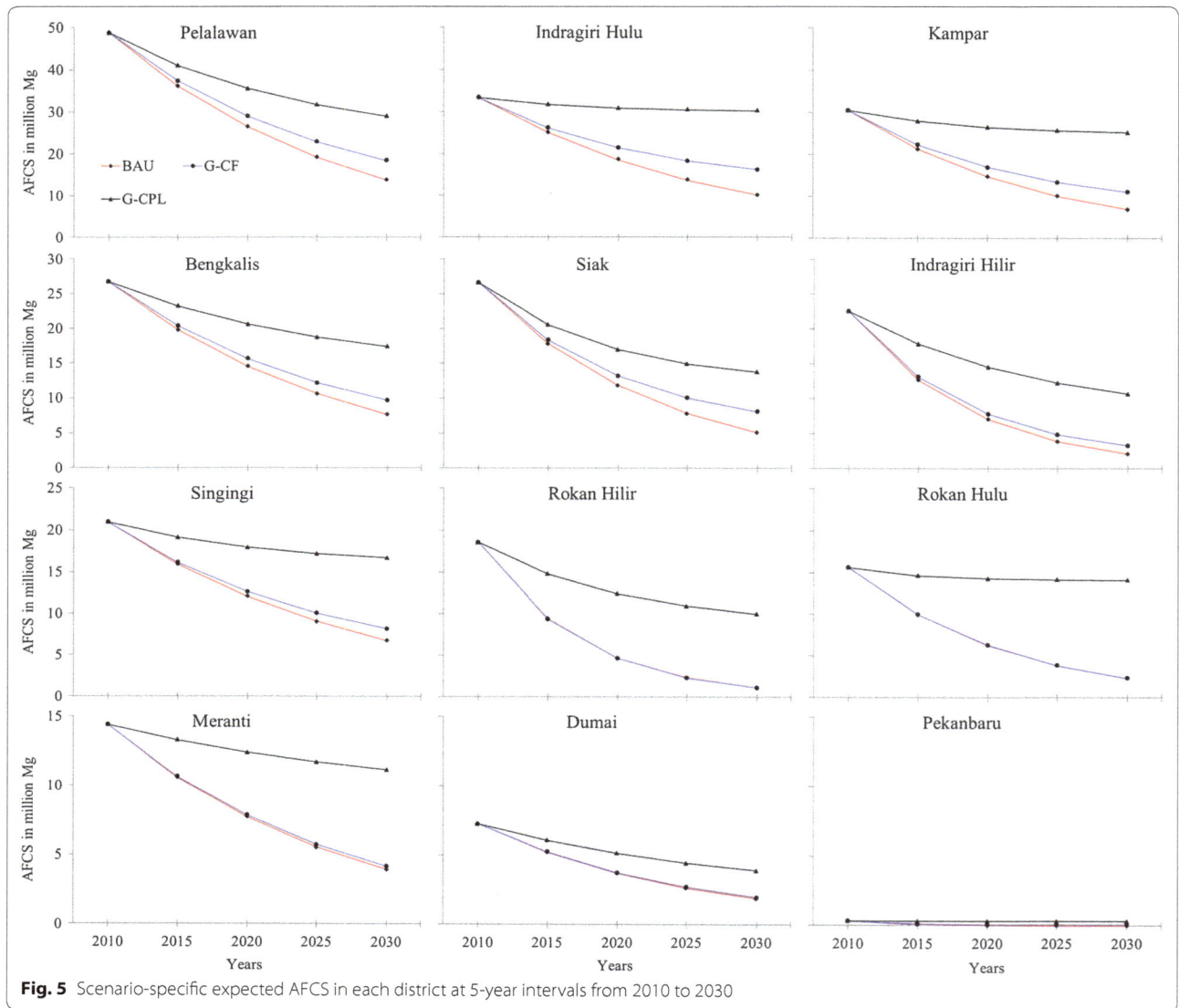

Fig. 5 Scenario-specific expected AFCS in each district at 5-year intervals from 2010 to 2030

2030, which is only 31 % of the current AFCS; this represents a reduction of 2.5 times compared with the BAU scenario. It is worth noting that the G-CPL scenario is likely to delay the carbon emissions by a further 15 years, whereas a similar amount is expected to be released by 2015 under the present conditions. If BAU is considered as a reference scenario, then the concession policy (G-CPL) scenario is likely to reduce the CO_2 emissions by 59.20 % (442.6 million Mg CO_2) through its expected deforestation by the end of 2030.

Despite the high spatial resolution of our AFCS map, which reveals the AFCS dynamics of natural forests on a local scale, information about the dynamics of belowground carbon stocks (BGCS) in natural forest areas is lacking in this study. Estimation of BGCS in natural

forest regions using remote sensing is extremely difficult. However, the general model [32, 33] for estimating the BGCS in tropical regions can be employed using the baseline map derived in this study if necessary. In addition to affecting the BGCS, the deforestation process also contributes to the release of carbon from other sources such as soils and peat lands. These additional sources may significantly increase the net carbon emissions, and the inclusion of forest fire parameters in the modeling process may improve the accuracy of future estimation. Furthermore, the majority of the natural forest land in this area has already undergone transformations to produce economically valuable industrial plantations, such as oil palm, acacia, coconut, and rubber trees. These plantation forests also store a significant amount of

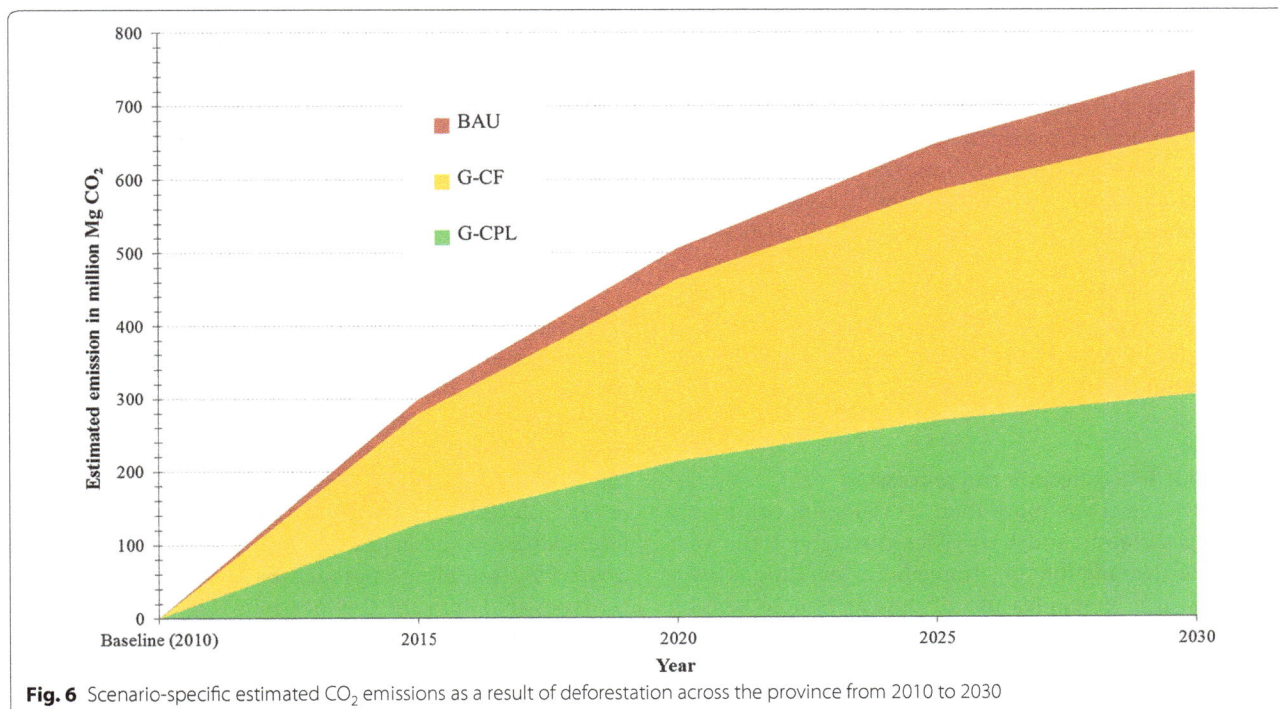

Fig. 6 Scenario-specific estimated CO_2 emissions as a result of deforestation across the province from 2010 to 2030

aboveground forest carbon, as reflected in baseline map (Fig. 3) and in Table 1. From a carbon stock perspective, a consideration of the AFCS dynamics in these forests may represent a trade off in overall carbon balance to some extent. However, there is an immediate risk of carbon emission when the natural forests are cleared, even when they are replaced by plantation forests. Additionally, these plantation forests store carbon only in the short term, as they are harvested within a certain period of time. For example, an acacia plantation in the study area is harvested every 4 years. The other plantations, such as oil palm, coconut, and rubber are often harvested every 20–30 years. The plantation forests maintain greenness in the province but still have a significant impact on carbon recycling and prevent the restoration of ecosystem services.

Conclusion

Through the integration of multiple remote sensing techniques, from airborne LiDAR to spaceborne SAR with field measurement data and a rule-based algorithm, an accurate baseline AFCS map with high spatial resolution was developed for one of the major tropical forests in Asia. This baseline map provides highly accurate, spatially explicit distributions and quantitative estimations of forest carbon stocks in the study area. The AFCS distribution varies geographically, indicating spatial variations in forest quality and vulnerability. The spatial modeling

technique provides an opportunity to extrapolate the spatial trends in AFCS and examine the implications of different forest management policies on carbon stocks and emissions over the next two decades. The inherent capability of the model to distinguish local variations in future AFCS trends under different scenarios is key to identifying the areas most vulnerable to high carbon emissions, which would require immediate mitigation measures to ensure forest conservation. The model was used to predict the spatiotemporal variations and associated quantities of remaining AFCS under different scenarios up until the year 2030. These predicted spatial patterns of AFCS indicate that the forest carbon emission rate is likely to be high in the coming decades across the province. Ongoing deforestation is expected to release around 747 million Mg CO_2 into the atmosphere by 2030. A forest conservation policy will slow the AFCS emissions, but the reduction will be insufficient. Among the scenarios tested, the concession scenario is the most promising, halving the expected emissions if it is implemented as planned. In addition to the high-resolution AFCS map, the modeling outcomes may provide opportunities for the identification of especially vulnerable localities and focuses for the implementation of REDD+ projects to obtain the greatest benefits based on the environmental settings. For the proper execution of the REDD+ project, it is important to understand how the expected trends in AFCS distribution are likely be affected in the long term by the implementation various plans, policies, and strategies. The AFCS under the BAU scenario may provide a reference emission scenario for REDD+, while the other scenarios can be used as examples in the initial exploration of the range of potential spatiotemporal issues and outcomes. These can provide important insights for preparedness activities that mitigate the problem of forest carbon emissions. The spatially explicit AFCS map and the modeled scenario results will therefore contribute to the sustainable management of forests in the study area and to the formulation of REDD+ projects, as well as representing a methodological reference for wider audiences in tropical regions and beyond.

Methods

L-band SAR data collection and processing

The study area spans more than 9 million ha commonly experiences cloudy and hazy skies throughout the year [20]. This degradation in atmospheric conditions over the study area precludes use of optical remote sensing techniques in assessing forest quality and AFCS. As a result, PALSAR mosaic data used for tropical forest monitoring as they are unaffected by atmospheric condition and are available for wall-to-wall mapping [1, 20]. The mosaic data are slope-corrected and orthorectified using the widely available SRTM 90 m digital elevation model without any alteration the image quality [21]. Currently, 25 m global mosaic data in two polarizations (HH and HV) are available as one set per year, from 2007 to 2010. The mosaic data are available at a downloadable size of 1-degree tiles, equivalent to approximately 111×111 km [34]. Those mosaic products covering the whole province of Riau for the years 2009 and 2010 were used. The mosaic data were converted into radar backscatter coefficients using gamma-naught ($\gamma°$) [21] due to high sensitivity to forest structure and its usefulness in forest cover analysis [20, 35].

To improve the confidence of the AFCS mapping, we also examined whether a particular filter or its size would affect the mapping results. Three filters were examined: the Lee, Frost, and median filters. These filters possess different formulations and assumptions for smooth speckled data in radar imagery. The Lee filter is a standard deviation-based filter, and filters data on the basis of statistics calculated within individual filtering windows. It conserves image sharpness and details while reducing speckle noise. The value of the pixel being filtered is replaced by a value computed using the neighboring pixels. In comparison, the Frost filter is an exponentially damped circularly symmetric filter, which utilizes local statistics. The value of the pixel being filtered is replaced by a value computed with a consideration of the damping factor, the local variance, and the distance from the

filter center. This filter is able to preserve the edges in the images. The median filter reduces the speckle noise in an image by conserving edges greater than the kernel dimensions. It replaces the value of each center pixel with the median value of the neighborhood specified by the filter size. In order to investigate the impact of speckle filtering sizes on the mapping, we also evaluated five different filtering window sizes ($3 \times 3, 5 \times 5, 7 \times 7, 9 \times 9$, and 11×11).

Field data collection and processing

The combination of LiDAR data and plot-based field measurements has emerged as a promising technique for accurately estimating AFCS [13, 17]. We conducted field measurements and airborne LiDAR surveys within the province during 2012 and 2013. Owing to the differences in forest structure and associated biomass in different land use and land cover (LULC) types, we adopted a stratified sampling approach based on the major forest types to determine the locations (Fig. 1) for field and LiDAR measurements. The major forest types in the study site were defined as natural forests, including peat swamps, dry moist, mangrove, and regrowth, and plantations including acacia, oil palm, rubber, and coconut. Based on these forest types, eight strata were created.

Across the field measurement campaigns, we made 87 biomass measurements within 1 ha-size plots that coincided with the LiDAR acquisition sites. Forest stands of all ages were inventoried, from mature to recent regrowth. Owing to the time and cost involved in conducting a census-based measurement of all trees in a 1-ha-sized plot, a sub-sampling approach was adopted using representative subplots. The sub-sampling methods differed between the natural and plantation forests.

To determine woody biomass, all living and standing deadwood trees with a diameter at breast height (at 1.3 m; DBH) ≥ 5 cm were measured in each subplot. We used allometric equations previously developed for the specific forest types: peat swamp forest [36], dry moist forest [37], mangrove [38], acacia [39], rubber [40], coconut and non-trees [41], oil palm [42], standing deadwood [43], lying deadwood [44], and bamboo [45]. The biomass of understory vegetation and litter was calculated by multiplying the mass of a fresh sample measured in the field by the ratio of sub-sample dry mass to sub-sample fresh mass. The plots within regrowth forests were all located in either peat swamps or dry moist forested areas. Therefore, the biomass for this class was calculated using the corresponding allometric equation, based on its location. Detailed descriptions of the field measurement method, plot specification, and allometric equations used to convert the field measured data to plot level aboveground biomass (AGB) are presented in Thapa et al. [13, 24]. In this study, the AFCS for each plot was considered to be 47 % of the field measured AGB [46].

Airborne LiDAR surveys were conducted at eight sites. LiDAR data were acquired on fine weather days in February 2012 and during November–December 2012. In February, we used an LM-5600 laser system operated at approximately 1000 m above the ground. This system captured first and last returns at a scan angle of ±20°. The average point density was 1.2 m^{-2}. During this measurement period, three of the eight sites were surveyed, covering 3600 ha of forested land. The remaining LiDAR data were collected using an Optech ALTM 3100EA laser system in the second survey period. This system was operated approximately 600 m above the ground, and captured full waveform data at a scan angle of ±32°. Discrete return data were recorded at average point density of 3.6 m^{-2}. At this time, 4472 ha of forested lands were surveyed, contributing to a total 8072 ha of LiDAR data acquired during the two periods combined. Further details of the LiDAR systems, data acquisition and processing, and the LiDAR-to-AFCS model calibration and validation are discussed in Thapa et al. [13]. In the present study, the LiDAR allometric model (Eq. 1) was used to create additional AFCS plots as the AFCS model was calibrated and validated over the same area [13].

$$
\begin{aligned}
\text{AFCS Mg C ha}^{-1} = {} & 259.488 - (146.373 \times \text{MCH}) \\
& + \left(4.738 \times \text{MCH}^2\right) \\
& - (4.881 \times \text{Cover}) \\
& + (3.513 \times \text{MCH\_cover}) \\
& - \left(0.0954 \times \text{MCH}^2\_\text{cover}\right) \\
& - (1.583 \times \text{QMCH\_cover}) \\
& + (22.568 \times P50) + (26.118 \times P90)
\end{aligned}
$$

$$(1)$$

where, MCH = mean canopy height, Cover = forest cover as a percentage of all returns above the MCH, MCH_cover = MCH × Cover, MCH^2_cover = MCH^2 × Cover, QMCH_cover = quadratic MCH × Cover, and P50 and P90 are the 50th and 90th percentiles of canopy height, respectively.

A total of 2716 1-ha LiDAR-based AFCS plots were created, avoiding an excess of path boundaries, field measurement plots, agricultural fields, clear cut areas, water areas, and built structures including buildings and roads. These LiDAR AFCS plots were combined with the 87 field measurement plots, resulting in a total of 2803 plots available for calibration and validation of the SAR-based AFCS baseline map.

Mapping and validating the AFCS

A supervised rule-based approach was adopted for the AFCS mapping, as the direct relationship between the PALSAR backscattered coefficients and field-measured AFCS in the study region is affected by saturation in higher biomass areas [24]. This approach analyzes the spatial patterns in the PALSAR mosaic data through the use of training samples and the rules formulated in a maximum likelihood algorithm (MLA) [47], as in Eq. 2, to determine the AFCS value for each PALSAR pixel. Half of the total AFCS plots were used to train the algorithm. The MLA quantitatively examines both the variance and the covariance of the patterns of corresponding backscatter in the training samples and provides the most probable AFCS values for the selected pixels. At first, the MLA was applied to the 2010 gamma naught image, which includes HH and HV polarizations. The different impacts of the three filters (Lee, Frost, and median) and five increasing window sizes on the mapping results were analyzed. Then, we investigated whether the addition of the gamma naught image of the previous year, in the form of mosaic data from 2009, would improve the mapping result. Layer stacking was performed while adding the 2009 image layers (HH and HV) to the 2010 image data set.

$$
g_i(x) = \ln p(\omega_i) - \frac{1}{2} \ln \left| \sum_i \right| - \frac{1}{2}(x - m_i)^t \sum_i^{-1} (x - m_i)
$$

$$(2)$$

where g_i is the AFCS Mg C ha^{-1} corresponding to the training sample, in which $i = 1...C$ (the number of AFCS training plots available); x is the position, in n-dimensional data where n is the number of image data layers (for example, 2 per year); $p(\omega_i)$ is the probability that the AFCS value ω_i occurs in the image; $|\Sigma_i|$ is the covariance matrix of the image data intersected with the spatial size of AFCS plot ω_i; t is the transposition of the matrix; Σ_i^{-1} is the inverse matrix; and m_i is the mean backscatter vector corresponding to AFCS plot ω_i.

A validation map was prepared using the remaining half of the sample plots. The AFCS map was compared with the validation map through the calculation of three statistical mapping uncertainty measures: root mean square error (RMSE, Eq. 3), bias (Eq. 4), and index of agreement (D, Eq. 5).

$$
\text{RMSE (Mg C ha}^{-1}) = \sqrt{\frac{\sum_{i=1}^{n}(P_i - O_i)^2}{n}}
$$

$$(3)$$

$$
\text{Bias (Mg C ha}^{-1}) = \frac{\sum_{i=1}^{n}(P_i - O_i)}{n}
$$

$$(4)$$

$$D = 1 - \frac{\sum_{i=1}^{n}(P_i - O_i)^2}{\sum_{i=1}^{n}(|P_i - \bar{O}| + |O_i - \bar{O}|)^2} \qquad (5)$$

where P_i represents predicted values, O_i represents observed values, $\bar{O}$ is the observed mean, and n is the number of observations. D is the index of agreement proposed by Willmott and Wicks [48]. D ranges from 0 to 1, which corresponds to disagreement or perfect agreement between the predicted and observed values, respectively.

Bias-adjusted RMSE (Eq. 6) was calculated for the final map to ensure the high accuracy of the AFCS estimation in the map.

$$\text{Adj. RMSE (Mg C ha}^{-1}) = \sqrt{\frac{\sum_{i=1}^{n}(P_i - O_i - Bias)^2}{n}}$$
$$(6)$$

Mapping of future expected AFCS footprints

In this study, we used the forest cover map derived from PALSAR mosaic data from 2010, in combination with the scenario maps for 2015, 2020, 2025, and 2030 from Thapa et al. [16] to generate the expected AFCS footprints for the various years in the future. Three policy scenarios were analyzed: BAU, corresponding to the 'business as usual policy', G-FC indicating the 'government-forest conservation policy', and G-CPL, representing the 'government-concession for plantations and logging policy'. The BAU policy scenario assumes that the deforestation process will continue with the same past trend everywhere in the province, and therefore, AFCS removal will occur in the corresponding deforested areas. The G-FC policy scenario assumes that the deforestation process does not follow the past trend and, in the future, will likely occur outside the designated forest conservation areas. In this case, the forest carbon stocks remain untouched within the conservation areas. For the G-CPL policy scenario, we assume that the concession areas are allotted for selective logging and industrial plantations, and imminent deforestation likely occurs only in the concession areas, therefore the AFCS will be untouched outside these areas. Scenario-wide AFCS maps were created at five-year intervals from 2015 to 2030. Using these maps, the AFCS was quantified at the district level to identify local variation in the carbon stocks. Furthermore, the expected CO_2 emissions for each scenario were computed at province level for each time interval using the approach of IPCC [46].

Authors' contributions

RBT, TM, MW, and MS conceived the research and analyzed the PALSAR data. RBT, TM and MW collected, processed, and analyzed the field data. RBT created the forest carbon map, performed spatial modeling, and wrote the manuscript. MS arranged necessary funds and coordinated the local collaborator for the research. All authors read and approved the final manuscript.

Acknowledgements

The authors would like to thank Prof. I. N. S. Jaya, E. S. Purnama and other colleagues and their students from IPB, Indonesia, Mr. K. Shono and his colleagues from Hatfield Consultants, Indonesia, Mr. Tomohiro Shiraishi from JAXA, and Mr. Takuya Itoh from RESTEC for their assistance with field data collection and for sharing their thoughts in this area of research.

Compliance with ethical guidelines

Competing interests

The authors declare that they have no competing interests.

References

1. Shimada M, Itoh T, Motohka T, Watanabe M, Shiraishi T, Thapa R, Lucas R. New global forest/non-forest maps from ALOS PALSAR data (2007–2010). Remote Sens Environ. 2014;155:13–31.
2. FAO. Forest resource assessment 2010. Rome: Food and Agricultural Organization of the United Nations; 2012.
3. Aguilar-Amuchastegui N, Riveros JC, Forrest JL. Identifying areas of deforestation risk for REDD+ using a species modeling tool. Carbon Bal Manag. 2014;9:10.
4. Canadell JG, Schulze ED. Global potential of biospheric carbon management for climate mitigation. Nat Commun. 2014;5:5282.
5. Grace J, Mitchard E, Gloor E. Perturbations in the carbon budget of the tropics. Glob Change Biol. 2014;20:3238–55.
6. Harris NL, Brown S, Hagen SC, Saatchi SS, Petrova S, Salas W, Hansen MC, Potapov PV, Lotsch A. Baseline map of carbon emissions from deforestation in tropical regions. Science. 2012;336:1573–6.
7. Strassburg B, Turner KR, Fisher B, Schaeffer R, Lovtt A. Reducing emissions from deforestation—The combined incentives mechanism and empirical simulations. Glob Environ Chang. 2009;19:265–78.
8. http://www.un-redd.org. Accessed 29 Sept 2014.
9. Carlson KM, Curran LM, Ratnasari D, Pittman AM, et al. Committed carbon emissions, deforestation, and community land conversion from oil palm plantation expansion in West Kalimantan, Indonesia. Proc Natl Acad Sci USA. 2012;109:7559–64.
10. Cutler M, Boyd D, Foody G, Vetrivel A. Estimating tropical forest biomass with a combination of SAR image texture and Landsat TM data: an assessment of predictions between regions. ISPRS J Photogram. 2012;70:66–77.
11. Leach M, Scoones I. Carbon forestry in West Africa: the politics of models, measures and verification processes. Glob Environ Chang. 2013;23:957–67.
12. Morel AC, Fisher JB, Malhi Y. Evaluating the potential to monitor aboveground biomass in forest and oil palm in Sabah, Malaysia, for 2000–2008 with Landsat ETM+ and ALOS-PALSAR. Int J Remote Sens. 2012;33:3614–39.
13. Thapa RB, Watanabe M, Motohka T, Shiraishi T, Shimada M. Calibration of aboveground forest carbon stock models for major tropical forests in central Sumatra using airborne LiDAR and field measurement data. IEEE J Sel Top Appl. 2015;8:661–73.
14. Asner GP, Knapp DE, Martin RE, Tupayachi R, Anderson CB, Mascaro J, Sinca F, Chadwick KD, Higgins M, Farfan W, Llactayo W, Silman MR. Targeted carbon conservation at national scales with high-resolution monitoring. Proc Natl Acad Sci USA. 2014;111:E5016–22.
15. Lin L, Sills E, Cheshire H. Targeting areas for reducing emissions from deforestation and forest degradation (REDD+) projects in Tanzania. Glob Environ Chang. 2014;24:277–86.
16. Thapa RB, Shimada M, Watanabe M, Motohka T, Shiraishi T. The tropical forest in South East Asia: monitoring and scenario modeling using Synthetic Aperture Radar data. Appl Geogr. 2013;41:168–78.
17. Asner GP, Powell GVN, Mascaro J, Knapp DE, Clark JK, Jacobson J, Kennedy-Bowdoin T, Balaji A, Paez-Acosta G, Victoria E, Secada L, Valqui M, Hughes RF. High-resolution forest carbon stocks and emissions in the Amazon. Proc Natl Acad Sci USA. 2010;107:16738–42.

18. Tsui OW, Coops NC, Wulder MA, Marshall PL. Integrating airborne LiDAR and space-borne radar via multivariate kriging to estimate above-ground biomass. Remote Sens Environ. 2013;139:340–52.

19. Mitchard ETA, Saatchi SS, Baccini A, Asner GP, Goetz SJ, Harris NL, Brown S. Uncertainty in the spatial distribution of tropical forest biomass: a comparison of pan-tropical maps. Carbon Bal Manag. 2013;8:10.

20. Thapa RB, Itoh T, Shimada M, Watanabe M, Motohka T, Shiraishi T. Evaluation of ALOS PALSAR sensitivity for characterizing natural forest cover in wider tropical areas. Remote Sens Environ. 2014;155:32–41.

21. Shimada M, Ohtaki T. Generating large-scale high-quality SAR mosaic datasets: application to PALSAR data for global monitoring. IEEE J Sel Top Appl. 2010;3:637–56.

22. Motohka T, Shimada M, Uryu Y, Setiabudi B. Using time series PALSAR gamma nought mosaics for automatic detection of tropical deforestation: a test study in Riau, Indonesia. Remote Sens Environ. 2014;155:79–88.

23. Saatchi S, Marlier M, Chazdon RL, Clark DB, Russell AE. Impact of spatial variability of tropical forest structure on radar estimation of aboveground biomass. Remote Sens Environ. 2011;115:2836–49.

24. Thapa RB, Watanabe M, Motohka T, Shimada M. Potential of high-resolution ALOS-PALSAR mosaic texture for aboveground forest carbon tracking in tropical region. Remote Sens Environ. 2015;160:122–33.

25. Dobson MC, Ulaby FT, Le Toan T, Beaudoin A, Kasischke ES, Christensen NC. Dependence of radar backscatter on conifer forest biomass. IEEE Trans Geosci Remote. 1992;30:412–5.

26. Lucas R, Armston J, Fairfax R, Fensham R, Accad A, Carreiras J, Kelley J, Bunting P, Clewley D, Bray S, Metcalfe D, Dwyer J, Bowen M, Eyre T, Laidlaw M, Shimada M. An evaluation of the ALOS PALSAR L-band backscatter—above ground biomass relationship Queensland, Australia: impact of surface moisture condition and vegetation structure. IEEE J Sel Top Appl. 2010;3:576–93.

27. Englhart S, Keuck V, Siegert F. Aboveground biomass retrieval in tropical forests—the potential of combined X- and L-band SAR data use. Remote Sens Environ. 2011;115:1260–71.

28. Luckman AJ, Baker JR, Honzák MH, Lucas RM. Tropical forest biomass density estimation using JERS-1 SAR: seasonal variation, confidence limits and application to image mosaics. Remote Sens Environ. 1998;62:126–39.

29. Watanabe M, Motohka T, Shiraishi T, Thapa RB, Kawano N, Shimada M. Dependency of forest biomass on full polarimetric parameters obtained from L-band SAR data for a natural forest in Indonesia. In: Proc Intl Geosci Remote Sens Symp 2013, 3919–3922.

30. MofF. IFCA consolidation report: reducing emissions from deforestation and forest degradation in Indonesia. Forestry Research and Development Agency, Jakarta, Indonesia; 2008.

31. BPS. Hasil Sensus Pnduduk 2010: data Agregat per Provinsi (in Indonesian language); 2010. Web: http://dds.bps.go.id/eng/download_file/SP2010_agregat_data_perProvinsi.pdf. Accessed 2012.3.7.

32. Cairns MA, Brown S, Helmer EH, Baumgardner GA. Root biomass allocation in the world's upland forests. Oecologia. 1997;111:1–11.

33. Pearson T, Walker S, Brown S. Sourcebook for land use, land-use change and forestry projects. Little Rock: Biocarbon Fund, Winrock International; 2005.

34. http://www.eorc.jaxa.jp/ALOS/en/dataset/dataset_index.htm. Accessed in 2015.6.30.

35. Shiraishi T, Motohka T, Thapa RB, Watanabe M, Shimada M. Comparative assessment of supervised classifiers for land use land cover classification in a tropical region using time-series PALSAR mosaic data. IEEE J Sel Top Appl. 2014;7:1186–99.

36. Murdiyarso D, Rosalina U, Hairiah K, Muslihat L, Suryadiputra INN, Jaya A. Petunjuk Lapangan: Pendugaan Cadangan Karbon pada Lahan Gambut. Proyek Climate Change, Forests and Peatlands in Indonesia. Wetlands International, Indonesia Programme and Wildlife Habitat Canada. Bogor, Indonesia; 2004.

37. Brown S. Estimating biomass and biomass change of tropical forests: a primer. UN FAO Forestry Paper, no. 134; 1997. p 55.

38. Komiyama A, Poungparn S, Kato S. Common allometric equations for estimating the tree weight of mangroves. J Trop Ecol. 2005;21:471–7.

39. Adiriono T. Measurement of carbon stock with carbonization method in forest plantation of acacia crassicarpa: a case study in PT Sebangun Bumi Andalas Wood Based Industries. MSc Thesis, Faculty of Forestry, Gadjah Mada University; 2009.

40. Schroth G, Angelo SAD, Teixeira WG, Haag D, Lieberei R. Conversion of secondary forest to agroforestry and monoculture plantations in Amazonia: consequences for biomass, litter and soil carbon stock after 7 years. For Ecol Manag. 2002;163:131–50.

41. Frangi JL, Lugo AE. Ecosystem dynamics of a subtropical floodplain forest. Ecol Monogr. 1985;55:351–69.

42. Yulianti N. Carbon stock of peatland in oil palm agroecosystem of PTPN IV Ajamu, Labuhan Batu, North Sumatra. MSc Thesis, Faculty of Agriculture, Bogor Agricultural University (IPB), Bogor; 2009.

43. Walker SM, Pearson TRH, Casarim FM, Harris N, Petrova S, Grais A, Swails E, Netzer M, Goslee KM, Brown S. Standard operating procedures for terrestrial carbon measurement. Little Rock: Winrock International; 2012.

44. Pearson T, Walker S, Brown S. Sourcebook for land use, land-use change and forestry (LULUCF) projects. Little Rock: BioCarbon Fund and Winrock International; 2005.

45. van Noordwijk M, Rahayu S, Hairiah K, Wulan YC, Farida A, Verbist B. Carbon stock assessment for a forest-to-coffee conversion landscape in Sumber-Jaya (Lampung, Indonesia): from allometric equations to land use change analysis. Sci China (Ser C). 2002;10:75–86.

46. IPCC. Agriculture, forestry and other land use. Guidelines for National Greenhouse Gas Inventories, 4; 2006. http://www.ipcc-nggip.iges.or.jp/public/2006gl/vol4.html. Accessed 2 Feb 2012.

47. Richards JA, Jia X. Remote sensing digital image analysis. Berlin: Springer; 2006.

48. Willmott CJ, Wicks DE. An empirical method for the spatial interpolation of monthly precipitation within California. Phys Geogr. 1980;1:59–73.

9

Allometric equations for estimating belowground biomass of *Androstachys johnsonii* Prain

Tarquinio Mateus Magalhães[*]

Abstract

Background: The belowground component of the trees is still poorly known because it needs labour- and time-intensive in situ measurements. However, belowground biomass (BGB) constitutes a significant share of the total forest biomass. I analysed the BGB allocation patterns, fitted models for estimating root components and root system biomasses, and called attention for its possible use in predicting anchoring functions of the different root components.

Results: More than half and almost one third of BGB is allocated to the lateral roots and to the root collar, respectively. More than 80% of the BGB is found at a depth range of 9.6–61.2 cm. As the tree size increased, the proportion of BGB allocated to taproots decreased and that allocated to lateral roots increased. All independent models performed almost equally, with the predictors explaining, on average, 98% of the variation in the BGB.

Conclusions: It was hypothesised that BGB allocation patterns are a response of the anchoring functions of the tap and lateral roots and therefore, root component biomass models can be used as a methodology to predict anchoring functions of the different root components. Based on the fact that all models performed almost equally, the models using either diameter at breast height (DBH) exclusively as a predictor should be preferred, as tree height is difficult to measure. Models using the root collar diameter (RCD) only should be preferred when the tree is found cut down, as sometimes the RCD is affected by root buttress. Given the large sample size, the validation results, and the coverage of a wide geographical, soil and climatic range, the models fitted can be applied in all *A. johnsonii* stands in Mozambique.

Keywords: Mecrusse, Anchorage, Additivity, Belowground biomass allocation patterns, Root components

Background

Androstachys johnsonii Prain (*A. johnsonii*) stands, known as mecrusse, are very important woodlands. Almost entirely restricted to Mozambique [1], it has an important socioeconomic value to local communities, that sell and use stakes and poles of *A. johnsonii* in the construction of homes, shelters, and furniture; and it is the main source of income in the Funhalouro and Mabote districts [2, 3]. On the global scale, mecrusse forests form part of the woodland belt that stretches over large portions of southern Africa and are reported to be a tipping point in regional ecological and socioeconomic development [4], hence, their importance in the mitigation of greenhouse gas emissions.

Forest biomass is a key variable employed when making estimates of carbon pools in forests, and for studying other biochemical cycles [5]. In the past, only the aboveground portion of trees was the desired products from forests [6]. However, with the increased significance of biomass estimation since the Kyoto Protocol was adopted in 1997 [7], and thus, the enhanced awareness of the sequestration functions of trees, climate change issues, have made belowground biomass (BGB) more relevant.

Despite the recent advances in examining root distribution and biomass with ground-penetrating radar

---
*Correspondence: tarqmag@yahoo.com.br
Departamento de Engenharia Florestal, Universidade Eduardo Mondlane, Campus Universitário, Edifício no.1, 257, Maputo, Mozambique

[8–12], the belowground component of trees is still poorly known because, traditionally, it requires labour- and time-intensive in situ measurements [13]. Yet, BGB constitutes a major share of total forest biomass. Cairns et al. [14] and Litton et al. [15] have maintained that BGB may represent up to 40% of the total biomass. In Mozambique, Magalhães and Seifert [16, 17] found that approximately 20% of the forest biomass of mecrusse woodlands was allocated to the root system, and so highlighting the need to study this carbon pool.

Besides the share of BGB in whole tree forest biomass, BGB happens to be a unique carbon pool because after exploitation, the root system, along with the stump, are left in the forest and, in some tree species, are then allowed to sprout, continuing the carbon sequestration process or decompose, releasing CO_2 and nutrients. Therefore, BGB can be used to estimate the carbon that will be transferred to the soil and the nutrients that will be reclaimed by the site.

BGB is often estimated indirectly, using root-to-shoot ratios (R/S) [17–21], root system biomass expansion factors (BEFs) [17], and by using regression equations of BGB versus aboveground biomass (AGB) [18, 22, 23] or versus easily measured variables (diameter at breast height (DBH) and tree height (TH)) [23, 24]. However, whatever the method utilised to estimate BGB (R/S, BEFs, equations), it is necessary that the root system is directly measured to develop those methods.

Based on the fact that measuring BGB is difficult and time-consuming, the root system is often partially removed from the soil [25–30], depths of excavation are predefined [29–31], and fine roots are excluded [19, 32, 33]. However, the depths of excavation and the definition of fine roots are not standardised [18, 34], but the depth selected in a given study is assumed to capture a large proportion of the roots [18]. Yet, according to Mokany et al. [20], sampling to what may be deemed an insufficient soil depth to capture the majority of the roots, while not sampling the root collar or fine roots, as well as sampling with inadequate replication, are a few of the methodological pitfalls associated with sampling root biomass, and can lead to underestimation.

In other cases, a root sampling procedure is applied where only a fraction of roots from each root system are fully excavated, and then the information from the excavated roots is employed to estimate biomass for the roots not excavated [6, 23, 35]. The disadvantage of relying on sampling procedures is that the observed biomass value for each individual root system is less accurately determined compared to excavating in full [6].

Very few allometric biomass models exist for Mozambican forests; exceptions include Magalhães and Seifert [16, 17], Ryan et al. [33], Mate et al. [36], and Sitoe

et al. [37]. As is best present known, the only studies that have included BGB are those by Magalhães and Seifert [16, 17], Ryan et al. [33], and Magalhães and Seifert [38]. However, the study by Ryan et al. [33] was based on only several sample trees (23) within a limited geographical range (27 ha) and the root system was not completely excavated (fine roots were not included). Although Magalhães and Seifert [16, 17, 38] considered relatively large sample trees and an expanded geographical area (93 trees harvested in 5 districts), besides including the entire root system, their allometric models were limited by not considering the different root components (e.g. taproot, root crown, lateral roots) and, therefore, the BGB allocation patterns were not analysed. Root component biomass models and BGB allocation patterns analyses are scarce worldwide, Litton et al. [15] being the only reference available in the literature.

Studying BGB allocation patterns is very important for understanding root anchorage as both root anchorage and BGB allocation patterns depend on root architecture, branching patterns, and size and depth of the roots [39–42]. In turn, those factors affecting tree anchorage and BGB allocation patterns depend on tree species and soil types [43] and resources [44–48]. The anchoring capacity of a tree is a critical factor for survival regarding external abiotic stresses [49].

It has been suggested by Herrel et al. [49] that for a fixed amount of biomass, a network of several small roots is more resistant to tension than a few large structural roots. This means that a root system with a larger taproot and several smaller lateral roots, as in the case of A. johnsonii trees [50], is more resistant to tension and therefore may have a better anchoring capacity than otherwise.

Parametric studies have shown that the number of lateral ramifications and their diameter were both major components affecting the resistance to pull-out for a given soil pressure [49]. On the other hand, the biomass allocated to a certain root component is also a function of the ramifications and/or their diameter, suggesting that studying biomass assigned to the different root components can help to identify which component affects the resistance to pull-out and anchorage more. Ennos and Fitter [51] showed that various anchorage strategies (plate-like and tap-like morphology) have an impact on biomass allocation patterns, therefore stressing the necessity of studying BGB allocation patterns to understand anchorage strategies.

Bila et al. [52] have demonstrated that silvicultural treatments positively influenced the health and growth of A. johnsonii and suggested that this species can be grown for commercial and urban forestry purposes. Knowledge of the extent and distribution of tree root systems is essential for managing trees in the constructed environment [53]. Conversely, the performance of urban trees depends upon the ability of their root systems to acquire resources and provide anchorage [53]. This emphasises the requirement to study BGB allocation patterns, though the anchorage of open grown trees and of those grown in the woods may be different.

Hence, as direct estimation of BGB is labour- and time-intensive, developing root component biomass models based on easily measurable variables is crucial for studying BGB allocation patterns and therefore anchoring functions of the different root components.

The present study was aimed at analysing the BGB allocation patterns, fitting and validating root component and root system models for *A. johnsonii*. A general model of the root system was also fitted using the best root component models using weighted non-linear seemingly unrelated regression (WNSUR) and critically compared against independent root system models. The excavation depth range that can capture more than 75% of the BGB was also estimated.

Results

Description of the data

Diameter distributions of the phase-1 and phase-2 trees are given in the Figure 1 and Table 1, respectively, and show that the phase-2 sample trees (outside the brackets in the Table 1) are representative of those from phase-1. On the other hand, it is also noted that the testing sample trees (inside the brackets in the Table 1) are also representative of those from phase-2.

The testing sample collected outside the study area (in Chicualacuala district) was distributed according to diameter classes in the Table 1, as follows: 4, 3, 3, 3, 2 and 1, respectively. Note that, although the study area comprised 5 districts (Chibuto, Madlakaze, Panda, Funhalouro and Mabote), during the randomization (Figure 2), none of the plots fell in Chibuto and Panda districts, and

Table 1 Diameter distribution of the training sample trees (outside the brackets) and of the testing sample trees felled inside the study area (inside the brackets)

Diameter class (cm)	Manjacaze/ Chibuto	Mabote	Funhalouro	Total
[05–10[	3 (1)	6 (2)	9 (2)	18 (5)
[10–15[	3 (1)	7 (2)	8 (2)	18 (5)
[15–20[	4 (1)	6 (0)	8 (2)	18 (3)
[20–25[	4 (1)	5 (1)	8 (1)	17 (3)
[25–30[	3 (0)	6 (2)	8 (2)	17 (4)
[30–35]+	3 (1)	1 (0)	1 (0)	5 (1)
Total	20 (5)	31(7)	42 (9)	93 (21)

those districts have almost a negligible share of mecrusse woodlands of the study area.

BGB allocation patterns

On average, the percentage of the root system biomass attributed to taproot, root collar, and lateral roots biomasses was 48.36, 30.79 and 51.64%, respectively; and the percentage of the taproot attributed to root collar was 64.89% (Table 2). The percentage of the root system biomass found at 20% of the taproot depth, which is equivalent to 9.6–61.2 cm in depth from the ground level, was 81.20%.

Table 3 shows that BGB allocation patterns vary with tree size (DBH, RCD, and TH), except the proportion of root system biomass allocated to the root collar (RC/RS), which is found to be independent on tree size by either Pearson´s correlation test or dcov test of independence.

Modelling

The laterals roots and root system biomass models showed that more than 99% of the BGB variation was explained by the predictor variables (Tables 4, 5). The root collar and taproot biomass models showed that more than 96 and 98% of the BGB variation, respectively, were explained by the predictor variables. The CVr varied from, approximately, 23 to 46%; the smallest and highest CVr values were verified for the root system and the root collar biomass models, respectively.

All the root component models presented statistically insignificant bias (MR) as tested by Student's *t*-test and their residuals showed homoscedasticity (*p*-value >> 0.05) and normal distribution (*p*-value > 0.05) (except for the model form 3 of the lateral roots) (Table 5). The plots of the residuals (not shown) presented no particular trend; the cluster of points was contained in a horizontal band, with the residuals evenly distributed under and over the axis of abscissas, meaning that there were not model defects. All the models performed almost equally.

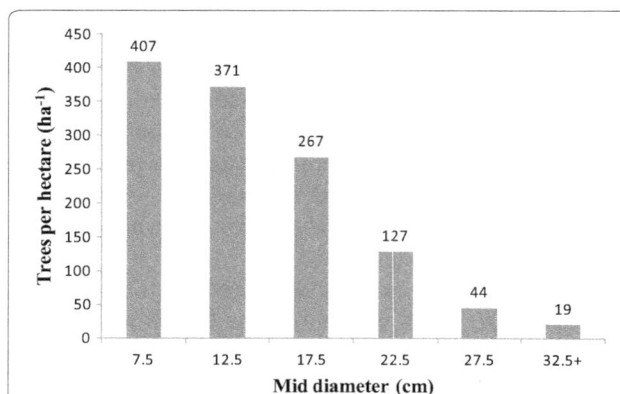

Figure 1 Diameter distribution histogram of phase-1 sampled *A. johnsonii* trees.

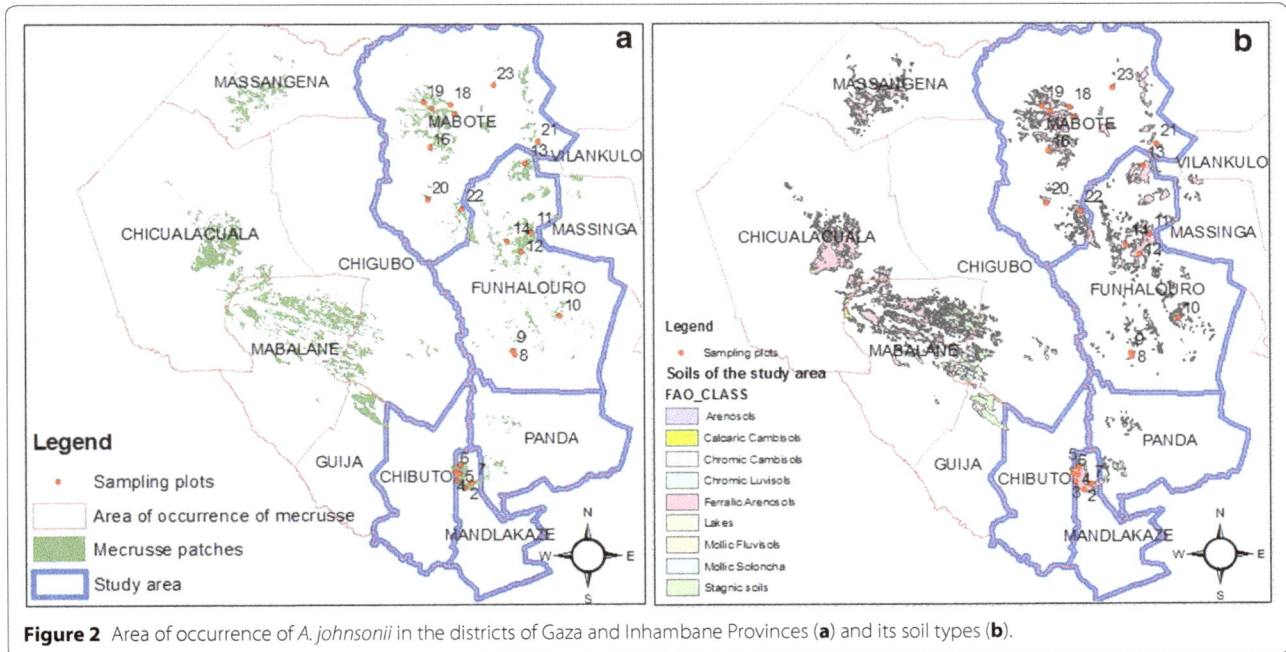

Figure 2 Area of occurrence of *A. johnsonii* in the districts of Gaza and Inhambane Provinces (**a**) and its soil types (**b**).

Table 2 Belowground biomass allocation patterns

Statistic	Taproot (TR)		Root collar (RC)			Lateral roots (LR)		Root system (RS) (Kg)	% of the RS found at 20% of the TR depth
	Kg	% RS$_{TR}$	Kg	% TR$_{RC}$	% RS$_{RC}$	Kg	% RS$_{LR}$		
Minimum	0.00	0.00	0.00	2.86	0.00	0.00	0.00	1.93	40.43
Average	18.64	48.36	13.32	64.92	30.79	24.05	51.64	42.69	81.20
Maximum	63.60	100.00	54.90	89.61	55.86	100.82	100.00	149.38	100.00
SD	15.78	15.34	12.29	16.64	11.22	23.98	15.34	37.97	11.16
CV	84.65	31.72	92.26	25.64	36.44	99.69	29.71	88.94	13.53

The last column represents the percentage of the root system biomass found at 20% of the taproot depth, which is equivalent to 9.6 to 61.2 cm in depth from the ground level.

SD standard deviation, CV coefficient of variation (%), % RS$_{TR}$ percentage of the root system biomass attributed to the taproot biomass, %TR$_{RC}$ percentage of the taproot biomass attributed to the root collar biomass, %RS$_{RC}$ percentage of the root system biomass attributed to root collar biomass, %RS$_{LR}$ percentage of the root system biomass attributed to lateral roots biomass.

Forcing additivity of the taproot and lateral roots biomasses into root system biomass

Because the different model forms performed almost equally (Table 5), the WNSUR with parameter restriction was applied to models using DBH and TH as predictors; and to those using either DBH or RCD only. This ensured that the additivity could be achieved either using two variables (DBH and TH) or using only one variable (DBH or RCD). The latter case was included because TH is difficult and time-consuming to measure in natural forests.

The relatively better taproot and lateral roots model forms were the model forms (5) and (6) (refer to "Methods"), respectively, as judged by Adj.R^2 and CVr, as other statistics were equally insignificant. Therefore, the root system model form is a function (sum) of the predictors of the models (5) and (6). The structural system of equations (including the root system biomass model) obtained by combining the best taproot and lateral roots model forms under parameter restriction is given in Eq. (1). Using the same principle for the model forms with either DBH or RCD only as predictors, the structural systems of equations for WNSUR are given in Eqs. (2) and (3).

$$\hat{Y}_{Taproot} = b_{10}DBH^{b_{11}}TH^{b_{12}}$$

$$\hat{Y}_{Lateral-roots} = b_{20}\left(DBH^2TH\right)^{b_{21}} \quad (1)$$

$$\hat{Y}_{Root-system} = b_{10}DBH^{b_{11}}TH^{b_{12}} + b_{20}\left(DBH^2TH\right)^{b_{21}}$$

Table 3 Dependence of BGB allocation patterns on tree size

No.	Pair of variables	Pearson's correlation test		Distance covariance test of independence		
		r	p value	dcov	dcor	p value
1	LR/RS vs. DBH	0.4120	4.08E−05	0.3388	0.4856	0.0150
2	TR/RS vs. DBH	−4.3129	4.08E−05	0.3388	0.4856	0.0150
3	RC/RS vs. DBH	0.0662	0.5285^{ns}	0.1118	0.1841	0.2200^{ns}
4	RC/TR vs. DBH	0.5816	9.7E−10	0.4451	0.5851	0.0150
5	LR/RS vs. RCD	0.4078	5.0E−05	0.3357	0.4722	0.0150
6	TR/RS vs. RCD	−0.4078	5.0E−05	0.3357	0.4722	0.0150
7	RC/RS vs. RCD	0.0942	0.3689^{ns}	0.1168	0.1888	0.1650^{ns}
8	RC/TR vs. RCD	0.6051	1.3E−10	0.4653	0.6002	0.0150
9	LR/RS vs. TH	0.2662	0.0099	0.1423	0.4040	0.0150
10	TR/RS vs. TH	−0.2662	0.0099	0.1423	0.4040	0.0150
11	RC/RS vs. TH	0.1451	0.1654^{ns}	0.0775	0.2528	0.0540^{ns}
12	RC/TR vs. TH	0.5418	0.0000	0.2093	0.5446	0.0150

r Pearson's correlation coefficient, dcov distance covariance, dcor distance correlation, ns not statistically significant at $\alpha = 0.05$.

Table 4 Coefficients of regression ($\pm$ standard error) for independently fitted models

Model form #	Weight function	b_0 ($\pm$SE)	b_1 ($\pm$SE)	b_2 ($\pm$SE)
Root collar				
1	$1/0.0002 \times DBH^{3.6194}$	0.0129 ($\pm$0.0040)**	2.3350 ($\pm$0.1024)***	
2	$1/0.0001 \times DBH^{3.6750}$	0.0035 ($\pm$0.0030)^{ns}	2.0979 ($\pm$0.1606)***	0.7807 ($\pm$0.4478)^{ns}
3	$1/0.0358 \times \exp(0.2527 \times DBH)$	0.0024 ($\pm$0.0010)*	1.0143 ($\pm$0.0497)***	
4	$1/0.0352 \times \exp(0.2219 \times RCD)$	0.0064 ($\pm$0.0024)**	2.4946 ($\pm$0.1216)***	
Taproot				
1	$1/0.0002 \times DBH^{3.7383}$	0.0427 ($\pm$0.0103)***	2.0594 ($\pm$0.0840)***	
2	$1/0.0002 \times DBH^{3.7521}$	0.0092 ($\pm$0.0066)^{ns}	1.7587 ($\pm$0.1447)***	0.9442 ($\pm$0.3989)*
3	$1/0.0001 \times DBH^{3.8793}$	0.0101 ($\pm$0.0029)***	0.8885 ($\pm$0.0353)***	
4	$1/0.0002 \times RCD^{3.6082}$	0.0269 ($\pm$0.0074)***	2.1454 ($\pm$0.0910)***	
Lateral roots				
1	$1/0.000008 \times DBH^{4.8739}$	0.0099 ($\pm$0.0021)***	2.6041 ($\pm$0.0748)***	
2	$1/0.00007 \times DBH^{4.0862}$	0.0045 ($\pm$0.0028)^{ns}	2.5257 ($\pm$0.1213)***	0.3983 ($\pm$0.3265)^{ns}
3	$1/0.00006 \times DBH^{4.1635}$	0.0012 ($\pm$0.0004)**	1.1585 ($\pm$0.0369)***	
4	$1/0.0189 \times \exp(0.2826 \times RCD)$	0.0038 ($\pm$0.0011)***	2.8352 ($\pm$0.0975)***	
Root system				
1	$1/0.00007 \times DBH^{4.3051}$	0.0405 ($\pm$0.0064)***	2.3402 ($\pm$0.0545)***	
2	$1/0.000008 \times DBH^{1.7645}$	0.0185 ($\pm$0.0064)**	2.1990 ($\pm$0.0793)***	0.4699 ($\pm$0.1898)*
3	$1/0.0003 \times DBH^{3.8639}$	0.0070 ($\pm$0.0015)***	1.0226 ($\pm$0.0251)***	
4	$1/0.00007 \times RCD^{4.2484}$	0.0230 ($\pm$0.0042)***	2.4519 ($\pm$0.0608)***	

SE standard error, ns not statistically significant at $\alpha = 0.05$.

* Significant at $\alpha = 0.05$.

** Significant at $\alpha = 0.01$.

*** Significant at $\alpha = 0.001$.

$$\hat{Y}_{Taproot} = b_{10}DBH^{b_{11}}$$
$$\hat{Y}_{Lateral-roots} = b_{20}DBH^{b_{21}}$$
$$\hat{Y}_{Root-system} = b_{10}DBH^{b_{11}} + b_{20}DBH^{b_{21}}$$

(2)

$$\hat{Y}_{Taproot} = b_{10}RCD^{b_{11}}$$
$$\hat{Y}_{Lateral-roots} = b_{20}RCD^{b_{21}}$$
$$\hat{Y}_{Root-system} = b_{10}RCD^{b_{11}} + b_{20}RCD^{b_{21}}$$

(3)

Table 5 Goodness of fit statistics, and heteroskedasticity and normality tests of residuals for independently fitted models

Model form #	Adj.R^2 (%)	CVr (%)	MR (%)	White,s test for heteroskedasticity		Lilliefors normality test	
				χ^2	p value	D	p value
Root collar							
1	98.22	42.91	-0.6582^{ns}	10.2673	0.5925	0.0917	0.0518
2	97.02	44.85	-0.9926^{ns}	9.4548	0.6637	0.0761	0.2048
3	97.43	44.80	-1.3605^{ns}	8.9433	0.7078	0.0783	0.1728
4	96.98	45.62	-1.4202^{ns}	11.0543	0.5243	0.0800	0.1508
Taproot							
1	98.55	37.71	-0.0437^{ns}	5.2695	0.9484	0.0420	0.9537
2	98.68	37.21	-0.1412^{ns}	3.5465	0.9903	0.0516	0.7859
3	98.16	38.85	-0.2415^{ns}	3.6553	0.9889	0.0413	0.9608
4	98.57	37.50	0.0035^{ns}	8.6969	0.7286	0.0621	0.5066
Lateral roots							
1	99.40	35.8213	0.0589^{ns}	7.9210	0.7913	0.0801	0.1504
2	99.44	33.8683	1.1136^{ns}	6.0923	0.9114	0.0873	0.0771
3	99.45	31.5039	0.1433^{ns}	7.1091	0.8503	0.1289	0.0006
4	99.26	38.5273	1.6894^{ns}	6.3422	0.8979	0.0755	0.2146
Root system							
1	99.73	24.6597	0.0182^{ns}	11.0001	0.5289	0.0562	0.6651
2	99.78	23.4060	0.1110^{ns}	9.9195	0.6230	0.0498	0.8249
3	99.80	23.1551	1.3768^{ns}	12.5144	0.4053	0.0470	0.8808
4	99.79	28.8790	0.1124^{ns}	16.2835	0.1786	0.0480	0.8620

D Lilliefors statistic, ns not statistically significant at $\alpha = 0.05$.

Note that, in Eqs. (1–3), the coefficients of regression of each regressor in each root component model (taproot or lateral roots) are forced (constrained, restricted) to be equal to coefficients of the equivalent regressor in the root system model, allowing additivity.

For the WNSUR in Eqs. (1–3) the range of Adj.R^2 is 83.30–93.46, 82.59–93.32, and 82.92–92.90%, respectively (Tables 6, 7). It was observed that by forcing additivity the Adj.R^2 decreased, and the normality of the residuals was lost (p-value < 0.05). However, the bias (MR) kept insignificant and the residuals showed homoscedasticity (p-value >> 0.05). The models in Eqs. (1–3), fitted under WNSUR, performed almost equally.

The t-test results for the restrictions imposed on WNSUR (Table 8) were insignificant (p-value $\approx$1), indicating that the data were consistent with the restriction and that the models fit as well with the restriction imposed.

Validation

The aggregate difference (AD) for independent models (Table 9) varied from −6.06 to 0.21% in the study area and from 0.98 to 6.0 outside the study area. For the whole testing sample (including both the inside and outside samples) the AD varied from −2.9 to 5.1%. For

simultaneous models the range of AD was from −5.4 to 0.68% and from 3.42 to 5.90%, inside and outside the study area, respectively (Table 10). For the whole testing sample the AD ranged from −2.87 to 5.1 %. The Wilcoxon signed rank test revealed that for both independent and simultaneous models (Tables 9, 10) the observed BGB did not differ statistically from the predicted BGB values (p-value >0.25); hence, the models can be used reliably inside and outside the study area.

Discussion
BGB allocation patterns
Considering the fact that 90% of the lateral roots of *A. johnsonii* trees are located in the first node, which is located close to the ground level [16, 17, 38, 50] it can be inferred that the 81.20% of the root system (found up to 61.2 cm in depth) is composed by root collar and lateral roots and therefore, the remaining portion of the taproot constitutes less than 20% of the root system biomass. This can be verified by summing the average taproot and lateral roots biomasses, which is equal to 82.43%, very close to the percentage of the root system biomass found at 20% of the taproot depth (81.20%). The difference of those percentages (1.13%) represents the lateral roots found at depths above 20% of the

Table 6 Coefficients of regression (± standard error) for simultaneously fitted models

Root component	Weight function	b_{10} (±SE)	b_{11} (±SE)	b_{12} (±SE)	b_{20} (±SE)	b_{21} (±SE)
Using DBH and TH as independent variables						
Taproot	$1/0.0002 \times DBH^{3.7421}$	0.0101 (±0.0072)^{ns}	1.7737 (±0.1439)***	0.8910 (±0.3949)*		
Lateral roots	$1/0.0002 \times DBH^{3.7421}$				0.0011 (±0.0004)*	1.1703 (±0.385)***
Root system	$1/0.0002 \times DBH^{3.7421}$	Taproot model + lateral roots model				
Using only DBH as independent variable						
Taproot	$1/0.0002 \times DBH^{3.6194}$	0.0430 (±0.0106)***	2.0567 (±0.0849)***			
Lateral roots	$1/0.0002 \times DBH^{3.6194}$				0.0090 (±0.0024)***	2.6368 (±0.0252)***
Root system	$1/0.0002 \times DBH^{3.6194}$	Taproot model + lateral roots model				
Using only RCD as independent variable						
Taproot	$1/0.0002 \times RCD^{3.6082}$	0.0269 (±0.0074)***	2.1454 (±0.0910)***			
Lateral roots	$1/0.0002 \times RCD^{3.6082}$				0.0048 (±0.0014)**	2.7624 (±0.0947)***
Root system	$1/0.0002 \times RCD^{3.6082}$	Taproot model + lateral roots model				

SE standard error, ns not statistically significant at $\alpha = 0.05$, * Significant at $\alpha = 0.05$, ** Significant at $\alpha = 0.01$, *** Significant at $\alpha = 0.001$.

Table 7 Goodness of fit statistics, and heteroskedasticity and normality tests of residuals for simultaneously fitted models

Root component	Adj.R² (%)	CVr (%)	MR (%)	White's test for heteroskedasticity		Lilliefors normality test		$\hat{\sigma}_{ii}$	$\hat{\sigma}^2_{WNSUR}$
				χ^2	p-value	D	p-value		
Using DBH and TH as independent variables									
Taproot	83.30	39.07	−0.2625^{ns}	13.7327	0.3181	0.1348	0.0003	3.2284	1.9462
Lateral roots	90.26	31.34	4.9056^{ns}	19.7679	0.0716	0.1299	0.0005	2.8909	
Root system	93.46	22.95	1.4469^{ns}	13.1392	0.3590	0.1450	5E−05	5.8425	
Using only DBH as independent variable									
Taproot	82.59	37.91	−0.0590^{ns}	15.1331	0.2342	0.1185	0.0026	4.9418	1.9570
Lateral roots	90.53	35.78	0.6559^{ns}	18.2943	0.1070	0.1537	1E−05	4.2161	
Root system	93.32	24.69	0.1103^{ns}	11.7599	0.4652	0.1403	1E−04	8.8881	
Using only RCD as independent variable									
Taproot	82.92	41.27	−0.0065^{ns}	18.3541	0.1054	0.1289	0.0006	3.5120	1.9570
Lateral roots	90.34	38.07	0.4631^{ns}	17.2169	0.1416	0.1836	3E−08	3.2507	
Root system	92.80	29.11	0.0843^{ns}	14.7963	0.2528	0.1696	6E−07	7.1575	

ns not statistically significant at $\alpha = 0.05$, D Lilliefors statistic, $\hat{\sigma}_{ii}$ the (i, i) element of the covariance matrix of the residuals $\hat{\Sigma}$ (error covariance matrix), it is the covariance error of the ith system equation, $\hat{\sigma}^2_{WNSUR}$ WNSUR system variance.

taproot depth (above the depth range of 9.6–61.2 cm from ground level).

Depths of excavation of the roots are not standardized [18, 34], but the depth selected in a given study is assumed to capture a large portion of the roots [18]. For example, Green et al. [19], Kuyah et al. [23], Sanquetta et al. [30], and Paul et al. [31], used excavation depths of 50–120, 40–200, 50, and <200 cm, respectively. Schenk and Jackson [54] found that globally, 95% of all roots are within the upper 200 cm of the soil profile. In this study, the excavation range (9.6–61.2 cm) that captured 81.20% of root biomass fall in the excavation range by those authors.

However, BGB estimates by Green et al. [19], Kuyah et al. [23], Sanquetta et al. [30], and Paul et al. [31] might have been underestimated, as according to Mokany et al. [20], sampling to insufficient soil depth to capture the majority of the roots, not sampling the root collar, not sampling fine roots, and sampling with inadequate replication are some methodological pitfalls associated with sampling root biomass, and can lead to underestimation. In this study, the root system was excavated to total depth and removed, including the root components generally ignored as stated by Mokany et al. [20].

The proportion of BGB allocated to the root collar (average = 30.79%, CV = 36.44%) is lower than

Table 8 *t* test for the restrictions imposed for WNSUR

Restriction	Parameter estimate	Standard error	t value	Pr > |t|
Using DBH and TH as independent variables				
Rest1	1.1859	30789.70	0.00	1.0000
Rest2	0.0102	287.40	0.00	1.0000
Rest3	1.7386	2664.90	0.00	0.9995
Rest4	0.0456	77.54	0.00	0.9995
Rest5	0.0451	68.64	0.00	0.9995
Using only DBH as independent variable				
Rest1	5.3731	3710.80	0.00	0.9989
Rest2	0.1517	102.70	0.00	0.9988
Rest3	0.1960	612.20	0.00	0.9987
Rest4	0.0215	76.40	0.00	0.9988
Using only RCD as independent variable				
Rest1	16.9897	6901.20	0.00	0.9981
Rest2	0.2539	104.50	0.00	0.9981
Rest3	0.1317	986.70	0.00	0.9999
Rest4	0.0112	80.31	0.00	0.9999

Rest1 to Rest5 are the restrictions imposed to each of the regression coefficients, as stated in Eqs. (1–3).

that reported by Mokany et al. [20] (average = 41%, CV = 3.10%). Inclusion of different terrestrial biomes by Mokany et al. [20] may justify this discrepancy.

It is noted from Table 3 that as the tree increased in DBH, RCD, and TH, the proportion of BGB allocated to taproots (TR/RS) decreased and that allocated to lateral roots (LR/RS) increased. This is presumably because as the trees grow the larger is the need for its anchorage in the soil; and that the anchorage function in larger *A. johnsonii* trees is much attributed to the lateral roots as they hold a larger amount of soil than the taproots. In fact, there are trees without taproots but hardly a tree can sustain itself in the soil without lateral roots, especially the larger ones. As maintained by Crook and Ennos [55] only relatively small tree species can rely solely on the taproot for anchorage and that this is the reason that most large trees develop thick lateral roots. Moreover, Bailey et al. [56] have verified that lateral roots play an important role in anchorage.

It has also been noted by Dupuy et al. [39] that heart like root systems (those root systems that possess large lateral roots originating from the centre of the bole) generally had the most efficient anchorage. The heart like root system is determined by lateral roots, implying that the lateral roots influence greatly the anchorage efficiency. *A. johnsonii* trees do not have larger lateral roots than the taproot; however, they have many lateral roots (up to 11 lateral roots per node) which make them to contain a larger proportion of biomass than the taproot

Table 9 Validation of independently fitted models

Model form #	Inside the study area (n = 21)			Outside the study area (n = 16)			Total (n = 37)		
	AD (%)	V	p-value	AD (%)	V	p-value	AD (%)	V	p-value
Root collar									
1	−3.8062	144.0	0.3377	0.9789	61.0	0.7436	−1.4127	399.0	0.6880
2	−2.9985	146.0	0.3038	2.6842	61.0	0.7436	−1.6542	418.0	0.4998
3	−2.2222	141.0	0.3926	3.6755	60.0	0.7057	−1.1550	409.0	0.5856
4	−6.0617	120.0	0.8917	2.8364	64.0	0.8603	−2.3628	380.0	0.8974
Taproot									
1	−2.3818	127.0	0.7079	4.1225	63.0	0.8209	−0.4035	392.0	0.7634
2	−1.7091	121.0	0.8649	5.7234	63.0	0.8209	−0.8994	378.0	0.9201
3	−1.6672	121.0	0.8649	5.7082	63.0	0.8209	−0.7959	378.0	0.9201
4	−5.3688	123.0	0.8117	3.5819	64.0	0.8603	−2.8804	397.0	0.7093
Lateral roots									
1	−1.1701	119.0	0.9187	2.7810	50.0	0.3755	4.3074	317.0	0.4465
2	0.1249	117.0	0.9729	4.4842	48.0	0.3225	5.0567	310.0	0.3885
3	0.2054	118.0	0.9457	5.4708	47.0	0.2979	4.2389	304.0	0.3425
4	−3.8743	139.0	0.4319	5.9664	62.0	0.7820	3.7639	375.0	0.9543
Root system									
1	−1.7047	131.0	0.6091	3.4307	52.0	0.4332	2.2047	346.0	0.7308
2	−1.2378	123.0	0.8117	4.4345	52.0	0.4332	2.0331	327.0	0.5371
3	−0.8691	113.0	0.9457	5.3854	48.0	0.3225	1.7691	318.0	0.4551
4	−5.4699	129.0	0.6578	3.6655	61.0	0.7436	−0.3734	366.0	0.9543

AD aggregate difference, *V* Wilcoxon statistic.

Table 10 Validation of simultaneously fitted models

Root component	Inside the study area (n = 21)			Outside the study area (n = 16)			Total (n = 37)		
	AD (%)	V	p-value	AD (%)	V	p-value	AD (%)	V	p-value
Using DBH and TH as independent variables									
Taproot	−1.7243	121.0	0.8649	5.6595	63.0	0.8209	−0.8602	378.0	0.9201
Lateral roots	0.6760	120.0	0.8917	5.8973	46.0	0.2744	4.7203	303.0	0.3352
Root system	−0.3711	117.0	0.9729	5.7933	49.0	0.3484	2.2976	319.0	0.4639
Using only DBH as independent variable									
Taproot	−2.4598	127.0	0.7079	4.0529	63.0	0.4037	−0.4861	393.0	0.7524
Lateral roots	−0.4410	119.0	0.9187	3.4277	51.0	0.8209	5.0618	319.0	0.4639
Root system	−1.3237	134.0	0.5392	3.7025	57.0	0.5966	2.6531	349.0	0.7634
Using only RCD as independent variable									
Taproot	−5.3576	123.0	0.8117	3.5923	64.0	0.8603	−2.8693	397.0	0.7093
Lateral roots	−5.3553	134.0	0.5392	4.3072	63.0	0.8209	2.0879	370.0	1.0000
Root system	−5.3563	131.0	0.6091	3.9955	61.0	0.7436	−0.0734	369.0	0.9886

AD aggregate difference, *V* Wilcoxon statistic.

(Table 2). In fact, Dupuy et al. [40] maintained that the number of lateral roots is determinant for anchorage, as they determine the root's ability to bear a large amount of soil and the area of soil mobilized during pull-out.

Trees are also known to respond to wind stress by increasing the number of lateral roots, which provide the most resistance to overturning [57]. This emphasizes the importance of lateral roots for tree anchorage, hence larger allocation of BGB biomass to them (Table 2), especially as trees grow (Table 3).

The decreasing BGB allocation to taproot with tree size might be related to decreasing anchorage function of the taproot with tree size. Khuder et al. [58] argued that "taproots may play an important role in anchoring young trees, but in adult trees, their growth is often impeded by the presence of a hard pan layer in the soil and the taproot becomes a minor component of tree anchorage".

Modelling, additivity, and validation
The homoscedasticity observed for the independent and simultaneous models implies that the derived weight functions were efficient in addressing the heteroskedasticity.

Because all the fitted independent models performed almost equally, any model can be used accurately to estimate BGB and carbon stocks in *A. johnsonii* stands (mecrusse woodlands), which, along with mopane and miombo are the most important woodlands in Gaza and Inhambane Provinces. However, as including tree height improved the fit statistics of the models negligibly, and sometimes worsened them, the models using either DBH or RCD only should be preferred, as tree height is difficult to measure in natural forests. The model using the

RCD only should be preferred when the tree is found cut down, as besides RCD being relatively difficult to measure compared to DBH, it is sometimes influenced by root buttress. The same holds true for simultaneous models, as they also performed equally.

The mean biomass per tree (MB) obtained using the WNSUR root system model based on DBH and TH [Eq. (1)], based on DBH [Eq. (2)], and based on RCD [Eq. (3)] was only 0.40, 0.31 and 0.23% larger than the MB obtained using the independent root system model based on DBH and TH (model form 6, found to be the best), based on DBH, and based on RCD, respectively. The differences between the MB of the root components (taproot and lateral roots) obtained using simultaneous and independent models were also negligible (>0.50%). These results are in line with those found by Repola [7].

As the differences between MB obtained using simultaneous and independent models were negligible, the independent models should be preferred to avoid unnecessary complexity of the models, and also because the residuals for simultaneously fitted models were not normally distributed.

This study is distinguished from other studies that include BGB (e.g. [6, 15, 19, 23, 29, 31, 33]) for five reasons: (1) in this study a very large sample size and geographical range was covered (93 trees of a single species harvested in 5 districts); (2) the root system was excavated to total depth and removed (including fine roots); (3) the root system was divided into three root components (taproot, root collar, and lateral roots) allowing, therefore, fitting the models for each root component and analysing BGB allocation patterns; (4) the error variance was modelled to derive the weight functions and address heteroskedasticity, which led to increases in Adj. R^2 and general improvement of the models' performance and; (5) the models were validated inside and outside the study area, therefore, the predictive capacity of the models was checked.

Mugasha et al. [6] used a large number of sample trees (80) distributed over 60 tree species, making up an average of 1.34 trees per species, which might not have been representative. Moreover, when modelling tree biomass, the use of species-specific equations are preferred because trees of different species may differ greatly in architecture and wood density [59], and architecture can influence biomass allocation and allometry [41, 42].

Mugasha et al. [6] and Kuyah et al. [23] did not fully excavated the root system, relied on sampling procedures, which might have led to less accurate estimates when compared to cases where all the root system is removed. Green et al. [19], Kuyah et al. [23], Ruiz-Peinado et al. [29], Paul et al. [31], and Ryan et al. [33] used predefined excavation depths and/or did not include fines roots; which leads to underestimation of the BGB [20].

Mugasha et al. [6], Litton et al. [15], Green et al. [19], Ruiz-Peinado et al. [29], and Ryan et al. [33] did not validate their models, therefore, the predictive capacity of the models was not checked. Paul et al. [31] checked the predictive capacity of the models only in the study area. Kuyah et al. [23] used a very small testing sample (6 trees) to validate the models which obviously did not cover all variation ranges.

The fit statistics for BGB model by Magalhães and Seifert [38] using the same model form as the one in Eq. (3) (Adj.R^2 = 94.94%; CVr = 21.79%) were different from those obtained here (Adj.R^2 = 99.80%; CVr = 23.16%). These differences might be because Magalhães and Seifert [38] considered the stump as part of the root system and because the weight functions were derived interactively; the weight functions were, therefore, just approximations.

Independent BGB (root system) models of this study (Adj.R^2 range 99.73–99.80%; bias range 0.02–1.38%) performed better than those by Kuyah et al. [23] (R^2 range 91.90–96.10%; bias range −49.60 to 35.40%) and Mugasha et al. [6] (R^2 range 89.00–94.00%; bias range 0.12–5.98%). Models of this study also showed superiority to the BGB models by Paul et al. [31] (R^2 range 64.00–95.00%) and Ryan et al. [33] (R^2 = 94.00%).

Husch et al. [5] suggested that the aggregate difference should not exceed $2 \times CV_r/\sqrt{n}$, where n is the number of trees used in the test. In this study the lowest value of $2 \times CV_r/\sqrt{n}$ was 11.96%, which was almost twice as large as the largest aggregate difference; therefore, the models can be applied inside and outside the study area without requiring any corrections, according to this criterion.

The dataset of this study comprised a training and testing sample of 93 and 37 trees, respectively, with DBHs varying from 5 to 32 cm and from 5.5 to 32 cm, respectively. *A. johnsonii* trees hardly can exceed 35 cm in DBH. Magalhães and Soto [60] (unpublished data) found only 13 *A. johnsonii* trees per ha with DBH ≥ 30 cm, corresponding to only 5% of the total trees per ha. Here, the number of trees with DBH ≥ 32.5 cm were only 19 per ha, equivalent to 1.54% of the total trees per ha. This implies that no serious bias will be added when extrapolating the models outside the DBH range used to fit the models since very few trees are found outside that DBH range.

A. johnsonii stands occur mainly in Ferralic Arenosols and Stagnic soils (Figure 2b) [61], which cover 410,144 ha (74%) and 108,960 ha (20%), respectively, of the total area of occurrence of *A. johnsonii* stands (Figure 2a, b) in Gaza and Inhambane Provinces; the remaining 6% are covered by other soil types [61]. Of the 23 sampled plots, 20 were located in Ferralic Arenosols and 3 (plots 8, 9 and 22 in Figure 2b) were located in Stagnic soils. The 20 plots from Ferralic Arenosols accounted with 81 felled trees of the training sample, the remaining 12 trees were from Stagnic soils. The testing sample trees from outside the study area (16 trees from Chicuacala district) were all from Ferralic Arenosols (Figure 2b). The climate in the districts where *A. johnsonii* occurs is dry and humid tropical [2, 3, 61–71], however, *A. johnsonii* occurs only in dry tropical climates [61]. Humid tropical climate occurs only along the coast, where *A. johnsonii* stands do not occur [2, 3, 61–71]. This implies that the fitted models can also be safely applicable and valid over a vast range of soils and regions where *A. johnsonii* occurs and outside the study area.

Therefore, besides the fact that the models were validated inside and outside the study area, the study area covered almost the entire range of soil and climate variations where *A. johnsonii* occurs (despite the apparent lack of large variations), and a wide range of DBHs and THs, therefore, the models can be applied in all *A. johnsonii* stands in Mozambique.

Conclusions

In this study, it was found that more than half and almost one-third of BGB is allocated to the lateral roots and to the root collar, respectively. More than 80% of the BGB was found at a depth range of 9.6–61.2 cm from the ground level. As the tree size increased, the proportion of BGB allocated to taproots decreased and that allocated to lateral roots increased. Consequently, it was hypothesised that BGB allocation patterns is a response of the anchorage functions of the tap and lateral roots and therefore, root component biomass models can be used as a methodology to predict anchoring functions of the different root components.

Because all fitted independent models performed almost equally, the models using either DBH or RCD exclusively are preferred as tree height is difficult to measure in natural forests. The model using RCD only as a predictor variable should be further preferred when the tree is found cut down, as sometimes the RCD is affected by root buttress. As a result of the differences between the mean biomasses obtained using independent and simultaneous models being negligible, the independent models should be preferred to avoid unnecessary complexity in the models.

The fitted independent models were based on a very large sample size (93 trees) and a wide geographical range (5 districts) and exhibited that, on average, 98% of the variation in BGB is explained by the predictor variables and were validated inside and outside the study area. Therefore, the models presented here could be applied to all *A. johnsonii* stands in Mozambique.

Methods
Study area
The study was conducted in Mozambique. The study area comprised 5 districts (Mabote, Funhalouro, Panda, Mandlakaze, and Chibuto) of 2 provinces (Inhambane and Gaza) with an extension of 4,502,828 ha [61], of which 226,013 ha (5%) were *A. johnsonii* stands (Mecrusse). Mecrusse is a forest type where the main tree species, many times the only one, in the upper canopy is *A. johnsonii*. Detailed description of the species, forest type and study area can be found in Magalhães and Seifert [16, 17, 38, 50].

Data acquisition
The data were collected in 2012 and 2014. In 2012, a two-phase sampling design was used to determine BGB. In the first phase, diameter at breast height (DBH), root collar diameter (RCD) and total tree height (TH) of 3574 trees were measured in 23 randomly located circular plots (20-m radius). Only trees with DBH ≥ 5 cm were considered. In the second phase, 93 trees (DBH range 5–32 cm; TH range 5.69–16 m), 2 to 6 per plot, were randomly selected from those analysed during the first phase for destructive measurement of BGB along with the variables from the first phase. In 2014, additional 37 trees (DBH range 5.5–32 cm; TH range 7.3–15.74 m) were felled outside sampling plots, 21 (DBH range 6.0–31 cm; TH range 9.37–15.74 m) inside and 16 (DBH range 5.5–32 cm; TH range 7.3–15.05 m) outside the study area (in Chicualacula district, Figure 2). The 93 trees collected in

2012 were used to fit BGB models (training sample) and those collected in 2014 (37 trees) were used to validate the models (testing sample).

Trees (both from 2012 and 2014) were cut down at 20 cm from the ground level. Thereafter, the root system was excavated and sampled as follows. First, the root system was partially excavated to the first node, using hoes, shovels, and picks; to expose the primary lateral roots. The primary lateral roots were numbered and separated from the taproot with a chainsaw and removed from the soil, one by one. This procedure was repeated in the subsequent nodes until all primary roots were removed from the taproot and the soil. Finally, the taproot was excavated and removed.

The removal of the root system to the total depth was relatively easy because 90% of the lateral roots of *A. johnsonii* are located in the first node, which is located close to ground level; the lateral roots grow horizontally to the ground level, do not grow downwards; and because the taproots had, at most, only 4 nodes and at least 1 node (at ground level). The root system was removed completely, so the depth of excavation depended on the depth of the taproot. For images illustrating the excavation process, refer to Magalhães and Seifert [16, 17, 38].

The root system was divided into following root components: lateral roots (fine and coarse), root collar, and taproot. These root components are not additive, as the taproot includes also the root collar (Figure 3); therefore, the root system is the sum of lateral roots and taproot. The remaining portion of the taproot, obtained after removing the root collar, was not considered as an independent component because it would be an artificial component, as the taproot includes the root collar as well; moreover, there is no name for such a portion (Figure 3). Lateral roots with diameters at the insertion point on the taproot <5 cm were considered as fine roots and those with diameters ≥5 cm were considered as coarse roots.

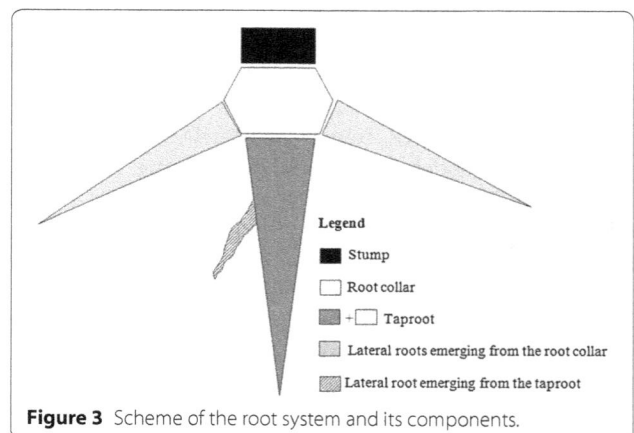

Figure 3 Scheme of the root system and its components.

Most of the studies on BGB (e.g. [15, 19, 23, 31, 72–75]) considered fine roots those with a diameter at insertion point on the taproot <2 mm and did not include them in the root samples. In this study, fine roots were defined as those with a diameter at insertion point on the taproot <5 cm and were included in the root samples. Therefore, although the definition of fine roots in this study is distinguished from that of most studies, the definition of fine roots by these authors is included in this study, and all dimensions of roots were considered here. Therefore, the definition of fine and coarse roots did not affect the estimates, as both definition categories were considered.

Fresh weight was obtained for the taproot, root collar, each coarse lateral root and for all fine lateral roots. A sample was taken from each root component, fresh weighed, marked, packed in a bag, and taken to the laboratory for oven drying. For the taproot, the samples were two discs, one taken on the top of the root collar and another from the middle of the taproot. For the coarse lateral roots, two discs were also taken, one from the insertion point on the taproot and another from the middle of it. For fine roots the sample was 5–10% of the fresh weight of all fine lateral roots. Oven drying of all samples was done at 105°C to constant weight, hereafter, referred to as dry weight.

Dry weights of the taproot, root collar, and lateral roots were determined by multiplying the ratio of oven-dry- to fresh-weight of each sample by the total fresh weight of the relevant component. The dry weight of the root system was obtained by summing the dry weights of taproot and lateral roots (fine and coarse ones). The disc taken on the top of the root collar was used to estimate its dry weight and the one taken in the middle of the taproot was used to estimate the dry weight of the remaining portion of the taproot.

Data analysis

Possible variations of BGB allocation patterns with tree size (DBH, RCD, and TH) were studied by investigating the dependence of the ratios of lateral roots to root system (LR/RS), taproot to root system (TR/RS), root collar to root system (RC/RS), and root collar to taproot (RC/TR) biomasses on tree size, using Pearson's correlation coefficient test of significance and distance covariance (dcov) test of independence. The first test was performed using the cor.test function of R software [76] and the second using the dcov.test function of energy package [77] in R software [76].

The Pearson's correlation coefficient measures only linear dependencies. Although, the dcov test of independence measures all types of dependences (linear, non-linear and non-monotone) between random vectors X and Y in arbitrary dimension [78], the Pearson's

correlation coefficient was also considered, because unlike dcov test, it shows the direction of variation between two variables.

Biomass models were fitted using weighted non-linear regression. Non-linear models were preferred over linear ones because biomass is a non-linear function of stem diameter and height [32, 79–82]. Weighted least squares (WLS) were preferred over ordinary least squares (OLS) to address the heteroskedasticity as, quite often, the error variance is functionally related to the independent variables in regression [83], that is, the variability of the biomass increases with tree independent variables [84]. The weight functions were obtained by modelling the error structure as described by Parresol [83, 85]. For that, the squares of the OLS residuals were fitted against the different combination of the independent variables. Thus, it was assumed that the squares of the OLS residuals are representative of the error variance.

The tested model forms for all root components are given below.

$$\hat{Y} = b_0 \times DBH^{b_1} \tag{4}$$

$$\hat{Y} = b_0 \times DBH^{b_1} \times TH^{b_2} \tag{5}$$

$$\hat{Y} = b_0 \times \left(DBH^2 \times TH\right)^{b_1} \tag{6}$$

$$\hat{Y} = b_0 \times RCD^{b_1} \tag{7}$$

where $\hat{Y}$ is expressed in Kg, DBH and TH are expressed in cm and m, respectively.

A model form using the RCD only as a predictor variable was also considered to allow the estimate of BGB when trees are found already cut down and only stump dimensions are available. Estimation of BGB of exploited trees is very important, as in that case, BGB can be used to estimate the carbon that will be transferred to the soil and/or the nutrients that will be reclaimed by the site.

However, the sum of the biomass predictions for the root components will not equal the biomass prediction for the root system, and a desired and logical feature of the root component regression equations is that the sum of biomass predictions of the root components equals the prediction for the root system. To cope with that, a new root system biomass model form was obtained as a function of the predictor variables of the best taproot and lateral roots biomass model forms. Then, the new root system model form and the best taproot and lateral roots biomass model forms were fitted again, simultaneously, using weighted non-linear seemingly unrelated regression (WNSUR) with parameter restriction, to achieve additivity.

Root component models were fitted with the statistical software R [76] and the function nls using the Gauss–Newton algorithm. The simultaneous models (WNSUR models) were fitted using PROC MODEL statement of SAS software [86], using the ITSUR option. Restrictions on the regression coefficients were imposed by using RESTRICT statement. The start values of the parameters in nls function of the R software and in PROC MODEL statement of the SAS were obtained by fitting the logarithmized models of each component in Microsoft Excel.

The following criteria were used to evaluate the models (independent and simultaneous ones): adjusted coefficient of determination (Adj.R^2), mean residual (MR), standard deviation of residuals ($S_{y.x}$), test for heteroskedasticity of residuals, test for normality of residuals, and graphical analysis of residuals. The mean residual and the standard deviation of residuals were expressed as relative values, hereafter referred to as percent mean residual [MR (%)] and coefficient of variation of residuals [CV_r (%)], respectively, which are more revealing. MR measures the average model bias, describing the directional magnitude, the size of expected under and overestimates. The ideal value is zero. CV_r measures the dispersion between the observed and the estimated values of the model. It indicates the error that the model is subject to when is used for predicting the dependent variable. The ideal value is zero.

The heteroskedasticity of residuals was evaluated using the White's test for heteroskedasticity, with the aid of the package het.test [87] of the R statistical software [76], under the null hypothesis of homoscedasticity. For that, the residuals were used as dependent variables and the predicted root component biomass as independent variable. The normality of residuals was evaluated using the Lilliefors normality test under the null hypothesis of normality, using the lillie.test function of R statistical software [76].

The models were then validated inside and outside the study area using aggregate difference in percentage (AD) [5, 88] and by comparing the observed and predicted BGB using Wilcoxon signed rank test in R [76] as recommended by Philip [89], under the null hypothesis of no difference between the observed and predicted BGB values. Aggregate difference is the prediction error of the models using an independent sample of trees (e.g. testing sample; trees not included in the sample used to fit the models).

All the statistical analyses were performed at $\alpha = 0.05$.

Abbreviations
AGB: aboveground biomass; BGB: belowground biomass; DBH: diameter at breast height; RCD: root collar diameter; TH: total tree height; WNSUR: weighted non-linear seemingly unrelated regression; AD: aggregate difference; MR: mean residual; CVr: coefficient of variation of residuals; Adj.R^2: adjusted coefficient of determination; SE: standard error; dcor: distance correlation; dcov: distance covariance; WLS: weighted least squares; OLS: ordinary least squares; MB: mean biomass per tree.

Acknowledgements
This study was funded by the Swedish International Development Cooperation Agency (SIDA). Thanks are extended to Professor Thomas Seifert for his contribution in data collection methodology and to Professor Almeida Sitoe for his advices during the preparation of the field work. I would also like to thank Professor Agnelo Fernandes and Madeirarte Lda for financial and logistical support.

Compliance with ethical guidelines

Competing interests
The author declares that he has no competing interests.

References
1. Cardoso GA (1963) Madeiras de Moçambique: *Androstachys johnsonii*. Serviços de agricultura e serviços de veterinária, Maputo
2. MAE (2005) Perfil do distrito de Funhalouro, província de Inhambane. Ministério da Administração Estatal, Maputo
3. MAE (2005) Perfil do distrito de Mabote, província de Inhambane. Ministério da Administração Estatal, Maputo
4. Leadley P, Pereira HM, Alkemade R, Fernandez-Manjarrés JF, Proença V, Scharlemann JPW et al. (2010) Biodiversity scenarios: projections of 21st century change in biodiversity and associated ecosystem services. Montereal: Secretariat of the Convention on Biological Diversity, Technical Series no. 50
5. Husch B, Beers TW, Kershaw JA Jr (2003) Forest mensuration, 4th edn. Wiley, Hoboken
6. Mugasha WA, Eid T, Bollandsås OM, Malimbwi RE, Chamshama SAO, Zahabu E et al (2013) Allometric models for prediction of above- and belowground biomass of trees in the miombo woodlands of Tanzania. For Ecol Manage 310:87–101
7. Repola J (2013) Modelling tree biomasses in Finland. Ph.D Thesis, University of Helsinki, Helsinki
8. Cui XH, Chen J, Shen JS et al (2010) Modeling tree root diameter and biomass by ground-penetrating radar. Sci China Earth Sci 54:711–719
9. Butnor JR, Doolittle JA, Johnsen KH, Samuelson L, Stokes T, Kress L (2003) Utility of ground-penetrating radar as a root biomass survey tool in forest systems. Soil Sci Soc Am J 67:1607–1615
10. Raz-Yaseef N, Kotten L, Baldocchi DD (2013) Coarse root distribution of a semi-arid oak savanna estimated with ground penetrating radar. J Geophys Res Biogeosci 118:1–13
11. Zhu S, Huang C, Su Y, Sato M (2013) 3D ground penetrating radar to detect tree roots and estimate root biomass in the field. Remote Sensing 6:5754–5773
12. Butnor JR, Doolittle JA, Kress L, Cohen S, Johnsen KH (2001) Use of ground-penetrating radar to study tree roots in the southeastern United States. Tree Physiol 21:1269–1278
13. GTOS (2009) Assessment of the status of the development of the standards for the terrestrial essential climate variables. FAO, Rome
14. Cairns MA, Brown S, Helmer EH, Baumgardner GA (1997) Root biomass allocation in the world's upland forests. Oecologia 111:1–11
15. Litton CM, Ryan MG, Tinker DB, Knight DH (2003) Belowground and aboveground biomass in young postfire lodgepole pine forests of contrasting tree density. Can J For Res 33:351–363
16. Magalhães T, Seifert T (2015) Estimation of tree biomass, carbon stocks, and error propagation in mecrusse woodlands. Open J For 5:471–488
17. Magalhães T, Seifert T (2015) Tree component biomass expansion factors and root-to-shoot ratio of Lebombo ironwood: measurement uncertainty. Carbon Balanc Manag 10:9

18. Brown S (2002) Measuring carbon in forests: current and future challenges. Environ Pollut 116:363–372
19. Green C, Tobin B, O'Shea M, Farrel EP, Byrne KA (2007) Above- and belowground biomass measurements in an unthinned stand of Sitka spruce (*Picea sitchensis* (Bong) Carr.). Eur J Forest Res 126:179–188
20. Mokany K, Raison RJ, Prokushkin AS (2006) Critical analysis of root: shoot ratios in terrestrial biomes. Global Change Biol 12:84–96
21. Carreiras JMB, Melo JB, Vasconcelos M (2013) Estimating the aboveground biomass in miombo savanna woodlands (Mozambique, East Africa) using L-band synthetic aperture radar data. Remote Sensing 5:1524–1548
22. Pearson TRH, Brown SL, Birdsey RA (2007) Measurement guidelines for the sequestration of forest carbon. United States Department of Agriculture, Forest Science, General Technical Report NRS-18
23. Kuyah S, Dietz J, Muthuri C, Jamnadass R, Mwangi P, Coe R et al (2012) Allometric equations for estimating biomass in agricultural landscapes: II. Belowground biomass. Agric Ecosyst Environ 158:225–234
24. Komiyama A, Poungparm S, Kato S (2005) Common allometric equations for estimating the tree weight of mangroves. J Trop Ecol 21:471–477
25. Levy PE, Hale SE, Nicoll BC (2004) Biomass expansion factors and root: shoot ratios for coniferous tree species in Great Britain. Forestry 77(5):421–430
26. Soethe N, Lehmann J, Engels C (2007) Root tapering between branching points should be included in fractal root system analysis. Ecol Model 207:363–366
27. Kalliokoski T, Nygren P, Sievänen R (2008) Coarse root architecture of three boreal tree species growing in mixed stands. Silva Fennica 42(2):189–210
28. Kalliokoski T (2011) Root system traits of Norway spruce, Scots pine, and silver birch in mixed boreal forests: an analysis of root architecture, morphology, and anatomy. PhD thesis, Department of Forest Sciences, University of Helsinki, Helsinki
29. Ruiz-Peinado R, del Rio M, Montero G (2011) New models for estimating the carbon sink of Spanish softwood species. For Syst 20(1):176–188
30. Sanquetta CR, Corte APD, Silva F (2011) Biomass expansion factors and root-to-shoot ratio for Pinus in Brazil. Carbon Balanc Manag 6:6
31. Paul KI, Roxburgh SH, England JR, Brooksbank K, Larmour JS, Ritson P et al (2014) Root biomass of carbon plantings in agricultural landscapes of southern Australia: Development and testing of allometrics. For Ecol Manag 318:216–227
32. Bolte A, Rahmann T, Kuhr M, Pogoda P, Murach D (2004) Gadow Kv. Relationships between tree dimension and coarse root biomass in mixed stands of European beech (*Fagus sylvatica* L.) and Norway spruce (*Picea abies* [L.] Karst.). Plant Soil 264:1–11
33. Ryan CM, Williams M, Grace J (2010) Above- and belowground carbon stocks in a Miombo woodland landscape in Mozambique. Biotropica 11(11):1–10
34. Miranda SC, Bustamante M, Palace M, Hagen S, Keller M, Ferreira LG (2014) Regional variations in biomass distribution in Brazilian Savanna woodland. Biotropica 46(2):125–138
35. Niiyama K, Kajimoto T, Matsuura Y, Yamashita T, Matsuo N, Yashiro Y et al (2010) Estimation of root biomass based on excavation of individual root systems in a primary dipterocarp forest in Pasoh Forest Reserve, Peninsular Malaysia. J Trop Ecol 26:271–284
36. Mate R, Johansson T, Sitoe A (2014) Biomass equations for tropical forest tree species in Mozambique. Forests 5:535–556
37. Sitoe AA, Mondlate LJC, Guedes BS (2014) Biomass and carbon stocks of Sofala bay mangrove forests. Forests 5:1967–1981
38. Magalhães T, Seifert T (2015) Biomass modelling of *Androstachys johnsonii* Prain: a comparison of three methods to enforce additivity. Int J For Res 2015:1–17
39. Dupuy L, Fourcaud T, Stokes A (2005) A numerical investigation into the influence of soil type and root architecture on tree anchorage. Plant Soil 278:119–134
40. Dupuy L, Fourcaud T, Stokes A (2005) A numerical investigation into factors affecting the anchorage of roots in tension. Eur J Soil Sci 56:319–327
41. Coll L, Potvin C, Messier C, Delagrange S (2008) Root architecture and allocation patterns of eight tropical native species with different successional status used in open-grown mixed plantations in Panama. Trees 22:585–596
42. Trubat R, Cortina J, Vilagrosa A (2012) Root architecture and hydraulic conductance in nutrient deprived *Pistacia lentiscus* L. seedlings. Oecologia 170(4):899–908
43. Nicoll BC, Gardiner BA, Rayner B, Peace AJ (2006) Anchorage of coniferous trees in relation to species, soil type, and rooting depth. Can J For Res 36:1871–1883
44. Fitter AH (1987) An architectural approach to the comparative ecology of plant root system. New Phytol 106(Suppl):61–77
45. Fitter AH, Stickland TR (1991) Harvey, GWW. Architectural analysis of plant root system 3.Architectural correlates of exploitation efficiency. New Phytol 118:375–382
46. Fitter AH, Stickland TR (1991) Architectural analysis of plant root system 2.Influence of nutrient supply on architecture in contrasting plant species. New Phytol 118:383–389
47. Malamy JE (2005) Intrinsic and environmental response pathways that regulate root system architecture. Plant Cell Environ 28:67–77
48. Echeverria M, Scambato AA, Sannazarro AI, Maiale S, Ruiz OA, Menéndez AB (2008) Phenotypic plasticity with respect to salt stress response by *Lotus glaber*: the role of its AM fungal and rhizobial symbionts. Mycorrhiza 18(6–7):317–329
49. Herrel A, Speck T, Rowe NP (2006) Ecology and biomechanics: a mechanical approach to the ecology of animals and plants. Taylor and Francis, Boca Raton
50. Magalhães T, Seifert T (2015) Below- and aboveground architecture of *Androstachys johnsonii* Prain: topological analysis of the root and shoot systems. Plant Soil. doi:10.1007/s11104-015-2527-0
51. Ennos AR, Fitter AH (1992) Comparative functional morphology of the anchorage systems of annual dicots. Funct Ecol 6:71–78
52. Bila JM, Chelene I, Manhiça G, Mabjaia N (2011) Efeito dos tratamentos silviculturais nos ecossistemas de mecrusse, em Mabote, província de Inhambane, Moçambique. PFP 31(65):63–67
53. Day SD, Wiseman PE, Dickinson SB, Harris JR (2010) Contemporary concepts of root system architecture of urban trees. Arboricult Urban For 36(4):149–159
54. Schenk HJ, Jackson RB (2002) The global biogeography of roots. Ecol Monogr 72:311–328
55. Crook MJ, Ennos AR (1998) The Increase in Anchorage with Tree Size of the Tropical Tap Rooted Tree *Mallotus wrayi*, King (Euphorbiaceae). Ann Bot 82:291–296
56. Bailey PHJ, Currey JD, Fitter AH (2002) The role of root system architecture and root hairs in promoting anchorage against uprooting forces in *Allium cepa* and root mutants of *Arabidopsis thaliana*. J Exp Bot 53(367):333–340
57. Stokes A (1994) Responses of young trees to wind: effects on root architecture and anchorage strength. Ph.D thesis, University of York, York
58. Khuder H, Stokes A, Danjon F, Gouskou K (2007) Langane. Is it possible to manipulate root anchorage in young trees? Plant Soil 294:87–102
59. Ketterings QM, Coe R, van Noordwijk M, Ambagau Y, Palm CA (2001) Reducing uncertainty in the use of allometric biomass equations for predicting above-ground tree biomass in mixed secondary forest. For Ecol Manage 146:199–209
60. Magalhães TM, Soto SJ (2005) Relatório de inventário florestal da concessão florestal de Madeirarte: bases para a elaboração do plano de maneio de conservação dos recursos naturais. Departamento de Engenharia Florestal (DEF), Universidade Eduardo Mondlane (UEM), Maputo
61. DINAGECA (1997) Mapa digital de uso e cobertura de terra. Cenacarta, Maputo
62. MAE (2005) Perfil do distrito de Chibuto, província de Gaza. Ministério da Administração Estatal, Maputo
63. MAE (2005) Perfil do distrito de Mandhlakaze, província de Gaza. Ministério da Administração Estatal, Maputo
64. MAE (2005) Perfil do distrito de Guijá, província de Gaza. Ministério da Administração Estatal, Maputo
65. MAE (2005) Perfil do distrito de Mabalane, província de Gaza. Ministério da Administração Estatal, Maputo
66. MAE (2005) Perfil do distrito de Chigubo, província de Gaza. Ministério da Administração Estatal, Maputo
67. MAE (2005) Perfil do distrito de Massangena, província de Gaza. Ministério da Administração Estatal, Maputo
68. MAE (2005) Perfil do distrito de Chicualacuala, província de Gaza. Ministério da Administração Estatal, Maputo

69. MAE (2005) Perfil do distrito de Panda, província de Inhambane. Ministério da Administração Estatal, Maputo
70. MAE (2005) Perfil do distrito de Vilankulo, província de Inhambane. Ministério da Administração Estatal, Maputo
71. MAE (2005) Perfil do distrito de Massinga, província de Inhambane. Ministério da Administração Estatal, Maputo
72. Ngo KM, Turner BL, Muller-Landau C, Davies SJ, Laryavaara M, Hassan NFBN, Lun S (2013) Carbon stocks in primary and secondary tropical forests in Singapore. For Ecol Manage 296:81–89
73. Baishya R, Barik SK (2011) Estimation of tree biomass, carbon pool and net primary production of an old-growth *Pinus kesiya* Royle ex. Gordon forest in north-eastern India. Ann For Sci 68:727–736
74. Tobin B, Nieuwenhuis M (2007) Biomass expansion factores for Sitka spruce (*Picea sitchensis* (Bong.) Carr.) in Ireland. Eur J Forest Res 126:189–196
75. Soares P, Tome M (2012) Biomass expansion factors for *Eucalyptus globulus* stands in Portugal. For Syst 21(1):141–152
76. R Core Team (2015) A language and environment for statistical computing. R Foundation for Statistical Computing, Vienna
77. Rizzo ML, Székelly GJ (2015) Energy: E-statistics (energy statistics). R package version 1.6.2. R Foundation for Statistical Computing, Vienna
78. Székelly GJ, Rizzo ML (2009) Brownian distance covariance. Ann Appl Stat 3(2):1236–1265
79. Ter-Mikaelian MT, Korzukhin MD (1997) Biomass equation for sixty five North American tree species. For Ecol Manage 97:1–27
80. Schroeder P, Brown S, Mo J, Birdsey R, Cieszewski C (1997) Biomass estimation for temperate broadleaf forest of the United States using inventory data. For Sci 43:424–434
81. de Jong TJ, Klinkhamer PGI (2005) Evolutionary Ecology of Plant reproductive strategies. Cambridge University Press, New York
82. Salis SM, Assis MA, Mattos PP, Pião ACS (2006) Estimating the aboveground biomass and wood volume of savanna woodlands in Brazil's Pantanal wetlands based on allometric correlations. For Ecol Manage 228:61–68
83. Parresol BR (1999) Assessing tree and stand biomass: a review with examples and critical comparisons. For Sci 45:573–593
84. Picard N (2012) Manual for building tree volume and biomass allometric equations: from field measurements to prediction. FAO, Rome
85. Parresol BR (2001) Additivity of nonlinear biomass equations. Can J For Res 31:865–878
86. SAS Institute Inc (1999) SAS/ETS User's Guide, Version 8. SAS Institute Inc, Cary
87. Andersson S (2015) het.test: White's Test for Heteroskedasticity. R package version 1.0-1. R Foundation for Statistical Computing, Vienna
88. de Gier IA (1992) Forest mensuration (fundamentals). International Institute for Aerospace Survey and Earth Sciences (ITC), Enschede
89. Philip MS (1984) Measuring trees and forests. University of Dar Es Salam, Dar Es Salam

10

Effects of harvest, fire, and pest/pathogen disturbances on the West Cascades ecoregion carbon balance

David P Turner[1*], William D Ritts[1], Robert E Kennedy[2], Andrew N Gray[3] and Zhiqiang Yang[1]

Abstract

Background: Disturbance is a key influence on forest carbon dynamics, but the complexity of spatial and temporal patterns in forest disturbance makes it difficult to quantify their impacts on carbon flux over broad spatial domains. Here we used a time series of Landsat remote sensing images and a climate-driven carbon cycle process model to evaluate carbon fluxes at the ecoregion scale in western Oregon.

Results: Thirteen percent of total forest area in the West Cascades ecoregion was disturbed during the reference interval (1991-2010). The disturbance regime was dominated by harvesting (59 % of all area disturbed), with lower levels of fire (23 %), and pest/pathogen mortality (18 %). Ecoregion total Net Ecosystem Production was positive (a carbon sink) in all years, with greater carbon uptake in relatively cool years. Localized carbon source areas were associated with recent harvests and fire. Net Ecosystem Exchange (including direct fire emissions) showed greater interannual variation and became negative (a source) in the highest fire years. Net Ecosystem Carbon Balance (i.e. change in carbon stocks) was more positive on public that private forestland, because of a lower disturbance rate, and more positive in the decade of the 1990s than in the warmer and drier 2000s because of lower net ecosystem production and higher direct fire emissions in the 2000s.

Conclusion: Despite recurrent disturbances, the West Cascades ecoregion has maintained a positive carbon balance in recent decades. The high degree of spatial and temporal resolution in these simulations permits improved attribution of regional carbon sources and sinks.

Keywords: Forests; Carbon; Net ecosystem production; Net ecosystem exchange; Net ecosystem carbon balance; Disturbance; West Cascades ecoregion

Background

Net uptake of carbon by forests provides a significant offset to anthropogenic carbon emissions at the global [1], national [2, 3], regional [4], and landscape [5] scales. However, forest carbon sinks are vulnerable to disturbances in the form of harvesting, fire, and pest/pathogen outbreaks. At the regional scale, we have a poor understanding of the relative contribution of these disturbances to overall carbon budgets [6], but such knowledge is important in understanding how the carbon cycle is responding to on-going management and climate change

[7, 8]. It is also critical for developing policies for greenhouse gas mitigation through altered land use [9].

Forest disturbances are a strong determinant of carbon stocks and fluxes on both managed and unmanaged landscapes [5, 10]. Clear-cut harvesting, as is commonly practiced in coniferous forests of western Oregon, shuts down the photosynthetic carbon sink and increases the carbon source from heterotrophic respiration of harvest residues. Partial harvests for thinning likewise induce a near term reduction in carbon sequestration [11]. Wildfire is similar to harvesting in reducing carbon uptake, but has a longer term impact on heterotrophic respiration because of the slow conversion of snags to more readily decomposed woody debris on the ground [12, 13]. Pest/pathogen outbreaks reduce leaf area and leave slow decomposing snags, thus altering ecosystem carbon flux for decades [14,

* Correspondence: david.turner@oregonstate.edu
[1]Department of Forest Ecosystems and Society, Oregon State University, 97331 Corvallis, OR, USA
Full list of author information is available at the end of the article

15]. Spatially-explicit carbon cycle assessments in the western U.S. have generally emphasized effects of harvests and fire [16–18] and not explicitly captured the impact of slow pest/pathogen disturbances.

Satellite remote sensing, particularly from the Landsat series of sensors (~30 m resolution), offers the opportunity to monitor forest disturbances [19, 20]. In combination with spatially-distributed ecosystem process models that simulate carbon cycle responses to specific disturbances, remote sensing data can be used to map and monitor forest carbon stocks and flux [21, 22]. Here, we take advantage of a new Landsat-based time series analysis of forest disturbance (LandTrendr) [23, 24] and a well-established modeling infrastructure for simulating regional carbon flux based on the Biome-BGC carbon cycle process model [25] to quantify carbon cycle impacts of harvesting, fire, and pest/pathogen outbreaks on forests of the West Cascades (WC) ecoregion in the Northwestern U.S. We make extensive use of plot scale and aggregated U.S.D.A. Forest Service Forest Inventory and Analysis (FIA) data [26, 27] for calibration and validation of both LandTrendr and Biome-BGC.

We quantify several distinct carbon fluxes [28] at the ecoregion scale. Net ecosystem production (NEP) is the balance of net primary production (NPP) and heterotrophic respiration (R_h). It reflects ecosystem metabolism as it responds to variation in weather, and to disturbance events. Net ecosystem exchange (NEE) is the absolute vertical flux of CO_2 over a given geographical domain. This is the flux as "seen" by a continental to global scale inversion, e.g. [29]. In addition to NEP and direct fire emissions, it includes river/stream evasion as well as emissions associated with harvested products [30]. Net ecosystem carbon balance (NECB) refers to the absolute change in carbon stocks, and is affected by NEP as well as removals in the form of harvested products, lateral transfers of dissolved organics, and by direct fire emissions. NECB is the equivalent of carbon sequestration as would be relevant to offsetting fossil fuel emissions. For the purposes of comparisons here, we report NEE using the same convention as with NEP and NECB, i.e. a positive value is a carbon sink. The capacity to isolate these fluxes is required to fully understand the role of forests and forest management in regulating the atmospheric CO_2 concentration.

Results

Domain characterization
Our study domain was the West Cascades ecoregion in western Oregon, U.S.A. It is characterized by a strong elevation gradient from west to east, with corresponding gradients in temperature and precipitation (Fig. 1). Land cover is predominantly conifer forest (Fig. 2a). Forest stand age tends to be <60 on the low elevation private lands that are managed for wood production (Figs. 2b,

and 3). On public lands at higher elevations, there is a broader range of stand ages.

Disturbance patterns
The total proportion of the study area that was disturbed during the 1991-2010 interval was 13 %. The proportion of total disturbed area attributed to harvest was 59 %. The corresponding proportion for fire was 23 %, and for pest/pathogen outbreak was 18 %. The location of the harvests have been predominantly on lower elevation private forestland (Fig. 4a). The time series of annual area harvested shows a decrease on public lands in the late 1980s and an increase on private lands in the most recent decade (Fig. 5). On public lands, there was a shift from stand replacing harvests to partial harvests whereas on private forestland stand replacing harvests were most common over the whole time series (Fig. 6).

The incidence of fire was low in the 1980s and 1990s but increased appreciably in the 2000s (Fig. 7), with fires located at both high and low elevations (Fig. 4b). The overall proportion of fires at high, medium, and low intensity was 28 %, 33 %, and 40 % respectively. These proportions did not change much from year to year.

Areas of pest/pathogen disturbance occurred primarily at high elevations (Fig. 6c). Intensity tended to be higher in the years with relatively large areas disturbed.

Carbon flux
In our stand-level simulations of pest/pathogen disturbance, NPP falls in parallel with the drop in stem and foliar biomass and begins to recover after the year of maximum intensity (e.g. Fig. 8). NEP correspondingly decreases, falling below zero in the case of a short, high intensity slow disturbance.

In the ecoregion-wide simulations, the majority of the surface area of the WC ecoregion had a positive NEP during the study period (e.g. Fig. 9). NEP tends to decrease as elevation increases (Fig. 10) because of decreasing rates of wood productivity associated with a shorter growing season, and a shift towards older, slower growing, age classes on public lands (Fig. 2b). Carbon sources (negative NEP) are indicated in areas recently burned or harvested.

The ecoregion total NEP was also generally positive (Fig. 11). Total direct fire emissions increased in the 2000s relative to the 1990s, but only exceeded NEP in 2003. Over the 1991-2010 interval, harvest removals offset 28 % of NEP and direct fire emissions offset 7 % of NEP.

The interannual variation in NEE is a function of both NEP and direct fire emissions. In 2003, which was a high fire year (Fig. 7), high temperatures and soil drought reduced NPP more than R_h, and hence reduced NEP. Direct fire emissions were also relatively high, causing a dip in NEE (Fig. 11).

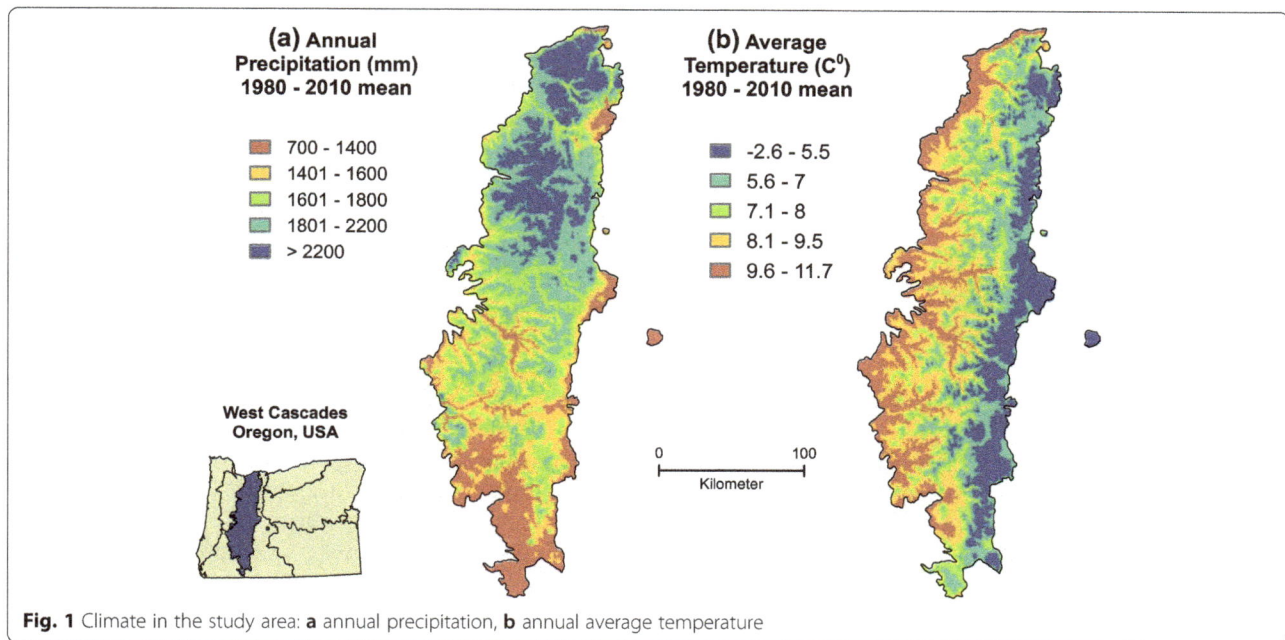

Fig. 1 Climate in the study area: **a** annual precipitation, **b** annual average temperature

The ecoregion total NECB averaged 1.9 (SD = 1.3) TgC yr^{-1} over the 1991-2010 interval. Only in 2003 did it fall below zero. Thus, the ecoregion has been a sustained sink for atmospheric CO_2 in recent decades. The area weighted mean NECB (1991-2010) for public lands was 65 gC m^{-2} year^{-1} compared to 8 gC m^{-2} year^{-1} on private forestland, reflecting a proportionally lower harvest rate on public lands in recent years.

Discussion

Harvest

Since much of the WC ecoregion is public forestland, the reduction in harvests associated with implementation of the Northwest Forest Plan (NWFP) in the early 1990s [31] had a notable impact on the overall rate of harvesting [24, 32]. The effect has been stabilization of a long-term decline in the proportion of public lands that is in the old-growth condition [33].

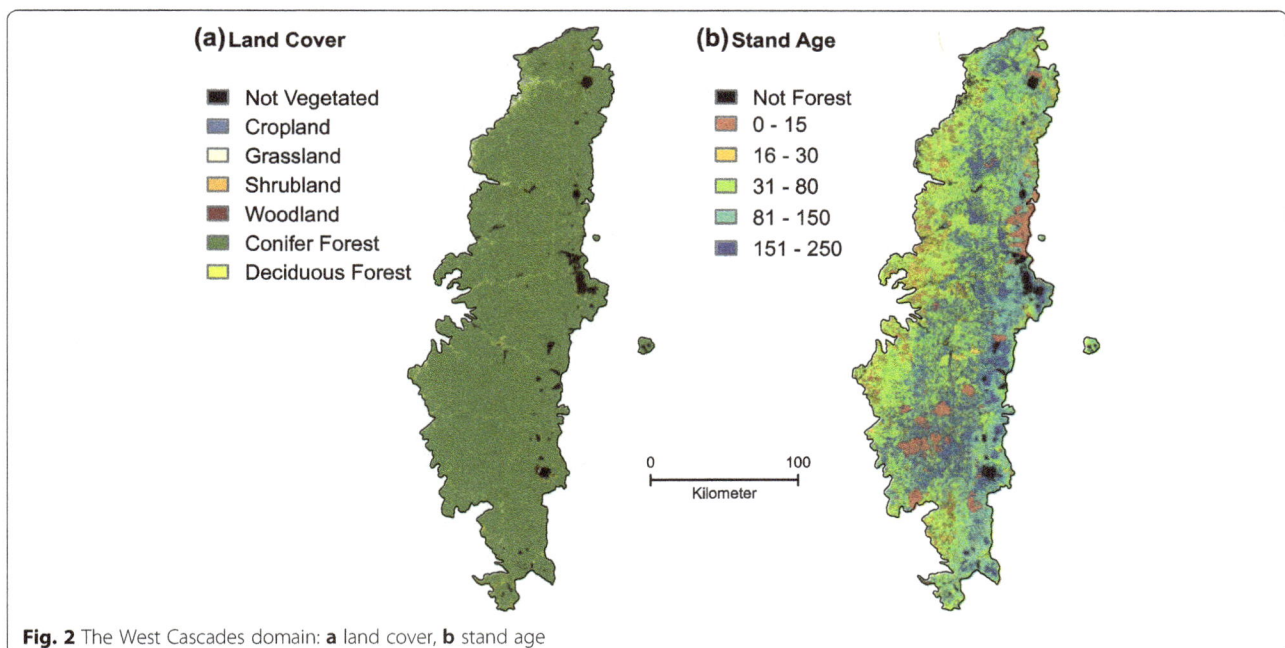

Fig. 2 The West Cascades domain: **a** land cover, **b** stand age

Land Ownership

- ■ Public
- ☐ Private

```
0        50
├──┬──┤
Kilometers
```

Fig. 3 The distribution of public and private forestland

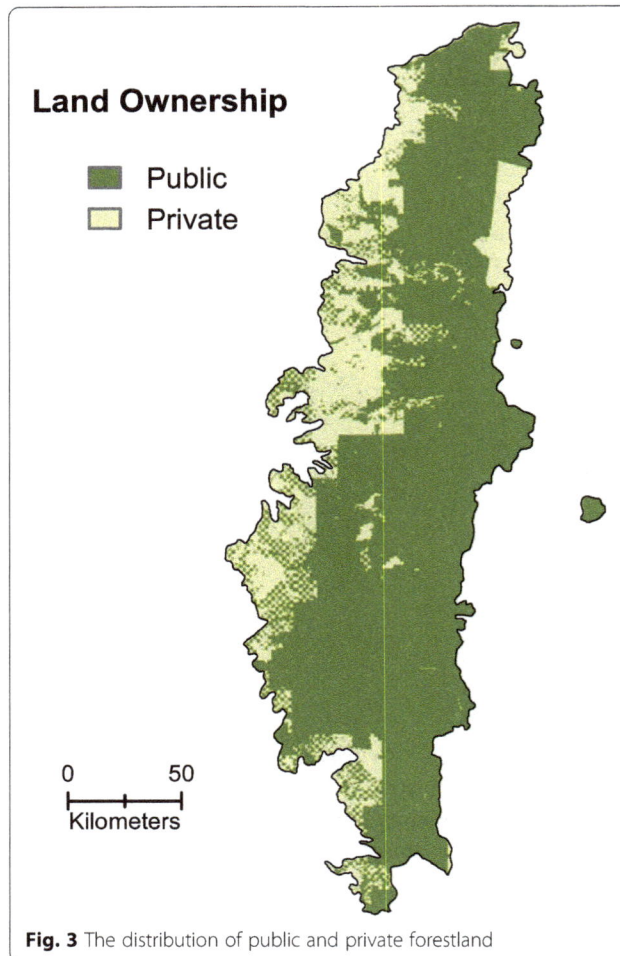

The rate of harvest on private forestland also shifted over the course of the study period, but in this case it was an increase that reflected a period of economic growth and high demand for wood products. The observed harvest rate of 1.5 % per year on private forestland over the 2000-2010 period is consistent with the 45–60 year rotation that is common in the area.

The notable shift on public lands from high intensity (clear-cut) harvesting to low and moderate intensities (thinning) reflects policy changes. Beginning in the 1990s, litigation largely prevented further harvesting of old growth stands. However, quotas established by the NWFP did allow for a low level of harvesting. Thinning of young to mature stands thus became the standard practice [34, 35], as reflected in the LandTrendr observations. On private lands that are managed primarily for wood production, the practice of clear-cut harvesting has remained the standard.

Fire

The fire regime in the WC ecoregion has varied over the last several hundred years. Fire scar data and tree age class distributions indicate a period of high fire about 1500, possibly associated with a relatively warm climate [36, 37]. During the settlement era in the late 19th century, the incidence of fire increased because of anthropogenic factors. That period was followed by a large decrease in the incidence of fire in the 20th century, associated with successful fire suppression. Most recently, an increase in the incidence of forest wildfire has been noted over much of the western U.S. and attributed in part to climate warming [38]. The post-2000 increase in area burned in this study is consistent with this broad pattern.

The sharp increase in the area burned in the WC ecoregion in the 2000s (3.6 % per year) compared to the 1990s (0.4 % per year) is associated with a 20 % decrease in mean May through September precipitation (Fig. 12). There was also a minor increase (0.2 °C) in mean May through September temperature. This natural experiment in interdecadal climate variation mimics to some degree the summer precipitation trends expected in the region for the 21st Century [39]. The observations here lend support to projections of an increased incidence of fire over the course of the 21st Century in the Cascade Mountains [40, 41]. Note that detecting relationships between temporal trends in burned area and climate is potentially confounded by changes in policy with respect to managing fire [42].

Pests/pathogens

Our trajectory-based LandTrendr algorithm is complimentary to the U.S. Forest Service airborne surveys for pests/pathogens in that it better resolves spatial heterogeneity and is sensitive to severity [43]. The duration of the slow disturbances is also informative with respect to identifying the relevant organisms: western spruce budworm show a more consistent long-term decline in the reference spectral vegetation index compared to mountain pine beetle [44].

The elevated incidence of pest/pathogen disturbances around 1990 is associated with a multiyear outbreak of western spruce budworm at relatively high elevations. This outbreak was widespread over the western U.S. [45, 44]. After 1995, the annual area disturbed became stable, presumably at an endemic population level characteristic of the native pests/pathogens. Meigs et al. [44] examined the incidence of area in the Cascade Mountains subject to pest/pathogen impacts and later burned. While that sequence was not uncommon, there did not appear to be a strong relationship. Indeed, mountain pine beetle outbreaks in lodgepole pine forests, may reduce the probability of active crown fire [46].

Carbon flux

NEP. In forest ecosystems, NEP is generally negative after a stand replacing harvest (because NPP is low and

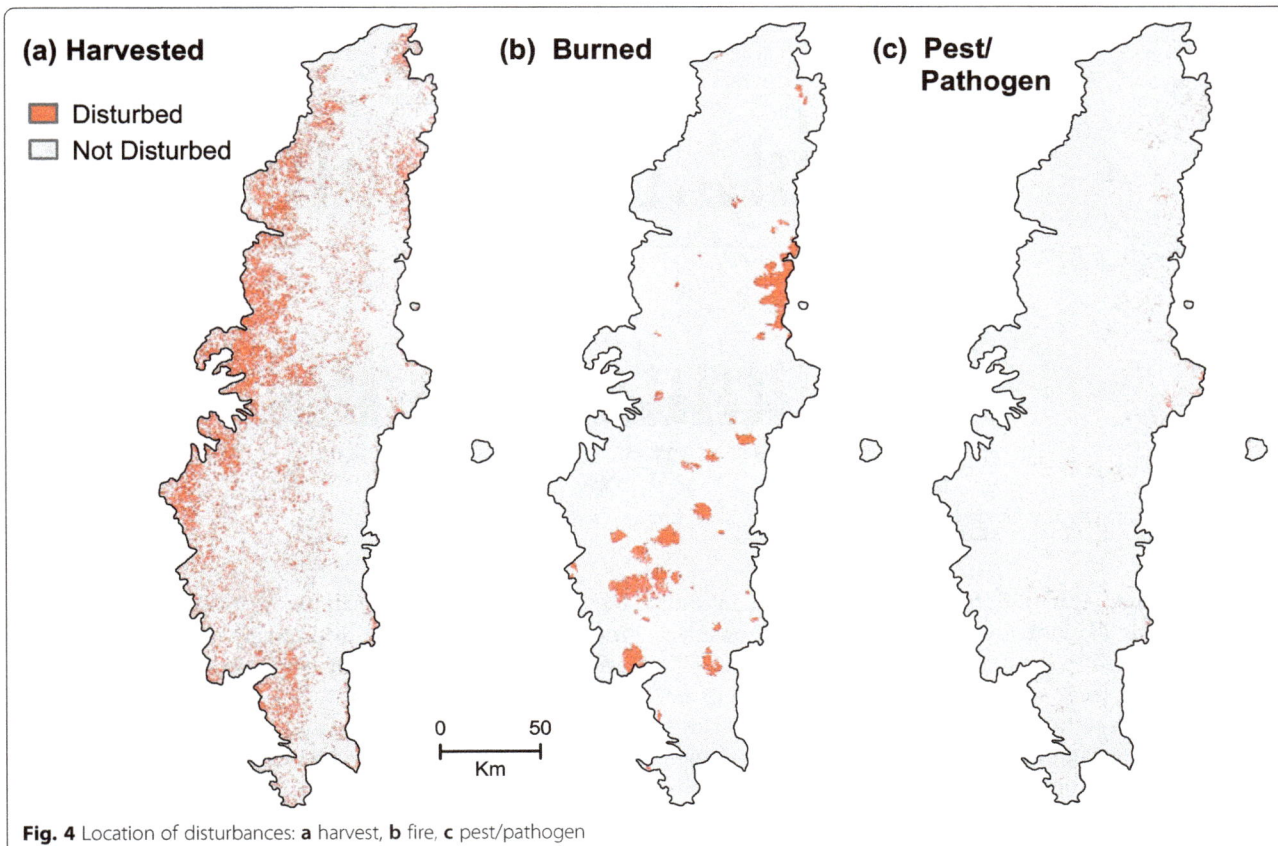

Fig. 4 Location of disturbances: **a** harvest, **b** fire, **c** pest/pathogen

R_h from decay of residues is high), becomes moderately positive for multiple decades while wood mass accumulates, and then falls to near zero in late succession [47, 48]. Stand-level simulations with Biome-BGC are in agreement with that pattern [17, 49].

The stand age for crossover from carbon source to carbon sink varies widely in the WC ecoregion because of differences in the amount of residues present at the time of stand initiation [50]. The post disturbance pulse of R_h and negative NEP is of greater magnitude in the

case of harvest but of greater duration in the case of fire because the proportion of the biomass actually burned in a wildfire is often small [51] and standing dead trees (snags) decay relatively slowly [13].

Moderate to low intensity abrupt disturbances (*e.g.* thinning) may also introduce periods of negative NEP [52]. Lost leaf area means a smaller fraction of photosynthetically active radiation is captured by the canopy, and heterotrophic respiration is boosted by the metabolism of the pests/pathogens themselves [53] or by heterotrophic respiration

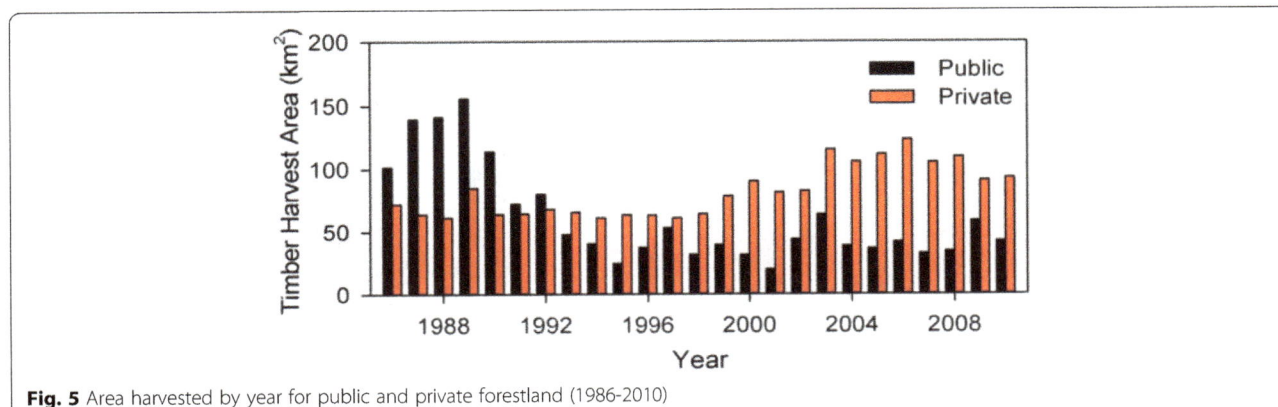

Fig. 5 Area harvested by year for public and private forestland (1986-2010)

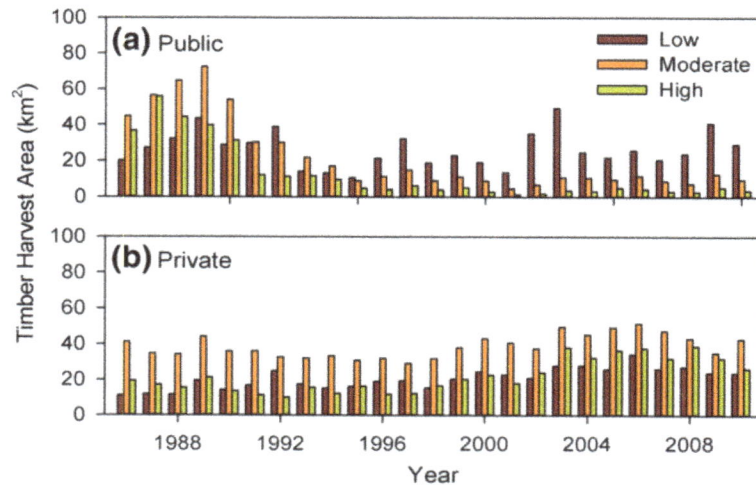

Fig. 6 Trends in harvest magnitude (1986-2010): **a** public forestland, **b** private forestland

of the increased leaf and fine root litter [54]. In our Biome-BGC simulations, the time to recovery of positive NEP is closely tied to the magnitude of the disturbance.

Here we also treat slow disturbances presumed to be caused by pests/pathogens. These disturbances are similar to fires in leaving large amounts of dead wood biomass behind that are slowly respired by heterotrophs and hence reduce NEP [14].

The most controversial aspect of the generalized trajectory for NEP over the course of succession in the WC ecoregion is the degree to which relatively old stands remain a carbon sink [55, 56]. Given that maritime Douglas-fir trees can survive for over 1000 years [57], and possibly benefit from increasing temperature and CO_2 concentration, a long term carbon sink might be expected. Mean simulated NEP (2001-2009) for old growth stands in the study area was 70 gC m^{-2} year^{-1}, which compares to and eddy covariance based mean of 49 gC m^{-2} year^{-1} (1998-2008) at the Wind River old growth Douglas-fir site [58].

Interannual variation in climate variables drives the large interannual variation in ecoregion-wide NEP in

our simulations. Observations at eddy covariance flux towers in the Cascade Mountains indicate that both gross primary production and ecosystem respiration contribute to interannual variation in NEP [58, 59], with smaller carbon sinks associated with warmer, drier years. Tree ring observations find that relatively warm years increase tree growth at high elevations [60, 61] and reduce it at mid to low elevations [62] in the Cascade Mountains. Thus, spatially distributed simulations are needed to evaluate if projected climate change driven warming and possible summer drying in the Pacific Northwest [39] would act to increase or decrease ecoregion mean NEP [7, 63].

NEE. The declining trend of NEE between the 1990s and 2000s was driven by both a decrease in NEP (dominant factor) and an increase in direct fire emissions. Schwalm et al. [7] also report a drop in regional NEP associated with turn of the century drought in western North America. Previous Biome-BGC simulations in Oregon suggest that both NPP and R_h have declined in recent decades, with a greater decline in NPP [64]. However, there remains considerable uncertainty about the

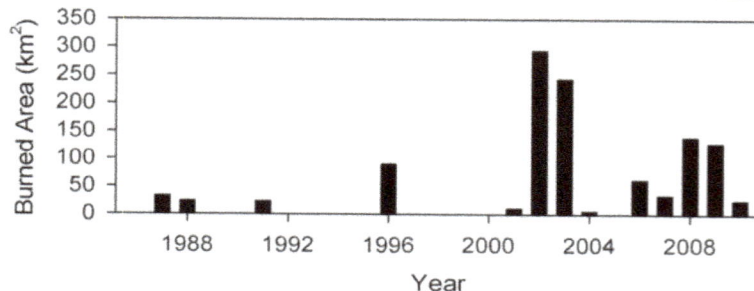

Fig. 7 Area burned by year for total forestland (1986-2010)

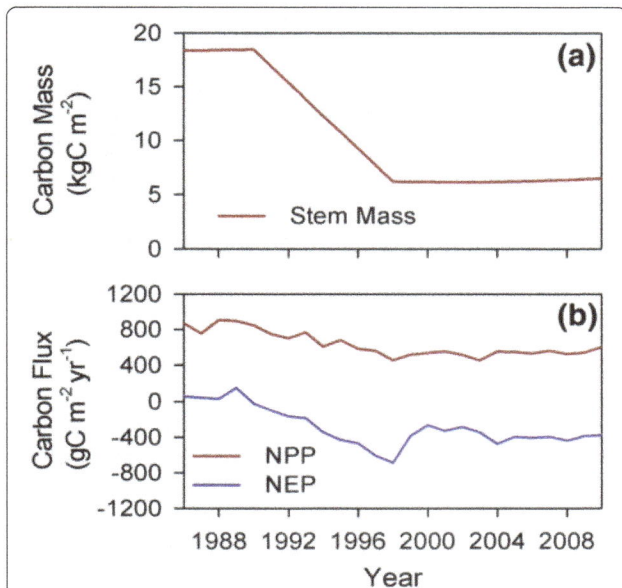

Fig. 8 Representative simulation results for a disturbed stand. Conditions are high intensity and short duration pest/pathogen disturbance beginning in 1990 at a mid-elevation site. **a** Stem mass, **b** Net Primary Production (NPP) and Net Ecosystem Production (NEP)

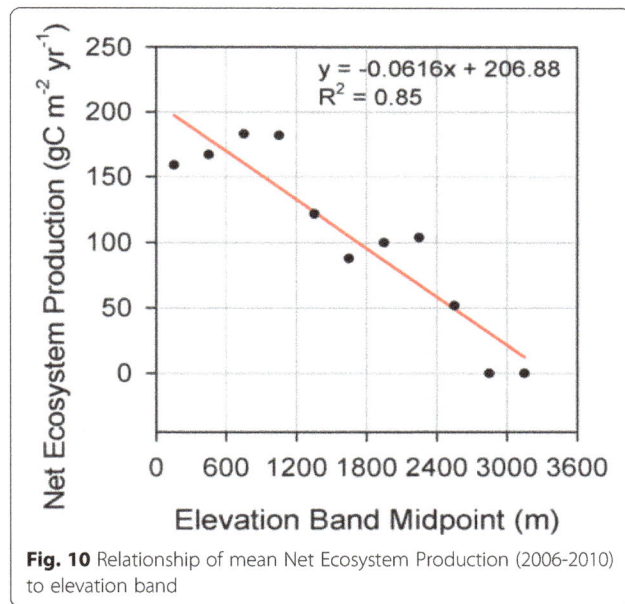

Fig. 10 Relationship of mean Net Ecosystem Production (2006-2010) to elevation band

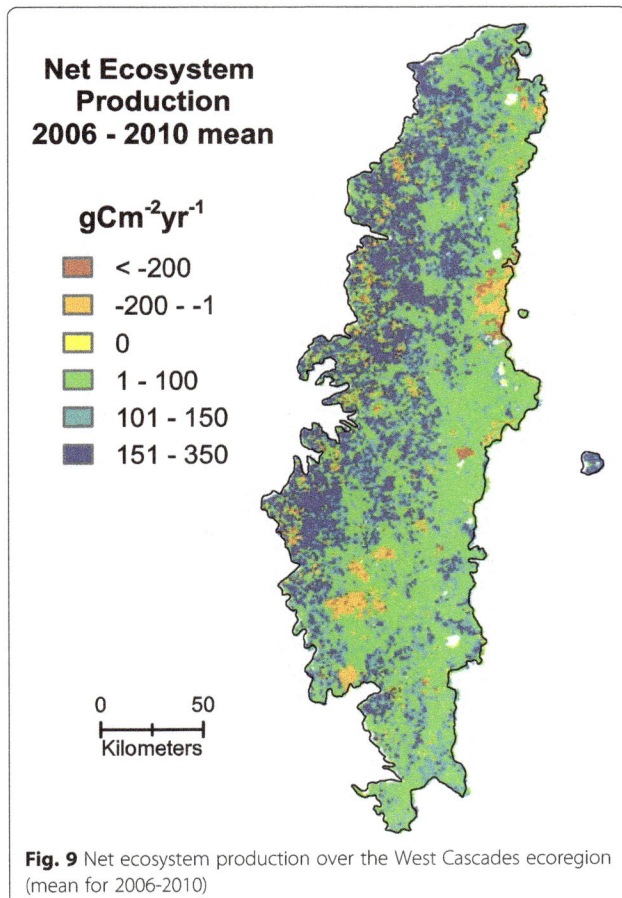

Fig. 9 Net ecosystem production over the West Cascades ecoregion (mean for 2006-2010)

impact of changing climate and CO_2 concentration on regional tree growth. Long term observations at 21 temperate and boreal zone eddy covariance flux towers support a broadly realized CO_2 benefit on forest water use efficiency [65], which may offset the impacts of lower summer precipitation.

By estimating NEE, we are in a position to compare our ecoregion fluxes with independent scaling approaches. Our mean NEE for 2004 was 148 gC m^{-2} years^{-1} (a carbon sink), much larger that an estimate from the CarbonTracker inversion [29, 66] of 18 gC m^{-2} years^{-1}. Differences of this magnitude have been observed in other NEE comparisons [67, 68] and point to the need to reconcile results from alternative scaling approaches [69]. LandTrendr could potentially deliver detailed information on the disturbance regime over much larger domains than the present study, thus opening the possibility for improved regional carbon budgets.

NECB. The positive NECB for the ecoregion using our flux scaling approach is consistent with carbon sinks in western Oregon reported using other modeling approaches [70] as well as inventory based approaches [35, 71, 72]. An additional forest sector sink is associated with harvested wood removals that accumulate in long-lived wood products [73] but is not estimated here. The high carbon accumulation rate on public lands, as driven by a policy-based lower harvest rate, supports the inclusion of carbon sequestration in the suite of ecosystem services to be considered in public forestland management.

Forest carbon stocks and flux are now widely recognized as highly relevant to mitigating the on-going rise in atmospheric CO_2 [74] and the combined remote sensing/modeling approach described here offers the opportunity

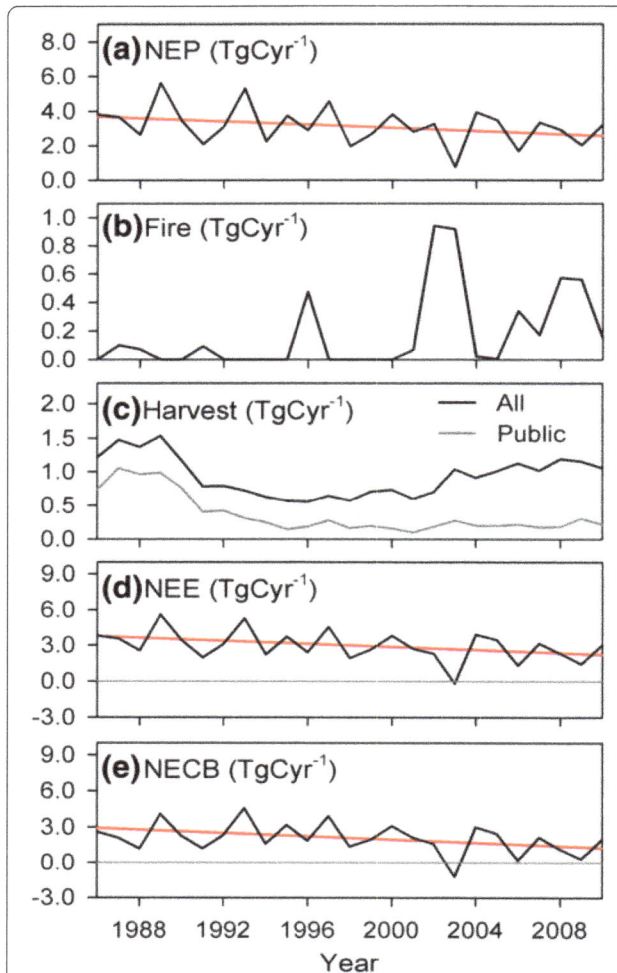

Fig. 11 Annual ecoregion totals for (a) net ecosystem production (NEP), (b) fire emissions, (c) harvest removals, (d) net ecosystem exchange (NEE), (e) net ecosystem carbon balance (NECB)

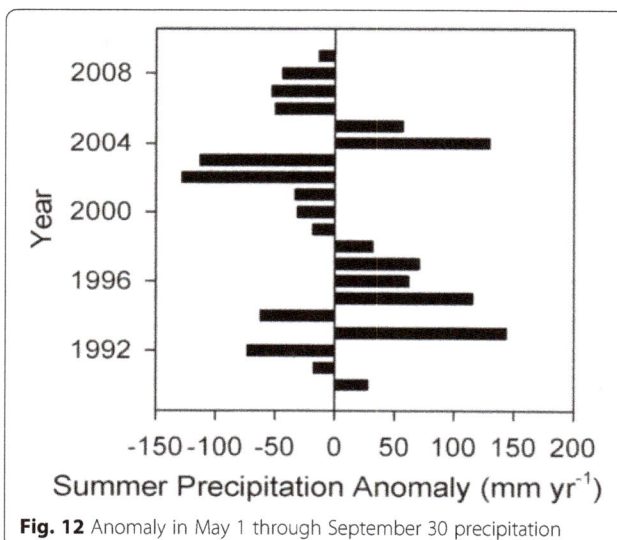

Fig. 12 Anomaly in May 1 through September 30 precipitation

to monitor forest carbon budgets at the fine spatial and temporal resolution that may be needed for quantifying carbon sinks. Ongoing efforts to characterize uncertainty associated with the remote sensing of disturbance and with stand-age-specific carbon fluxes (*e.g.* [75]) will increase the policy relevance of these carbon flux maps. The end points of our analysis are also the prerequisites for a realistic landscape simulation into the future that accounts for climate change and land use [76].

Conclusions

Harvests are the dominant form of disturbance in the West Cascades ecoregion, followed by fire, and pests/pathogens. The majority of the WC landscape has a positive NEP most of the time, and annual total NEP has been positive in recent decades. In high fire years, ecoregion total NEE can fall below zero (become a source) because of low NEP and high direct fire emissions. Harvest removals offset 4 times more NEP than do direct fire emissions. The sustained carbon accumulation on the West Cascades land base contributes to the regional forest carbon sink.

Methods
Overview

Our NEP/NECB scaling methods and validation results for applications in Oregon, Washington, and California have been reported in Turner et al. [18, 21, 22, 64] and Law et al. [77, 78]. These previous studies considered only stand-replacing disturbances and employed simple Landsat-based change detection algorithms to specify the approximate year of disturbance. Here we added treatment of pest/pathogen outbreaks (slow disturbance), allowed for multiple disturbance intensity classes, and accounted for sequential disturbance events based on an annual time series of Landsat data. To quantify NEE, we added a previously derived estimate of emissions from harvested products (wood and crops) to NEP and fire emissions.

NEP modeling

Our primary tool for scaling carbon flux across the ecoregion is the Biome-BGC carbon cycle process model [25]. The model has a daily time step and is run over multiple years to simulate succession. Simulated carbon cycle processes include photosynthesis, autotrophic respiration, heterotrophic respiration, plant C allocation, and mortality. Simulated C pools include stemwood, coarse roots, fine roots, foliage, litter, coarse woody debris, snags, and soil organic matter. A water balance is calculated based on the Penman-Montieth formulation of evapotranspiration. For each model run, there is a spin-up using a 25-year repeating loop of climate data to bring the soil carbon pool into near equilibrium with the climate. At its end, one or two disturbance events are prescribed by year, type, duration, and intensity to bring the simulation up to the current condition.

A look-up table determines the partitioning of the extant carbon stocks existing at the time of disturbance into removals, direct fire emissions, and transfers of necromass from one ecosystem component to another, thus maintaining ecosystem mass balance (Table 1). The disturbance history of a stand is prescribed in terms of one or two disturbances based on the record of Landsat imagery.

The version of Biome-BGC used here was adapted from version 4.1.2 [25] to simulate stand-replacing disturbances [77], dynamic allocation over the course of succession [78], and mixed severity fire [17]. To accommodate partial disturbances, the proportional disturbance intensity from satellite data (see below) is applied directly to the live carbon pools (Table 1).

The initiation year, duration, and maximum magnitude are prescribed by LandTrendr in the case of pest/pathogen disturbances. A linear ramp from initiation year to the maximum magnitude year is used to prescribe an annual mortality amount such that the prescribed maximum mortality is achieved in the maximum mortality year (Fig. 8). The stand recovers from disturbance prognostically. Transfers of dead foliar and fine root C are made to the litter pools and transfers of tree C to a standing dead (snag) pool. There is subsequent transfer to coarse woody debris on the ground [79–81]. Partitioning factors at the time of disturbance are based on observations within the region [51, 82–84].

Biome-BGC has 20 cover-type specific ecophysiological parameters that must be specified. For the most part, we adopted the recommendations in White et al. [85] and a representative set of parameters is given in Turner et al. [64]. In the case of the evergreen needle leaf cover type, we did an ecoregion specific adjustment on two of the parameters (the fraction of leaf nitrogen as rubisco, and the mortality fraction) that had been identified in sensitivity analyses as strongly impacting wood mass and NEP. Reference data for these parameter selections were site- and age-specific estimates of net stem growth and wood mass from USDA Forest Service FIA plots [35]. We ran the model at all FIA plot locations (approximated within 500 m to preclude potential disclosure of confidential locations) within the ecoregion to the stand age in the plot data, and compared predicted and observed net stem production using a range of possible parameter values. The optimal value for the mortality parameter was 0.0125 (proportion of biomass per year) and for the fraction of leaf nitrogen as rubisco parameter was 0.035 (unitless).

Model inputs

Our base land cover dataset (Fig. 2a) is the 2011 National Land Cover Database (NLCD) [86] which is derived from Landsat data. Areas that had been harvested and were classified by NLCD as shrubland were reclassified as forest. The distribution of public and private land (Fig. 3) was from U.S. Geological Survey [87].

The attributes of the disturbance regime were from the LandTrendr analysis of Landsat Thematic Mapper time series data [23, 24]. It captures both abrupt events, such as fire and harvest, and slow (multiyear)

Table 1 Partitioning of biomass pools at the time of disturbance

Disturbance Type	Disturbance Magnitude	Leaf/fine root	Stem	Coarse root	Litter	CWD	Snag
Harvest	Low	0.25[a]	0.25[b]	0.25[c]			
	Medium	0.55[a]	0.55[b]	0.55[c]			
	High	1.00[a]	1.00[b]	1.00[c]			
Fire[d]	Low	0.125	0.02[e]	0.02[f]	0.66	0.17	0.11
	Medium	0.50	0.03[e]	0.03[f]	0.66	0.22	0.14
	High	1.00	0.05[e]	0.05[f]	1.00	0.39	0.18
Pest/Pathogen	Low	0.30[a]	0.30[g]	0.30[c]			
	High	0.75[a]	0.75[g]	0.75[c]			

Values are proportions of total biomass. CWD = coarse woody debris
[a]Transferred to litter
[b]Stem transferred off-site, associated branches to go coarse woody debris (CWD)
[c]Transferred to coarse woody debris
[d]All transfers are to the atmosphere
[e]Residual mortality is transferred to snag
[f]Residual mortality is transferred to coarse woody debris
[g]Transferred to snag

disturbance processes caused by pests/pathogens. Each disturbance event is characterized by year, type, magnitude, and duration. The year is indicated by an inflection point in the trajectory of a spectral vegetation index in each pixel-level Landsat time series. The type of disturbance can be harvest (thinning), fire, or pest/pathogen driven. The distinction for abrupt disturbances between fire and harvest is based on reference to the Monitoring Trends in Burn Severity (MTBS) dataset [88]. Variables in the LandTrendr output are reported in a continuous format but here we binned the outputs into 3 magnitude classes for fire and harvest (see Table 1), and 2 magnitude classes (see Table 1), each with 3 possible durations (9, 17, 25 years), for slow disturbances. Bin midpoints are assigned to represent stand age, disturbance magnitude, and disturbance duration.

The LandTrendr analysis covered the interval from 1985 to 2012, but here we used multiyear buffers at the beginning and end when reporting carbon fluxes to limit artifacts associated with identifying inflection points in slow disturbances.

Disturbances previous to 1986 were prescribed on the basis of mapped stand age class from gradient nearest neighbor analysis (GNN) [89]. GNN integrates data from Landsat and Forest Inventory and Analysis plots [26] to estimate stand age. Conifer stands not having LandTrendr-based disturbance since 1985 were binned into 4 age classes and assigned the age class midpoint (45, 80, 150, 250 years). The binning of disturbance magnitude and duration as well as stand age class was necessary to constrain the number of unique disturbance histories within each 1 km climate grid cell. Model runs were made for the 10 most frequent combinations of cover type and disturbance history in each 1 km grid cell, and fluxes were reported as the weighted mean per 1 km^2 cell. Using this approach, we covered >90 % of the forest area under consideration.

The meteorological inputs to Biome-BGC are daily minimum and maximum temperature, precipitation, humidity, and solar radiation. We obtained 25 years (1986-2010) of climate data at 1 km resolution over our study ecoregion through the North American Carbon Program [90]. The distributed climate data (Fig. 1) are based on interpolations of meteorological station observations using a digital elevation model and general meteorological principles [91–93]. Uncertainty in the interpolations has previously been evaluated in our region and elsewhere [94, 95].

Soil Data Input. Soil texture and depth are specified from US Geological Survey soil maps [96].

Carbon flux reporting

NEP was calculated as NPP minus R_h. NEE was NEP minus direct fire emissions and an estimate for emissions from harvested products [97, 68]. This later estimate was made by reference to the change in the stock of previously harvested wood and crop products (based on inputs associated with harvests and product-specific turnover rates) and population distribution [97]. River stream evasion is also a significant term in regional NEE, however much of it derives from CO_2 originating in R_h of soil organic matter that is flushed into streams with the flow of the soil solution [98]. Since our Biome-BGC simulation includes R_h of soil organic matter, we did not attempt to additionally account for river/stream evasion here. NECB was calculated as NEP minus fire emissions and harvest removals (from our own simulations). Our NECB did not explicitly account for land use change, but the rate of land use conversion is quite low in the WC ecoregion [72].

Abbreviations
FIA: Forest inventory and analysis; GNN: Gradient nearest neighbor analysis; MTBS: Monitoring trends in burn severity; NEP: Net ecosystem production; NECB: Net ecosystem carbon balance; NEE: Net ecosystem exchange; R_h: Heterotrophic respiration; NPP: Net primary production; NWFP: Northwest forest plan; U.S.: United States; USDA: United States Department of Agriculture; WC: West Cascades.

Competing interests
The authors declare that they have no competing interests.

Authors' contributions
DT designed the study and contributed to data analysis and writing; WR contributed to assembling the model inputs and implementing model runs; RK provided the stand age map and developed the disturbance maps; AG organized and analyzed the FIA data; and ZY contributed to model coding and implementing the model runs. All authors read and approved the final manuscript.

Acknowledgments
Support was provided by the NASA Terrestrial Ecology Program (NNX12AK59G). Climate data from the North American Carbon Program and data from the U.S.D.A. Forest Service Forest Inventory and Analysis Program were essential to this study.

Author details
[1]Department of Forest Ecosystems and Society, Oregon State University, 97331 Corvallis, OR, USA. [2]College of Earth, Ocean, and Atmospheric Sciences, Oregon State University, 97331 Corvallis, OR, USA. [3]USDA Forest Service, Pacific Northwest Station, 97331 Corvallis, OR, USA.

References
1. Pan YD, Birdsey RA, Fang JY, Houghton R, Kauppi PE, Kurz WA, et al. A large and persistent carbon sink in the world's forests. Science. 2011;333(6045):988–93.
2. Stinson G, Kurz WA, Smyth CE, Neilson ET, Dymond CC, Metsaranta JM, et al. An inventory-based analysis of Canada's managed forest carbon dynamics, 1990 to 2008. Global Change Biol. 2010;17(6):2227–44.
3. McKinley DC, Ryan MG, Birdsey RA, Giardina CP, Harmon ME, Heath LS, et al. A synthesis of current knowledge on forests and carbon storage in the United States. Ecol Appl. 2011;21(6):1902–24.
4. Lu XL, Kicklighter DW, Melillo JM, Yang P, Rosenzweig B, Vorosmarty CJ, et al. A contemporary carbon balance for the Northeast Region of the United States. Environ Sci Technol. 2013;47(23):13230–8.
5. Chen B, Arain MA, Khomik M, Trofymow JA, Grant RF, Kurz WA, et al. Evaluating the impacts of climate variability and disturbance regimes on the historic carbon budget of a forest landscape. Agr Forest Meteorol. 2013;180:265–80.

6. Liu SG, Bond-Lamberty B, Hicke JA, Vargas R, Zhao SQ, Chen J, et al. Simulating the impacts of disturbances on forest carbon cycling in North America: processes, data, models, and challenges. J Geophys Res Biogeosci. 2011;116:22.

7. Schwalm CR, Williams CA, Schaefer K, Baldocchi D, Black TA, Goldstein AH, et al. Reduction in carbon uptake during turn of the century drought in western North America. Nat Geosci. 2012;5(8):551–6.

8. Williams CA, Collatz GJ, Masek J, Goward SN. Carbon consequences of forest disturbance and recovery across the conterminous U.S. Global Biogeochem Cy. 2012;26:GB1005.

9. USCCSP. United States Carbon Cycle Science Plan, http://downloads.globalchange.gov/carbon-cycle/us-carbon-cycle-science-plan.pdf. 2011.

10. Densai AR, Moorcroft PR, Bolstad PV, Davis KJ. Regional carbon fluxes from an observationally constrained dynamic ecosystem model: impacts of disturbance, CO2 fertilization, and heterogeneous land cover. J Geophys Res Biogeosci. 2007;112:G1.

11. Campbell J, Alberti G, Martin J, Law BE. Carbon dynamics of a ponderosa pine plantation following a thinning treatment in the northern Sierra Nevada. Forest Ecol Manag. 2009;257(2):453–63.

12. Harmon ME, Bond-Lamberty B, Tang JW, Vargas R. Heterotrophic respiration in disturbed forests: a review with examples from North America. J Geophys Res Biogeosci. 2011;116:17.

13. Acker SA, Kertis J, Bruner H, O'Connell K, Sexton J. Dynamics of coarse woody debris following wildfire in a mountain hemlock (Tsuga mertensiana) forest. Forest Ecol Manag. 2013;302:231–9.

14. Edburg SL, Hicke JA, Lawrence DM, Thornton PE. Simulating coupled carbon and nitrogen dynamics following mountain pine beetle outbreaks in the western United States. J Geophys Res Biogeosci. 2011;116:15.

15. Kurz WA, Dymond CC, Stinson G, Rampley GJ, Neilson ET, Carroll AL, et al. Mountain pine beetle and forest carbon feedback to climate change. Nature. 2008;452:987–90.

16. Cohen WB, Harmon ME, Wallin DO, Fiorella M. Two decades of carbon flux from forests of the Pacific Northwest. Bio Sci. 1996;46:836–44.

17. Meigs GW, Turner DP, Ritts WD, Yang ZQ, Law BE. Landscape-scale simulation of heterogeneous fire effects on pyrogenic carbon emissions, tree mortality, and net ecosystem production. Ecosystems. 2011;14(5):758–75.

18. Turner DP, Ritts WD, Yang ZQ, Kennedy RE, Cohen WB, Duane MV, et al. Decadal trends in net ecosystem production and net ecosystem carbon balance for a regional socioecological system. Forest Ecol Manag. 2011;262:1318–25.

19. Cohen WB, Goward SN. Landsat's role in ecological applications of remote sensing. Bio Sci. 2004;54(6):535–45.

20. Goward SN, Masek JG, Cohen WB, Moisen G, Collatz GJ, Healey SP, et al. Forest disturbance and North American carbon flux. EOS. 2008;89:105–16.

21. Turner DP, Ollinger SV, Kimball JS. Integrating remote sensing and ecosystem process models for landscape to regional scale analysis of the carbon cycle. Bio Sci. 2004;54:573–84.

22. Turner DP, Guzy M, Lefsky M, Ritts W, VanTuyl S, Law BE. Monitoring forest carbon sequestration with remote sensing and carbon cycle modeling. Environ Manage. 2004;4:457–66.

23. Kennedy RE, Yang ZG, Cohen WB. Detecting trends in forest disturbance and recovery using yearly Landsat time series: 1. LandTrendr - Temporal segmentation algorithms. Remote Sens Environ. 2010;114(12):2897–910.

24. Kennedy RE, Yang ZQ, Cohen WB, Pfaff E, Braaten J, Nelson P. Spatial and temporal patterns of forest disturbance and regrowth within the area of the Northwest Forest Plan. Remote Sens Environ. 2012;122:117–33.

25. Thornton PE, Law BE, Gholz HL, Clark KL, Falge E, Ellsworth DS, et al. Modeling and measuring the effects of disturbance history and climate on carbon and water budgets in evergreen needleleaf forests. Agr Forest Meteorol. 2002;113:185–222.

26. Bechtold WA, Patterson PL. The enhanced Forest Inventory and Analysis program - national sampling design and estimation procedures, USDA Forest Service General Technical Report SRS-GTR-80. 2005.

27. Woudenberg SW, Conkling BL, O'Connell BM, LaPoint EB, Turner JA, Waddell KL. The Forest Inventory and Analysis database: Description and user manual version 4.0 for Phase 2, USDA Forest Service, General Techncial Report RMRS-GTR-245. 2010.

28. Chapin FS, Woodwell GM, Randerson JT, Rastetter EB, Lovett GM, Baldocchi DD, et al. Reconciling carbon-cycle concepts, terminology, and methods. Ecosystems. 2006;9(7):1041–50.

29. Peters W, Jacobson AR, Sweeney C, Andrews AE, Conway TJ, Masarie K, et al. An atmospheric perspective on North American carbon dioxide exchange: CarbonTracker. Proc Natl Acad Sci U S A. 2007;104(48):18925–30.

30. Hayes DJ, Turner DP. The need for "apples-to-apples" comparisons of carbon dioxide source and sink estimates. EOS. 2012;93:404–5.

31. USDA. Record of Decision for amendments to Forest Service and Bureau of Land Management planning documents within the range of the northern spotted owl. 74 p. (plus Attachment A: Standards and guidelines). 1994.

32. Cohen WB, Spies TA, Alig RJ, Oetter DR, Maiersperger TK, Fiorella M. Characterizing 23 years (1972-95) of stand replacement disturbance in western Oregon forests with Landsat imagery. Ecosystems. 2002;5:122–37.

33. Moeur M, Ohmann JL, Kennedy RE, Cohen WB, Gregory MJ, Yang Z, et al. Status and trends of late-successional and old-growth forests, General Technical Report PNW-GTR-853. 2011.

34. Franklin JF, Johnson KN. A restoration framework for federal forests in the Pacific Northwest. J Forest. 2012;110(8):429–39.

35. Gray AN, Whittier TR. Carbon stocks and changes on Pacific Northwest national forests and the role of disturbance, management, and growth. Forest Ecol Manag. 2014;328:167–78.

36. Weisberg PJ, Swanson FJ. Regional synchroneity in fire regimes of western Oregon and Washington. USA Forest Ecol Manag. 2003;172(1):17–28.

37. Weisberg PJ. Historical fire frequency on contrasting slope facets along the McKenzie River, Western Oregon Cascades. West North Am Naturalist. 2009;69(2):206–14.

38. Westerling AL, Hidalgo HG, Cayan DR, Swetnam TW. Warming and earlier spring increases western U.S. forest wildfire activity. Science. 2006;313:940–3.

39. Mote PW, Salathe EP. Future climate in the Pacific Northwest. Clim Change. 2010;102(1-2):29–50.

40. Littell JS, Oneil EE, McKenzie D, Hicke JA, Lutz JA, Norheim RA, et al. Forest ecosystems, disturbance, and climatic change in Washington State, USA. Climatic Change. 2010;102(1-2):129–58.

41. Rogers BM, Neilson RP, Drapek R, Lenihan JM, Wells JR, Bachelet D, et al. Impacts of climate change on fire regimes and carbon stocks of the U.S. Pacific Northwest. J Geophys Res Biogeosci. 2011;116:13.

42. NCWFMS. The National Strategy (National Cohesive Wildland Fire Management Strategy). United States Department of Interior and Department of Agriculture http://www.forestsandrangelands.gov/strategy/thestrategy.shtml, accessed: May 1, 2015. 2014.

43. Meigs GW, Kennedy RE, Gray AN, Gregory MJ. Spatiotemporal dynamics of recent mountain pine beetle and western spruce budworm outbreaks across the Pacific Northwest region, USA. Remote Sens Environ. 2015;339:71–86.

44. Meigs GW, Kennedy RE, Cohen WB. A Landsat time series approach to characterize bark beetle and defoliator impacts on tree mortality and surface fuels in conifer forests. Remote Sens Environ. 2011;115(12):3707–18.

45. Hummel S, Agee JK. Western spruce budworm defoliation effects on forest structure and potential fire behavior. Northwest Sci. 2003;77(2):159–69.

46. Simard M, Romme WH, Griffin JM, Turner MG. Do mountain pine beetle outbreaks change the probability of active crown fire in lodgepole pine forests? Ecol Monogr. 2011;81(1):3–24.

47. Grant RF, Barr AG, Black TA, Margolis HA, McCaughey JH, Trofymow JA. Net ecosystem productivity of temperate and boreal forests after clearcutting-a Fluxnet-Canada measurement and modelling synthesis. Tellus B. 2010;62(5):475–96.

48. Kasischke ES, Amiro BD, Barger NN, French NHF, Goetz SJ, Grosse G, et al. Impacts of disturbance on the terrestrial carbon budget of North America. J Geophys Res Biogeosci. 2013;118(1):303–16.

49. Turner DP, Guzy M, Lefsky MA, Van Tuyl S, Sun O, Daly C, et al. Effects of land use and fine-scale environmental heterogeneity on net ecosystem production over a temperate coniferous forest landscape. Tellus B. 2003;55:657–68.

50. Janisch JE, Harmon ME. Successional changes in live and dead wood carbon stores: implications for net ecosystem productivity. Tree Physiol. 2002;22:77–89.

51. Campbell J, Donato D, Azuma DL, Law B. Pyrogenic carbon emission from a large wildfire in Oregon United States. J Geophys Res Biogeosci. 2007;12:G04014.

52. Harmon ME, Moreno A, Domingo JB. Effects of partial harvest on the carbon stores in Douglas-fir/Western Hemlock forests: a simulation study. Ecosystems. 2009;12(5):777–91.

53. Schmitz OJ, Raymond PA, Estes JA, Kurz WA, Holtgrieve GW, Ritchie ME, et al. Animating the carbon cycle. Ecosystems. 2014;17(2):344–59.

54. Dore S, Montes-Helu M, Hart SC, Hungate BA, Koch GW, Moon JB, et al. Recovery of ponderosa pine ecosystem carbon and water fluxes from thinning and stand-replacing fire. Global Change Biol. 2012;18(10):3171–85.

55. Harmon ME, Bible K, Ryan MG, Shaw DC, Chen H, Klopatek J, et al. Production, respiration, and overall carbon balance in an old-growth Pseudotsuga-Tsuga forest ecosystem. Ecosystems. 2004;7(5):498–512.

56. Luyssaert S, Schulze ED, Borner A, Knohl A, Hessenmoller D, Law BE, et al. Old-growth forests as global carbon sinks. Nature. 2008;455(7210):213–5.

57. Waring RH, Franklin JF. Evergreen forests of the Pacific Northwest. Science. 1979;204:1380–6.

58. Wharton S, Falk M, Bible K, Schroeder M, Paw KT. Old-growth CO2 flux measurements reveal high sensitivity to climate anomalies across seasonal, annual and decadal time scales. Agr Forest Meteorol. 2012;161:1–14.

59. Falk M, Wharton S, Schroeder M, Ustin SL, Paw UKT. Flux partitioning in an old-growth forest: seasonal and interannual dynamics. Tree Physiol. 2008;28:509–20.

60. Case MJ, Peterson DL. Fine-scale variability in growth-climate relationships of Douglas-fir, North Cascade Range, Washington. Can J For Res-Rev Can Rech For. 2005;35(11):2743–55.

61. Case MJ, Peterson DL. Growth-climate relations of lodgepole pine in the North Cascades National Park. Washington Northwest Sci. 2007;81(1):62–75.

62. Beedlow PA, Lee EH, Tingey DT, Waschmann RS, Burdick CA. The importance of seasonal temperature and moisture patterns on growth of Douglas-fir in western Oregon. USA Agr Forest Meteorol. 2013;169:174–85.

63. Wharton S, Chasmer L, Falk M, U KTP. Strong links between teleconnections and ecosystem exchange found at a Pacific Northwest old-growth forest from flux tower and MODIS EVI data. Global Change Biol. 2009;15(9):2187–205.

64. Turner DP, Ritts WD, Law BE, Cohen WB, Yang Z, Hudiburg T, et al. Scaling net ecosystem production and net biome production over a heterogeneous region in the western United States. Biogeosci. 2007;4:597–612.

65. Keenan TF, Hollinger DY, Bohrer G, Dragoni D, Munger JW, Schmid HP, et al. Increase in forest water-use efficiency as atmospheric carbon dioxide concentrations rise. Nature. 2013;499(7458):324.

66. CT. CarbonTracker. http://www.esrl.noaa.gov/gmd/ccgg/carbontracker/. 2014.

67. Desai AR, Helliker BR, Moorcroft PR, Andrews AE, Berry JA. Climatic controls of interannual variability in regional carbon fluxes from top-down and bottom-up perspectives. J Geophys Res Biogeosci. 2010;115:15.

68. Turner DP, Jacobson AR, Ritts WD, Wang WL, Nemani R. A large proportion of North American net ecosystem production is offset by emissions from harvested products, river/stream evasion, and biomass burning. Global Change Biol. 2013;19(11):3516–28.

69. Turner DP, Gockede M, Law BE, Ritts WD, Cohen WB, Yang Z, et al. Multiple constraint analysis of regional land-surface carbon flux. Tellus. 2011;63B:207–21.

70. Zhu Z, Reed BC, editors. Baseline and projected future carbon storage and greenhouse-gas fluxes in ecosystems of the Western United States. U.S. Geological Survey Professional Paper 1797. 2012.

71. Alig RJ, Krankina ON, Yost A, Kuzminykh J. Forest carbon dynamics in the Pacific Northwest (USA) and the St. Petersburg region of Russia: comparisons and policy implications. Clim Change. 2006;79:335–60.

72. Gray AN, Whittier TR, Azuma DL. Estimation of aboveground forest carbon flux in Oregon: adding components of change to stock-difference assessments. Forest Sci. 2014;60(2):317–26.

73. Malmsheimer RW, Bowyer JL, Fried JS, Gee E, Izlar RL, Miner RA, et al. Managing forests because carbon matters: integrating energy, products, and land management policy. J Forest. 2011;109(7):S7–48.

74. Olander LP, Cooley DM, Galik CS. The potential role for management of U.S. public lands in greenhouse gas mitigation and climate policy. Environ Manage. 2012;49(3):523–33.

75. Williams CA, Collatz GJ, Masek J, Huang CQ, Goward SN. Impacts of disturbance history on forest carbon stocks and fluxes: Merging satellite disturbance mapping with forest inventory data in a carbon cycle model framework. Remote Sens Environ. 2014;151:57–71.

76. Metsaranta JM, Dymond CC, Kurz WA, Spittlehouse DL. Uncertainty of 21st century growing stocks and GHG balance of forests in British Columbia, Canada resulting from potential climate change impacts on ecosystem processes. Forest Ecol Manag. 2011;262(5):827–37.

77. Law BE, Turner D, Campbell J, Van Tuyl S, Ritts WD, Cohen WB. Disturbance and climate effects on carbon stocks and fluxes across Western Oregon USA. Global Change Biol. 2004;10:1429–44.

78. Law BE, Turner DP, Lefsky M, Campbell J, Guzy M, Sun O, et al. Carbon fluxes across regions: observational constraints at multiple scales, Scaling

79. Busse MD. Downed bole-wood decomposition in Lodgepole pine forests of central Oregon. Soil Sci Soc Am J. 1994;58(1):221–7.

80. Mitchell RG, Priesler HK. Fall rate of lodgepole pine killed by the mountain pine beetle in Central Oregon. West J Appl For. 1998;13:23–6.

81. Brown M, Black TA, Nesic Z, Foord VN, Spittlehouse DL, Fredeen AL, et al. Impact of mountain pine beetle on the net ecosystem production of lodgepole pine stands in British Columbia. Agr Forest Meteorol. 2011;150(2):254–64.

82. Turner DP, Koerper GJ, Harmon ME, Lee JJ. A carbon budget for forests of the conterminous United States. Ecol Appl. 1995;5:421–36.

83. Law BE, Sun OJ, Campbell J, Van Tuyl S, Thornton PE. Changes in carbon storage and fluxes in a chronosequence of ponderosa pine. Global Change Biol. 2003;9(4):510–24.

84. Sun OJ, Campbell J, Law BE, Wolf V. Dynamics of carbon stocks in soils and detritus across chronosequences of different forest types in the Pacific Northwest. USA Global Change Biol. 2004;10(9):1470–81.

85. White MA, Thornton PE, Running SW, Nemani RR. Parameterization and sensitivity analysis of the BIOME-BGC terrestrial ecosystem model: net primary production controls. Earth Interact. 2000;4:1–85.

86. Jin S, Yang L, Danielson P, Homer C, Fry J, Xian G. A comprehensive change detection method for updating the National land cover database to circa 2011. Remote Sens Environ. 2013;132:159–75.

87. GAP. US Geological Survey, Gap Analysis Program (GAP). National Land Cover, Version 2. http://gapanalysis.usgs.gov/gaplandcover/data/. 2014.

88. Eidenshink J, Schwind B, Brewer K, Zhu ZL, Quayle B, Howard SM. A project for monitoring trends in burn severity. Fire Ecol. 2007;3:3–21.

89. Ohmann JL, Gregory MJ. Predictive mapping of forest composition and structure with direct gradient analysis and nearest-neighbor imputation in coastal Oregon. USA Can J For Res. 2002;32:725–41.

90. Daymet: Daily Surface Weather Data on a 1-km Grid for North America, Version 2. Data set. Available on-line [http://daac.ornl.gov] from Oak Ridge National Laboratory Distributed Active Archive Center, Oak Ridge, Tennessee, USA [database on the Internet] 2014. Available from: http://dx.doi.org/10.3334/ORNLDAAC/1219. Accessed: January 1, 2015.

91. Thornton PE, Running SW, White MA. Generating surfaces of daily meteorological variables over large regions of complex terrain. J Hydrol. 1997;190(190):214–51.

92. Thornton PE, Running SW. An improved algorithm for estimating incident daily solar radiation from measurements of temperature, humidity, and precipitation. Agr Forest Meteorol. 1999;93:211–28.

93. Thornton PE, Hasenauer H, White MA. Simultaneous estimation of daily solar radiation and humidity from observed temperature and precipitation: an application over complex terrain in Austria. Agr Forest Meteorol. 2000;104:255–71.

94. Hasenauer H, Merganicova K, Petritsch R, Pietsch SA, Thornton PE. Validating daily climate interpolations over complex terrain in Austria. Agr Forest Meteorol. 2003;119:87–107.

95. Daly C, Halbleib M, Smith JI, Gibson WP, Doggett MK, Taylor GH, et al. Physiographically sensitive mapping of climatological temperature and precipitation across the conterminous United States. Int J Climatol. 2008;28(15):2031–64.

96. CONUS. Conterminous United States multi-layer soil characteristics data set for regional climate and hydrology modeling. 2007. http://www.soilinfo.p-su.edu/index.cgi?soil_data&conus.

97. Hayes DJ, Turner DP, Stinson G, McGuire AD, Wei YX, West TO, et al. Reconciling estimates of the contemporary North American carbon balance among terrestrial biosphere models, atmospheric inversions, and a new approach for estimating net ecosystem exchange from inventory-based data. Global Change Biol. 2012;18(4):1282–99.

98. Butman D, Raymond PA. Significant efflux of carbon dioxide from streams and rivers in the United States. Nat Geosci. 2011;4(12):839–42.

and uncertainty analysis in ecology: methods and applications. New York: Columbia University Press; 2006.

Integrating forest inventory and analysis data into a LIDAR-based carbon monitoring system

Kristofer D Johnson[1*], Richard Birdsey[1], Andrew O Finley[2], Anu Swantaran[3], Ralph Dubayah[3], Craig Wayson[1] and Rachel Riemann[4]

Abstract

Background: Forest Inventory and Analysis (FIA) data may be a valuable component of a LIDAR-based carbon monitoring system, but integration of the two observation systems is not without challenges. To explore integration methods, two wall-to-wall LIDAR-derived biomass maps were compared to FIA data at both the plot and county levels in Anne Arundel and Howard Counties in Maryland. Allometric model-related errors were also considered.

Results: In areas of medium to dense biomass, the FIA data were valuable for evaluating map accuracy by comparing plot biomass to pixel values. However, at plots that were defined as "nonforest", FIA plots had limited value because tree data was not collected even though trees may be present. When the FIA data were combined with a previous inventory that included sampling of nonforest plots, 21 to 27% of the total biomass of all trees was accounted for in nonforest conditions, resulting in a more accurate benchmark for comparing to total biomass derived from the LIDAR maps. Allometric model error was relatively small, but there was as much as 31% difference in mean biomass based on local diameter-based equations compared to regional volume-based equations, suggesting that the choice of allometric model is important.

Conclusions: To be successfully integrated with LIDAR, FIA sampling would need to be enhanced to include measurements of all trees in a landscape, not just those on land defined as "forest". Improved GPS accuracy of plot locations, intensifying data collection in small areas with few FIA plots, and other enhancements are also recommended.

Keywords: Aboveground biomass; Carbon; Inter-comparison; LIDAR; Forest inventory and analysis

Background

Accurate, high resolution Light Detection and Ranging (LIDAR) biomass maps facilitate decision making to sequester C, for example, by identifying areas for protecting existing C stocks or planning for additional C accumulation in other areas. However, biomass maps modeled from LIDAR returns have uncertainty that should be assessed for the maps to be more useful. Forest inventories such as the U.S. Forest Service Forest Inventory and Analysis (FIA) program can be valuable for evaluating LIDAR-based and other remotely sensed biomass maps. FIA plots are systematically arranged to provide spatially unbiased estimates of forest biomass over an area, follow well-documented measurement protocols, and are quality

* Correspondence: kristoferdjohnson@fs.fed.us
[1]USDA Forest Service, Northern Research Station, Newtown Square, Pennsylvania, USA
Full list of author information is available at the end of the article

controlled. Thus, FIA plots have been successfully used to calibrate remote sensing-based models [1-3] and provide independent estimates of biomass stocks [4,5], and biomass change [6]. The FIA plot design has also been used specifically in calibrating and validating LIDAR-derived biomass maps [7,8] and in optimizing sampling strategies to train LIDAR biomass models [9].

There are also challenges with using FIA data for biomass map evaluation since the program was not specifically designed for this purpose. First, FIA defines forest land based on both tree stocking or land use[a], and does not usually sample areas that are considered to be "nonforest" (e.g. pastures, roads, suburban areas, parks and rights-of-way) even if trees are present [10]. In Maryland, about 25% of the aboveground carbon was estimated to be stored in "nonforest land" [11] and this discrepancy alone could account for considerable disagreement between FIA data and LIDAR mapped results. Another issue is uncertainty in the estimation of biomass from field measurements, specifically because of allometric model error and choice of allometric model to apply [12]. Finally, the FIA plot design and geolocation errors of FIA plots complicate comparisons with biomass map pixels.

Maryland is one of several U.S. states with statewide LIDAR available, and since 2011 the use of LIDAR-derived biomass maps has been explored for carbon monitoring purposes [13]. Unlike many large-area remote sensing biomass mapping efforts [2,3], FIA plots in Maryland were not used for the development of the LIDAR biomass prediction models in this study. Instead the FIA data in Maryland was used as an independent comparison to LIDAR-based biomass maps trained from a separate field inventory. In the current analysis we report key results about comparing FIA data with two high resolution (30-m) biomass maps, one using random forest and one using Bayesian spatial regression (see Methods) at both the plot and county scales in a case study of the Anne Arundel and Howard counties in Maryland (Figure 1). Although we include standard comparison statistics (R^2, RMSE, etc.), the purpose was not to determine which biomass map was "better" for the two counties. Rather, we investigated issues with integrating FIA data with LIDAR-based maps by analyzing the consequences of incomplete tree data (i.e. no measurements of "nonforest" trees) and measurement error (i.e. allometric model choice and allometric

model prediction error). Finally, we recommend ways that the FIA protocol could be enhanced for integration with LIDAR-based carbon monitoring, and suggest some approaches to efficiently combine the two observation systems.

Results
Allometric model and model choice errors
When simulated allometric model errors were propagated to each plot, they were relatively small with an average 95% confidence interval of only 5 Mg/ha (11% of the total biomass) for all the plots. One plot's confidence interval was 93% of its total biomass, though this plot also had low biomass in absolute terms (22 Mg/ha). The mean biomass calculated from the three different allometric model choices was somewhat variable, although none of the estimates was significantly different from the others ($P < 0.05$). The highest estimate resulted from the Species Specific approach (mean: 208 Mg/ha, std dev: 147 Mg/ha), which was 31% higher than the CRM (mean: 159 Mg/ha, std dev: 98 Mg/ha). The Jenkins equations also produced estimates that were higher than the CRM, by 16% (mean: 184 Mg/ha, std dev: 119 Mg/ha). For the following comparisons to the LIDAR-modeled biomass map products, the Jenkins estimate was used since the Jenkins equations were applied to estimate biomass for the plot data used in the training of the LIDAR-models.

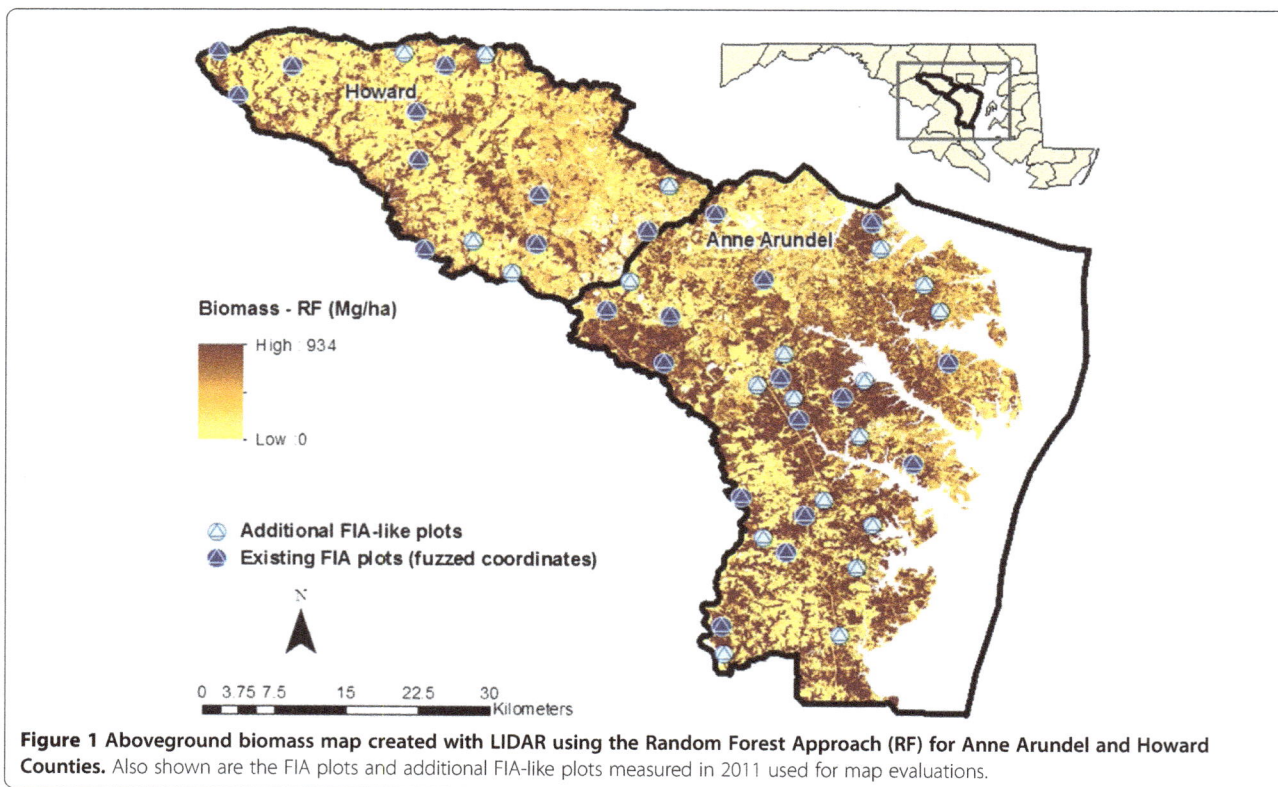

Figure 1 Aboveground biomass map created with LIDAR using the Random Forest Approach (RF) for Anne Arundel and Howard Counties. Also shown are the FIA plots and additional FIA-like plots measured in 2011 used for map evaluations.

Plot and pixel comparisons

Overall, the biomass estimates from the FIA + NFI and FIA-like plots were moderately correlated with both LIDAR maps. For the RF model, the R^2 was 0.49 with a slope of 0.94 (RMSE − 91.5 Mg/ha). For the BAY model, the R^2 was 0.52 with a slope of 1.34 (RMSE − 89.0 Mg/ha) (Figure 2a,b). Both LIDAR maps predicted higher biomass in areas where the plot biomass measured in the field was very low or zero and yet also tended to predict lower biomass for plots with very high biomass. For example, disagreements were reflected in the comparisons of cumulative distributions (Figure 2c). Half of the field measured observations had biomass less than 1 Mg/ha, whereas half of the predicted values at the same locations were less than 68 and 81 Mg/ha for the RF and BAY models, respectively. In contrast, the mean biomass of the 5 highest biomass plots was 434 Mg/ha, compared to 228 and 240 Mg/ha for the RF and BAY models, or about half of the ground measurement. There was greater disagreement between the distributions of the BAY map (KS = 0.55) and the plot estimate than the RF map (KS = 0.34).

When purely "nonforest" plots were removed, so that only traditionally field-measured plots were included in the regressions (n = 42), the agreement was poor. The RF map R^2 was 0.27 with a slope of 0.91 and the BAY model R^2 was 0.43 with a slope of 1.28. The BAY map also included pixel-level 95% confidence intervals which allowed for the comparison of 95% confidence intervals (with propagated error) of individual plot measurements (Figure 3). The mean confidence interval at the pixel level for the LIDAR model was 246 Mg/ha and 84% of the plots had a confidence interval that overlapped the confidence intervals of the corresponding pixels.

Forest and nonforest county level estimates

County-level biomass estimated from inventory data was higher in the FIA + NFI inventory than the FIA-only inventory (Table 1a). In Anne Arundel biomass in "nonforest" conditions accounted for 27%, or 1.42 Tg, of the total. Similarly, "nonforest" biomass was 21%, or 1.35 Tg, of the total in Howard County. The biomass in Howard County was also higher than Anne Arundel for both estimates (Table 1a).

The mean and total biomass was also calculated from pixel values at the plot locations for county level estimates (Table 1a). In Anne Arundel County, the LIDAR-derived value was 53 and 51% higher (about 3 Tg) than the FIA + NFI estimate for the RF and BAY maps, respectively. In contrast, for Howard County, the difference between the LIDAR-derived valued and FIA + NFI estimate was small.

When all biomass map pixels were summed and compared to total biomass estimated from field data, there were even larger discrepancies in Anne Arundel County,

being well outside the 95% intervals of the FIA + NFI estimate (Table 1b). The LIDAR maps were more than twice as high in biomass, a difference of 6.95 and 5.99 Tg biomass for the RF and BAY models, respectively. In contrast, in Howard County all the LIDAR-derived county biomass estimates were well within the FIA + NFI confidence interval. Additionally, the summed pixel estimates were lower than the FIA + NFI estimate for maps in Howard, 0.94 and 1.09 Tg for the RF and BAY maps, respectively.

Discussion
Anne Arundel and Howard county case study

When FIA data were combined with a "nonforest" inventory, the plot data proved to be valuable for evaluation of LIDAR biomass maps in Anne Arundel and Howard Counties. Despite the low R^2's of both models, the comparisons still revealed that the RF model seemed to be less biased with a slope closer to 1, but tended to more severely underestimate plots with very high biomass (the reason for the lower R^2) compared to the BAY model. Plots traditionally measured by the FIA program (i.e. no "nonforest" inventory enhancement) were also useful to evaluate LIDAR maps at the plot scale, but only for densely forested plots not confounded by plots that had both "forest and "nonforest" conditions. This comparison was aided by including the 95% confidence intervals of the LIDAR model for each pixel and the propagated allometric model and sampling errors of the plots (Figure 3).

There were significant discrepancies at the county scale, indicating that the biomass maps are predicting low biomass in areas where little or no biomass is measured. The consequence of predicting low biomass instead of none for landcovers with no trees results in comparatively larger total biomass for the counties when the pixels are summed because these areas are proportionately very large. It is unclear why the difference in Anne Arundel was so much greater than in Howard, though we note the higher proportion of agricultural landcover in Howard (30% v. 12%, determined from NLCD 2006 data). It is possible that the LIDAR biomass maps at 30-m resolution may be more successful at delineating tree v. tree-less areas in counties with higher agricultural landcover like Howard, as opposed Anne Arundel that perhaps has landcover with more fragmented tree canopies.

Using the same allometric model for both inventory and map estimates (the Jenkins equations [14]) resulted in relatively small errors compared to the choice of the LIDAR biomass model in this study. At the same time, the different allometric models led to significantly variable estimates. The CRM method has been shown to produce substantially lower biomass estimates in a number of studies due to the incorporation of tree height. For example, the 16% difference between the Jenkins and CRM methods

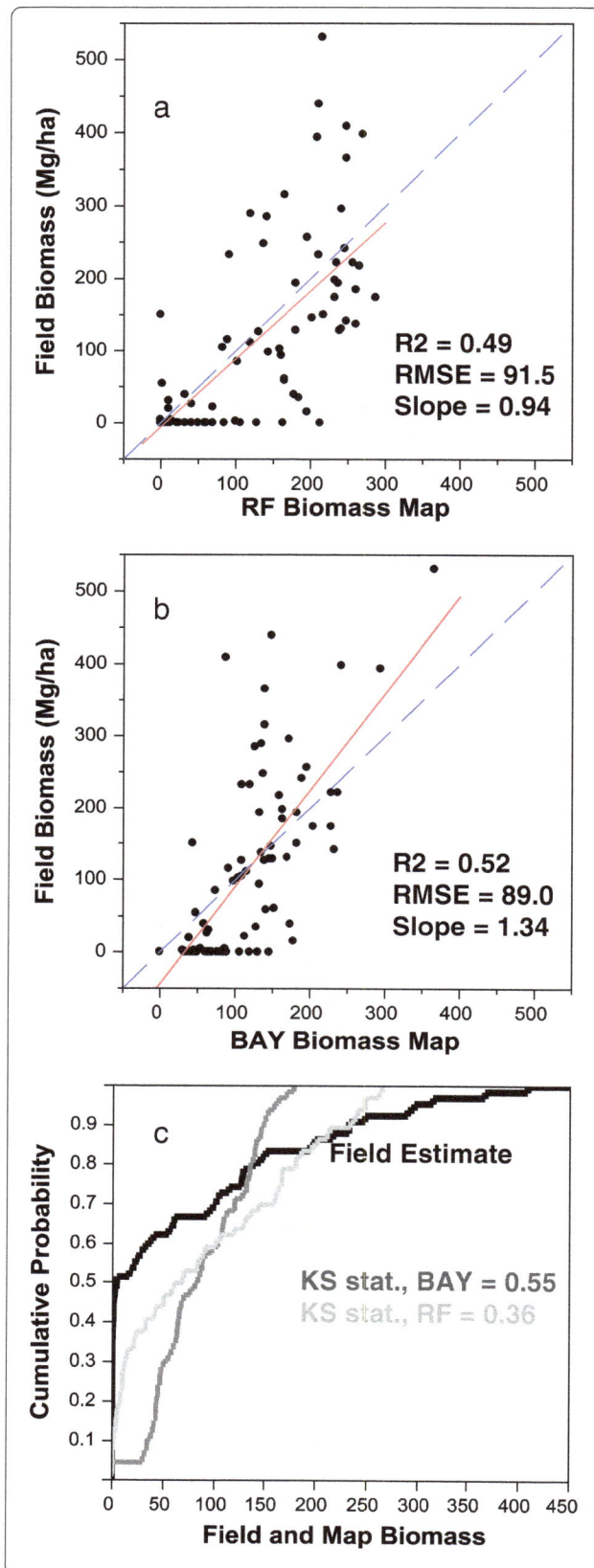

Figure 2 Comparisons of biomass map pixels and field plots for the (a) RF and (b) BAY biomass maps. (c) Comparisons of the cumulative distribution functions and respective Kolmogorov-Smirnov statistics (KS stat.) for both maps. High KS stat indicates a higher maximum difference between the distributions.

found in this study was the same as that found on average nationally [15], but lower than the 8% difference found for Northeastern forests [16]. [12] suggested that model selection error introduced 20 to 40% to live biomass uncertainty, a range that captures the 31% difference in mean biomass between the CRM and Species Specific estimates of this study. However, these differences are less important for the purpose of map evaluation here, given that the maps used the same allometric models as the inventory for their training data.

In terms of whether the biomass maps are "accurate enough" to be recommended for carbon management purposes in these counties, it appears that one can obtain reasonable biomass values in many, but not all, areas at the plot scale (roughly 1.5 acres). Furthermore, as mentioned above, county scale estimates were only useful for Howard County, but not Anne Arundel, where more work is needed. The current evaluations have already been considered in the process of designing more effective field collection strategies and modeling approaches for developing improved biomass maps in Maryland counties. For example, newer random forest models exclude variable radius plot locations that had biomass detected by LIDAR over a 30-m area (the pixel size) but that had no trees measured in them. This can occur when trees are at the edge of a pixel, too far away to be included in the variable radius plot measurement, but still being observed by the LIDAR. When these locations were excluded the resulting model had better agreement with the FIA data because there were fewer instances where biomass was predicted in the FIA plot but there was no biomass measured ($R^2 = 0.59$, RMSE = 82.4 Mg/ha, slope = 1.1; compare with Figure 2a). Another issue contributing to the poor agreement was probably our combination of a single plot design of the NFI and the regular FIA plot design, resulting in inconsistent plot-pixel comparisons throughout the sample. As "nonforest" biomass is important to consider in Maryland and elsewhere, plot designs and overall strategies for addressing the "nonforest" biomass gap, are discussed below.

Conclusions

Enhancing the FIA protocol by sampling trees on nonforest land

It is critical that field biomass data be both accurate and complete for evaluating biomass maps in order to improve the maps. Despite the uncertainty estimates and inconsistencies revealed by this case study, there are good reasons

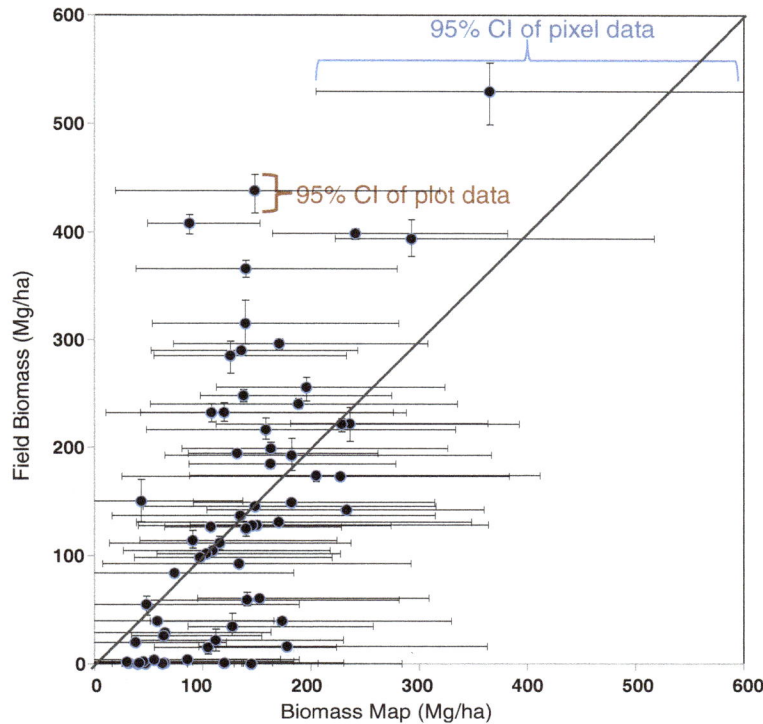

Figure 3 Comparison of field measured biomass (FIA and FIA-like) and mapped biomass at the plot level, including only plots with "forest" conditions according to the FIA definition (i.e. purely "nonforest" plots were excluded). The vertical bars are the 95% confidence interval of the mean of the field biomass values after propagating allometric and sampling errors. The horizontal bars are the mean 95% confidence levels of the LIDAR biomass map pixels for the BAY model, sampled from the posterior predictive distribution that acknowledges spatial dependence (see Methods).

for integrating FIA data with LIDAR biomass maps in an aboveground carbon monitoring system. A consistent analysis would require an all-tree inventory enhancement to the current FIA protocol. This enhancement greatly facilitates comparisons at both the plot and county scales, especially in fragmented canopy landscapes that are common throughout the eastern United States. If both the maps and FIA are composed of "wall-to-wall" biomass estimates, then there is no need to distinguish "forest" from "nonforest" areas for estimating total land biomass.

The main barrier to enhancing FIA data collection to include "nonforest" trees is the additional cost. An all-tree inventory would require field crews to sample trees in "nonforest" areas that are currently monitored mostly with aerial imagery. However, the cost may be lower than expected because the pre-field work imagery analysis that is already performed by FIA could screen out many plots that have essentially no chance of having tree biomass (e.g. plots located in agricultural fields). Furthermore, FIA crews already visit many "mixed condition" plots that

Table 1 County level comparisons of mean and total aboveground biomass

	Anne Arundel		Howard	
	Mean biomass	Total biomass	Mean biomass	Total biomass
a. ESTMATES FROM SAMPLED DATA	Mg/ha (95% CI)	Tg (95% CI)	Mg/ha (95% CI)	Tg (95% CI)
FIA (2006-2010)	41.4 (18.0, 65.2)	3.90 (1.66, 6.14)	74.9 (26.4, 123.5)	5.24 (1.85, 8.64)
FIA (2006-2010) + NFI (1999)	56.5 (30.5, 82.5)	5.32 (2.87, 7.76)	94.1 (41.1, 147.1)	6.59 (2.88, 10.29)
LiDAR-RF sample	86.5 (62.3, 110.7)	8.14 (5.87, 10.42)	93.5 (60.6, 126.5)	6.54 (4.24, 8.85)
LiDAR-BAY sample	85.2 (70.8, 99.6)	8.02 (6.67, 9.38)	89.4 (74.4, 104.3)	6.25 (5.21, 7.30)
b. ESTIMATES BY SUMMING PIXELS				
LiDAR-RF		12.89		5.65
LiDAR-BAY		11.93		5.5

have "nonforest" trees and so the extra time spent could be minimal, especially if only a subset of typical tree measurements are needed. The FIA program would also need to consider availability of capacity to accommodate this demand for more detailed information, but we note that cost-sharing agreements with other entities to this end have already occurred [17,18]. When a current all-tree inventory is cost prohibitive, another approach is to use previous all-tree inventories, recognizing limitations as was done in this study. For example, urban tree inventories are already available in many areas [19].

When designing an all-tree inventory to integrate into the FIA protocol, there are several alternatives to consider, each with their own set of limitations. One option is to measure the trees in all the "nonforest" conditions within the actual FIA subplots, without modifying the plot design [18]. The advantage to this approach is that newly collected data from purely nonforest plots can be easily combined with existing FIA plot data. Nonetheless, a major disadvantage to this approach is that in residential areas the four subplots will commonly cover multiple properties with different owners. Obtaining permission to visit all the subplots would therefore be more difficult and increase the chances of denied access and potentially bias the study. An alternative plot design like the one used to collect the NFI dataset of this study [20] reduces this impact on field time by sampling one larger plot instead of four. However, this design makes the total area sampled smaller and less compatible with existing FIA measurements and for relating to map pixels. A compromise between the two options is one that FIA is currently implementing in urban forest inventories, where every tree is measured in a single circular plot, located at the center of current FIA plots, and has the same area as four FIA subplots (670 m^2) (James Westfall, personal communication). An advantage to the large continuous area is that it is much more useful for comparing to map pixels, though the design does not strictly complement the original FIA design.

Additional enhancements and modifications

Geolocation error was not evaluated in this study but also contributes to confounding plot and pixel comparisons, especially near forest and agricultural field interfaces. For example, GPS error with the current units used by FIA is between 1 and 13 meters in heavy canopy in northeastern forests (Richard McCullough, personal communication). Thus, another enhancement to the FIA protocol would be to obtain more accurate coordinates. Though survey-grade GPS units would be ideal, even submeter accuracy obtained from relatively inexpensive units would be a great improvement.

In some situations it may be useful to intensify the sample size to obtain more information in areas where

biomass is highest or lowest relative to the average. From our experience, it is more useful to locate the additional plots in a manner similar to the FIA design, so that the additional data are complementary for county level estimates [17]. Instead, we somewhat opportunistically located supplemental FIA-like plots in pixels indicated as forest by NLCD maps, though its stratification is not fully compatible with FIA definitions of "forest" and "nonforest". The unintended result was that the additional FIA-like plots were located in homogeneous areas that were higher in biomass than the average FIA sample. Thus, to obtain the most information from plot intensification, a systematic design throughout the area of interest should be maintained.

Another common issue is the disparity of collection years of the different types of data. Though the error resulting from the difference in years is probably small compared to, for example, the LIDAR-biomass model error, efforts should be made to harmonize the date of LIDAR collection and the date of field data collection. Practically speaking, in the current study this would have been difficult since we were using data available to us at the time, but this should be considered in planning FIA-LIDAR data integration.

For carbon monitoring purposes, it is important to consider the discrepancies in biomass estimates from different allometric model choices [21]. The impact of allometric model choice depends on the objective for making the biomass estimate. If the estimate is used to quantify absolute biomass stocks for comparison to other counties and states, then the same allometric approach should be used in all cases. When biomass maps are used as tools for estimating biomass change in a single county, the negative consequence of choosing allometric models that are different than neighboring areas is less serious, though model selection will still have an impact. There is also unknown error when applying allometric equations developed for forestland trees, to trees located in yards and parking lots that may have different growth forms [22]. Thus, it is difficult to recommend one approach, but it is important to recognize that different allometric models can produce significantly different results, and therefore it would be useful to report estimates from more than one method or validate the selection of an allometric model with some additional field measurements of tree biomass.

Another way to improve the comparability of FIA and LIDAR estimations is to design mapping approaches that are more consistent with the ground data. For example, being careful to mimic the distribution of field measured biomass at point locations will result in a greater chance that the total biomass predicted by maps will have better agreement. Furthermore, since FIA has committed to providing biomass estimates using the

CRM allometric approach, training data for making LIDAR relationships should also use this method. Additionally, providing meaningful pixel level confidence intervals (e.g. the BAY model of this study), are useful for analyzing agreement. Finally, when an all tree forest inventory is not practical, a serviceable but less ideal alternative is to exclude residential areas from LIDAR biomass maps so that they are more comparable with FIA measurements.

Finally, to achieve a robust and spatially explicit carbon monitoring system, it is most ideal for comparison purposes to have independently sampled model training and model evaluation field datasets, as was done in this study. Nevertheless, we think it is worthwhile to examine other approaches that could represent a fully integrated biomass inventory system, including assessing the uncertainties and costs. For example, it could be significantly less costly to collect all the field data needed for training and verification of biomass maps at the same time, rather than supporting two independent field efforts.

Methods
Study area and datasets
The study area includes the Anne Arundel and Howard counties composed mostly of oak-hickory forest [23]. The counties are almost cleanly divided by two different physiographic regions. Anne Arundel belongs to the Coastal Plain Province principally containing sandy soils at low elevation (100 ft). In contrast, Howard belongs to the Piedmont Province containing loamy and clayey soils at somewhat higher elevations (100–500 ft).

There were three field inventory datasets used to evaluate LIDAR biomass maps. Two of the inventories followed the conventional Forest Inventory and Analysis ("FIA") plot design, that is four clustered subplots, each 168 m^2, and spaced 7-m apart [10] (Figure 4). FIA tree level data for 64 plots within the Anne Arundel and Howard counties were downloaded from the FIA Data-Mart website for the 2006 to 2010 cycle period. There were a total of 72 forest plot locations, but 8 of these plots were not visited due to denied access. Of the 64 visited plots, only 9 were recorded to have purely "forest" conditions; that is, some proportion of the sample-plot area was determined to be "nonforest". Therefore, to augment the dataset for plot-level comparisons in forested areas, an additional 20 forest plots of the same dimensions were measured in the two counties in 2011 ("FIA-like") (Figure 1). The FIA-like plot locations were placed within forest landcover indicated by National Land Cover Database 2006 (NLCD; [24]). Finally, we took advantage of a previously collected dataset - a Nonforest Inventory ("NFI") collected by [20] in Maryland in 1999 at FIA plots. An important nuance of the NFI dataset is that

only the center subplot was measured, sampling a larger subplot area, but overall the sampled area per plot changed from 670 m^2 to 400 m^2. Due to the disparity in inventory years between the NFI and FIA inventories, the locations of each plot were checked with imagery one by one for evidence of clearing or forest ingrowth, but none was found. Despite the difference in inventory years, and recognizing the potential errors of combining different plot designs, for some analyses we used the NFI dataset to fill the "nonforest" biomass data gap when trees were present but not measured in the regular FIA data collection ("FIA + NFI").

LIDAR-derived biomass maps
Leaf-off LIDAR data collected by the Maryland Department of Natural Resources (DNR) over Anne Arundel and Howard Counties in 2004 were used to derive biomass maps for this study. LIDAR first and last returns were interpolated and differenced to obtain a normalized difference surface model (nDSM) with a resolution of 2 m. Next, a high resolution tree cover map was created by segmenting the LIDAR data and NAIP imagery [25,26] and further used to mask out everything but tree crowns on the nDSM. The resulting canopy height model (CHM) was used to calculate height percentiles, density metrics, canopy cover and other LIDAR metrics describing the vertical and spatial distribution of vegetation structure within 30 m pixels [13].

Field biomass data for developing LIDAR biomass models were collected in 300 new variable radius plots in the two counties, independently of the FIA program. Variable radius sampling is typically used to estimate basal area of a forested tract by sampling trees with probability proportional to tree basal area and is known to be a quick and accurate method for estimating stand basal area and volume [27].We collected tree measurements over variable radius plots using a model-based stratified sampling approach based on the NLCD land cover class and LIDAR height class. Field based allometric estimates of biomass, calculated using equations from [14], were then related to LIDAR variables to predict biomass using Bayesian model averaging and Random Forests regression (Figure 1).

Random forest model
Random Forests, (RF) [28,29] is a machine learning algorithm in which a large number of regression trees are fit to a dataset (~500). Bootstrap samples are used from the data to construct each tree and at each node, a random subset of predictors are tested. Response values from all trees are averaged to provide accurate predictions and "out-of-bag" error estimates are calculated using 37% of the data in each regression tree, thus avoiding over fitting and reducing the need for cross validation. Predictions

Figure 4 An example of the size of FIA subplots overlaid onto imagery and a biomass map of 30-m pixel resolution.

from RF regression can be used to model linear/non-linear relationships using a large number of predictor variables. The RF model of this study, using the 300 variable radius plots, had an R^2 of 0.67 and RMSE of 73.5 Mg/ha and, similar to findings in other studies for mixed forests [7,30].

Bayesian spatial regression model

Given ground data locations and coinciding LIDAR height metrics, we used a Bayesian spatial regression model (BAY) to make pixel-level biomass predictions. Exploratory variogram analysis showed that a non-spatial LIDAR height metric regression model did not adequately explain the spatial dependence in biomass observations, i.e., there was spatial autocorrelation among the model residuals. The presence of spatial dependence among residuals violates model assumptions which can lead to incorrect parameter and prediction inference [31]. The spatial regression model includes spatial random effects that estimate, and accommodate, this residual structure. Here, the random effects arise from a spatial Gaussian process with a covariance matrix constructed using an exponential spatial correlation function. In addition to the slope coefficients associated with the LIDAR metrics and an intercept, this model estimates a spatial correlation function decay and variance parameter, as well as the non-spatial residual variance parameter. The analysis was conducted in the spBayes R package using the spLM function [32]. This modeling framework uses a Markov chain Monte Carlo approach to generate samples from parameters' posterior distributions. Given these posterior samples, composition sampling is used to sample from the posterior predictive distribution of biomass at unobserved locations (pixels) [33]. From these pixel-level posterior distributions any error statistic can be created by simply summarizing the sets of posterior samples. In the current study, the 95% confidence levels were used to map pixel-level uncertainty in Anne Arundel and Howard Counties.

Fitted values for the BAY model yield RMSE of 34.67 and an R^2 of 0.91. Note that these values are not strictly comparable to those of the RF model because they reflect the highly flexible Gaussian process used to specify the BAY model's random effects for accurate interpolation of the observed data.

Analysis of measurement error

To investigate allometric errors, functions for standard errors were derived by simulating a population of 10,000 data points around the regression lines of each species group published by [14]. For each species group, populations were created until the R^2 from the regression line from the simulated points matched the R^2 from the original equation. Next, points equaling the number of observations used in the original equations were randomly drawn from a Weibull distribution and a new standard error function was fit to the subset, where at least 100 subsets and associated standard error functions were generated. Tests of this method with actual destructive harvest data from Canada's Energy from the Forest (ENFOR) dataset [34] showed consistent results and reflected increasing uncertainty in biomass estimates of larger trees (Figure 5). The mean of all the standard error functions for each species group was then applied on a tree by tree basis using a Monte Carlo simulation technique to calculate plot level 95% confidence intervals of the plot mean (see [35] for further details). Thus, the final plot level 95% confidence interval depended on the mixture of species groups found on the plot and their diameters.

For investigating differences in mean biomass for different allometric approaches, three sets of equations relating biomass to diameter at breast height (DBH) were applied. One set of equations was derived using the Component Ratio Method (CRM), the method used by the FIA program to report biomass stocks. The CRM equations calculate bole volume as a first step and so require "bole height" (height of the stem to 4 in diameter) in

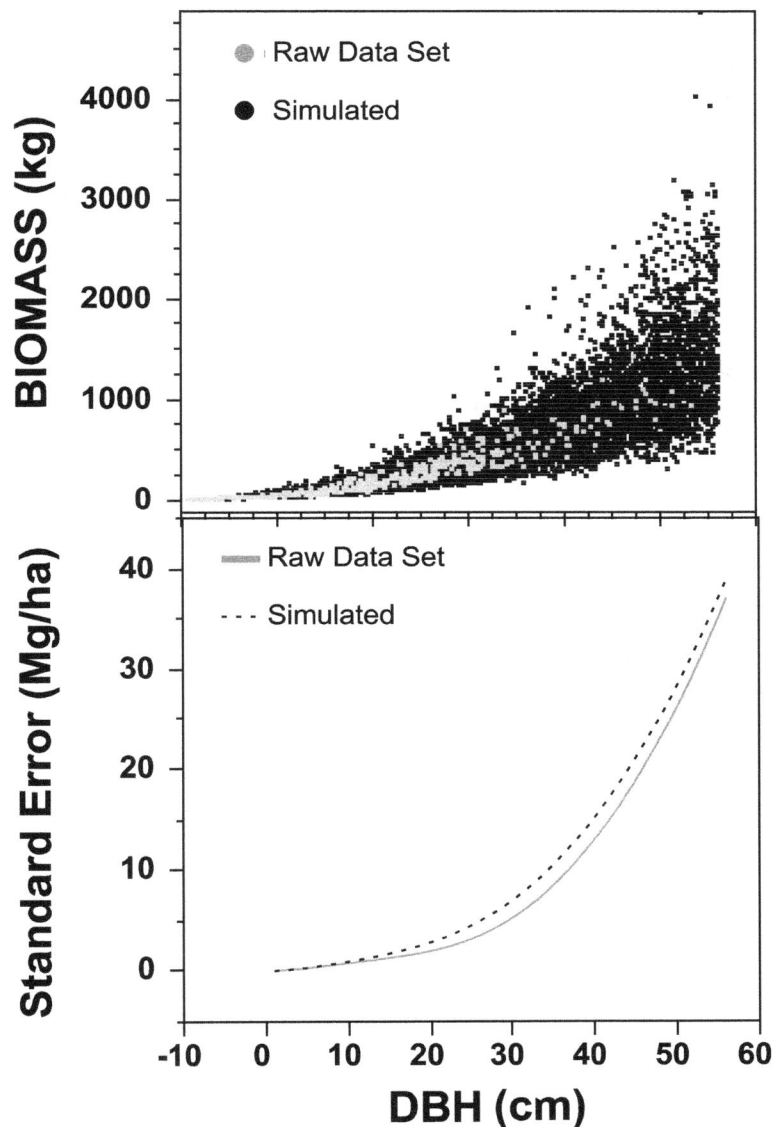

Figure 5 The comparison of raw destructive harvest data from the ENFOR dataset and simulated biomass results (top panel) and the associated standard error function (bottom panel). This example is for the "mixed hardwood" species group from [10].

addition to DBH measurements, and the relationships are region-specific [36]. In contrast, equations applied from [14] require only a DBH measurement, are generalized for 10 species groups, and are not region-specific. Finally, yet another set of local equations, not volume-based, for species found in Maryland was used ("Species Specific") [37-39]. In the case that no specific equation was available for a species in the Species Specific approach, the general Jenkins equation was substituted.

For comparing FIA plot measurements to mapped biomass, we chose to use the biomass equations from [14] because they represented mid-level biomass values (of the three equation types we tested) and because they were

also used in the separate inventory used for the LIDAR biomass models. At the plot level, biomass map values were extracted for the coordinates of each FIA subplot from which an average of the four pixel values was calculated and compared to the ground measurement (Figure 4). The cumulative distribution functions and Kolmogorov-Smirnov (KS) statistic were calculated and compared for both the FIA observations and the biomass map observations. The KS statistic a metric of the maximum distance between the field and mapped cumulative distribution functions, where higher values reflect poorer agreement [40]. At the county level, we used two comparison approaches with the FIA + NFI data. First, we calculated the mean biomass from the pixel values extrapolated from the plot locations (in units of Mg/ha), and then multiplied the mean by the area of the county (ha) to get total biomass. This approach allowed us to mimic the FIA sample design to investigate disagreement at the plot locations. In the second approach, we calculated total biomass by simply summing the map pixels and then multiplying by 0.09 to adjust for the 30-m and 1 ha difference. We did not include the "FIA-like" plots in county level comparisons in order to maintain a systematic random sample. We also note, but do not consider in this analysis, the discrepancy between the area sampled in the field (670 m^2) and the pixel area extracted for 4 subplots (3600 m^2) which leads to additional errors [40]. All statistical analyses were performed using JMP [41].

Endnote

[a]Land at least 120 feet wide and 1 acre in size with at least 10 percent cover (or equivalent stocking) by live trees of any size, including land that formerly had such tree cover and that will be naturally or artificially regenerated. Tree-covered areas in agricultural production settings, such as fruit orchards, or tree-covered areas in urban settings, such as city parks, are not considered forest land.

Competing interests

The authors declare that they have no competing interests.

Authors' contributions

The study was devised by KJ, RB, AS and RD. KJ and CW performed the allometric uncertainty analysis. Comparisons were performed by KJ with ground data from RR and LIDAR biomass maps from AF, AS, and RD. All authors read and approved the final manuscript.

Acknowledgements

We thank Chhun-Huor Ung for facilitating use of Canada's Energy from the Forest program data (ENFOR) which was valuable for comparing our simulations of standard errors. Matt Patterson collected the field data for the FIA-like plots and provided technical assistance. We thank James Westfall and Tanya Lister of FIA who both reviewed the paper and provided useful comments. Other U.S. Forest Service personnel we would like to thank are Thomas Willard, Stephen Potter and Robert Ilgenfritz for help extracting information for new plots, and general guidance. Research was supported by NASA grant # NNX12AN07G (Dubayah, Principal Investigator). Support for the additional plots and analysis was funded by NASA's Carbon Monitoring System through grant # NNH12AU32I.

Author details

[1]USDA Forest Service, Northern Research Station, Newtown Square, Pennsylvania, USA. [2]Departments of Forestry and Geography, Michigan State University, East Lansing, Michigan, USA. [3]Department of Geographical Sciences, University of Maryland, College Park, Maryland, USA. [4]USDA Forest Service, Northern Research Station, Troy, New York, USA.

References

1. Blackard J, Finco M, Helmer E, Holden G, Hoppus M, Jacobs D: Mapping US forest biomass using nationwide forest inventory data and moderate resolution information. *Remote Sens Environ* 2008, 112:1658–1677.

2. Kellndorfer J, Walker W, LaPoint E, Bishop J, Cormier T, Fiske G, Hoppus M, Kirsch K, Westfall J: *NACP Aboveground Biomass and Carbon Baseline Data (NBCD 2000), U.S.A., 2000, Dataset*; 2012. Http://daac.ornl.gov from ORNL DAAC, Oak Ridge, Tennessee, U.S.A.

3. Wilson B, Lister A, Riemann R: A nearest-neighbor imputation approach to mapping tree species over large areas using forest inventory plots and moderate resolution raster data. *For Ecol Manage* 2012, 271:182–198.

4. Nelson R, Short A, Valenti M: Measuring biomass and carbon in Delaware using an airborne profiling LIDAR. *Scand J For Res* 2004, 19:500–511.

5. Zhang X, Kondragunta S: Estimating forest biomass in the USA using generalized allometric models and MODIS land products. *Geophys Res Lett* 2006, 33.

6. Powell S, Cohen W, Healey S: Quantification of live aboveground forest biomass dynamics with Landsat time-series and field inventory data: A comparison of empirical modeling approaches. *Remote Sens Environ* 2010, 114:1053–1068.

7. Popescu S, Wynne R, Scrivani J: Fusion of small-footprint lidar and multispectral data to estimate plot-level volume and biomass in deciduous and pine forests in Virginia, USA. *For Sci* 2004, 50:551–565.

8. Gonzalez P, Asner G, Battles J, Lefsky M, Waring K, Palace M: Forest carbon densities and uncertainties from Lidar, QuickBird, and field measurements in California. *Remote Sens Environ* 2010, 114:1567–1575.

9. Junttila V, Finley A, Bradford J, Kauranne T: Strategies for minimizing sample size for use in airborne LiDAR-based forest inventory. *For Ecol Manage* 2013, 292:75–85.

10. Bechtold WA, Patterson PL: *The Enhanced Forest Inventory and Analysis Program - National Sampling Design and Estimation Procedures*. US Department of Agriculture Forest Service: Ashville, NC: Southern Research Station; 2005:85.

11. Jenkins J, Riemann R: What does nonforest land contribute to the global C balance? In *Proc third Annu For Invent Anal Symp*. Edited by McRoberts R, Reams GA, Van Dousen PC, Mosor JW. U.S. Department of Agriculture: Forest Service, North Central Station; 2003.

12. Melson SL, Harmon ME, Fried JS, Domingo JB: Estimates of live-tree carbon stores in the Pacific Northwest are sensitive to model selection. *Carbon Balance Manag* 2011, 6:1–16.

13. Dubayah R: *County- Scale Carbon Estimation in NASA's Carbon Monitoring System, NASA CMS Pilot Projects. Biomass and Carbon Storage*. 2012.

14. Jenkins J, Chojnacky D, Heath L, Birdsey R: National-scale biomass estimators for United States tree species. *For Sci* 2003, 49:12–35.

15. Domke G, Woodall C, Smith J, Westfall J, McRoberts R: Consequences of alternative tree-level biomass estimation procedures on US forest carbon stock estimates. *For Ecol Manage* 2012, 270:108–116.

16. Westfall J: A Comparison of Above-Ground Dry-Biomass Estimators for Trees in the Northeastern United States. *North J Appl For* 2012, 29:26–34.

17. Lister AJ, Scott CT, Rasmussen S: Inventory methods for trees in nonforest areas in the Great Plains States. *Environ Monit Assess* 2012, 184:2465–74.

18. Nowak D, Cumming A, Twardus D, Hoehn R, Oswalt C, Brandeis T: *Urban forests of Tennessee, 2009*. U.S. Department of Agriculture, Forest Service: General Technical Report, SRS-149; 2012:52.

19. Nowak D, Greenfield E, Hoehn R, Lapoint E: Carbon storage and sequestration by trees in urban and community areas of the United States. *Environ Pollut* 2013, 178:229–236.

20. Riemann R: *Pilot Inventory of FIA Plots Traditionally Called "Nonforest."*. US Department of Agriculture, Forest Service: Northeastern Research Station; 2003.

21. Zhao F, Guo Q, Kelly M: Allometric equation choice impacts lidar-based forest biomass estimates: A case study from the Sierra National Forest, CA. *Agric For Meteorol* 2012, 165:64–72.

22. Russo A, Escobedo F, Timilsina N, Schmitt A, Varela S, Zerbe S: Assessing urban tree carbon storage and sequestration in Bolzano, Italy. *Int J Biodivers Sci Ecosyst Serv Manag* 2014, 10:54–70.

23. Miles P: *Forest Inventory EVALIDator web-application version 1.5.1.06*. St. Paul, MN: U.S. Department of Agriculture, Forest Service, Northern Research Station; 2014 [http://apps.fs.fed.us/Evalidator/evalidator.jsp].

24. Fry J, Xian G, Jin S, Dewitz J, Homer C, Limin Y, Barnes C, Herold N, Wickham J: Completion of the 2006 national land cover database for the conterminous United States. *Photogramm Eng Remote Sensing* 2011, 77:858–864.

25. O'Neil-Dunne JPM, Pelletier K, MacFaden S, Troy AR, Grove JM: Object-Based High-Resolution Land-Cover Mapping: Operational Considerations. In *Proc 17th Int Conf Geoinformatics*. Virginia, USA: Fairfax; 2009.

26. O'Neil-Dunne JPM, MacFaden S, Pelletier KC: **Incorporating Contextual Information into Object-Based Image Analysis Workflows**. In *Proc 2011 ASPRS Annu Conf.* Wisconsin: Milwaukee; 2011.

27. Radtke P, Packard K: **Forest sampling combining fixed- and variable-radius sample plots.** *Can J For Res* 2007, **37**:1460–1471.

28. Breiman L: **Random forests.** *Mach Learn* 2001, **45**:5–32.

29. Prasad P, Iverson L, Liaw A: **Newer Classification and Regression Tree Techniques: Bagging and Random Forests for Ecological Prediction.** *Ecosystems* 2006, **9**:181–199.

30. Skowronski N, Clark K, Gallagher M, Birdsey R, Hom J: **Airborne laser scanner-assisted estimation of aboveground biomass change in a temperate oak–pine forest.** *Remote Sens Environ* 2014. http://dx.doi.org/10.1016/j.rse.2013.12.015.

31. Hoeting J, Madigan D, Raftery A, Volinsky C: **Bayesian model averaging: a tutorial.** *Stat Sci* 1999, **14**:382–401.

32. Finley A, Banerjee S, Gelfand A: *spBayes for large univariate and multivariate point-referenced spatio-temporal data models.* ; 2013. arXiv Prepr arXiv13108192.

33. Banerjee S, Gelfand A, Carlin B: *Hierarchical Modeling and Analysis for Spatial Data.* 2004. Crc Press.

34. Ung C, Bernier P, Guo X: **Canadian national biomass equations: new parameter estimates that include British Columbia data.** *Can J For Res* 2008, **38**:1123–1132.

35. Yanai R, Battles J, Richardson A, Blodgett C, Wood C, Rastetter E: **Estimating uncertainty in ecosystem budget calculations.** *Ecosystems* 2010, **13**:239–248.

36. Heath L, Hansen M, Smith J, Miles P, Smith W: **Investigation into calculating tree biomass and carbon in the FIADB using a biomass expansion factor approach.** In *Proc FIA [Forest Invent Anal Symp 2008.* Edited by McWilliams W, Moisen G, Czaplewski R. Fort Collins, Colorado, USA: Park City, Utah: USDA Forest Service, Rocky Mountain Research Station; 2008.

37. Siccama T, Hamburg S, Arthur M, Yanai R, Bormann F, Likens G: **Corrections to allometric equations and plant tissue chemistry for Hubbard Brook Experimental Forest.** *Ecology* 1994, **75**:246–248.

38. Martin J, Kloeppel B, Schaefer T, Kembler D, McNulty S: **Aboveground biomass and nitrogen allocation of ten deciduous southern Appalachian tree species.** *Can J For Res* 1998, **28**:1648–1659.

39. Naidu S, DeLucia E, Thomas R: **Contrasting patterns of biomass allocation in dominant and suppressed loblolly pine.** *Can J For Res* 1998, **28**:1116–1124.

40. Riemann R, Wilson B, Lister A, Parks S: **An effective assessment protocol for continuous geospatial datasets of forest characteristics using USFS Forest Inventory and Analysis (FIA) data.** *Remote Sens Environ* 2010, **114**:2337–2352.

41. *JMP, Version 9.* NC: Cary; 2007.

Rapid forest carbon assessments of oceanic islands: a case study of the Hawaiian archipelago

Gregory P. Asner[1]* , Sinan Sousan[1], David E. Knapp[1], Paul C. Selmants[2], Roberta E. Martin[1], R. Flint Hughes[3] and Christian P. Giardina[3]

Abstract

Background: Spatially explicit forest carbon (C) monitoring aids conservation and climate change mitigation efforts, yet few approaches have been developed specifically for the highly heterogeneous landscapes of oceanic island chains that continue to undergo rapid and extensive forest C change. We developed an approach for rapid mapping of aboveground C density (ACD; units = Mg or metric tons C ha^{-1}) on islands at a spatial resolution of 30 m (0.09 ha) using a combination of cost-effective airborne LiDAR data and full-coverage satellite data. We used the approach to map forest ACD across the main Hawaiian Islands, comparing C stocks within and among islands, in protected and unprotected areas, and among forests dominated by native and invasive species.

Results: Total forest aboveground C stock of the Hawaiian Islands was 36 Tg, and ACD distributions were extremely heterogeneous both within and across islands. Remotely sensed ACD was validated against U.S. Forest Service FIA plot inventory data ($R^2 = 0.67$; RMSE = 30.4 Mg C ha^{-1}). Geospatial analyses indicated the critical importance of forest type and canopy cover as predictors of mapped ACD patterns. Protection status was a strong determinant of forest C stock and density, but we found complex environmentally mediated responses of forest ACD to alien plant invasion.

Conclusions: A combination of one-time airborne LiDAR data acquisition and satellite monitoring provides effective forest C mapping in the highly heterogeneous landscapes of the Hawaiian Islands. Our statistical approach yielded key insights into the drivers of ACD variation, and also makes possible future assessments of C storage change, derived on a repeat basis from free satellite data, without the need for additional LiDAR data. Changes in C stocks and densities of oceanic islands can thus be continually assessed in the face of rapid environmental changes such as biological invasions, drought, fire and land use. Such forest monitoring information can be used to promote sustainable forest use and conservation on islands in the future.

Keywords: Carbon stocks, Carnegie Airborne Observatory, Forest inventory, Invasive species, LiDAR, Random Forest Machine Learning

Background

Aboveground carbon (C) stock assessments have become a mainstay of forest management [1]. In the past decade, the importance of such assessments has also grown in the climate change mitigation arena [2]. In step with these efforts, there has been increasing focus on developing quantitative methods to monitor forest C stocks over time, as a means to support policies that reduce emissions from deforestation and forest degradation, and increase C storage in existing forests (REDD+) [3]. C storage has also become an important metric for assessing forest habitat and condition in the broader conservation arena [4, 5].

Based on the increasing value in understanding the geography of forest C stocks, both field-based and remote sensing-assisted C assessments have been

*Correspondence: gpa@carnegiescience.edu
[1] Department of Global Ecology, Carnegie Institution for Science, 260 Panama St, Stanford, CA 94305, USA
Full list of author information is available at the end of the article

undertaken over larger and larger geographic areas [6, 7]. Far less attention, however, has been given to oceanic islands, likely due to their relatively small land area. Oceanic islands provide model socio-ecological systems with which to examine spatial patterns in forest C stocks, because islands are often comprised of highly heterogeneous ecosystems, where many of the drivers of C storage (e.g., vegetation types, climate, fire, and land use) vary strongly over short distances [8, 9]. While C stocks on oceanic islands may be small in a global context, they provide unique opportunities to test fundamental concepts on the landscape ecology, sociology, economics and management of forest C sequestration. Further, forests on oceanic islands are quite important to the provisioning of ecosystem goods and services, including fresh water supply, prevention and mitigation of soil erosion that can deplete upland soil resources and pollute aquatic ecosystems including coral reefs [10], and both timber and non-timber forest products. Island forests also play a strong cultural role as a locus of subsistence and recreational activities [11, 12]. However, relative to continental ecosystems, forests on oceanic islands continue to undergo a much greater proportional extent and rate of change in cover and composition, which threatens the sustainability of forest-based good and services including C stocks [13, 14]. Not only have islands been heavily deforested in some regions of the world, they have also undergone enormous change via introduced disturbance regimes, such as fire, and alien invasive species [15, 16]. The effects of these and other changes on forest C stocks remain poorly understood, despite numerous local- to landscape-scale assessments [17]. Without continuous and spatially extensive forest monitoring, patterns of change and/or opportunities for recovery of island forests will remain a challenge to incorporate into conservation, management and resource policy initiatives.

Like most oceanic islands, aboveground forest C stocks within and across the Hawaiian Islands are poorly known, owing to extreme environmental heterogeneity combined with local inaccessibility and complex terrain. This has greatly limited efforts to develop and maintain operational, repeat forest inventory on the ground. Global remote sensing-based carbon mapping approaches generally yield lower spatial resolutions and C stock sensitivities [18–21], which are difficult to apply in regions of high ecological heterogeneity like islands. While high-resolution remote sensing methods, such as airborne Light Detection and Ranging (LiDAR) [22], are suitable for such settings [23], mapping remote or difficult-to-access areas with aircraft can be expensive. In particular, cloud cover is often persistent over higher-elevation forests of key interest in forest C and watershed assessments. As a result, airborne campaigns can

be prolonged and accumulate costs. An added challenge is that island forest assessments are needed on a repeat basis in response to the inherent vulnerability of many island landscapes to rapid change driven by land use, fire, storms (e.g., hurricanes), biological invasions and sea level rise. The issue of rapid change calls for the development of a low-cost, repeatable forest monitoring method for island forests. Such rapid, high-resolution assessment capabilities must be sensitive to the drivers of forest C change, not only as a metric for climate change mitigation, but also as a measure of forest health and provisioning services.

While mapping of forest C stocks has been challenged by uncertainty and cost [7], recent progress at subnational to national levels indicates that significant methodological hurdles can be overcome at larger scales, especially through the fusion of ground, aircraft and satellite based measurements [21, 24]. These approaches can simultaneously increase map resolution in ways that benefit forest managers, while reducing uncertainty to levels acceptable to policy makers. Despite these advances, important methodological questions remain regarding how to provide high resolution, low uncertainty monitoring at low cost in heterogeneous landscapes. A further need is the simultaneous assessment of the drivers of spatial variation in C storage.

We developed an approach for monitoring forest aboveground carbon density (ACD; units = Mg or metric tons C ha^{-1}) across island archipelagos at a spatial resolution of 30 m (0.09 ha) using a combination of airborne LiDAR and freely available satellite data (Fig. 1). The approach involves initial use of high-resolution LiDAR sampling of a selected island within an archipelago to derive vegetation canopy height data. These data from the sampled island are then used to train a geospatial model that incorporates maps of multiple environmental factors, as well as forest canopy structural metrics derived from Landsat or comparable satellite imagery [25]. The resulting model is applied to all islands within the archipelago using as input the same portfolio of environmental and satellite-based canopy structural maps used on the model-training island, thereby yielding a multi-island map of canopy height at 30-m spatial resolution. Finally a regionally-tuned equation is applied to relate mapped canopy height to ACD [26], resulting in a carbon density map at 30-m resolution for the entire island chain. Critically, once the model is built for an archipelago, subsequent changes in ACD can be detected using only Landsat imagery, thereby greatly reducing longer-term monitoring costs [24].

For this study, we first sampled Hawaii Island, by far the largest island in the Hawaiian archipelago, with airborne LiDAR to assess forest top-of-canopy height (TCH) responses to natural environmental gradients

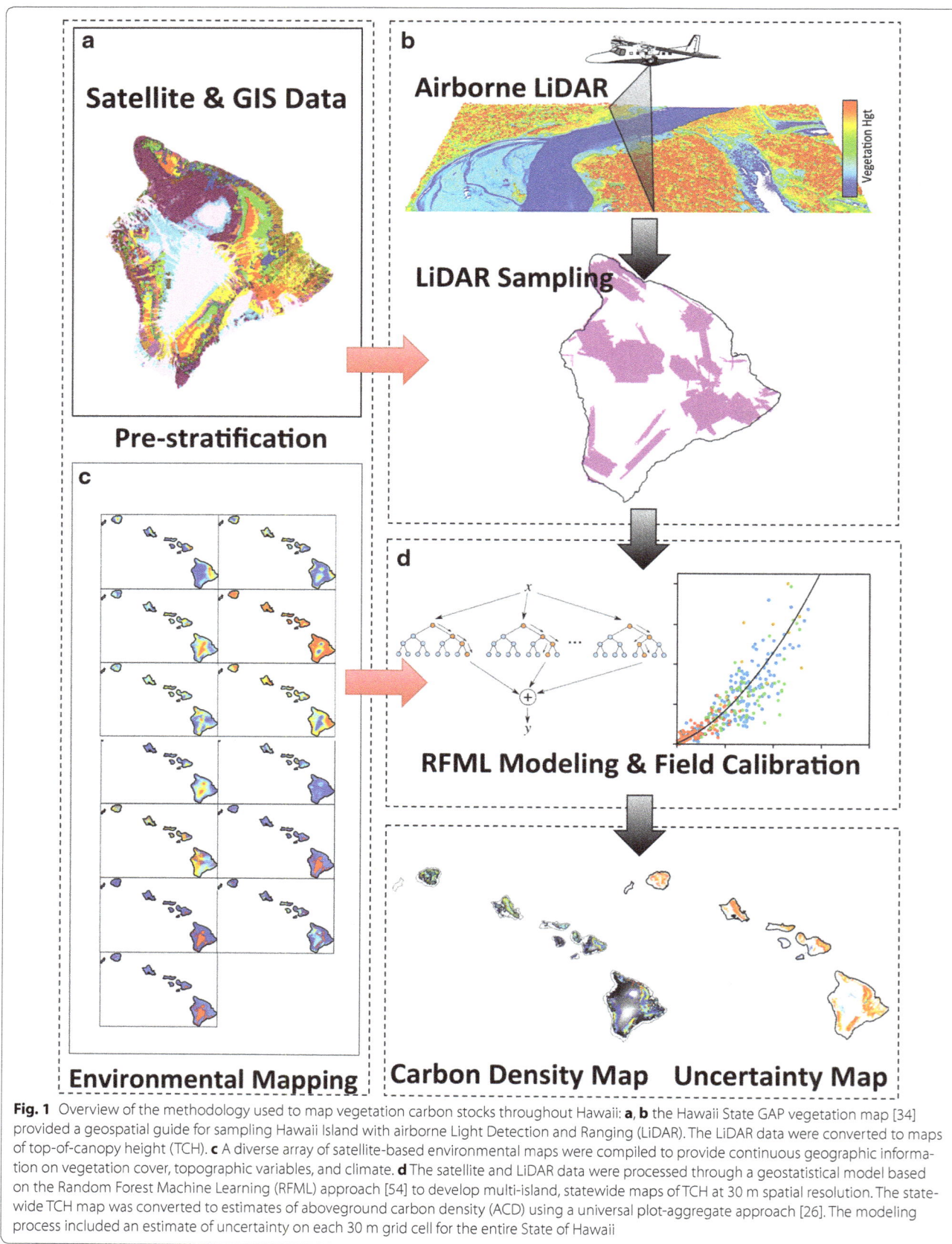

Fig. 1 Overview of the methodology used to map vegetation carbon stocks throughout Hawaii: **a, b** the Hawaii State GAP vegetation map [34] provided a geospatial guide for sampling Hawaii Island with airborne Light Detection and Ranging (LiDAR). The LiDAR data were converted to maps of top-of-canopy height (TCH). **c** A diverse array of satellite-based environmental maps were compiled to provide continuous geographic information on vegetation cover, topographic variables, and climate. **d** The satellite and LiDAR data were processed through a geostatistical model based on the Random Forest Machine Learning (RFML) approach [54] to develop multi-island, statewide maps of TCH at 30 m spatial resolution. The statewide TCH map was converted to estimates of aboveground carbon density (ACD) using a universal plot-aggregate approach [26]. The modeling process included an estimate of uncertainty on each 30 m grid cell for the entire State of Hawaii

and land use (Additional file 1: Figure S1). These LiDAR TCH data from Hawaii Island were used to calibrate a Random Forest Machine Learning (RFML) model, which was subsequently used to predict TCH at 30 m resolution on all islands from a portfolio of spatially explicit predictor maps (Additional file 1: Figures S2-S4). The resulting statewide model of forest TCH was then used to estimate forest ACD via a conversion equation developed for the Hawaiian Islands (Additional file 1: Figure S5). The resulting map was compared to US Forest Service Forest Inventory and Analysis (FIA—http://www.fia.fs.fed.us/) plot data for evaluation of mapped ACD precision. Finally, we used the new ACD map to assess aboveground forest C stocks within and among islands, in protected and unprotected areas, and among forests dominated by native and invasive plant species.

Results and discussion
Island carbon stocks and distributions
Total forest cover and aboveground carbon stock for seven main Hawaiian Islands was estimated at 550,065 ha and 36.0 Tg (million metric tons), respectively (Fig. 2; Table 1). A map of estimated uncertainty indicated greatest absolute uncertainties of 20–40 % in very high-biomass forests, with much lower uncertainties in low-to-moderate biomass conditions (Additional file 1: Figure S6). Forest ACD varied widely by island (Fig. 3). Hawaii Island contained 57 % of the total forest cover of the State, and almost 20 Tg of the State's forest carbon. Kauai, Maui and Oahu islands collectively accounted for 36 % of the total forest cover and 14.7 Tg of aboveground C. Molokai, Lanai, and Kahoolawe together accounted for only 7 % of the State's forest cover and less than 1.4 Tg C. The small northwest-most island of Niihau was not considered in this study.

The highest forest ACDs were found on Hawaii Island, reaching 537 Mg C ha^{-1}. Maui supported the next highest ACDs, reaching 294 Mg C ha^{-1}. We also found extremely variable C stocks on each island (Additional file 1: Figures S7-S10). Aboveground forest C density varied up to three fold among State Districts, which are the minimum State-level political units of civil governance

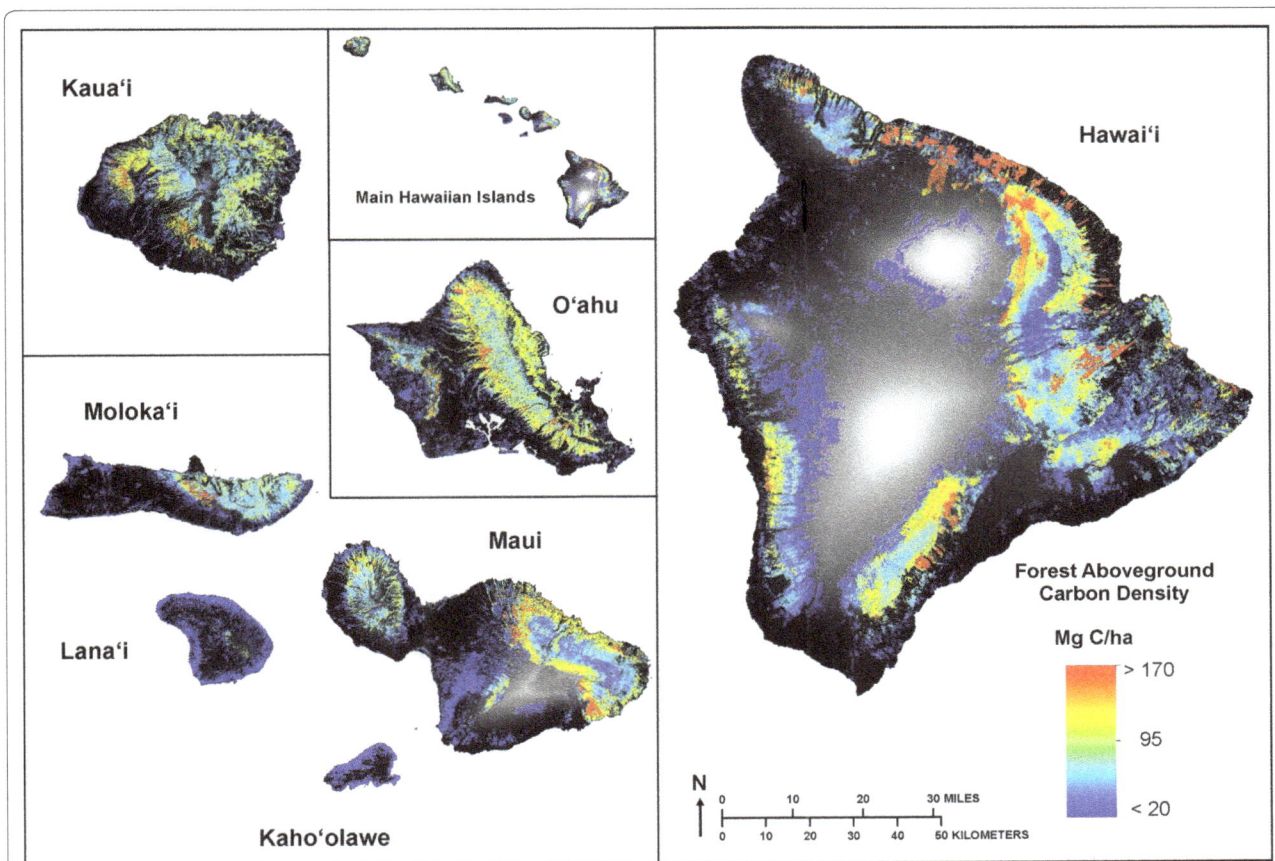

Fig. 2 Spatial distribution of forest aboveground carbon density (ACD; Mg C ha^{-1}) for the State of Hawaii at 30-m mapping resolution. A map of estimated uncertainty is provided in Additional file 1: Figure S6. The islands are displayed so that their relative sizes are preserved

Table 1 Forest cover and aboveground carbon stock and density for each island and the State's Districts

Island	Counties and districts	Forest cover (ha)	Aboveground carbon density (Mg C ha^{-1})	Aboveground carbon stock (Tg C)
Hawaii		311,977.0	64.0 + 43.7	20.0
	Hawaii County			
	Hamakua	23,391.8	51.4 + 47.4	1.2
	Kau	63,204.2	67.0 + 43.3	4.2
	North Hilo	18,598.8	93.3 + 49.3	1.7
	South Hilo	67,056.8	72.8 + 41.7	4.9
	North Kohala	8341.1	65.9 + 47.9	0.6
	South Kohala	7057.3	47.7 + 31.9	0.3
	North Kona	28,391.2	30.0 + 35.7	0.9
	South Kona	30,635.5	60.8 + 40.8	1.9
	Puna	65,302.4	66.3 + 36.8	4.3
Maui		75,532.9	67.1 + 47.0	5.1
	Maui County			
	Hana	29,763.6	78.7 + 41.4	2.3
	Lahaina	22,113.8	33.5 + 40.7	0.7
	Makawao	34,542.2	47.5 + 52.7	1.6
	Wailuku	7543.5	61.1 + 39.5	0.5
	Molokai	23,018.2	54.9 + 37.3	1.3
Molokai		23,018.2	54.9 + 37.3	1.3
Lanai		13,048.5	7.6 + 15.1	0.1
Kahoolawe		5391.1	3.0 + 2.5	0.02
Oahu		64,673.4	78.3 + 40.2	5.1
	Honolulu County			
	I	3562.3	77.4 + 39.2	0.3
	II	1648.4	92.4 + 33.3	0.2
	III	2803.9	63.4 + 47.0	0.2
	IV	6669.2	87.9 + 32.9	0.6
	V	10,988.6	95.5 + 32.0	1.1
	VI	14,449.8	80.2 + 38.9	1.2
	VII	4812.6	85.3 + 36.3	0.4
	VIII	5047.9	34.3 + 37.3	0.2
	IX	14,407.3	74.0 + 38.8	1.1
Kauai		56,424.0	80.4 + 35.5	4.5
	Kauai County			
	Hanalei	16,033.8	82.8 + 32.0	1.3
	Kawaihau	9020.1	83.0 + 32.1	0.8
	Koloa	3024.7	78.5 + 43.2	0.2
	Lihue	8886.3	79.2 + 32.7	0.7
	Waimea	19,449.3	78.1 + 39.2	1.5

(Table 1). On Hawaii Island, for example, forest ACD values varied from means of 30–93 Mg C ha^{-1} across Districts, yet within Districts, spatial variation in forest ACD ranged from 50 to 111 % of their District means. Moreover, three of nine Districts on Hawaii Island contained two-thirds of the entire island's forest C stock. The island with the most variable inter-District forest C stocks was Maui.

Model comparison to FIA plots

Comparison of modeled ACD to values estimated from FIA plot inventory indicated good precision ($R^2 = 0.67$) and accuracy (average root mean squared error or RMSE = 30.4 Mg C ha^{-1}) (Fig. 4). Bias was just 11.2 Mg C ha^{-1}, and heteroscedasticity was similar to that derived in plot-inventory comparison studies [27]. These map performances were particularly strong relative

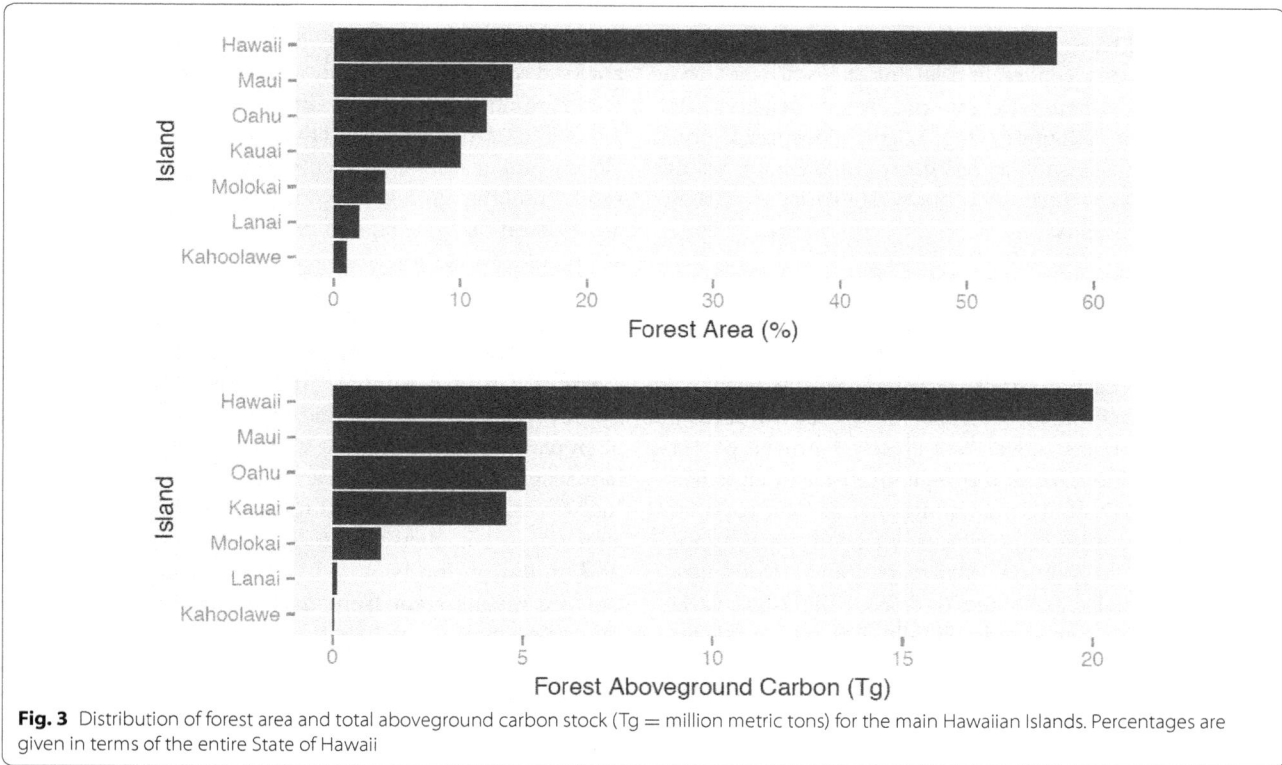

Fig. 3 Distribution of forest area and total aboveground carbon stock (Tg = million metric tons) for the main Hawaiian Islands. Percentages are given in terms of the entire State of Hawaii

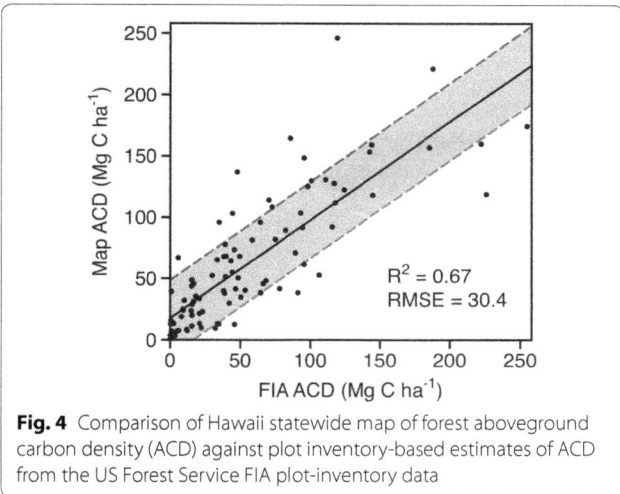

Fig. 4 Comparison of Hawaii statewide map of forest aboveground carbon density (ACD) against plot inventory-based estimates of ACD from the US Forest Service FIA plot-inventory data

non-differentially corrected global positioning system (GPS) instruments. This leads to plot location uncertainties of up to 30 m. The combination of relatively small size (18 m radius), circular shape, and non-contiguity of the FIA plots (see "Methods"), explains higher uncertainty when comparing to ACD estimates in 30 m × 30 m mapping cells. Asner et al. [28] found that mismatches in location and plot shape alone account for up to 15 % uncertainty in field validation studies. Additionally, the allometric scaling applied to the FIA field measurements can result in additional uncertainties of up to 50 % of the plot mean value [29, 30].

Given these, and other sources of uncertainty, we contend that the verification step undertaken here was successful in validating the map results. Nonetheless, validation with FIA or other plots could be significantly improved by more accurate GPS measurements of plot locations, and by employing plot and sampling design that is better suited to validating remotely-sensed estimates of ACD. Specifically, plots should be similar in area to the final grid size and all trees >5 cm dbh should have height and diameter measured in each plot. Better allometry would also decrease uncertainty. Currently, we employ species-specific allometric equations only for the two most dominant native woody tree species (*Metrosideros polymorpha* and *Acacia koa*) and for four

to the accuracy of the equation used for estimating ACD from canopy height (Additional file 1: Figure S5).

Here we note the challenges involved in comparing the FIA plot data to mapped C densities based on remote sensing. First, there was an offset of about 6 years between the time the LiDAR flights were completed and the time the FIA measurements were taken in the field. Second, the FIA data in Hawaii were geo-located using

non-native tree species. Aboveground biomass for the remaining 114 tree species encountered in FIA plots was estimated using a general model for tropical trees that incorporates diameter, height and wood density [31]. Species-specific allometry for large, widespread non-native tree species, such as *Falcataria moluccana*, would almost certainly reduce uncertainty in estimates of their aboveground biomass.

Factors affecting carbon stocks

The geospatial analysis indicated that fractional canopy cover (FC) was the principal driver of spatial variation in forest carbon stocks throughout the Hawaiian archipelago, accounting for 27 % of the total variance in ACD (Fig. 5). Forest cover was closely followed by forest type, as defined using the vegetation-cover classification, which accounted for an additional 24 % of variation in ACD. Other important factors included mean annual precipitation, vegetation structure, and cloudiness, which individually explained 6–8 % of the ACD variation throughout the islands. Finally, fire return factors, elevation and additional climate variables individually explained 1–4 % of the variability in carbon density.

Note that while the results presented in Fig. 5 account for co-variation in explanatory factors, many of them are ecologically and/or geospatially convolved with one another. For example, forest FC is broadly related to elevation and topographic aspect, with less forest cover often observed at high elevations and on leeward aspects, although low forest FC was also observed in deforested zones at lower elevations on windward aspects. Thus the factor rankings presented here indicate an additional effect of elevation and aspect not already explained by FC alone. Similar inter-factor co-variances occur among the model rankings in Fig. 5. Nonetheless, it is clear that FC and vegetation type explain much of the geographic variation in forest carbon stocks.

Effects of biological invasion on forest carbon

Although this study is limited to a single time step, the current Hawaii vegetation map allowed us to conduct the first statewide assessment of the large-scale effects of alien plant species on forest C stocks. Numerous plot- to landscape-scale studies have reported on this issue, with highly variable outcomes ranging from no effect of invasion on carbon densities, to increases and decreases in ACD following invasion [17, 23, 28, 32, 33]. Such wide-ranging results stem from underlying variability in the mediating factors, such as time-since-introduction, rates of invasion, relative changes in plant functional and structure types, and environmental filters such as soils and climate. There is thus a general need for large-scale, high-resolution assessments that go beyond local contextual results.

The Hawaii State vegetation map was generated using manual and automated classification of Landsat imagery

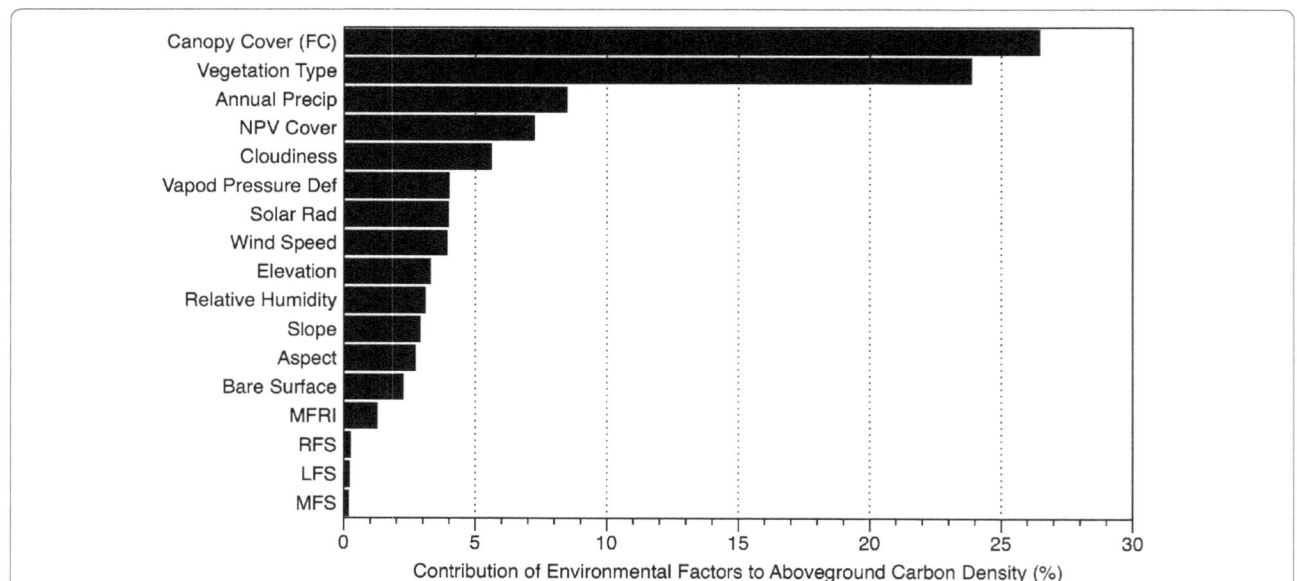

Fig. 5 Contribution of each potential explanatory factor determining aboveground carbon density (ACD) in the Hawaiian Islands. Fractional canopy cover (FC), non-photosynthetic vegetation (NPV) cover, and bare surface cover (soils, rock, infrastructure) were derived from sub-30 m resolution Landsat-based satellite mapping of the islands (see [25]). Vegetation type was provided by the Hawaii State GAP vegetation map [34]. *MFRI* mean fire return interval; *RFS* replacement fire severity; *LFS* low fire severity; *MFS* mixed fire severity

against aerial photography [34]. Experience with this map in field studies indicates that the "alien-dominated" classes are comprised of mature stands of non-native species, while "native-dominated" classes are comprised of mature stands of native species, particularly dominated by the keystone canopy species *Metrosideros polymorpha* and *Acacia koa*. We focused our analysis on these two groups because the Hawaii State vegetation map alone does not provide sufficient detail to partition the mapped C results into finer levels of invasion, particularly since the invasion process is ongoing and highly dynamic (in favor of alien invasive species dominance). We further partitioned the native- and alien-dominated groups by three major environmental filters known to mediate C stocks: annual precipitation, elevation and substrate age (from volcanic activity dating back to the early Pliocene) (Additional file 1: Table S1).

Our results show that, on medium-to-older substrates in both drier and wetter conditions, the total area of alien-dominated forest exceeds that of native-dominated forest in lower-elevation zones (Fig. 6a). In contrast, the majority of wetter, higher-elevation and/or older-substrate conditions remain dominated by native forest cover. Critically, however, we found that ACD is greater in native-dominated forests in low-to-medium elevation, dry-to-mesic regions of the islands, whereas alien-dominated forests tend to have slightly higher ACD levels in wetter environments across the board (Fig. 6b). At these broad multi-island scales, substrate age played only a small role in determining the *relative* difference in alien- and native-dominated forest ACD. This suggests strong limiting effects of nutrient-poor soils on growth and biomass accumulation for all species, independent of origin [35]. In contrast, higher biomass of native forest canopies in drier zones on older substrates may reflect evolutionary adaptation to these environments, as well as a lack of analog tree taxa in the current alien species pool on the islands.

Our results are also suggestive of how native biological diversity intersects with C storage, and how alien invasive species alter those relationships. For example, higher-elevation, drier forests on older substrates may be dominated by alien forest cover (smallest solid green dot; Fig. 6a), but native-dominated forests in similar environments support twice the stored C on a per-area basis (Fig. 6b). Thus actions to conserve and restore high-elevation native ecosystems yield a co-benefit of increased C storage. On the other hand, higher-elevation, drier conditions on younger substrates are areas currently dominated by native forest cover (open small red dot; Fig. 6a), but alien species can double the ACD levels in these environments (Fig. 6b). Forest managers and conservationists can use these landscape-scale relationships as trade-offs

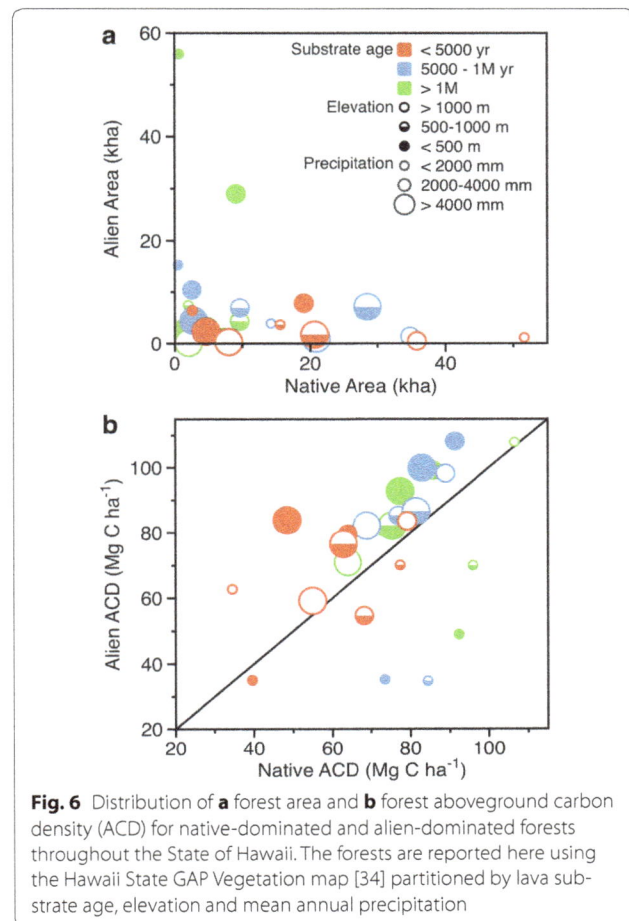

Fig. 6 Distribution of **a** forest area and **b** forest aboveground carbon density (ACD) for native-dominated and alien-dominated forests throughout the State of Hawaii. The forests are reported here using the Hawaii State GAP Vegetation map [34] partitioned by lava substrate age, elevation and mean annual precipitation

in planning efforts to increase C storage while managing for biological diversity [36, 37].

Forest carbon protections and opportunities

High-resolution C mapping also affords a way to assess current protections, threats and opportunities for sequestered carbon and generating healthy forests via land-use allocation and management [21]. Using land tenure data provided by the State of Hawaii, we quantified C stocks and densities on State, federal and private reserves. Of the total aboveground forest C stock found on the islands (36 Tg C), about 18.5 Tg C or 51 % is officially protected on State (e.g., Natural Area Reserves; Forest Reserves), federal (National Parks; Wildlife Refuges) and private (The Nature Conservancy; Kamehameha Schools lands) lands covering 257,691 ha (Fig. 7a, Additional file 1: Table S2). This is almost equally matched by forests outside of protected reserves, which in total cover more land area at 292,374 ha, but which contain 17.5 Tg of aboveground C. This finding indicates that a large amount of forest C could be incorporated into more formal reserve

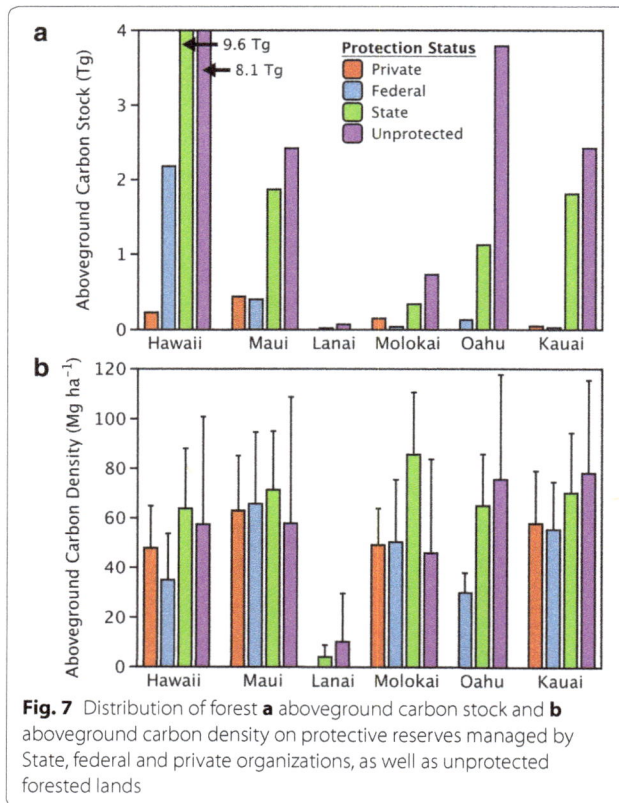

Fig. 7 Distribution of forest **a** aboveground carbon stock and **b** aboveground carbon density on protective reserves managed by State, federal and private organizations, as well as unprotected forested lands

Replication on oceanic island chains

The approach we have developed and tested here for high-resolution mapping of aboveground forest carbon density is intended for replication on oceanic islands worldwide, but also any set of highly heterogeneous landscapes. The methodology is based on a previously established strategy that relies on airborne LiDAR sampling of forests found across a range of ecological conditions, but limited to one island [23]. Here we greatly advanced the approach by extending the initial LiDAR sampling of a single island, via a machine-learning algorithm [38, 39], to the multi-island or archipelago scale using a suite of environmental maps and satellite data that is, in combination, sufficiently sensitive to variation in the LiDAR-based estimates of canopy height. Shared environmental characteristics among neighboring islands usually include geology, climatic zones, and dominant vegetation types. Satellite-based metrics of forest structure, derived from Landsat-based spectral mixture analysis, are time-variant and key to the linkage with the LiDAR data. Strategically, these Landsat-based metrics can be updated through time using the fully automated CLASlite software [25].

The conversion of either LiDAR-scale or modeled canopy height to estimates of ACD requires plot-aggregate allometric equations [40]. This worked well in Hawaii, relative to plot-estimated ACD from U.S. Forest Service inventory data. The universal or regional plot-aggregate allometric equations proposed by Asner et al. [40] have also worked reasonably well in other regions [17, 41, 42], and they tend to result in mismatches between LiDAR-based and field-based estimates of ACD of 10–15 % when applied at 1-ha spatial resolution [26]. Nonetheless, application of these conversion equations to oceanic islands requires further validation, particularly for isolated islands in which vegetation types (and thus allometrics) may diverge from general databases.

There is an initial cost for installing a forest C monitoring program on any given island chain or archipelago. It includes an initial airborne LiDAR survey of one island or part of the archipelago, which varies widely in cost depending upon whether the data are sourced from non-profit, government, or commercial organizations. Our LiDAR data collection and processing cost was approximately $150,000 for the Island of Hawaii, but costs have greatly declined since the data acquisition was made for this study [43]. The LiDAR component was followed by personnel and computing costs required to link the LiDAR data to the satellite imagery and for validation work. However, the satellite imagery was free of charge, and CLASlite is also currently available at no charge [44], thereby providing us with a low-cost way to complete the initial carbon map. Moreover, the free imagery

protections. Moreover, we found that reserve ACD averages 61.8 ± 22.3 Mg C ha^{-1}, whereas non-reserve forests have carbon densities of 59.6 ± 34.2 3 Mg C ha^{-1} (Fig. 7b). Combined, these results underscore the C-storage benefit of adding long-term protection status to remaining island forests; Total forest aboveground C stock increases linearly with increasing reserve area (Additional file 1: Figure S11).

On all islands, 189 State-managed reserves hold the vast majority of protected carbon stocks—14.8 Tg C, while 25 federal and 14 private reserves contain just 2.8 and 0.9 Tg C, respectively (Additional file 1: Table S2). Carbon densities are highest in State reserves (66.3 ± 23.2 Mg C ha^{-1}), followed by private (56.1 ± 19.3 Mg C ha^{-1}) and federal reserves (41.4 ± 17.6 Mg C ha^{-1}). Differences in forest carbon densities are reflective of the location of the reserves (lowland vs. montane, wet windward vs. dry leeward) as well as species composition and management. A desired outcome of this work is to provide forest managers and the public with information to compare, for example, carbon stocks on a reserve-by-reserve basis against environmental maps, to identify opportunities for increasing C densities through conservation and management actions.

and software makes updates to the map extremely cost-efficient, likely requiring the effort of a single geospatial technician for the State of Hawaii. Even if field inventories could be done at large geographic scales on a spatially contiguous basis, which is not possible, the recurring costs would be extremely high for each monitoring step through time.

Conclusions

We have shown that a combination of one-time airborne LiDAR data acquisition, and freely available satellite data with automated analysis, can provide effective forest C mapping and monitoring of oceanic islands. The method is highly replicable and cost-effective. From the first map generated, and with regular updates using satellite data over time, assessments of C storage can be derived by political entity (e.g., State Districts), land-use allocation (e.g., protected vs. unprotected areas), or any other unit of governance or management. Moreover, changes in C stocks and densities can be continually assessed in the face of rapid environmental changes, such as climate, fire and biological invasion. The resulting information is spatially explicit, allowing for actions that promote sustainability of forests and the services they provide to island biodiversity and societies. High-resolution monitoring approaches also provide a geography of forest C stock that facilitates the inclusion of multiple stakeholders ranging from individual landowners to national governments. The resulting empowerment afforded by this type of ecological information will be important to the protection, enhancement and/or restoration of island ecosystems in the future.

Methods

Our mapping approach is summarized in Fig. 1. The necessary technologies are airborne Light Detection and Ranging (LiDAR), which yields highly detailed measurements of forest canopy height and vertical canopy profile, and satellite-derived maps of environmental variables and forest canopy fractional cover. A second component relies on machine learning algorithms to scale airborne LiDAR samples of one island up to multi-island or archipelago maps. Several studies have employed a Random Forest Machine Learning (RFML) algorithm to model the relationship between LiDAR-based estimates of forest structure or biomass and a suite of satellite data sets [19, 21, 45, 46]. RFML fits multiple environmental datasets (predictors) to estimates of vegetation structure or biomass (response), as described later. In doing so, a direct scaling of LiDAR samples to full-coverage maps can be derived without artificial boundaries between ecosystems that often occur using traditional stratification approaches.

Random Forest Machine Learning also provides quantitative information on which predictors (e.g., satellite data) are most important in determining the response variable (LiDAR-derived canopy height) [47]. Here the importance of a predictor to the RFML model was assessed by randomly permuting the values of the factor within a validation dataset, and processing the validation data through the regression trees. In our implementation of RFML, a temporary validation dataset is created to build each regression tree, and is chosen as a randomly selected set of 250 samples left out of the full training dataset. To assess the importance of a single factor, we compared the mean square error (MSE) values of the validation data both before (MSE_i) and after (MSE^*_i) randomly permuting the values of the factor for each tree [48]. For each tree i, the difference between MSE_i and MSE^*_i, divided by MSE_i, was collected. The importance of the given factor was then taken to be the mean of these relative difference values across all trees. By repeating the above procedure for each explanatory factor, the relative importance of each factor could be compared.

LiDAR data acquisition and analysis

LiDAR data were collected using the Carnegie Airborne Observatory [49]. Flights covered 379,337 ha of Hawaii Island (Additional file 1: Figure S1) including all major forest types (Additional file 1: Figure S2b) [23]. LiDAR data were collected at 1000 or 2000 m above ground level, using two corresponding configurations: higher resolution with 0.56 m on-the-ground laser spot spacing, 24° field of view (FOV) and a 70 kHz pulse repetition frequency; low resolution with a 1.12 m spot spacing, 30° FOV and a 50 kHz pulse repetition frequency, respectively. Ground cover was sampled along parallel flight lines with 50 % overlap to ensure LiDAR coverage of no less than 4 laser shots m^{-2}.

Mean top-of-canopy height (TCH) was calculated for each 30 m × 30 m grid cell of LiDAR coverage on Hawaii Island (Additional file 1: Figure S1). To create this layer, the laser range measurements from the LiDAR were combined with the embedded high resolution Global Positioning System-Inertial Measurement Unit (GPS-IMU) data to determine the 3-D locations of the laser returns. This calculation produced a 'cloud' of LiDAR data. The LiDAR data cloud was processed to identify where the laser pulses penetrated the canopy volume, reaching the ground surface, from which a digital terrain model (DTM) was produced. This was achieved using a 10 m × 10 m filter kernel throughout the LiDAR coverage, and the lowest elevation in each kernel was deemed as possible ground detection. These filtered points were then evaluated by fitting a horizontal plane through each point. If the closest unclassified point was <1.5 m higher in elevation, the pre-filtered point was finalized as a ground-classified surface point. This process was repeated until all potential ground points within the LiDAR coverage were evaluated. A digital surface model (DSM), which is essentially the top-most surface (e.g., canopies, buildings, exposed ground), was also generated based on interpolations of all first-return points at 1.12 m spatial resolution. The DTM and DSM were combined as a tightly matched pair of data layers. The vertical difference between them resulted in a model of top-of-canopy height (TCH) at 1.12 m spatial resolution throughout the 379,337 ha LiDAR sampling coverage. Validation studies of this CAO LiDAR TCH estimation approach have shown it to be highly accurate across a wide range of forests including extremely densely foliated, tall tropical forests exceeding 60 m in height [28, 42].

Environmental predictor variables

We used 17 environmental predictor variables from co-aligned spatial datasets covering State of Hawaii to model canopy height based on the LiDAR TCH measurements made on Hawaii Island (Additional file 1: Figure S2-S4). All predictor variables were gridded at 30-m spatial resolution. Three predictor variables were fractional cover of forest canopy (FC), non-photosynthetic vegetation (NPV), and bare surfaces. These were determined from nine primary Landsat-8 images collected in 2013 and 2014. The mosaic of nine images included a few small cloud-covered areas, so those areas were backfilled with Landsat-7 and Landsat-8 data going back to 2010. The Landsat mosaic was run through a probabilistic spectral mixture analysis algorithm embedded in the CLASlite forest monitoring software package [25]. These fractional cover images have been validated and used in numerous studies in Hawaii and elsewhere [e.g., 23, 50].

An important additional predictor variable was the Hawaii State GAP vegetation map, which provides the highest resolution and most widely used vegetation cover type information for the State of Hawaii. The version used was based on Gon et al. [34], with improvements based on high-resolution satellite images and other more recent vegetation mapping information [51]. Three additional predictor variables were derived from 30-m Shuttle Radar Topography Mapping (SRTM) mission data: elevation, slope and aspect. In addition, mean annual precipitation (MAP), mean wind speed at 2 m above ground, vapor pressure deficit, total solar radiation, mean relative humidity, and cloud frequency data were acquired from http://climate.geography. http://hawaii.edu/downloads. html. Finally, we used four fire-related predictor variables: low fire severity (LFS), mixed fire severity (MFS), replacement fire severity (RFS), and mean fire return interval (MFRI) provide by http://www.landfire.gov/fireregime.php.

These 17 predictor maps and the 30-m LIDAR-derived TCH map were applied to the RFML model for Hawaii Island to develop the prediction-based regression trees. The regression trees were then used to predict TCH values across the entire State of Hawaii using the 17 predictor maps as input.

Estimating aboveground carbon density

We estimated ACD from the statewide TCH map using a plot-aggregate allometric scaling approach [26]. A biophysical link was previously developed to quantitatively link mapped TCH to field estimates of ACD by applying regional plot-aggregated estimates of vegetation wood density and diameter-to-height relationships. To develop a TCH-to-ACD calibration for Hawaiian forests and other vegetation types throughout the State, we used 209 field plots located on Hawaii Island for which ACD was measured using field plot-based inventory measurements as detailed by Asner et al. [23]. The resulting calibration between TCH and ACD is shown in Additional file 1: Figure S5, with in $R^2 = 0.82$ and RMSE $= 78.7$ Mg C ha^{-1}. The final calibration equation for relating TCH to ACD was: $ACD = 3.744 * TCH^{1.391}$.

Uncertainty map

The uncertainty of the mapped ACD estimates was estimated by developing a relationship between the mapped ACD values and the RMSE of ACD for those areas on Hawaii Island covered by the LiDAR data [21]. These RMSE values were partitioned into 30 bins across the range of RFML-modeled ACD values. A polynomial was fit to model the RMSE of an ACD estimate as a function of its predicted ACD value. The polynomial was then applied to the ACD map to produce an estimate of ACD uncertainty (Additional file 1: Figure S6).

Map validation

To evaluate the accuracy of the final carbon map, we compared data from the map to georeferenced plots surveyed across the Hawaiian Islands in 2011 and 2012 by the United States Department of Agriculture Forest inventory and Analysis (FIA) Program. The FIA Program is a national network of plots designed to represent all forest conditions across the United States [52]. Each FIA plot is a cluster of four circular 7.32-m radius subplots arranged in a fixed pattern. All trees and tree ferns ≥12.7 cm diameter at breast height (dbh; 1.37 m above the ground) had diameter, height, and species recorded in each subplot. Trees and tree ferns <12.7 cm dbh had diameter, height, and species were recorded in microplots, which are 2.07 m radius plots located within each subplot. Macroplots, which are 17.95 m radius and immediately surround each subplot, are usually reserved for destructive sampling. However, FIA plots sampled in Hawaii in 2011–2012 using the 'experimental forest' (EXPFOR) protocol (n = 96) had all trees $\geq$12.7 cm dbh measured in Macroplots as well, greatly enlarging the sample footprint of each plot. We used data from these 96 EXPFOR FIA plots to validate the accuracy of the final carbon map.

We estimated ACD for each tree measured in the 96 FIA plots using a combination of species-specific and general diameter-to-ACD and height-to-diameter models. We used locally derived, species-specific diameter to ACD models for eight species, including the two most common species in the FIA dataset: *Metrosideros polymorpha* and *Acacia koa* (Additional file 1: Table S3). For all other species, and for large trees that exceeded the diameter range of species-specific diameter-to-ACD models, we used a general allometric model for tropical trees developed by Chave et al. [31] that uses diameter, height, and wood density to estimate ACD (Additional file 1: Table S4). When the Chave model was employed, we used species-specific wood density values from Hawaii [23] and a global wood density database [53]. If a species-specific wood density value was unavailable, we used a mean value for the genus, and if this was not available we used a default value of 0.5 (Additional file 1: Table S5). We note here that wood densities are difficult to find for some commonly occurring oceanic island species, and thus we encourage research and measurement in this area. Occasionally, a height measurement was lacking for trees requiring the general Chave model. In these instances, we used locally derived, species-specific diameter to height models from Asner et al. [23]. When no species-specific diameter-to-height model was available, we used a general diameter-to-height model developed by Chave et al. for tropical trees that incorporates an environmental stress *E* parameter. Plot-level ACD was estimated by (1) estimating aboveground biomass (AGB) per unit area of microplots and macroplots within each FIA plot; (2) summing AGB per unit area within each FIA plot (n = 96); and (3) multiplying plot-level AGB per unit area by 0.48 to estimate ACD. The ACD of the 96 FIA plot locations were extracted from the statewide carbon map and averaged in a 3 × 3 pixel window (~1 ha) centered on each plot location.

Additional file

Additional file 1. Supporting figures and tables.

Abbreviations
ACD: aboveground carbon density; AGB: aboveground biomass; C: carbon; FC: fractional cover; FIA: Forest Inventory and Analysis; LFS: low fire severity; LiDAR: Light Detection and Ranging; MFRI: mean fire return interval; MFS: mixed fire severity; NPV: non-photosynthetic vegetation; PV: photosynthetic vegetation; RFML: Random Forest Machine Learning; RFS: replacement fire severity; TCH: top of canopy height.

Authors' contributions
GA designed the study, led the airborne remote sensing data collection, analyzed data, and wrote the paper. SS analyzed satellite remote sensing data, and carried out the modeling analyses. DK analyzed field and airborne remote sensing data, and carried out the modeling analyses. PS analyzed field data, provided GIS data analyses, and contributed to the writing of the paper. RM, FH, and CG assisted with study design, acquisition of funding, data interpretation, and writing of the paper. All authors read and approved the final manuscript.

Author details
[1] Department of Global Ecology, Carnegie Institution for Science, 260 Panama St, Stanford, CA 94305, USA. [2] Department of Natural Resources and Environmental Management, University of Hawaii at Manoa, 1910 East–West Rd., Honolulu, HI 96822, USA. [3] USDA Forest Service, Pacific Southwest Research Station, Institute of Pacific Islands Forestry, 60 Nowelo Street, Hilo, HI 96720, USA.

Acknowledgements
We thank Lori Tango for assistance with interpreting FIA plot data, and Tom Thompson and Jane Reid with the USDA Forest Service FIA Program for access to Hawaii field plot data. We thank past and current Carnegie Airborne Observatory team members for assistance with data collection and processing. The USGS Biological Carbon Sequestration Program funded this project, a part of LandCarbon Carbon Assessment of Hawaii initiative. The Carnegie Airborne Observatory is made possible by the Avatar Alliance Foundation, John D. and Catherine T. MacArthur Foundation, Mary Anne Nyburg Baker and G. Leonard Baker Jr., and William R. Hearst III.

Competing interests
The authors declare that they have no competing interests.

References
1. Cannell M. Woody biomass of forest stands. For Ecol Manage. 1984;8(3–4):299–312.
2. Gibbs HK, Brown S, Niles JO, Foley JA. Monitoring and estimating tropical forest carbon stocks: making REDD a reality. Environ Res Lett. 2007;2:1–13.
3. Angelsen A. Moving Ahead with REDD: issues, options and implications. Bogor, Indonesia: Center for International Forestry Research (CIFOR); 2008.
4. Lindenmayer DB, Laurance WF, Franklin JF, Likens GE, Banks SC, Blanchard W, et al. New policies for old trees: averting a global crisis in a keystone ecological structure. Conserv Lett. 2013;7(1):61–9. doi:10.1111/conl.12013.
5. Berenguer E, Ferreira J, Gardner TA, Aragão LEOC, De Camargo PB, Cerri CE, et al. A large-scale field assessment of carbon stocks in human-modified tropical forests. Glob Change Biol. 2014;20(12):3713–26. doi:10.1111/gcb.12627.
6. Mitchard ETA, Feldpausch TR, Brienen RJW, Lopez-Gonzalez G, Monteagudo A, Baker TR, et al. Markedly divergent estimates of Amazon forest carbon density from ground plots and satellites. Glob Ecol Biogeogr. 2014;23(8):935–46. doi:10.1111/geb.12168.
7. Goetz S, Baccini A, Laporte N, Johns T, Walker W, Kellndorfer J, et al. Mapping and monitoring carbon stocks with satellite observations: a comparison of methods. Carbon Balance Manag. 2009;4(1):2. doi:10.1186/750-0680-4-2.
8. Vitousek PM. The Hawaiian Islands as a model system for ecosystem studies. Pac Sci. 1995;49:2–16.

9.   Loope LL, Hamman O, Stone CP. Comparative conservation biology of oceanic archipelagoes: Hawaii and the Galapagos. Bioscience. 1988;38:272–82.

10. Maina J, de Moel H, Zinke J, Madin J, McClanahan T, Vermaat JE. Human deforestation outweighs future climate change impacts of sedimentation on coral reefs. Nature Commun. 2013;4:1986. doi:10.1038/ncomms2986

11. Ticktin T, Whitehead AN, Fraiola HA. Traditional gathering of native hula plants in alien-invaded Hawaiian forests: adaptive practices, impacts on alien invasive species and conservation implications. Environ Conserv. 2006;33(03):185–94.

12. Berkes F, Colding J, Folke C. Rediscovery of traditional ecological knowledge as adaptive management. Ecol Appl. 2000;10(5):1251–62.

13. Fordham D, Brook B. Why tropical island endemics are acutely susceptible to global change. Biodivers Conserv. 2010;19(2):329–42.

14. D'Antonio CM, Dudley TL. Biological invasions as agents of change on islands versus mainlands. Ecol Stud. 1995;115:103–19.

15. D'Antonio CM, Vitousek PM. Biological invasions by exotic grasses, the grass/fire cycle, and global change. Annu Rev Ecol Syst. 1992;23:63–87.

16. Loope LL, Mueller-Dombois D. Characteristics of invaded islands, with special reference to Hawaii. In: Drake J, DiCastri F, Groves R, Kruger F, Mooney HA, Rejmanek M, et al., editors. Biological invasions: a global perspective. Chichester: Wiley and Sons; 1989. p. 257–80.

17. Hughes RF, Asner GP, Mascaro J, Uowolo A, Baldwin J. Carbon storage landscapes of lowland Hawaii: the role of native and invasive species through space and time. Ecol Appl. 2014;24(4):716–31.

18. Harris NL, Brown S, Hagen SC, Saatchi SS, Petrova S, Salas W, et al. Baseline map of carbon emissions from deforestation in tropical regions. Science. 2012;336(6088):1573–5. doi:10.1126/science.1217962.

19. Baccini A, Goetz SJ, Walker WS, Laporte NT, Sun M, Sulla-Menashe D, et al. Estimated carbon dioxide emissions from tropical deforestation improved by carbon-density maps. Nat Clim Change. 2012;. doi:10.1038/nclimate1354.

20. Mitchard E, Saatchi S, Baccini A, Asner G, Goetz S, Harris N, et al. Uncertainty in the spatial distribution of tropical forest biomass: a comparison of pan-tropical maps. Carbon Balance Manage. 2013;8(1):10.

21. Asner GP, Knapp DE, Martin RE, Tupayachi R, Anderson CB, Mascaro J, et al. Targeted carbon conservation at national scales with high-resolution monitoring. Proc Natl Acad Sci. 2014;111(47):E5016–22.

22. Lefsky MA, Cohen WB, Parker GG, Harding DJ. Lidar remote sensing for ecosystem studies. Bioscience. 2002;52(1):19–30.

23. Asner GP, Hughes RF, Mascaro J, Uowolo AL, Knapp DE, Jacobson J, et al. High-resolution carbon mapping on the million-hectare Island of Hawaii. Front Ecol Environ. 2011;9(8):434–9.

24. Asner GP. Tropical forest carbon assessment: integrating satellite and airborne mapping approaches. Environ Res Lett. 2009;3:1748–9326.

25. Asner GP, Knapp DE, Balaji A, Paez-Acosta G. Automated mapping of tropical deforestation and forest degradation: CLASlite. J Appl Remote Sens. 2009;3:033543.

26. Asner GP, Mascaro J. Mapping tropical forest carbon: Calibrating plot estimates to a simple LiDAR metric. Remote Sens Environ. 2014;140:614–24. doi:10.1016/j.rse.2013.09.023.

27. Brown S, Gillespie AJR, Lugo AE. Biomass estimation methods for tropical forests with application to forest inventory. Forest Sci. 1989;35:881–902.

28. Asner GP, Hughes RF, Varga TA, Knapp DE, Kennedy-Bowdoin T. Environmental and biotic controls over aboveground biomass throughout a tropical rain forest. Ecosystems. 2009;12:261–78.

29. Chave J, Andalo C, Brown S, Cairns MA, Chambers JQ, Eamus D, et al. Tree allometry and improved estimation of carbon stocks and balance in tropical forests. Oecologia. 2005;145:87–99. doi:10.1007/s00442-005-0100-x.

30. Keller M, Palace M, Hurtt G. Biomass estimation in the Tapajos National Forest, Brazil: examination of sampling and allometric uncertainties. For Ecol Manage. 2001;154:371–82.

31. Chave J, Réjou-Méchain M, Búrquez A, Chidumayo E, Colgan MS, Delitti WBC, et al. Improved allometric models to estimate the aboveground biomass of tropical trees. Glob Change Biol. 2014;20(10):3177–90. doi:10.1111/gcb.12629.

32. Asner GP, Martin RE, Knapp DE, Kennedy-Bowdoin T. Effects of *Morella faya* tree invasion on aboveground carbon storage in Hawaii. Biol Invasions. 2010;12:477–94. doi:10.1007/s10530-009-9452-1.

33. Mascaro J, Hughes RF, Schnitzer SA. Novel forests maintain ecosystem processes after the decline of native tree species. Ecol Monogr. 2011;82(2):221–8. doi:10.1890/11-1014.1.

34. Gon SM, Allison A, Cannarella RJ, Jacobi JD, Kaneshiro KY, Kido MH et al. A GAP analysis of Hawaii: Final report. US Department of the Interior. US Geological Survey, Washington, DC. 2006.

35. Funk JL, Cleland EE, Suding KN, Zavaleta ES. Restoration through reassembly: plant traits and invasion resistance. Trends Ecol Evol. 2008;23(12):695–703. doi:10.1016/j.tree.2008.07.013.

36. Stone CP, Cuddihy LW, Tunison JT. Responses of Hawaiian ecosystems to removal of feral pigs and goats. In: Stone CP, Smith CW, Tunison JT, editors. Alien Plant Invasions in Native Ecosystems of Hawaii: Management and Research. Honolulu: University of Hawaii Cooperative National Park Resources Study Unit; 1992. p. 666–704.

37. Loope LL, Scowcroft PG. Vegetation response within exclosures in Hawaii: A review. In: Stone CP, Scott JM, editors. Hawaii's terrestrial ecosystems: preservation and Management. Honolulu: University of Hawaii Cooperative National Park Resources Study Unit; 1985. p. 377–402.

38. Mascaro J, Asner GP, Knapp DE, Kennedy-Bowdoin T, Martin RE, Anderson C, et al. A tale of two "forests": Random Forest machine learning aids tropical forest carbon mapping. PLoS One. 2014;9(1):e85993. doi:10.1371/journal.pone.0085993

39. Breiman L. Random forests. Mach Learn. 2001;45:5–32.

40. Asner GP, Mascaro J, Muller-Landau HC, Vieilledent G, Vaudry R, Rasamoelina M, et al. A universal airborne LiDAR approach for tropical forest carbon mapping. Oecologia. 2012;168(4):1147–60. doi:10.1007/s00442-011-2165-z.

41. Mascaro J, Detto M, Asner GP, Muller-Landau HC. Evaluating uncertainty in mapping forest carbon with airborne LiDAR. Remote Sens Environ. 2011;115(12):3770–4. doi:10.1016/j.rse.2011.07.019.

42. Taylor PG, Asner GP, Dahlin K, Anderson CB, Knapp DE, Martin RE, et al. Landscape-scale controls on aboveground forest carbon stocks on the Osa Peninsula, Costa Rica. PLoS One. 2015;10(6):e0126748.

43. Mascaro J, Asner G, Davies S, Dehgan A, Saatchi S. These are the days of lasers in the jungle. Carbon Balance Manage. 2014;9(1):1–3. doi:10.1186/s13021-014-0007-0.

44. Asner GP. Satellites and psychology for improved forest monitoring. Proc Natl Acad Sci. 2014;111(2):567–8.

45. Baccini A, Asner GP. Improving pantropical forest carbon maps with airborne LiDAR sampling. Carbon Manag. 2013;4(6):591–600. doi:10.4155/cmt.13.66.

46. Mascaro J, Asner GP, Knapp DE, Kennedy-Bowdoin T, Martin RE, Anderson C et al. A tale of two "forests": Random Forest machine learning aids tropical forest carbon mapping. PLoS One. 2014;e85993.

47. Asner G, Mascaro J, Anderson C, Knapp D, Martin R, Kennedy-Bowdoin T, et al. High-fidelity national carbon mapping for resource management and REDD+. Carbon Balance Manage. 2013;8(1):7.

48. Gromping U. Variable importance assessment in regression: linear regression versus random forest. Am Stat. 2009;63(4):308–19. doi:10.2307/25652309.

49. Asner GP, Knapp DE, Kennedy-Bowdoin T, Jones MO, Martin RE, Boardman J, et al. Carnegie airborne observatory: in-flight fusion of hyperspectral imaging and waveform light detection and ranging for three-dimensional studies of ecosystems. J Appl Remote Sens. 2007;1:013536.

50. Reimer F, Asner GP, Joseph S. Advancing reference emission levels in subnational and national REDD + initiatives: a CLASlite approach. Carbon Balance Manag. 2015;10:5. doi:10.1186/s13021-015-0015-8

51. Jacobi JD, Price JP, Fortini LB, Berkowitz P. Baseline Land Cover. In: Z. Zhu e, U.S. Department of Interior, U.S. Geological Survey, editor. Baseline and projected future carbon storage and greenhouse-gas fluxes in ecosystems of Hawaii. 2015.

52. Woudenberg SW, Conkling BL, O'Connell BM, LaPoint EB, Turner JA, Waddell KL. The Forest Inventory and Analysis Database: Database description and users manual version 4.0 for Phase 2. 2010.

53. Chave J, Coomes D, Jansen S, Lewis SL, Swenson NG, Zanne AE. Towards a worldwide wood economics spectrum. Ecol Lett. 2009;12:351–66.

54. Liaw A, Wiener M. Classification and regression by randomForest. R News. 2002;2:18–22.

13

An assessment of the carbon stocks and sodicity tolerance of disturbed *Melaleuca* forests in Southern Vietnam

Da B Tran[1*], Tho V Hoang[1] and Paul Dargusch[2]

Abstract

Background: In the lower Mekong Basin and coastal zones of Southern Vietnam, forests dominated by the genus *Melaleuca* have two notable features: most have been substantially disturbed by human activity and can now be considered as degraded forests; and most are subject to acute pressures from climate change, particularly in regards to changes in the hydrological and sodicity properties of forest soil.

Results: Data was collected and analyzed from five typical *Melaleuca* stands including: (1) primary *Melaleuca* forests on sandy soil (VS1); (2) regenerating *Melaleuca* forests on sandy soil (VS2); (3) degraded secondary *Melaleuca* forests on clay soil with peat (VS3); (4) regenerating *Melaleuca* forests on clay soil with peat (VS4); and (5) regenerating *Melaleuca* forests on clay soil without peat (VS5). Carbon densities of VS1, VS2, VS3, VS4, and VS5 were found to be 275.98, 159.36, 784.68, 544.28, and 246.96 tC/ha, respectively. The exchangeable sodium percentage of *Melaleuca* forests on sandy soil showed high sodicity, while those on clay soil varied from low to moderate sodicity.

Conclusions: This paper presents the results of an assessment of the carbon stocks and sodicity tolerance of natural *Melaleuca cajuputi* communities in Southern Vietnam, in order to gather better information to support the improved management of forests in the region. The results provide important information for the future sustainable management of *Melaleuca* forests in Vietnam, particularly in regards to forest carbon conservation initiatives and the potential of *Melaleuca* species for reforestation initiatives on degraded sites with highly sodic soils.

Keywords: Carbon sequestration, Climate change, *Melaleuca*, REED+, Sodicity

Background

Numerous studies have shown that tropical wetlands typically contain large carbon stocks [1–7]. Protecting and restoring tropical coastal wetlands is considered a critical part of how society adapts to and mitigates global climate change [8].

Large areas of *Melaleuca* forests in Vietnam are disturbed ecosystems that experience extreme conditions, and are associated with floods and/or sodic soils. They mostly occur in the lower Mekong Basin, which has been severely impacted by climate change [9–12]. Little is known about the carbon sequestration potential of disturbed *Melaleuca* forests in Australasia and South-East Asia where the genus occurs. Carbon stocks of *Melaleuca* forests are generally considered to be low (i.e. about 27.8 tC/ha estimated by Australian Government Office [13]). However, Tran et al. [14] suggested that this has been grossly under-estimated and that *Melaleuca cajuputi* forests on peatland soils in Vietnam, Indonesia and Malaysia are likely to have a high potential for carbon sequestration.

Sea level rise has significant impacts on the coastal zone, where soils will become saline and/or highly sodic [15]. Sodic soils are distinguished by an excessively high concentration of Sodium (Na) in their cation exchange complex. High sodicity causes soil instability due to poor physical and chemical properties, which affects plant growth and can have a more significant impact than excessive salinity growth [16, 17]. Sodicity impacts plant

*Correspondence: tranbinhda@gmail.com
[1] The Vietnam Forestry University, Hanoi, Vietnam
Full list of author information is available at the end of the article

growth in three ways, including: soil dispersion, specific ion effects, and nutritional imbalance in plants [18, 19]. Excessive sodium concentrations cause clay dispersion which is the primary physical effect of the sodic soil. Sodium-induced dispersion can reduce water infiltration, decrease hydraulic conductivity, and increase soil surface crusting that strongly affect roots such as root penetration, root development, and blocking plant uptake of moisture and nutrients [19].

Except for those containing mangroves and other halophytes, most ecosystems are severely affected by salinity and/or sodicity. A few studies have examined saline-sodic soils in shrimp farming areas in the coastal regions of Vietnam (i.e. ECe = 29.25 dS/m and exchangeable sodium percentage ranged from 9.63 to 72.07%, which had a big impact on plant cultivation systems [20]).

Several studies (such as Dunn et al. [21], Niknam and McComb [22], van der Moezel et al. [23, 24]) have examined the tolerance of woody species such as *Acacia*, *Eucalyptus*, *Melaleuca*, and *Casuarina* species to salinity and/or sodicity, but more research is required. This paper examines the carbon stocks of disturbed *Melaleuca* forests and the sodicity tolerance of *M. cajuputi* forests in Southern Vietnam.

Results and discussion
Characteristics of the typical *Melaleuca* forests in the study areas
The major characteristics of five *Melaleuca* forests types examined include standing trees, an understory, and saturated conditions (Table 1). The variation in these characteristics not only distinguishes the different stands but also improves understanding of their carbon stocks.

The stand densities of the five typical *Melaleuca* forest types varied considerably: they were 2,330, 10,950, 980, 9,833, and 6,867 trees/ha for VS1, VS2, VS3, VS4, and VS5, respectively (Table 1). Within each study site, the tree densities of regenerating forests (VS2, VS4, and VS5) were significantly higher than primary forests (VS1) and secondary forests (VS2) (Figure 1a). The increased stand densities of types VS2, VS4, and VS5 were mostly comprised of trees with a diameter at breast height (DBH) <10 cm. In contrast, VS1 was dominated by trees with DBH < 20 cm (accounting for 84.3%), with the balance of trees having a DBH $\geq$ 20 cm (including 4.2% of trees with DBH $\geq$ 30), while VS3 was mostly dominated by trees with a 5 cm $\leq$ DBH < 20 cm (accounting for 96%), with the balance having a 20 cm $\leq$ DBH < 40 cm (accounting for 4%) (Table 1).

Average DBH of all stand classes were 16.71, 5.36, 12.93, 5.88, and 6.20 for VS1, VS2, VS3, VS4, and VS5, respectively (Figure 1b). There was a significant difference in DBH in the five *Melaleuca* forest types (χ^2 = 446.86,

$p = 2.2e^{-16}$). However, post hoc test shows that there is no significant difference in tree DBH between VS1 and VS3, and between VS2, VS4, and VS5 (Additional file 1: 2b).

Average total height of all stand classes were 14.69, 7.11, 9.69, 5.68, and 7.50 m for VS1, VS2, VS3, VS4, and VS5, respectively (Figure 1c). There was a significant difference in the total height of the five *Melaleuca* forest types (χ^2 = 11.616, p = 0.0088) (Additional file 1: 2c). Furthermore, the tree density of the five forest types was generally very high, especially of VS2, VS4 and VS5 (over 2,000 individuals/ha), which can contribute to a large biomass. The basal areas shown in Figure 1d further confirm the potential high biomass of VS2, VS4 and VS5 (BA = 28.41, 30.14, and 23.14 m²/ha, respectively). Furthermore, the basal area of VS1 is significantly greater than VS3, accounting for 41.45 and 10.29 m²/ha, respectively (F = 3.341, p = 0.0423) (Additional file 1: 2d).

Different species were found in the understorey of the various *Melaleuca* forest types. Key species for VS1 and VS2 include *Leptocarpus* sp., *Lepironia* sp., *Hanguana* sp., *Eleocharis* sp., *Euriocaulon* sp., *Xyris* sp., *Stenochlaena* sp., *Melastoma* sp., and *Imperata cylindrica*. For VS3, VS4, VS5, the following species dominate the understorey: *Stenochlaenapalustris* sp., *Phragmitesvallatoria* sp., *Melastomadodecandrum* sp., *Diplaziumesculentum* sp., *Lygodiumscandens* sp., *Aspleniumnidus* sp., *Scleriasumatrensis*, *Cassia tora*, *Paederiafoetida* sp., *Flagellariaindica* sp., and *Cayratiatrifolia* sp. (Table 1).

Carbon stocks of *Melaleuca* forests
The carbon densities of five typical *Melaleuca* forests in Southern Vietnam were 275.98, 159.36, 784.68, 544.28, and 246.96 tC/ha, respectively, for primary *Melaleuca* forests on sandy soil (VS1), regenerating *Melaleuca* forests on sandy soil (VS2), degraded secondary *Melaleuca* forests on clay soil with peat (VS3), regenerating *Melaleuca* forests on clay soil with peat (VS4), and regenerating *Melaleuca* forests on clay soil without peat (VS5) (Figure 2a). There is significant difference in carbon densities between the forest types (χ^2 = 10.419, p = 0.0339) (Additional file 1: 2e). On sandy soils, the carbon density of VS1 was significantly greater (1.7 times) than VS2. The carbon density of *Melaleuca* forests on clay soil with peat was still high after disturbance (VS3 was 1.4 times higher than VS4). The carbon density of VS5 was lower than VS3 and VS4 because there was no peat layer.

On sandy soil, the stands and soil layers were the highest contributors to carbon density of VS1 (accounting for 41.34 and 29.11%, respectively), while VS2 has a high contribution from the soil layer, then stands (soil and stand categories contribute for carbon density of 56.15 and 28.53%, respectively) (Figure 2b). However, in

Table 1 Major characteristics of five typical *Melaleuca* forests in the study areas

Forest types	Tree classes	Code	Stand trees Density (trees/ha) Mean	SE	DBH (cm) Mean	SE	BA (m²/ha) Mean	SE	Height (m) Mean	SE	Understory	Saturation levels
Primary *Melaleuca* on sandy soil	DBH < 5 cm	VS1C0	800	248.3	3.87	0.11	na	na	6.00	0.28	*Leptocarpus sp.* *Lepironia sp.* *Hanguana sp.* *Eleocharis sp.* *Euriocaulon sp.* *Xyris sp.* *Stenochlaena sp.* *Melastoma sp.* *Imperata sp.*	Including non-inundated, seasonal, and permanent inundation
	5 cm ≤ DBH < 10 cm	VS1C1	400	100.0	7.18	0.36	na	na	9.81	0.68		
	10 cm ≤ DBH < 20 cm	VS1C2	750	273.8	14.63	0.22	na	na	14.80	0.26		
	20 cm ≤ DBH < 30 cm	VS1C3	285	34.0	24.33	0.49	na	na	18.44	0.40		
	30 cm ≤ DBH < 40 cm	VS1C4	80	28.3	34.37	0.90	na	na	20.17	0.97		
	DBH ≥ 40 cm	VS1C5	20	10.0	48.73	3.75	na	na	22.20	1.77		
	All classes	VS1	2,330	558.0	16.71	0.55	41.54	6.16	14.69	0.30		
Regenerating *Melaleuca* on sandy soil	DBH < 5 cm	VS2C0	5,450	2,850.0	3.63	0.07	na	na	6.13	0.16		Including non-inundated and seasonal inundation
	5 cm ≤ DBH < 10 cm	VS2C1	5,500	700.0	7.07	0.14	na	na	8.08	0.15		
	DBH ≥ 10 cm	na	na	na	na	na	na	na	na	na		
	All classes	VS2	10,950	3,550.0	5.36	0.14	28.41	3.14	7.11	0.13		
Degraded secondary *Melaleuca* on clay soil with peat	DBH < 5 cm	VS3C0	150	na	4.41	0.23	na	na	5.00	0.29	*Stenochlaena palustris* *Phragmites vallatoria* *Melastoma dodecandrum* *Diplazium esculentum* *Lygodium scandens* *Asplenium nidus* *Scleria sumatrensis* *Cassia tora* *Paederia foetida* *Flagellaria indica* *Cayratia trifolia*	Including seasonal and permanent inundation
	5 cm ≤ DBH < 10 cm	VS3C1	350	na	7.12	0.68	na	na	4.57	0.38		
	10 cm ≤ DBH < 20 cm	VS3C2	440	20.0	13.11	0.36	na	na	10.44	0.41		
	20 cm ≤ DBH < 30 cm	VS3C3	30	na	25.00	1.20	na	na	14.33	0.17		
	30 cm ≤ DBH < 40 cm	VS3C4	10	na	35.35	na	na	na	12.50	na		
	DBH ≥ 40 cm	VS3C5	na	na	na	na	na	na	na	na		
	All classes	VS3	980	560.0	12.93	0.71	10.29	4.74	9.69	0.45		
Regenerating *Melaleuca* on clay soil with peat	DBH < 5 cm	VS4C0	3,867	2,258.6	3.84	0.06	na	na	4.15	0.11		Including seasonal and permanent inundation
	5 cm ≤ DBH < 10 cm	VS4C1	5,967	176.4	7.20	0.12	na	na	6.68	0.17		
	DBH ≥ 10 cm	na	na	na	na	na	na	na	na	na		
	All classes	VS4	9,833	2,265.9	5.88	0.12	30.14	1.46	5.68	0.13		
Regenerating *Melaleuca* on clay soil without peat	DBH < 5 cm	VS5C0	2,133	592.6	3.82	0.09	na	na	4.95	0.17		Including seasonal and permanent inundation
	5 cm ≤ DBH < 10 cm	VS5C1	4,733	1,560.3	7.27	0.13	na	na	8.65	0.31		
	DBH ≥ 10 cm	na	na	na	na	na	na	na	na	na		
	All classes	VS5	6,867	1,970.1	6.20	0.14	23.02	8.53	7.50	0.25		

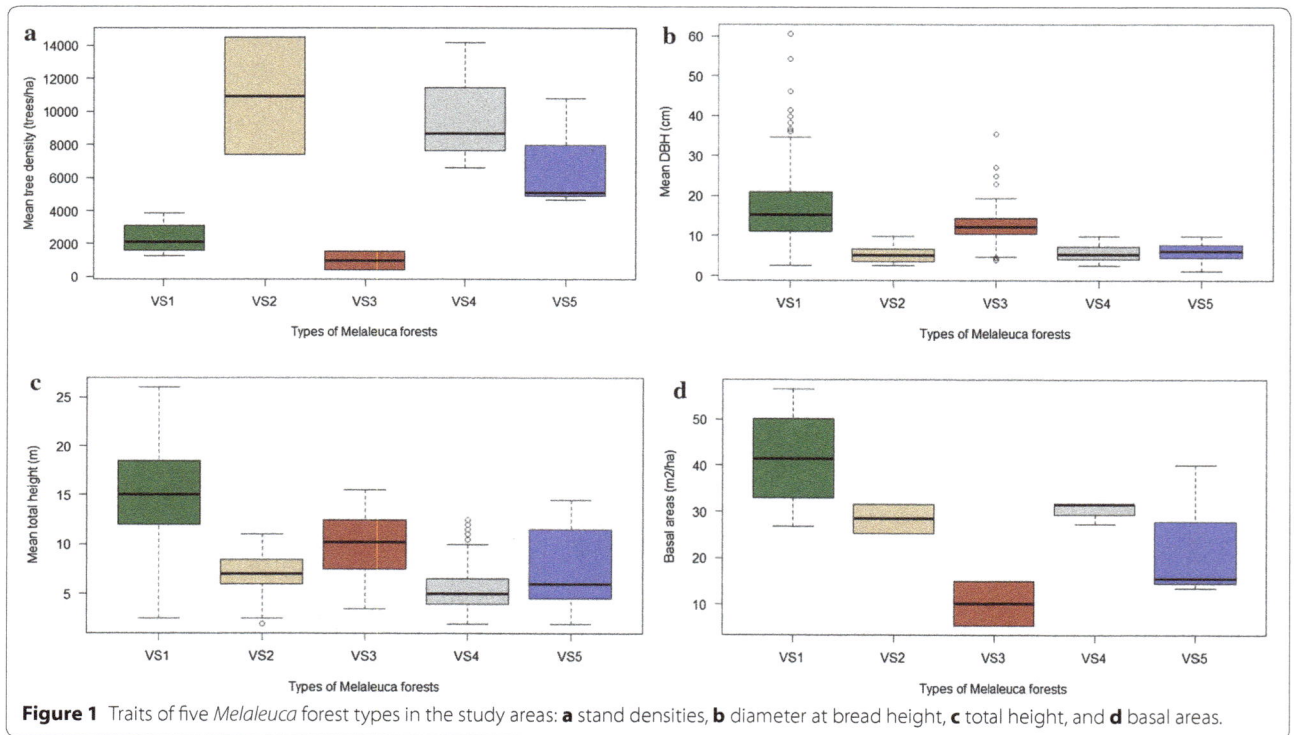

Figure 1 Traits of five *Melaleuca* forest types in the study areas: **a** stand densities, **b** diameter at bread height, **c** total height, and **d** basal areas.

the peat land, the greatest contribution of carbon densities for VS3 and VS4 are the peat and soil categories (accounting for 61.41%, 22.10% of VS3, and 57.66, and 16.72% of VS4, respectively). Separately, carbon density of VS5 is mostly linked to the soil, deadwood, and stand categories (accounting for 33.54, 32.16, and 14.66%, respectively) (Figure 2b).

Variability of carbon stocks in different types of *Melaleuca* forests

This study investigated the carbon stocks of six categories: stands, understory, deadwood, litter, root, and soil for five types of *Melaleuca* forests in Southern Vietnam (Figure 3).

The carbon densities of stands of the various forest types were 110.67, 44.27, 22.79, 48.25, and 37.20 tC/ha for VS1, VS2, VS3, VS4, and VS5, respectively (Figure 3a). There was a significant difference in stand carbon density between the forest types ($\chi^2 = 48.3184$, $p = 8.1e^{-10}$) (Additional file 1: 2f). The carbon density of the stand VS1 is the highest and is 2.5, 4.9, 2.3, and 3.0 times higher than VS2, VS3, VS4, and VS5. Surprisingly, there is no statistical difference in stand carbon densities between secondary forests (VS3) and regenerating forests (VS2, VS4 and VS5) (Additional file 1: 2f). These carbon stocks were lower those from other studies of different forests (e.g. 144 tC/ha for Asian tropical forests [25]; 200.23 tC/ha and 92.34 tC/ha of primary and secondary

swamp forests in Indonesia (involving *Melaleuca* vegetation), respectively [26]).

The carbon densities of the understory in the *Melaleuca* forests of Vietnam were 2.45, 2.48, 6.23, 1.65, and 5.27 tC/ha for VS1, VS2, VS3, VS4, and VS5, respectively (Figure 3b). There was a statistically significant difference in understory carbon density between the forest types ($\chi^2 = 30.7189$, $p = 3.49e^{-6}$) (Additional file 1: 2g). However, there was no significant difference in understory carbon density between *Melaleuca* forest types on sandy soils (VS1 and VS2). On clay soils, the understory carbon densities of VS3 and VS5 were significantly higher than VS4.

The carbon densities of deadwood in the forest types were 30.47, 0, 67.90, 45.06, and 74.59 tC/ha for VS1, VS2, VS3, VS4, and VS5, respectively (Figure 3c). There was a statistically significant difference in deadwood carbon density between the *Melaleuca* forest types ($\chi^2 = 3.0978$, $p = 0.5416$), but pairwise comparisons show no significant differences (Additional file 1: 2 h). Surprisingly, deadwood was not present in regenerating forests in the study sites on Phu Quoc Island. This is probably due to frequent forests fires and/or fuelwood collection by people associated crop cultivation.

Some of the carbon stock of *Melaleuca* forests is contributed by layers of coarse and fine litter. The carbon densities of the total litter layer of the forest types were

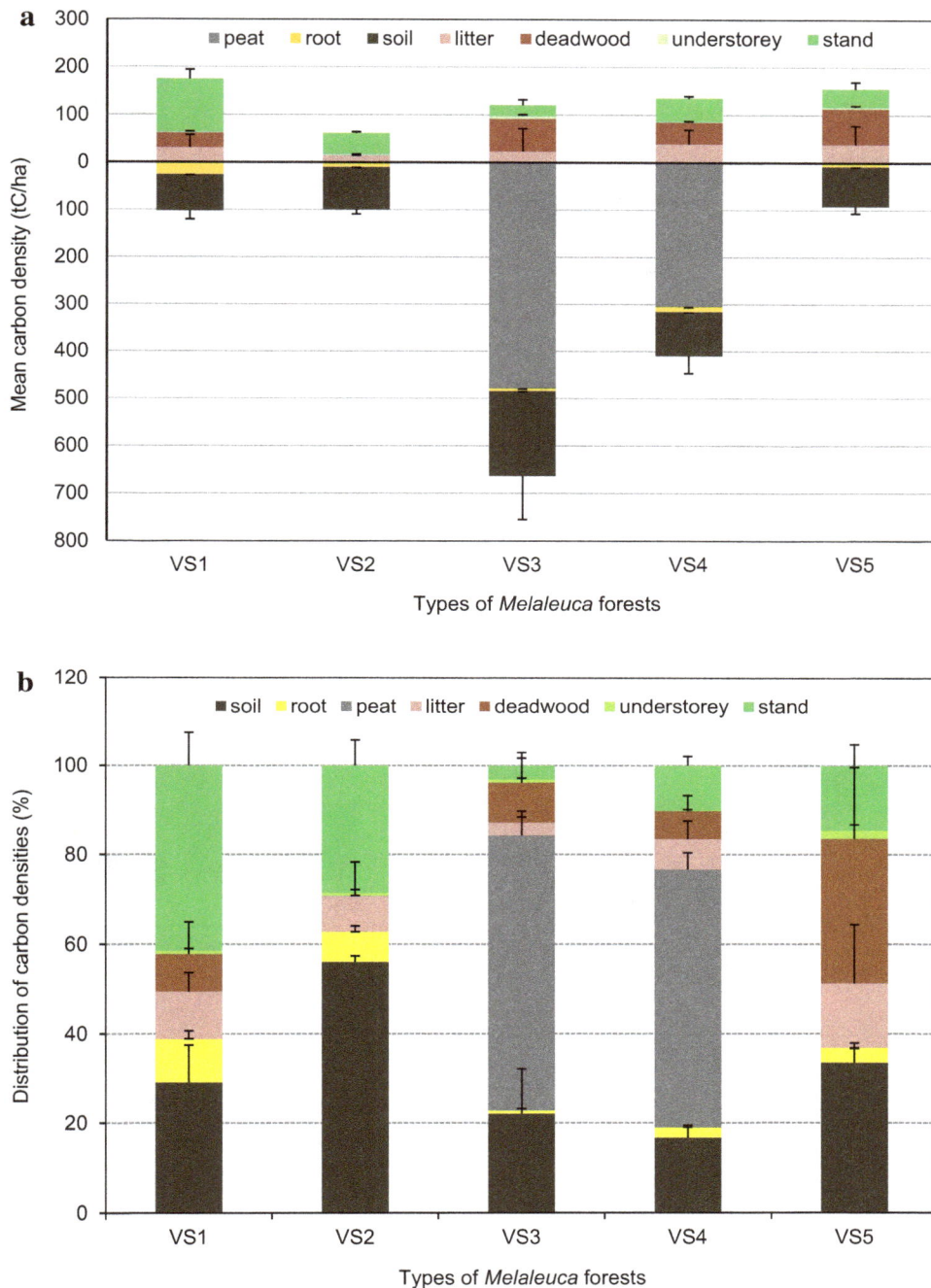

Figure 2 Carbon densities of five typical *Melaleuca* forests in the study areas: **a** mean carbon density, and **b** distribution of carbon densities.

31.03, 14.45, 23.76, 57.35, and 39.23 tC/ha for VS1, VS2, VS3, VS4, and VS, respectively (Figure 3d). There was a statistically significant difference in overall litter carbon density between these forest types ($\chi^2 = 1.5619$, $p = 0.08156$), but pairwise comparisons show no significant differences (Additional file 1: 2i).

The carbon densities from peat of the *Melaleuca* forests were 479.62 and 294.57 tC/ha for secondary forests (VS3) and regenerating forests (VS4), respectively (Figure 3e). The carbon density from peat of VS3 is significantly greater than that of VS4 ($\chi^2 = 5.2359$, $p = 0.0221$) (Additional file 1: 2j). This is almost certainly due to

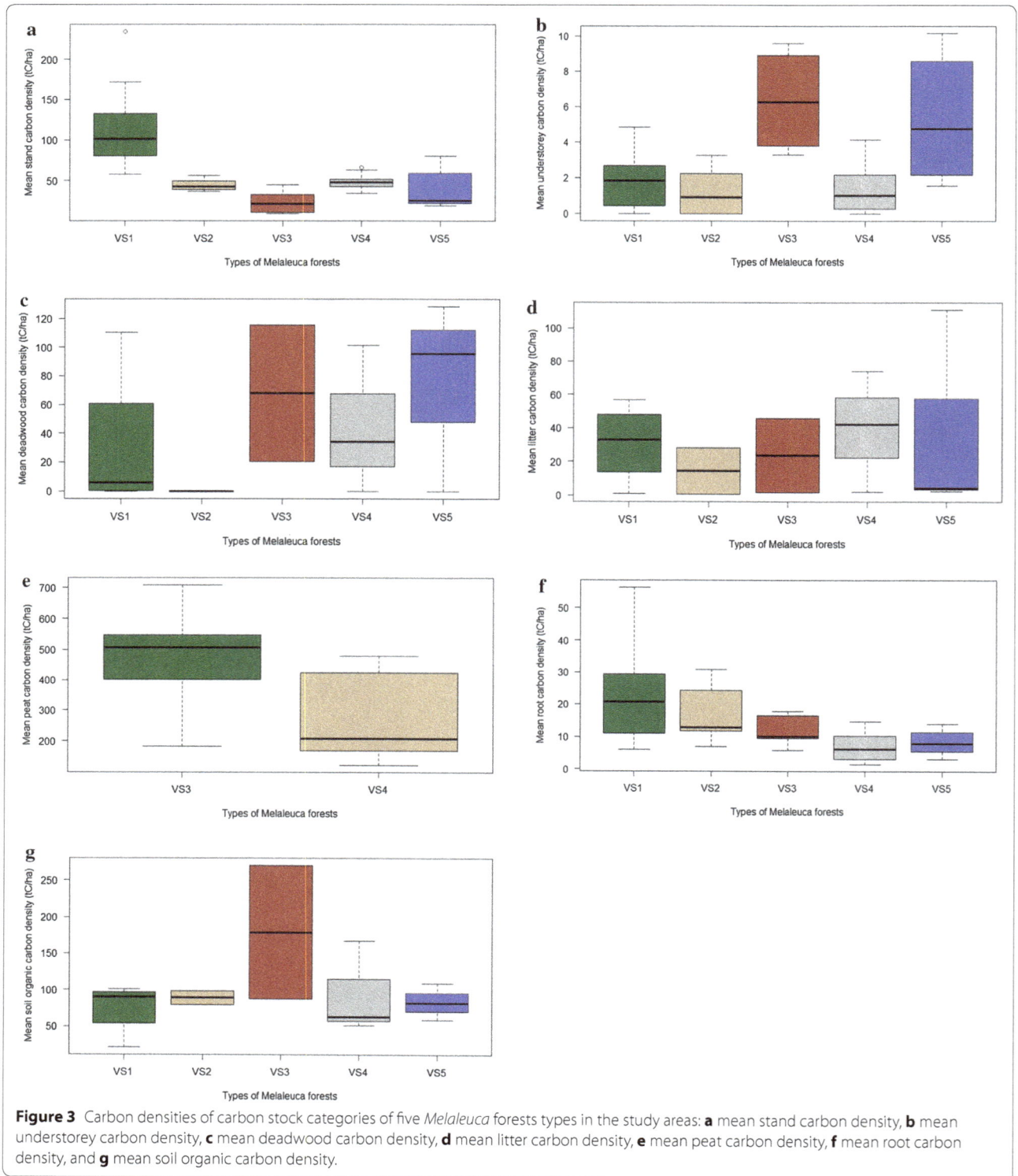

Figure 3 Carbon densities of carbon stock categories of five *Melaleuca* forests types in the study areas: **a** mean stand carbon density, **b** mean understorey carbon density, **c** mean deadwood carbon density, **d** mean litter carbon density, **e** mean peat carbon density, **f** mean root carbon density, and **g** mean soil organic carbon density.

peat being partly burned in the regenerating forest by the severe fire of 2002. In U Minh Thuong National Park, peat comprises the top soil layer, with a deep layer of clay below. The depth of the peat layer ranged from 15 to 62 cm in 18 soil cores, and the peat bulk density ranged from 0.19 to 0.3. The depths of the peat layer in this study were much thinner than in other forests (i.e. primary peat layer in U Minh Thuong was over 90 cm depth [27], and the thick peat layer in U Minh Ha was over 120 cm depth [28]).

The carbon densities of roots in the *Melaleuca* forests were 22.75, 16.97, 11.97, 6.99, and 8.35 tC/ha for VS1, VS2, VS3, VS4, and VS5, respectively (Figure 3f). There was a statistically significant difference in root carbon density between the forest types ($\chi^2 = 22.437$, $p = 0.00016$). The carbon densities of roots in *Melaleuca* forests in sandy soil were higher than those in clay soil, in particular, the root carbon density of VS2 was significant higher than that of VS4 (Additional file 1: 2k).

Organic soil carbon densities to a 30 cm depth in the study areas were 75.81, 89.22, 178.93, 93.94, and 83.58 tC/ha for VS1, VS2, VS3, VS4, and VS5, respectively (Figure 3 g). There was a statistically significant difference in organic soil carbon density between the forest types ($\chi^2 = 1.7333$, $p = 0.230$), but pairwise comparisons showed no significant differences (Additional file 1: 2k). These results are consistent with those of other studies of soil carbon stocks in wetlands (e.g. organic soil carbon stocks in swamp forests in Indonesia (with *Melaleuca* vegetation) were 106.00 and 135.63 tC/ha in the top 30 cm of soil of primary and secondary forests, respectively [29]).

Overall, the carbon density of *Melaleuca* forests on sandy soil in Southern Vietnam ranged from 159.36 tC/ha for regenerating forests to 275.98 tC/ha for primary forests. The carbon densities of forests on clay soil ranged from 246.96 tC/ha for regenerating forests without peat to 784.68 tC/ha of secondary forests with peat. Compared with the carbon stocks of other forests on peatland (e.g. the carbon density of mangrove forests in the Indo-Pacific region was 1,030 tC/ha [30]), the carbon density of disturbed *Melaleuca* forests on the peatland of Southern Vietnam is about one half, but the results are consistent with other studies on peat swamp forests (e.g. the carbon density of undisturbed swamp forests in South-East Asia ranged from 182 to 306 tC/ha [31]). Despite this, *Melaleuca* forests in the peatlands of Vietnam still have high

potential as carbon stores. The case of U Minh Thuong National Park is an example. The total carbon stock of 8,038 ha of *Melaleuca* forests in the park is about 2.69 M tC (Table 2), which is equivalent 9.43 M tCO$_2$e. Furthermore, there were 8,576 hectares of *Melaleuca* forested peatland in U Minh Ha National Park that have peat layers ranging from 40 cm to over 120 cm deep [32], which provides an even higher potential carbon store.

Sodicity tolerance of *Melaleuca cajuputi* forests toward the adaptation to global climate change

Sea-level rise is a consequence of global climate change that will severely affect coastal and wetland ecosystems. *Melaleuca* forests are largely located in coastal and wetland areas that may be affected by climate change [33], so the risk of salinization of the region will increase. Salinity in soils can damage woody plant species by stunting buds, reducing leaf size and causing necroses in buds, roots, leaf margins and shoot tips [34]. Salinity can also inhibit seed germination, and can even kill non-halophytic species [35]. Both vegetative and reproductive growth of woody species are also reduced by high concentrations of sodium chloride in soil [35, 36]. The combination of flooding and salinity can create a more pronounced effect on growth and survival of plants than either stress alone [35]. High concentrations of sodium can affect the structure of sodic soils [37–39]. In contrast, low sodium concentration, soil structure is not affected by salinity in saline soil [40]. Sodicity and salinity always occur together and coming to have negative impacts on soil properties and plants [38, 41], but sodic soils may be either non-saline or saline [17].

The lower Mekong Basin and coastal regions of southern Vietnam are highly vulnerable to global climate change impacts [9, 33, 42, 43]. Most of Vietnam's *Melaleuca* forests occur in these areas and will be affected projected sea-level rise. Fortunately, this study has shown that *M. cajuputi* has the ability to tolerant increase in sodic soils.

About 28 soil samples collected from *Melaleuca* forests in Southern Vietnam were examined and all were

Table 2 Potential carbon storage in *Melaleuca* peat-swamp forests: case in U Minh Thuong National Park

Land cover type	Area (ha)	Carbon density (tC/ha)	Carbon storage tC
Mature *Melaleuca* forests on clay soil without peat	1,765	305.06	538,431
Mature *Melaleuca* forests on clay soil with peat	601	784.68	471,593
Regenerating *Melaleuca* on clay soil with peat	2,106	544.28	1,146,254
Regenerating *Melaleuca* on clay soil without peat	1,106	246.96	273,138
Others (open water, reeds and grasses)	2,460	107.91	265,459
Total	8,038		2,694,874

The areas of *Melaleuca* forests in U Minh Thuong National Park are taken from a Vietnam Environment Protection Agency report [48].

Table 3 Chemical element concentration and sodicity levels of the *Melaleuca* forest soils in the study areas

Forest types	Soil layers (cm)	$pH_{(KCl)}$		Ca^{2+} (meq/100 g)		Mg^{2+} (meq/100 g)		Na^+ (meq/100 g)		K^+ (meq/100 g)		Al^{3+} (meq/100 g)		Fe^{3+} (mg/100 g)		ESP (%)		Sodicity
		Mean	SE	Mean	SE	Mean	SE	Mean	SE	Mean	SE	Mean	SE	Mean	SE	Mean	SE	
Primary *Melaleuca* on sandy soil (VS1)	0–10	3.97	0.15	1.413	0.75	1.783	1.58	1.790	1.56	0.600	0.47	0.910	0.29	3.303	1.20	32.05	4.28	High
	10–30	4.12	0.17	1.065	0.41	1.138	1.02	1.708	1.58	0.383	0.33	0.660	0.24	7.310	2.23	39.78	7.90	High
Regenerating *Melaleuca* on sandy soil (VS2)	0–10	3.68	0.03	0.690	0.10	0.310	0.10	0.310	0.15	0.155	0.02	1.860	0.14	1.615	0.36	21.16	7.82	High
	10–30	3.86	0.04	0.645	0.03	0.175	0.00	0.150	0.02	0.065	0.02	1.280	0.24	1.810	0.56	14.49	2.28	Moderate
Degraded secondary *Melaleuca* on clay soil with peat (VS3)	0–10	4.12	0.25	7.585	1.82	6.320	1.81	1.705	0.81	0.455	0.19	0.100	0.10	37.155	17.63	10.61	2.37	Moderate
	10–30	4.07	0.32	7.585	2.96	5.795	2.59	1.470	0.41	0.705	0.07	1.680	1.64	48.245	7.19	9.45	1.09	Low
Regenerating *Melaleuca* on clay soil with peat (VS4)	0–10	4.67	0.19	8.845	0.55	6.685	0.58	1.760	0.51	0.585	0.27	0.00	0.00	47.550	16.06	9.85	3.05	Low
	10–30	5.00	0.04	5.855	0.33	4.860	0.42	1.320	0.19	0.575	0.18	0.00	0.00	54.825	36.49	10.47	1.27	Moderate
Regenerating *Melaleuca* on clay soil without peat (VS5)	0–10	4.16	0.26	11.580	4.19	5.557	0.55	1.330	0.24	0.663	0.07	5.533	5.53	67.433	9.03	6.95	0.49	Low
	10–30	3.91	0.40	8.603	1.65	5.170	0.08	1.507	0.24	0.717	0.09	8.000	7.02	78.440	10.37	9.42	0.37	Low

shown to be sodic (Table 3). While the exchangeable sodium percentage (ESP) of soil layers of Melaleuca forests on clay soil (VS3, VS4, and VS5) ranges from low to moderate sodicity, those of Melaleuca forests on sandy soil (VS1 and VS2) were significantly higher, particularly VS1, which had an ESP of up to 39.78% in soil taken from depths of 10–30 cm (Table 3). This indicates that both mature and young M. cajuputi forests have a high tolerance of sodic soils. Furthermore, M. cajuputi seeds can germinate and grow in highly sodic soil [e.g. M. cajuputi in forest type VS2 was able to grow in highly sodic soil with ESP up to 21.16% in the top 0–10 cm (Table 3)].

With the exception of mangroves, few woody species can tolerate saline and/or sodic soils. Many woody species have been examined for their tolerance of salinity and/or sodicity. For example, Eucalyptus, Melaleuca, Acacia, Casuarina [21–24], Grevillea robusta, Lophostemon confertus and Pinus caribea [44], and Moringa olifera [45] have been examined and their tolerance to salinity assessed in the field and in glasshouses. In extremely saline soils in Australia, Niknam and McComb [22] suggested that the land care benefit of establishing species such as Melaleuca or Casuarina is more important than their commercial value. As well as the land care value, this study has shown that M. cajuputi forests in Vietnam can adapt to climate change through their tolerance to sodicity, and other harsh conditions [33], and can help to mitigate climate change through their carbon storage abilities.

Conclusion

By undertaking original field data, this study examined the carbon sequestration potential of five types of Melaleuca forests including 'Primary Melaleuca forests on sandy soil' (VS1), 'Regenerating Melaleuca forests on sandy soil' (VS2), 'Degraded secondary Melaleuca forests on clay soil with peat' (VS3), 'Regenerating Melaleuca forests on clay soil with peat' (VS4), and 'Regenerating Melaleuca forests on clay soil without peat' (VS5). The study also assessed the sodicity tolerance of M. cajuputi forests in coastal and wetland regions of Vietnam.

The carbon densities of VS1, VS2, VS3, VS4, and VS5 were 275.98 (±38.62) tC/ha, 159.36 (±21.01) tC/ha, 784.68 (±54.72) tC/ha, 544.28 (±56.26) tC/ha, and 246.96 (±27.56) tC/ha, respectively. Most carbon stocks were contributed from the soil (including peat) and stands.

The exchangeable sodium percentage (ESP) of soil from Melaleuca forests on clay soil (VS3, VS4, and VS5) ranged from low to moderate sodicity, but those from Melaleuca forests on sandy soil (VS1 and VS2) were highly sodic.

The results provide important information for the future sustainable management of Melaleuca forests in Vietnam, particularly in regards to forest carbon conservation initiatives and the potential of Melaleuca species for reforestation initiatives on degraded sites with highly sodic soils. In Vietnam, forest carbon conservation initiatives such as REDD+ have hereto, in our view, not placed appropriate priority or consideration on the protection of carbon stocks of Melaleuca forests. The results presented in this paper suggest that Melaleuca forests in Vietnam, particularly those on peatland areas, hold globally significant carbon stocks—arguably greater than those found in upland rainforest ecosystems, which have so far been given higher priority in REDD+ planning in Vietnam. Furthermore, the results presented in this paper suggest that some Melaleuca forest species in Vietnam, particularly those on sandy soils, exhibit a tolerance for highly sodic soils. This suggests that those species might be useful in reforestation initiatives on degraded sites with highly sodic soils. As degradation pressures including climate change continue to alter the hydrological features of soil systems in areas such as the Mekong Delta in Vietnam, and the sodicity of soils in some areas increases, Melaleuca species could offer a useful option for reforestation and rehabilitation initiatives.

The results in this research provide further scientific information to support better Melaleuca ecosystem management. The results should help policy makers make better decisions in an era of global change. The results have particular relevance for the application of REED+ in the Southeast Asia.

Methods
Study sites and disturbance context

Melaleuca cajuputi is naturally distributed as scattered shrub populations along the coastal regions in the middle Provinces and up to the Northern hilly regions, and as tall forests in the Mekong Delta of Vietnam [46]. Thus, the study focussed on the sites in Southern Vietnam (involving Mekong Delta). The study investigated two sites: the Phu Quoc National Park and U Minh Thuong National Park, which both contain extensive Melaleuca forests in coastal wetlands (Figure 4). A total of 14 plots were randomly selected for carbon storage assessment, covering five types of Melaleuca stands: 'Primary Melaleuca forests on sandy soil'(VS1), 4 plots; 'Regenerating Melaleuca forests on sandy soil' (VS2), 2 plots; 'Degraded secondary Melaleuca forests on clay soil with peat' (VS3), 2 plots; 'Regenerating Melaleuca forests on clay soil with peat' (VS4), 3 plots; and 'Regenerating Melaleuca forests on clay soil without peat' (VS5), 3 plots.

Phu Quoc National Park is located on the northern Phu Quoc Island of Vietnam (at N 10°12′07″–N 10°27′02″, E 103°50′04″–E 104°04′40″) (Figure 4). Melaleuca forest areas cover 1,667.50 ha out of the total area

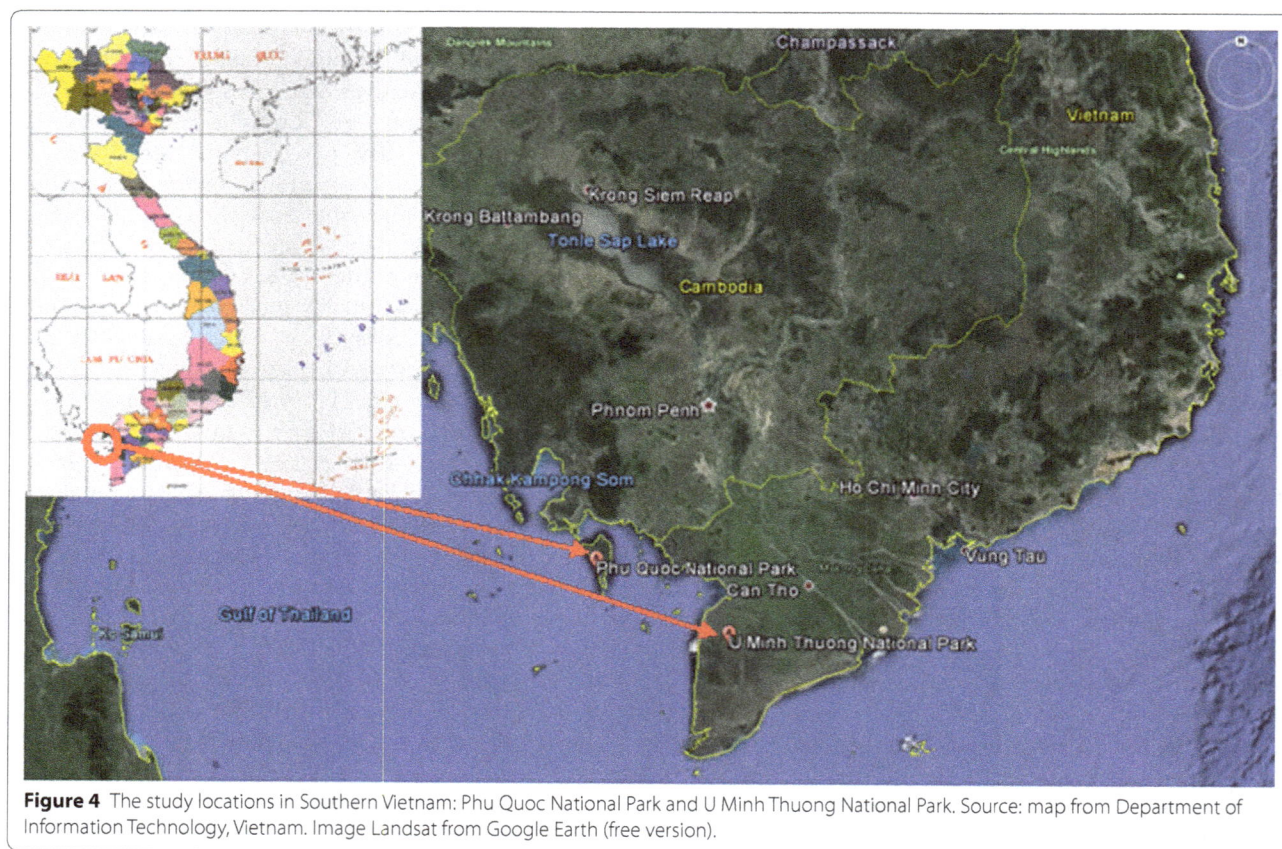

Figure 4 The study locations in Southern Vietnam: Phu Quoc National Park and U Minh Thuong National Park. Source: map from Department of Information Technology, Vietnam. Image Landsat from Google Earth (free version).

of 28,496.90 ha. These *Melaleuca* forests naturally occur on lowland regions of the island where they are seasonally inundated and/or permanent saturated, and also on permanent sand bars where no inundation occurs [47]. The rest areas of the park are hilly and mountainous forests. Two *Melaleuca* forest types were found in the park: primary *Melaleuca* forest (VS1); and regenerating *Melaleuca* forest (VS2). Before the park was established in 2001, key disturbance included forest fires and human intrusion for crop cultivation. The regenerating *Melaleuca* forests were up to 10–12 years of age at the time this study was conducted.

U Minh Thuong National Park is located in the Kien Giang Province (at N 9° 31′–N 9° 39′, E 105° 03′–E 105° 07′) (Figure 4). *Melaleuca* forest on swamp peatland is an endemic ecosystem in the lower Mekong Basin of Vietnam. The core area of the park is 8,038 ha, which is surrounded by a buffer zone of 13,069 ha. Here, the key disturbance is fire, with the last major fire occurring in April 2002, which burnt the primary vegetation as well as the peat soil. The Vietnamese Environment Protection Agency [48] reported that 3,212 hectares of *Melaleuca* forests was almost destroyed, so a canal system was built as a key management solution

to increase water inundation of the forest to prevent fires. Currently, there are three *Melaleuca* forest types in U Minh Thuong National Park: VS3, VS4, and VS5. At the time of this study, the VS4 and VS5 areas were up to 10 years old.

Field sampling and data collection

The major plots were set out as 500 m^2 quadrats (20 m × 25 m), and all trees with a DBH $\geq$ 10 cm were measured and recorded. Sub-plots also were set out as 100 m^2 quadrats (20 m × 5 m) within the major plots to measure all trees with DBH < 10 cm and a total height of >1.3 m (modified from Van et al. [49]). Data on DBH, alive or dead, and height were recorded for all standing trees.

Deadwood (dead fallen trees) with a diameter $\geq$10 cm were measured within the major plots (500 m^2), while deadwood with 5 cm $\leq$ diameter < 10 cm were measured within the sub-plots (100 m^2). Diameters at both ends of the trunk (D1 and D2), length (if $\geq$50 cm length), and the decay classes (involved sound, intermediate, and rotten [50, 51]) were recorded for all deadwood.

Seventy random quadrats (1 m × 1 m) were located in the main plots to collect and record the 'fresh weight' of the understory. Samples of all species from the understory were collected in each major plot and taken back to the Vietnam Forestry University laboratory for drying.

Seventy random coarse litter samples and seventy random fine litter samples were collected in the major plots. The fresh weight of each litter sample was recorded. Each litter type (coarse litter and fine litter) collected in every major plot were well mixed and taken to the laboratory for drying.

Two soil samples, one from the upper (0–10 cm) soil layer and one from the lower (10–30 cm) soil layer, were taken from each of 14 plots, giving a total of 28 soil samples. The 28 soil samples were taken back to the National Institute of Agricultural Planning and Projection laboratory for further analysis. Various soil chemical properties of the 28 samples were tested including: pH_{KCl}, total C, total N, Ca^{2+}, Mg^{2+}, Na^+, K^+, Al^{3+}, and Fe^{3+}. Twenty-eight duplicate soil samples were collected and analyzed for bulk density.

Sample analysis

Each understory and litter sample was divided into three sub-samples and dried in a drying oven at 60°C to measure the moisture content, based on the Eq. (1) below:

$$R_{moist} = \frac{\sum_{i=1}^{n} \frac{W_{fi} - W_{di}}{W_{fi}}}{n}. \tag{1}$$

where R_{moist} = moist ratio [0:1], W_{fi} = fresh weight of sub-sample i, W_{di} = dry weight of sub-sample i, n = number of sub-samples. The scales used to weight sub-samples were accurate to ± 0.01 g.

Total organic carbon (C%) was measured using the Walkley–Black method, which is commonly used to examine soil organic carbon via oxidation with $K_2Cr_2O_7$ [52, 53]. Total nitrogen was measured using the Kjeldahl method, which is the standard way to determine the total organic nitrogen content of soil [54]. A standard bulk density test was used to analyze all soil bulk samples in a dryven. Bulk density was calculated using Eq. (2):

$$BD = \frac{Ms}{V}. \tag{2}$$

where BD = the bulk density of the oven-dry soil sample (g/cm³), Ms = the oven dry-mass of the soil sample (gram), V = the volume of the ring sample (cm³).

Exchangeable sodium percentage (ESP) was calculated using Eq. (3) [55–57], and classified with four sodic levels as non-sodic soil (ESP < 6), low sodic soil (ESP = 6–10), moderately sodic soil (ESP = 10–15), and highly sodic soil (ESP > 15) [55–57].

$$ESP = \frac{Na^+}{\Sigma [Na^+][K^+][Mg^{2+}][Ca^{2+}]} \times 100. \tag{3}$$

Basal area (BA) was calculated with Eq. (4) (modified from Jonson and Freudenberger [58]):

$$BA = \frac{\sum_{n}^{1} [\pi \times (DBH_i/200)^2]}{S_{plot}} \times 10,000 \tag{4}$$

where BA = basal area (m²/ha), DBH_i = diameter at bread height of tree i (cm), i = stand individual (i = [1:n]), n = number of trees of sample plot, S_{plot} = area of the sample plot (m²).

Biomass allometric computation

Nine allometric equations, which are most common way to measure forest carbon stocks, were applied to calculate the above-ground and root biomass of the stands (Table 4). The selected allometric equations were tested for statistical significance using the R Statistic Program (Additional file 1: 1). Using these equations, the average biomass was analyzed for five typical *Melaleuca* stands (VS1, VS2, VS3, VS4, and VS5). To convert from fresh to dry biomass, a moisture rate of 0.5 was applied as suggested by Van et al. [49] for the allometric equation of Finlayson et al. [59]. According to the Global Wood Density Database, the density of *M. cajuputi* timber ranges from 0.6 to 0.87 g/cm³ [60], so 0.6 g/cm³ was applied for the above-ground biomass allometric equation of Chave et al. [61].

The fallen deadwood biomass were calculated using Eq. (5) ([62], p 12):

$$B = \pi \times r^2 \times L \times \delta \tag{5}$$

where B = biomass (kg), r = ½ diameter (cm), L = length (m), and δ = wood density (= 0.6 g/cm³).

Then, the biomass of the fallen deadwood was determined using the IPCC [50, 51] density reduction factors (sound = 1, intermediate = 0.6, and rotten = 0.45). The biomass of standing dead trees was measured using the same criteria as live trees, but a reduction factor of 0.975 is applied to dead trees that have lost leaves and twigs, and 0.8 for dead trees that have lost leaves, twigs, and small branches (diameter <10 cm) ([51], p 4.105).

To convert biomass to carbon mass for all categories (stands, roots, deadwood, understory, and litter), a factor of 0.45 was applied.

Soil organic carbon (SOC) was calculated using Eq. (6) [50, 51]:

$$SOC = Dep \times BD \times C_{sample} \times 100 \tag{6}$$

where SOC = Soil organic carbon, Dep = depth of soil layer (m), BD = bulk density (g/cm³), C_{sample} = organic

Table 4 List of allometric equations applied to examine stand biomass of the *Melaleuca* forests

Allometric equations	R^2	Vegetation	Sites	References
$\log_{10}(FW) = 2.266\log_{10}(D) - 0.502$ where FW = fresh above-ground biomass (kg/tree), D = diameter at breast height (cm)	0.98	*Melaleuca* spp.	Northern Territory	Finlayson et al. [59]
$y = 0.124 \times DBH^{2.247}$ where y = above-ground biomass (kg/tree), DBH = diameter at breast height (cm)	0.97	*Melaleuca cajuputi*	Vietnam	Le [63]
$y = \exp[-2.134 + 2.53\ln(D)]$ where y = above-ground biomass (kg/tree), D = diameter at breast height (cm)	0.97	Mixed species	Tropical, moist forest	IPCC [51] or Brown [64]
$\ln(y) = 2.4855\ln(x) - 2.3267$ where y = above-ground biomass (kg/tree), x = diameter at breast height (cm)	0.96	Native sclerophyll forest	NSW, ACT, VIC, TAS, and SA	Keith et al. [65]
$\ln(AGB) = -1,554 + 2.420\ln(D) + \ln(\rho)$ where AGB = above-ground biomass (kg/tree), D = diameter at breast height (cm), ρ = wood density (g/cm^3)	0.99	Tropical forests	America, Asian and Oceania	Chave et al. [61]
$\ln(RBD) = -1,085 + 0.926\ln(ABD)$ where RBD = root biomass density (tons/ha), ABD = above-ground biomass density (tons/ha)	0.83	Upland forests	Worldwide	IPCC [51] or Cairn et al. [66]
$y = 0.27x$ where y = total root biomass (tons/ha), x = total shoot biomass (tons/ha)	0.81	Natural forests	Worldwide	Mokany et al. [67]
$Wr = 0.0214 \times D^{2.33}$ where W_r = coarse root biomass (kg/tree), D = diameter at breast height (cm)	0.94	Tropical secondary forests	Sarawak, Malaysia	Kenzo et al. [68]
$W_r = 0.023 \times D^{2.59}$ where W_r = coarse root biomass (kg/tree), D = diameter at breast height (cm)	0.97	Tropical secondary forests	Sarawak, Malaysia	Niiyama et al. [69]

NSW New South Wales, *ACT* Australian Capital Territory, *VIC* Victoria, *TAS* Tasmania, *SA* South Australia.

carbon content of soil sample (%), and 100 is the default unit conversion factor.

Statistical analysis

One-way ANOVA tests were applied to compare stand densities, DBH, height classes, basal areas, and six categories of carbon stocks of the five *Melaleuca* forest types. LSD post hoc tests were also used for all pairwise comparisons between group means. Statistical analysis was undertaken using Microsoft Excel 2010 and the R Statistic Program.

Additional file

> **Additional file 1:** Data analysis.

Authors' contributions
DBT conducted design of the study, field data collection, carried out all analyses and drafted the manuscript. TVH and PD helped field data collection, guided the research, and assisted with the writing. All authors read and approved the final manuscript.

Author details
[1] The Vietnam Forestry University, Hanoi, Vietnam. [2] School of Geography, Planning and Environmental Management, The University of Queensland, Brisbane, QLD, Australia.

Acknowledgements
This study was authorized to access and collect vegetation and soil samples by the director boards of two national parks including the Phu Quoc National Park and U Minh Thuong National Park. All work was approved by the Vietnam Forestry University. We would like to thank the staffs of Phu Quoc National Park; U Minh Thuong National Park; the National Institute of Agricultural Planning and Projection; and the Vietnam Forestry University for their association of doing fieldwork and laboratory work. We also specially thank the anonymous reviewers for their excellent comments on the earlier version of this manuscript. We gratefully thank International Foundation for Science (IFS) for research funds.

Compliance with ethical guidelines

Competing interests
The authors declare that they have no competing interests.

References
1. Mitsch W, Bernal B, Nahlik A, Mander Ü, Zhang L, Anderson C et al (2013) Wetlands, carbon, and climate change. Landsce Ecol 28(4):583–597. doi:10.1007/s10980-012-9758-8
2. Bernal B, Mitsch WJ (2012) Comparing carbon sequestration in temperate freshwater wetland communities. Glob Change Biol 18(5):1636–1647. doi:10.1111/j.1365-2486.2011.02619.x
3. Mitsch W, Nahlik A, Wolski P, Bernal B, Zhang L, Ramberg L (2010) Tropical wetlands: seasonal hydrologic pulsing, carbon sequestration, and methane emissions. Wetlands Ecol Manag 18(5):573–586. doi:10.1007/s11273-009-9164-4
4. Bernal B, Wolski P, Nahlik A, Ramberg L, Zhang L, Mitsch WJ (2010) Tropical wetlands: seasonal hydrologic pulsing, carbon sequestration, and methane emissions. Wetlands Ecol Manag 18(5):573–586
5. Mitsch WJ, Tejada J, Nahlik A, Kohlmann B, Bernal B, Hernández CE (2008) Tropical wetlands for climate change research, water quality management and conservation education on a university campus in Costa Rica. Ecol Eng 34(4):276–288. doi:10.1016/j.ecoleng.2008.07.012
6. Bernal BS (2008) Carbon pools and profiles in wetland soils: the effect of climate and wetland type. The Ohio State University, Ohio
7. Bernal B, Mitsch WJ (2008) A comparison of soil carbon pools and profiles in wetlands in Costa Rica and Ohio. Ecol Eng 34(4):311–323. doi:10.1016/j.ecoleng.2008.09.005

8. Irving AD, Connell SD, Russell BD (2011) Restoring coastal plants to improve global carbon storage: reaping what we sow. PLoS One 6(3):e18311

9. Erwin K (2009) Wetlands and global climate change: the role of wetland restoration in a changing world. Wetlands Ecol Manag 17(1):71–84. doi:10.1007/s11273-008-9119-1

10. Renaud FG, Kuenzer C (2012) Climate and environmental change in River Deltas globally: expected impacts, resilience, and adaptation. In: Renaud FG, Kuenzer C (eds) Mekong delta system: interdisciplinary analyses of a River Delta, vol Book. Springer Netherlands, Whole

11. Bastakoti RC, Gupta J, Babel MS, van Dijk MP (2014) Climate risks and adaptation strategies in the Lower Mekong River basin. Reg Environ Change 14(1):207–219. doi:10.1007/s10113-013-0485-8

12. Le TVH, Nguyen HN, Wolanski E, Tran TC, Haruyama S (2007) The combined impact on the flooding in Vietnam's Mekong River delta of local man-made structures, sea level rise, and dams upstream in the river catchment. Estuar Coast Shelf Sci 71(1):110–116. doi:10.1016/j.ecss.2006.08.021

13. MIG (2008) Australia's State of the forests report: five-yearly report 2008. Montreal process implementation group for Australia, Bureau of Rural Sciences, Canberra

14. Tran DB, Dargusch P, Herbohn J, Moss P (2013) Interventions to better manage the carbon stocks in Australian Melaleuca forests. Land Use Policy 2013(35):417–420. doi:10.1016/j.landusepol.2013.04.018

15. Renaud FG, Le T, Lindener C, Guong V, Sebesvari Z (2014) Resilience and shifts in agro-ecosystems facing increasing sea-level rise and salinity intrusion in Ben Tre Province, Mekong Delta. Clim Change 1–16. doi:10.1007/s10584-014-1113-4

16. Rengasamy P, Olsson K (1991) Sodicity and soil structure. Soil Res 29(6):935–952. doi:10.1071/SR9910935

17. Bernstein L (1975) Effects of salinity and sodicity on plant growth. Annu Rev Phytopathol 13:295–312

18. Mahmood K (2007) Salinity, sodicity tolerance of Acacia ampliceps and identification of techniques useful to avoid early stage salt stress. Kassel Univ. Press, Kassel

19. Warrence NJ, Bauder JW, Pearson KE (2002) Basics of salinity and sodicity effects on soil physical properties. Land Resources and Environmental Sciences Department, Montana State University, Bozeman

20. Tho N, Vromant N, Hung NT, Hens L (2008) Soil salinity and sodicity in a shrimp farming coastal area of the Mekong Delta, Vietnam. Environ Geol 54(8):1739–1746. doi:10.1007/s00254-007-0951-z

21. Dunn GM, Taylor DW, Nester MR, Beetson TB (1994) Performance of twelve selected Australian tree species on a saline site in southeast Queensland. For Ecol Manag 70(1–3):255–264. doi:10.1016/0378-1127(94)90091-4

22. Niknam SR, McComb J (2000) Salt tolerance screening of selected Australian woody species: a review. For Ecol Manag 139(1–3):1–19. doi:10.1016/S0378-1127(99)00334-5

23. van der Moezel PG, Pearce-Pinto GVN, Bell DT (1991) Screening for salt and waterlogging tolerance in Eucalyptus and Melaleuca species. For Ecol Manag 40(1–2):27–37. doi:10.1016/0378-1127(91)90089-E

24. van der Moezel P, Watson L, Pearce-Pinto G, Bell D (1988) The response of six Eucalyptus species and Casuarina obesa to the combined effect of salinity and waterlogging. Austr J Plant Physiol 15(3):465–474

25. VCS (2011) Methodology for sustainable grassland management (SGM). Verified Carbon Standard-A global Benchmark for Carbon

26. Taiyab N (2006) Exploring the market for voluntary carbon offsets. International Institute for Environment and Development (IIED), London

27. Polglase PJ, Reeson A, Hawkins CS, Paul KI, Siggins AW, Turner J et al (2013) Potential for forest carbon plantings to offset greenhouse emissions in Australia: economics and constraints to implementation. Clim Change 121(2):161–175. doi:10.1007/s10584-013-0882-5

28. Singh A, Nigam PS, Murphy JD (2011) Renewable fuels from algae: an answer to debatable land based fuels. Bioresour Technol 102(1):10–16. doi:10.1016/j.biortech.2010.06.032

29. Rahayu S, Harja D (2012) A study of rapid carbon stock appraisal: average carbon stock of various land cover in Merauke, Papua Province. World Agroforestry Centre (ICRAF-SEA)

30. Donato DC, Kauffman JB, Murdiyarso D, Kurnianto S, Stidham M, Kanninen M (2011) Mangroves among the most carbon-rich forests in the tropics. Nat Geosci 4:293–297

31. Verwer CC, Meer PJVD (2010) Carbon pool in tropical peat forest: toward a reference value for forest biomass carbon in relatively undisturbed peat swamp forests in Southeast Asia. Wageningen, Allterra Wageningen UR

32. Le PQ (2010) Inventory of peatlands in U Minh Ha Region, Ca Mau Province, Vietnam. Institute for Environment and Natural Resources, National University, HCM City

33. Tran DB, Dargusch P, Moss P, Hoang TV (2013) An assessment of potential responses of Melaleuca genus to global climate change. Mitig Adapt Strat Glob Change 18(6):851–867. doi:10.1007/s11027-012-9394-2

34. Larcher W (1980) Physiological plant ecology. vol Book, Whole. Springer, Berlin

35. Kozlowski TT (1997) Responses of woody plants to flooding and salinity. Tree Physiol 17(7):490. doi:10.1093/treephys/17.7.490

36. Greenway H, Munns R (1980) Mechanisms of salt tolerance in nonhalophytes. Annu Rev Plant Physiol 31(1):149–190. doi:10.1146/annurev.pp.31.060180.001053

37. Wong VL, Dalal R, Greene RB (2008) Salinity and sodicity effects on respiration and microbial biomass of soil. Biol Fertil Soils 44(7):943–953. doi:10.1007/s00374-008-0279-1

38. Department of Primary Industries (2008) Identifying, understanding and managing hostile subsoils for cropping. University of Adelaide-South Australian Research and Development Institute

39. Mavi MS, Marschner P, Chittleborough DJ, Cox JW, Sanderman J (2012) Salinity and sodicity affect soil respiration and dissolved organic matter dynamics differentially in soils varying in texture. Soil Biol Biochem 45:8–13. doi:10.1016/j.soilbio.2011.10.003

40. Howat D (2000) Acceptable salinity, sodicity and pH values for Boreal forest reclamation: Alberta Environment, Environmental Sciences Division, Edmonton Alberta. Report # ESD/LM/00-2. ISBN 0-7785-1173-1 (printed edition) or ISBN 0-7785-1174-X (on-line edition)

41. Nuttall JG, Armstrong RD, Connor DJ, Matassa VJ (2003) Interrelationships between edaphic factors potentially limiting cereal growth on alkaline soils in north-western Victoria. Soil Res 41(2):277–292. doi:10.1071/SR02022

42. Nicholls RJ, Wong PP, Burkett VR, Codignotto JO, Hay JE, McLean RF et al (2007) Coastal systems and low-lying areas. In: Parry ML, Canziani OF, Palutikof JP, Linden PJVD, Hanson CE (eds) Climate change 2007: impacts, adaptation and vulnerability. Contribution of working group II to the fourth assessment report of the intergovernmental panel on climate change (IPCC). Cambridge University Press, Cambridge, pp 315–356

43. Toan TL (2009) Impacts of climate change and human activities on environment in the Mekong Delta, Vietnam. Centre d'Etudes Spatiales de la Biosphère (CESBIO), Toulouse

44. Sun D, Dickinson G (1993) Responses to salt stress of 16 Eucalyptus species, Grevillea robusta, Lophostemon confertus and Pinus caribea var. hondurensis. For Ecol Manag 60(1–2):1–14. doi:10.1016/0378-1127(93)90019-J

45. Paul KI, Roxburgh SH, England JR, Ritson P, Hobbs T, Brooksbank K et al (2013) Development and testing of allometric equations for estimating above-ground biomass of mixed-species environmental plantings. For Ecol Manag 310:483–494. doi:10.1016/j.foreco.2013.08.054

46. Cuong NV, Quat HX, Chuong H (2004) Some comments on indigenous Melaleuca of Vietnam. Sci Technol J Agric Rural Dev (Vietnam). (11/2004)

47. Hoover CM, Smith JE (2012) Site productivity and forest carbon stocks in the United States: analysis and implications for forest offset project planning. Forests 3(4):283–299. doi:10.3390/f3020283

48. Vietnam Environment Protection Agency (2003) Report on peatland management in Vietnam. Ministry of Natural Resources and Environment

49. Van TK, Rayachetry MB, Center TD (2000) Estimating above-ground biomass of Melaleuca quinquenervia in Florida, USA. J Aquat Plant Manag 38:62–67

50. IPCC (2006) Good practice guidance for land use, land-use change and forestry. Institute for Global Environmental Strategies (IGES) for the IPCC, Kanagawa

51. IPCC (2003) Good practice guidance for land use, land-use change and forestry. Institute for Global Environmental Strategies (IGES) for the IPCC, Kanagawa

52. Walkley A (1947) A critical examination of a rapid method for determination of organic carbon in soils—effect of variations in digestion conditions and of inorganic soil constituents. Soil Sci 63:251–257

53. Schumacher BA (2002) Methods for the determination of total organic carbon (TOC) in soils and sediments. Ecological Risk Assessment Support Center Office of Research and Development, US Environmental Protection Agency

54. LABCONCO (1998) A guide to Kjeldahl nitrogen determination methods and apparatus. An Industry Service Publication, Houston

55. Rengasamy P, Olsson KA (1991) Sodicity and soil structure. Aust J Soil Res 29(6):935–952. doi:10.1071/SR9910935

56. Ford G, Martin J, Rengasamy P, Boucher S, Ellington A (1993) Soil sodicity in Victoria. Soil Res 31(6):869–909. doi:10.1071/SR9930869

57. Gj C (1999) Cation exchange capacity, exchangeable cations and sodicity. vol Book, Whole

58. Jonson JH, Freudenberger D (2011) Restore and sequester: estimating biomass in native Australian woodland ecosystems for their carbon-funded restoration. Aust J Bot 59(7):640–653. doi:10.1071/BT11018

59. Finlayson CM, Cowie ID, Bailey BJ (1993) Biomass and litter dynamics in a *Melaleuca* forest on a seasonally inundated floodplain in tropical, Northern Australia. Wetlands Ecol Manag 2(4):177–188

60. Thomas S, Hoegh-Guldberg OOHG, Griffiths A, Dargusch P, Bruno J (2010) The true colours of carbon. Nat Preced. http://precedings.nature.com/documents/5099/version/1

61. Chave J, Andalo C, Brown S, Cairns MA, Chambers JQ, Eamus D et al (2005) Tree allometry and improved estimation of carbon stocks and balance in tropical forests. Oecologia 145(1):87–99

62. Hairiah K, Sitompul S, Noordwijk MV, Palm C (eds) (2001) Methods of sampling carbon stocks above and below ground. ASB Lecture Note. International Centre for Research in Agroforestry (ICRAF)-Southeast Asian Regional Research Program, Bogor

63. Le ML (2005) Phương pháp đánh giá nhanh sinh khối và Ảnh hưởng của độsâu ngập lên sinh khối rừng Tràm (*Melaleuca cajuputi*) trên đất than bùn và đất phèn khu vực U Minh Hạ tỉnh Cà Mau (Evaluation biomass and Effect of submergence depth on growth of *Melaleuca* planting on peat soil and acid sulfate Soil in U Minh Ha area—Ca Mau Province). Nong Lam University, Ho Chi Minh City

64. Brown S (1997) Estimating biomass and biomass change of tropical forests: a primer. FAO, Quebec City

65. Keith H, Barrett D, Keenan R (2000) Review of allometric relationships for estimating woody biomass for New South Wales, the Australian Capital Territory, Victoria, Tasmania and South Australia

66. Cairns MA, Brown S, Helmer EH, Baumgardner GA (1997) Root biomass allocation in the world's upland forests. Oecologia 111(1):1–11

67. Mokany K, Raison RJ, Prokushkin AS (2006) Critical analysis of root:shoot ratios in terrestrial biomes. Glob Change Biol 12:84–96

68. Kenzo T, Ichie T, Hattori D, Itioka T, Handa C, Ohkubo T et al (2009) Development of allometric relationships for accurate estimation of above- and below-ground biomass in tropical secondary forests in Sarawak, Malaysia. J Tropic Ecol 25(4):371–386

69. Niiyama K, Kajimoto T, Matsuura Y, Yamashita T, Matsuo N, Yashiro Y et al (2010) Estimation of root biomass based on excavation of individual root systems in a primary dipterocarp forest in Pasoh Forest Reserve, Peninsular Malaysia. J Tropic Ecol 26(3):271–284

14

Analysis of biophysical and anthropogenic variables and their relation to the regional spatial variation of aboveground biomass illustrated for North and East Kalimantan, Borneo

Carina Van der Laan[1*], Pita A Verweij[1], Marcela J Quiñones[2] and André PC Faaij[1]

Abstract

Background: Land use and land cover change occurring in tropical forest landscapes contributes substantially to carbon emissions. Better insights into the spatial variation of aboveground biomass is therefore needed. By means of multiple statistical tests, including geographically weighted regression, we analysed the effects of eight variables on the regional spatial variation of aboveground biomass. North and East Kalimantan were selected as the case study region; the third largest carbon emitting Indonesian provinces.

Results: Strong positive relationships were found between aboveground biomass and the tested variables; altitude, slope, land allocation zoning, soil type, and distance to the nearest fire, road, river and city. Furthermore, the results suggest that the regional spatial variation of aboveground biomass can be largely attributed to altitude, distance to nearest fire and land allocation zoning.

Conclusions: Our study showed that in this landscape, aboveground biomass could not be explained by one single variable; the variables were interrelated, with altitude as the dominant variable. Spatial analyses should therefore integrate a variety of biophysical and anthropogenic variables to provide a better understanding of spatial variation in aboveground biomass. Efforts to minimise carbon emissions should incorporate the identified factors, by 1) the maintenance of lands with high AGB or carbon stocks, namely in the identified zones at the higher altitudes; and 2) regeneration or sustainable utilisation of lands with low AGB or carbon stocks, dependent on the regeneration capacity of the vegetation. Low aboveground biomass densities can be found in the lowlands in burned areas, and in non-forest zones and production forests.

Keywords: Aboveground biomass; Tropical forest landscapes; Disturbance; Spatial analysis; Multiple regression; Geographically weighted regression; Biophysical and anthropogenic variables; East Kalimantan; North Kalimantan

Background

More insights into the spatial variation of aboveground biomass (AGB) are crucial to minimise carbon emissions and global climate change from tropical deforestation, forest degradation and agricultural expansion. According to van der Werf *et al.* globally, approx. 12% of anthropogenic carbon emissions in 2008 were caused by deforestation and forest degradation [1]. During the period 1973–2010, Kalimantan, Indonesian Borneo, has lost ~31% of the total forest area [2]. With regard to land use changes, according to the Governors' Climate and Forests Task Force Indonesia [3], the recently merged provinces North and East Kalimantan are when combined the third largest carbon emitting provinces in Indonesia, with 255 Mt CO_2e yr^{-1}, after Central Kalimantan (324 Mt CO_2e yr^{-1}) and Riau (258 Mt CO_2e yr^{-1}). According to their 'business as usual' scenarios, land use change will cause carbon emissions in North and East Kalimantan to increase to 331 Mt CO_2e yr^{-1} by 2030 [3]. Mechanisms such as Reducing Emissions from Deforestation and forest

* Correspondence: C.Vanderlaan@uu.nl
[1]Copernicus Institute of Sustainable Development, Group Energy and Resources, Utrecht University, Heidelberglaan 2, 3584 CS Utrecht, The Netherlands
Full list of author information is available at the end of the article

Degradation + (REDD+) [4] have been developed to halt such emissions by maintaining lands with high carbon stocks contained in living forest biomass, such as secondary and undisturbed forests. Meanwhile, expansion of low carbon stock agricultural lands can be instead shifted towards areas with already low carbon stocks or AGB, such as abandoned agricultural or restored degraded lands [5] by implementation of sustainable land zoning tools [6].

AGB is not static, but rather spatially and temporally highly variable, particularly in the tropics [7-11]. This makes its quantification and the avoidance of high AGB densities or high carbon stocks challenging (it is generally assumed that about half of AGB consists of carbon). As in other tropical forest landscapes, complex matrices of low to high AGB densities can be expected in North and East Kalimantan, due to varying biophysical conditions present, such as terrain and soil types, and anthropogenic disturbances such as fire or logging. For example, forest fires can cause substantial losses in carbon by the emission of large quantities of CO_2 by the burning of biomass [12,13], and via logging by the extraction of timber [14,15]. However, after a fire or logging activities, regeneration can occur, resulting in an increasing sensitivity of the remaining live and dead vegetation to subsequent disturbance events [13,16-19]. Additionally, the type and severity of the disturbance and local biophysical conditions, such as altitude, soil type and the presence of pioneer species, influence the carbon accumulation potential [18,20]. Therefore, in this paper we address the question of how such biophysical and anthropogenic variables are related to AGB, and contribute to the spatial variation of AGB in a disturbed tropical forest landscape.

AGB can be estimated at forest stand to landscape scale by plot-based measurements [21,22]. Several existing plot-based studies in tropical forest landscapes have statistically analysed the relationships between AGB and multiple biophysical variables including soil factors [7,11,23,24], altitude [11,25-27] and slope [11,28]. Anthropogenic variables, however, are usually not considered, while specifically in highly disturbed tropical areas like North and East Kalimantan, these factors are expected to strongly affect AGB. Additionally, anthropogenic variables are important and useful to support the management of, and decision-making on, maintaining carbon stocks in disturbed areas. For these reasons, our analyses include both biophysical and anthropogenic variables.

Other plot-based studies have focused on the impacts of e.g. logging [14,15] and fire [13,29], by comparing AGB between undisturbed and disturbed land classes. The relationship between forest cover change and anthropogenic variables has also been analysed [16,30,31]. These studies have instead focused on discrete land use and forest classes and therefore have not accounted for local scale AGB variation. Furthermore, the reviewed studies were not spatially explicit or conducted over larger spatial scales, thereby limiting a landscape scale view on the factors that influence the spatial variation in AGB or forest cover.

A variety of spatially explicit data and methods exist to map and monitor land with high and low AGB or carbon over large spatial scales, such as extrapolating plot-based field AGB estimates to vegetation types with remotely-sensed reflectance data and spatial data of biophysical variables [32,33]. For example, optical data can be used for mapping forest cover, such as Landsat [34]. However, in areas with frequent cloud cover such as the tropics, radar technologies such as ALOS (Advanced Land Observing Satellite) PALSAR (Phase Arrayed L-band SAR) are more suitable [35,36]. Additionally, the integration of optical and/or radar technologies, including LiDAR (Light Detection And Ranging), has the potential to improve AGB estimates because it may reduce data saturation and mixed pixel problems [35,37-39]. Although the output maps of the aforementioned studies have visualised the spatial distribution of AGB at high resolutions and over large spatial scales, these did not include the effects of biophysical or anthropogenic factors on AGB.

Changes in AGB or carbon stocks have also been modelled at different spatial and temporal scales and resolutions [33,40-43]. Additionally, studies using spatial data for AGB have compared AGB between forest types with different levels of degradation or disturbances, e.g. by logging or fire [36,44-47]. However, the focus was mostly on a single anthropogenic variable, e.g. logging or fire, and interrelationships between or interaction effects amongst variables were not investigated.

The aforementioned studies are useful for the mapping and monitoring of AGB and carbon stocks, for e.g. REDD + mechanisms. To monitor and quantify AGB whilst taking into consideration the high spatial variation, and additionally to enable the modelling of carbon stocks, further analysis of the underlying biophysical and anthropogenic conditions and processes, using a multi-variable approach, is essential. An improved level of information quality, that considered a broader set of variables and their interactions, would allow decision-making to focus on manageable factors in support of land use allocation that minimises carbon emissions and maximises carbon uptake in support of climate change mitigation.

The aim of this study is to define which of a preselected set of biophysical and anthropogenic variables contribute significantly to the spatial variation of AGB. To this end, statistical analyses were conducted, including analysis of variance (ANOVA), non-spatial multiple linear regression and spatial geographically weighted regression (GWR). An AGB map based on radar remote sensing data and plot-based measurements were utilised, plus landscape scale data on terrain, soil types, land allocation zoning, fires, roads, rivers and cities, covering North and East Kalimantan, Indonesian Borneo (see Figure 1). The results are shown in the Results section and can support the quantification and maintenance of living AGB and carbon stocks. In the Discussion section, the results are discussed in terms of their scientific and societal contribution, followed by the Conclusions and an extensive description of the data and analyses in the Methods section.

Results

Relationships between AGB and the continuous explanatory variables

The distribution of AGB was negatively skewed (Skewness: −0.852, st. error: 0.113, Kurtosis: 0.207, st. error: 0.226), which can be expected in a disturbed tropical forest landscape (Figure S1, in Additional file 1). AGB varied between 2 and 480.0 t ha^{-1} with an overall mean of 213.6 ± 80.1 t ha^{-1} (for descriptive statistics, see Table S2, in Additional file 1). AGB and the selected continuous explanatory variables altitude, slope, and distance to the nearest fire, road, river and city (logarithmically transformed) appeared to have a strong, positive correlation (Table S3, in Additional file 1). All relationships are plotted in Figure 2. The Pearson's correlation coefficients (r)

indicated the strongest relationships between AGB and the terrain variables, altitude (r = 0.740, $P < 0.001$) and slope (r = 0.563, $P < 0.001$), and between AGB and distance to the nearest fire (r = 0.607, $P < 0.001$) and city (r = 0.478, $P < 0.001$). Moderately positive relationships were found between AGB and distance to the nearest river and road (r ~ 0.335, $P < 0.001$). Altitude and distance to the nearest city were strongly related to all other explanatory variables (r > 0.400, $P < 0.001$), but not to distance to the nearest river. Distance to the nearest fire was related to the distance to the nearest river and the nearest city. No strong multicollinearity was found (Tolerance >0.200).

Variation in AGB between altitude ranges and soil types

The ANOVA on the categorised altitude variable revealed significant differences in mean AGB between the categories lowlands (<750 m), midlands (750–1,500 m) and highlands (>1,500 m), F (2,462) = 32.85, $P < 0.001$. The lowlands (M = 201) had significantly lower AGB than the midlands (M = 276, $P < 0.001$) and the highlands (M = 282, $P < 0.05$). A boxplot of altitude and AGB is shown in Figure 3a. An ANOVA was used to test for mean differences in AGB among four soil types. Means in AGB for soil types differed significantly across the four types (F(3,461) = 14.88, $P < 0.001$). Bonferroni's post-hoc comparisons on the four soil types indicate that AGB on peatland (M = 142) gave significantly lower

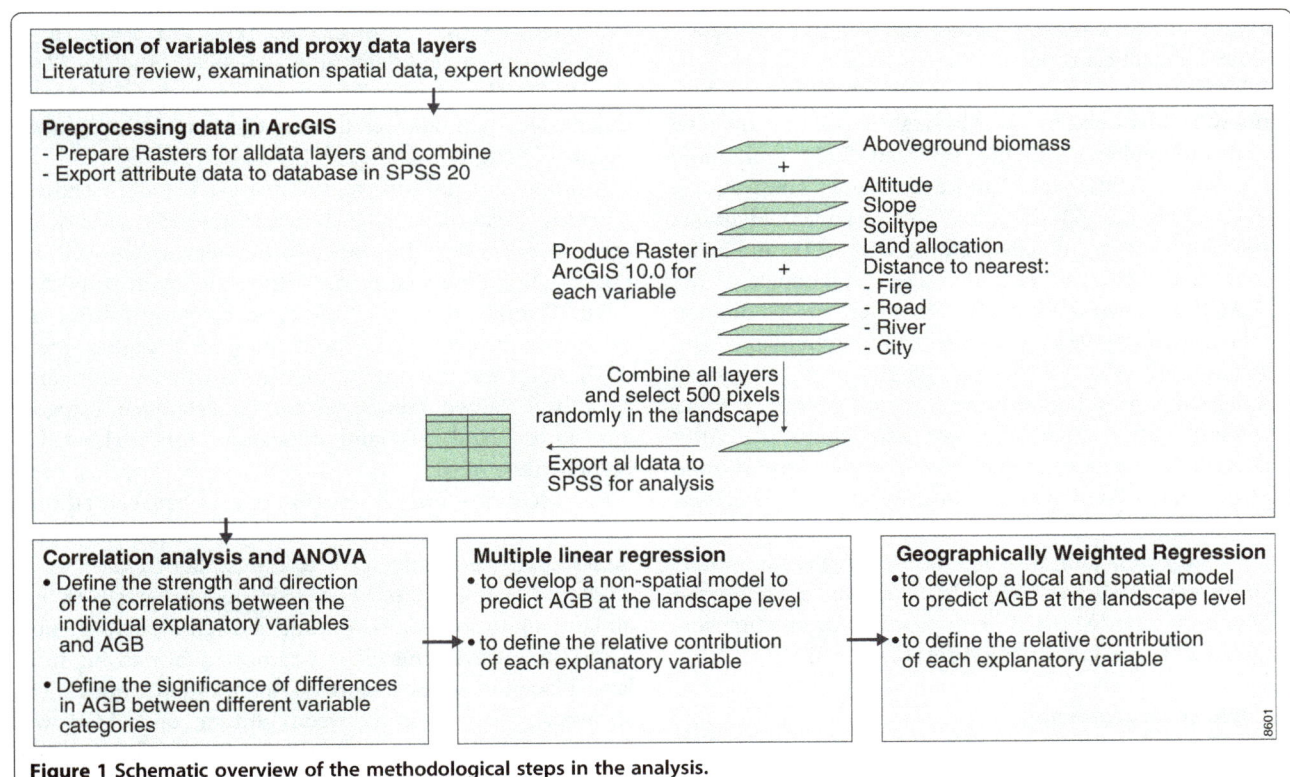

Figure 1 Schematic overview of the methodological steps in the analysis.

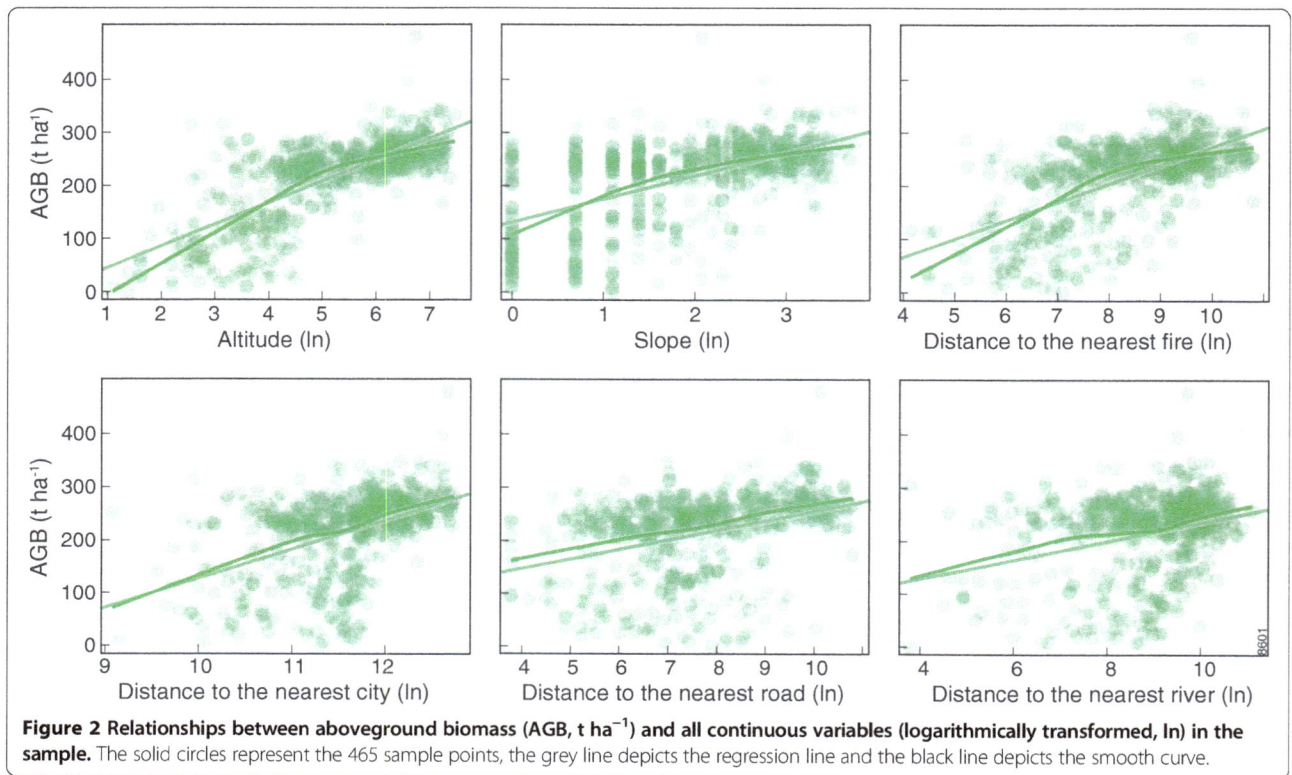

Figure 2 Relationships between aboveground biomass (AGB, t ha⁻¹) and all continuous variables (logarithmically transformed, ln) in the sample. The solid circles represent the 465 sample points, the grey line depicts the regression line and the black line depicts the smooth curve.

means than on karst ($M = 261$, $P = 0.001$) and volcanic soils ($M = 282$, $P < 0.001$) (Figure 3b).

Variation in AGB between burned and non-burned areas and land allocation zones

AGB in non-burned areas (i.e. areas where no fire hotspots were identified by the Moderate-resolution Imaging Spectroradiometer (MODIS) between 2000 and 2008) ($M = 223$, $P < 0.001$) was significantly higher compared to burned areas (i.e. MODIS fire hotspots were identified within 500 m from the data point between 2000 and 2008) ($M = 114$, F $(1,463) = 79.22$, $P < 0.001$). A boxplot of fire and AGB is shown in Figure 3c. Fires were more common in the lowlands (98%) compared to the midlands and highlands. An ANOVA showed significant differences in the mean for AGB between the five land allocation zones (F $(4,460) = 56.06$, $P < 0.001$) (see also Figure 3d). After Bonferroni's correction, pairwise comparisons showed that the mean AGB was significantly lower in the non-forest land zone ($M = 152$, $P < 0.001$) compared to the other categories, and was significantly higher in watershed protection forest ($M = 272$) and the forest limited production zone ($M = 253$), compared to production forest ($M = 193$) and conservation forest ($M = 211$).

Multiple linear regression

After removal of the non-significant explanatory variables via conducting a backward multiple linear regression, the variables altitude, distance to the nearest fire, and the categorical variables land allocation zoning and soil type significantly contributed to predicting AGB, and combined explained approx. 59% of the observed variance in AGB in the sample (Adjusted $R^2 = 0.589$, $F(9,455) = 72.46$, $P < 0.001$). The standardised coefficients showed that in this analysis, altitude was the most important explanatory variable (Table 1).

Altitude and distance to the nearest fire both showed a positive relation with AGB, which means that with increasing altitude and distance to the nearest fire AGB increased. Soil type also made a difference with respect to AGB. Compared to the reference category 'other', the categories volcanic and karst showed a higher mean AGB. Karst was the only significant coefficient compared to 'other'. When compared to the reference category 'non-forest land', all land allocation zones showed a higher mean AGB.

In a second model, interaction effects between altitude and land allocation zoning were added (Table 1). These interaction effects added 2% to the explained variance of AGB (R^2 change $= 0.02$, $F(4, 451) = 5.29$, $P < 0.001$). For all land allocation zones, altitude showed a positive effect on AGB; however, this effect was not equally strong in all land allocation zones (Figure S4, in Additional file 1). The strongest relationship between altitude and AGB was found in conservation forests, where altitude explained the AGB variance with about 86% ($R^2 = 0.860$, $P < 0.001$).

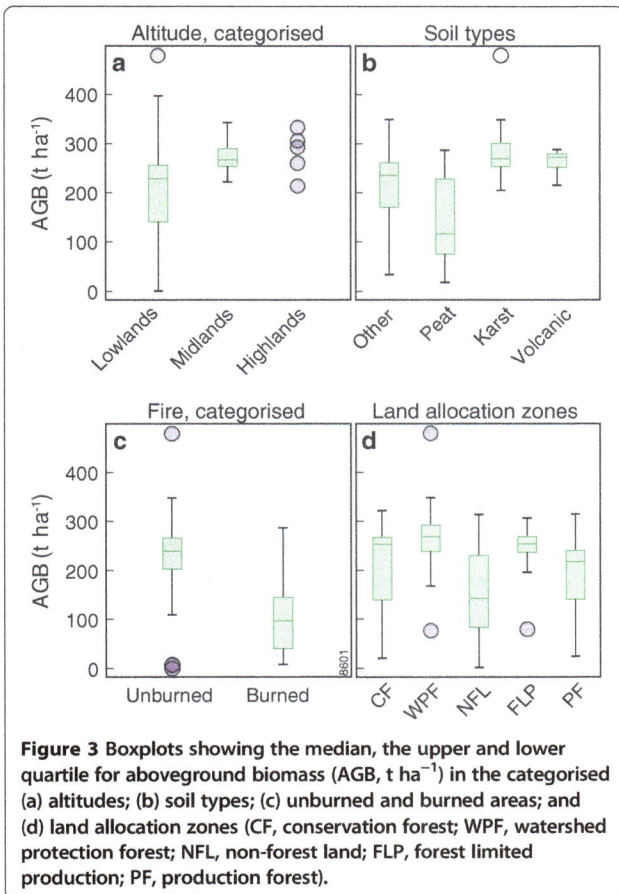

Figure 3 Boxplots showing the median, the upper and lower quartile for aboveground biomass (AGB, t ha^{-1}) in the categorised (a) altitudes; (b) soil types; (c) unburned and burned areas; and (d) land allocation zones (CF, conservation forest; WPF, watershed protection forest; NFL, non-forest land; FLP, forest limited production; PF, production forest).

Table 1 Unstandardised coefficients resulting from the non-spatial multiple regression, without (Model 1) and with interaction terms (Model 2)

Variable	Coefficients Model 1	Coefficients Model 2
Altitude (ln)	32.8 (0.60)**	40.1 (0.73)**
Distance to the nearest fire (ln)	8.1 (0.14)**	8.4 (0.15)**
Soil type		
Karst	43.2*	42.1*
Peat	19.2	25.7
Volcanic	16.8	22.7*
Land allocation zone		
Forest limited production	19.4*	168.8**
Conservation forests	−14.5	−23.1
Production forests	16.5*	41.7
Watershed protection forests	17.1	114.3*
Interaction effects		
Altitude (ln) x Forest limited production		−27.9**
Altitude (ln) x Conservation forests		−1.0
Altitude (ln) x Production forests		−6.6
Altitude (ln) x Watershed protection forests		−18.3*

Standardised coefficients are indicated between brackets; **$P < 0.001$, *$P < 0.05$; ln, logarithmically transformed.

The weakest relationship with altitude was found in watershed protection forests, where altitude explained only 17% of the variance in AGB.

No multicollinearity was present and the standardised residuals showed a normal distribution (Range = [−2.92; 3.96]), meeting the important assumptions of normality and multicollinearity underlying multiple linear regression (see Figure S5, in Additional file 1). However, the Breusch-Pagan test exposed the presence of spatial non-stationarity or heteroscedasticity (Chi-square df = 88.381, $P \leq 0.05$), invalidating the significance of the statistical tests. The Moran's I test (Index = 0.147, z-score = 12.02, $P \leq 0.05$) showed spatial autocorrelation of the standardised residuals, which can cause an unexplained shift in the regression coefficients and can thus influence the output of the model.

Geographically weighted regression (GWR)
Because of the presence of spatial autocorrelation in the standardised residuals of the non-spatial multiple linear regression, geographically weighted regression (GWR) was conducted, producing for each sample point a local relationship between AGB and the explanatory variables. The variables; distance to the nearest road, city and river showed strong multicollinearity with altitude. Finally, three GWR models (Table 2) were computed that did not show multicollinearity; however, Moran's I test of two of these models showed spatial autocorrelation. In the best model (R^2 Adjusted = 0.641, $P \leq 0.05$), the explanatory variables; altitude, distance to the nearest fire, and land allocation zoning were significant, and explained the variation of AGB in the sample with approx. 64% (Table 2). The presence of spatial autocorrelation was unlikely (Index = 0.02, z-score = 1.8, $P \leq 0.1$). In Figure 4 the AGB values observed on the AGB map are plotted against the AGB values predicted by the model. The standardised residuals showed a normal distribution (Range ~ [−3.80; 4.80]), indicating that the normality assumptions underlying multiple regression were met (see Figure S6, in Additional file 1).

Discussion
In this study, we combined ANOVA, multiple regression and GWR and used multiple thematic spatial data layers to define which biophysical and anthropogenic variables contributed significantly to the spatial AGB variation in a disturbed tropical forest landscape. Altitude showed the strongest relationship with AGB; individually, and in both regression analyses. This strong positive correlation with altitude is supported for other areas by previous studies e.g. [11,27]. In our study, the mean AGB was highest in the higher altitudes where volcanic soils are

Table 2 Output of the spatial GWR model computed in ArcGIS ($P < 0.05$); * for each variable (ln, logarithmically transformed) the mean of the coefficients is indicated: the GWR produced for each sample point a local model and variable coefficient

	Model
R^2	0.660
R^2 Adjusted	0.637
Response variable:	
Observed mean AGB (t ha^{-1})	213.6
Predicted mean AGB (t ha^{-1})	211.9
Explanatory variables:	
Altitude (ln) (mean coefficient)	31.7*
Distance to the nearest fire (ln) (mean coefficient)	8.6*
Land allocation zoning (mean coefficient)	0.17*
Residuals	1.7
Standard Error	46.6
Standardised Residual	0.03

present, and most of the land was allocated to zones where land clearing is not allowed. Moreover, these higher altitudes are less suitable for agriculture and poorly accessible by road. However, by taking into account interaction effects, we found that the influence of altitude on AGB was highest in conservation forest where AGB variation was explained with approx. 86%. It is likely that the forests in these higher, and thus more remote, conservation areas are less impacted by anthropogenic variables, leading to higher AGB densities.

The mean AGB was significantly lower in the lowlands [11,27]. Lowlands are more susceptible to timber extraction, agricultural expansion and mining because of better

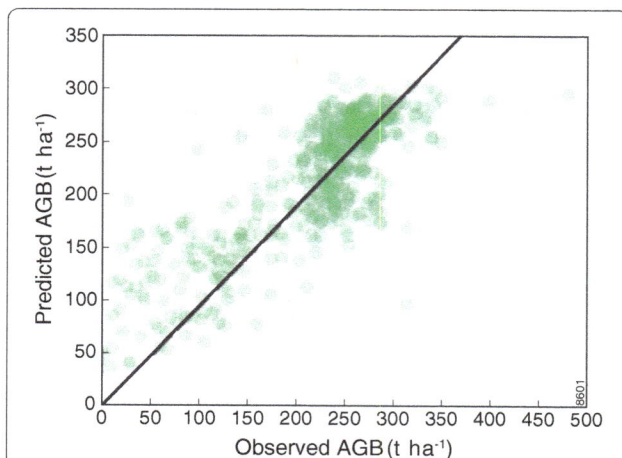

Figure 4 Relationship between the observed aboveground biomass (AGB, t ha^{-1}) on the AGB map and the predicted AGB generated by the GWR model for the 465 sample points. The line corresponds to a perfect fit.

accessibility, and typical land allocation zoning where such activities are allowed. Fire is a commonly-used land-clearing method in Indonesia, and has severely affected the study area [16,48]. Combined with regularly occurring ENSO events, fires that start very locally may quickly spread. In our study, between 2000 and 2008 significantly more fires occurred in the lowlands, and in these burned areas we observed significantly lower AGB [13]. These findings support the results of Fuller et al. who found that lowland forests and areas designated for forest conversion in East Kalimantan were more threatened than upland forests and areas not designated for conversion, with slope, elevation, and fire being important factors in determining the threat to forest cover [49].

The contribution of distance to the nearest fire and land allocation zoning is in line with the observations of e.g. Siegert et al. [12,16] and Broich et al. [31] that fires, logging, and land clearing contribute substantially to forest cover loss and thus to lower AGB values in North and East Kalimantan. Additionally, Broich et al. [31] identified a relatively lower forest cover in 2000 within production forest and non-forest-use zones, and the greatest decrease in forest cover between 2000 and 2008, compared to other land allocation zones. In this study, these differences between the land allocation zones are now also shown for AGB, which may support carbon stock estimations.

Previous studies have focused on the effects of biophysical or anthropogenic factors on forest cover in North and East Kalimantan or Borneo. With the exception of Fuller et al. [49], most of these studies, however, included only a limited number of variables, and highlighted the influence of land allocation zoning [31]; land use and fire [30]; or logging and logging roads [2]. In this paper, we provided a reconciliation of these seemingly contradicting results and showed that multiple explanatory variables had a significant effect on AGB and were interrelated to one another. Because of these interrelationships, we underline the importance of including a wide range of biophysical and anthropogenic variables.

Most previous studies have incorporated mainly biophysical variables [e.g. 10, 24–26]. In this study we have demonstrated that in disturbed tropical forest landscapes it is essential to include biophysical and anthropogenic variables in AGB quantification and modelling. Also, the identification of anthropogenic factors that negatively impact AGB can support projections about if and where AGB losses may occur in an area.

We found substantially lower mean AGB on peatland soils (142 t ha^{-1}), than was found by Budiharta et al. [50] for undisturbed peat swamp forests in Borneo (348.7 t ha^{-1}). This may indicate strong anthropogenic disturbances in our study area by for example fires, logging or conversion. Next to aboveground biomass, peatlands also store large amounts of biomass belowground [51,52]. In regions with large areas of peatland, such as Indonesia [52], the inclusion of belowground biomass in addition to AGB may improve carbon stock predictions.

The GWR showed improved model accuracy (~64%) and a similar ranking of variables compared to the multiple regression; however, soil type did not significantly contribute to this model. This may be caused by an underrepresentation of the limited number of sample points across all soil types.

The variables were selected based on expected relations with AGB, but also by data availability. In order to refine this model for applicability in other landscapes, climate variables such as precipitation and temperature, but also infrastructure data such as logging roads and settlements, may be included, if data is available at an appropriate level of detail.

The root mean squared error between the field-estimated AGB and the AGB from the radar map was 10 t ha^{-1}. Although generally this is considered low, it may influence the resulting accuracy of the regression models.

AGB does not only vary spatially, but also temporally because of disturbance and regeneration processes. Temporal analysis of AGB could add a valuable dimension to the present approach, by providing insight into the potential increase or decrease of AGB and carbon stocks.

For maintenance and identification of lands with high AGB, many studies have focused on AGB and carbon mapping and modelling using plot-based and remote sensing data [33]. Many analyses, however, use maps with discrete classes of AGB e.g. [12,30], and consequently AGB quantification, or modelling, may not be covering the existing high spatial variation sufficiently. Therefore, we recommend using to use AGB maps with a continuous scale.

Methods such as multiple regression and GWR have been utilised for the analysis of, for example; forest attributes [53], the NDVI–rainfall relationship [54], and even for estimating AGB in a tropical forest area [55]. However, using GWR for generating a better understanding of the biophysical and anthropogenic variables that contribute to the regional spatial variation of AGB, by using an extensive spatial dataset, is a relatively new approach. The advantage of using GWR, compared to non-spatial multiple regression, is that it produces a local model, thus accounting for the spatially varying relationships between AGB and the explanatory variables. Moreover, by using GWR the effects of spatial autocorrelation were minimised.

Conclusions

Better insights into the spatial variation of AGB are needed to support the maintenance of carbon stocks in disturbed tropical forested landscapes. In this paper, we analysed how a set of biophysical and anthropogenic variables were related to, and contributed to the spatial variation of AGB in such a landscape. As was expected for disturbed forest conditions, mean AGB was relatively low and varied strongly throughout the landscape.

Through non-spatial and spatial multiple linear regression, we were able to explain this high spatial variation with, respectively, about 59% and 64%. Because of spatial autocorrelation in the standardised residuals of the non-spatial multiple regression, we conducted GWR. The GWR showed that altitude, distance to the nearest fire and land allocation zoning had the largest significant effect on AGB. Our study showed that mean AGB was relatively higher at the higher altitudes, on karst and volcanic soils, with increasing distance from fire hotspots, in limited production forest, and in watershed protection and conservation forests.

Because of the strong effects of these factors on variation in AGB, efforts to minimise carbon emissions, such as REDD+, should incorporate these factors. This can be implemented through maintenance of lands with high carbon stocks or through the utilisation or regeneration of lands with low carbon stocks or AGB. For example, the maintenance of high carbon stocks should be a priority in the aforementioned zones with high AGB values at higher altitudes. Low AGB or carbon stock lands, such as found in the lowlands in burned areas and in non-forest lands and production forests, should be considered for either regeneration or utilisation purposes, dependent on the regeneration capacity of the vegetation. In these lowland areas, the use of fire should be prevented as much as possible. The utilisation of peatlands should be avoided, especially because of the presence of high belowground carbon stocks. In our study, the variation in AGB was less affected by proximity of roads, rivers and cities.

The high correlations between the explanatory variables showed that the variables were interrelated and thus that AGB variation cannot be explained by one single variable. Instead, spatial analyses should integrate a variety of biophysical and anthropogenic variables to provide a better understanding of spatial variation in AGB.

Methods
Study area
The natural resource-rich provinces of North and East Kalimantan (North Kalimantan was established on 25 October 2012 and was previously part of East Kalimantan) have a high spatial variation in biophysical and anthropogenic conditions and processes. For use in this study, the provinces are regarded as one case study region. The terrain consists of undulating slopes and altitudes up to about 2,200 m. Karst and peatlands occur

mainly in the lowlands (respectively, ~2% and ~4). The remaining landscape consists mainly of volcanic soils and other soil types (respectively, ~7% and 87%). This landscape is highly dynamic with regard to its past, current and expected land use changes. Until the early 1970s, the original land cover in the lowlands of North and East Kalimantan consisted of extensive dipterocarp forests with high AGB and species richness [18], but driven by forest and land development policies in the 1980s, large-scale degradation, deforestation and conversion to agricultural land have taken place [48]. The main activities were high intensity logging [48], but also large-scale forest fires occurred that were often initiated for land clearing purposes [13,16,48,56], and events associated with El Niño Southern Oscillation (ENSO) [57]. In 1997−98, again very destructive fires related to ENSO occurred, burning 5.2 million ha of North and East Kalimantan's pristine and logged forests. Hoffmann et al. [58] have found that approx. 75% of the burned forests were allocated for logging, timber or oil palm concessions. The frequency and spatial extension of fires have increased over the last few decades in North and East Kalimantan because of deforestation and degradation processes associated with logging, mining and agriculture, and intensifying droughts related to ENSO events [13].

Selection of variables and proxy data layers

An overview of the method is given in Figure 1. We included multiple biophysical and anthropogenic variables in the analyses, based on data at a regional scale so that the interrelationships between the explanatory variables could be accounted for. In Figure 5, a landscape-scale view on the data layers is presented in which the landscape-scale pattern for each variable is visible. The use of spatial data enabled the analyses of AGB and several explanatory variables on a continuous scale. The initial selection of the explanatory variables was based on a literature review, field visits, visual examination of spatial data and data availability (for Data sources, see Table S7 and Appendix S8, in Additional file 1).

The AGB map (Figure 5a) is based on ALOS PALSAR-LiDAR data and plot-based measurements [38]. Disturbed tropical forest landscapes such as North and East Kalimantan are often covered by clouds or haze. Radar remote sensing is not affected by clouds and has proven to be a remote sensing system responsive to AGB [36,59]. Saturation of the radar signal at medium AGB levels (150 t ha^{-1}) restricts the use of radar remote sensing for a direct radar image inversion into AGB maps. A radar-based forest type map is used in combination with estimated vegetation heights per land cover type, derived from Geoscience Laser Altimeter System (GLAS) LiDAR data, to overcome such saturation effects. This resulted in an AGB map with a resolution of 50 m. Available

vegetation height-AGB allometric equations were used to invert heights into AGB values per pixel, overcoming the effect of radar saturation. An accuracy assessment of the AGB map was conducted using field measurements over 54 plots of 0.2 ha over a range of degraded forest types in the study area. AGB values were estimated using the allometric equation developed by Saatchi et al. [38,39]. The accuracy of the AGB map is estimated as 10 t ha^{-1}, using the root mean squared error between the field-estimated AGB and the AGB from the radar map for the same location. For more information see Quiñones et al. [38].

Altitude (Figure 5b) varied from −90 m − 2,230 m in the landscape and was selected because a relationship with AGB is expected [11,25-27]. *Slope* (Figure 5c) was found to have a positive relationship with AGB [8]. Altitude and slope were derived from the Digital Elevation Model (90 m) by the Shuttle Radar Topography Mission (SRTM-DEM) [60]. For the multiple linear regression and GWR, altitude was included as a continuous variable. For the ANOVA, altitude was categorised into several altitude ranges (Lowlands <750 m; Midlands 750−1,500 m; Highlands >1,500 m) (see Appendix S9 in Additional file 1).

Logging and land conversion decreases AGB substantially by the harvesting and loss of especially rare tree species [13]. Therefore a relationship is expected between AGB and logging intensity, and thus differences in AGB between protected forested areas, areas allocated for timber or forest concessions, and non-forest land. The data source selected is the land (use) allocation zoning data (Figure 5d) classified by WRI and originally produced by the Ministry of Forestry of Indonesia within the Tata Guna Hutan Kesepakatan (TGHK) mapping program (Ministry of Forestry Indonesia, year unknown) as a proxy for logging and land clearance intensity. The classes designated within this data layer and present in the study area are: 'forest limited production', where logging is accompanied by measures to reduce impacts on soil erosion; 'conservation forest', which is conservation forest for protected areas; 'watershed protection forest', which is intended for watershed protection; 'non-forest land' has the status of non-forest use; and 'production forest', which is intended for commercial logging [31,62].

Relationships were reported between AGB and soil drainage [26], soil texture [11], and soil fertility [23,24]. *Soil type* (Figure 5e) was selected as a proxy, and was included by reclassifying the improved reproduction of the RePPProT land systems map [63] into the categories 'karst', 'peat', 'volcanic', and 'other' (for details see Appendices S8 and S9 in Additional file 1).

Forest *fires* (Figure 5f) can occur multiple times at the same spot and in this way can cause substantial losses in AGB [12,13,64]. Also the rate of post-fire regeneration depends, amongst others, on the frequency and age of the fire [13,18]. The proxy for fire included in the multiple

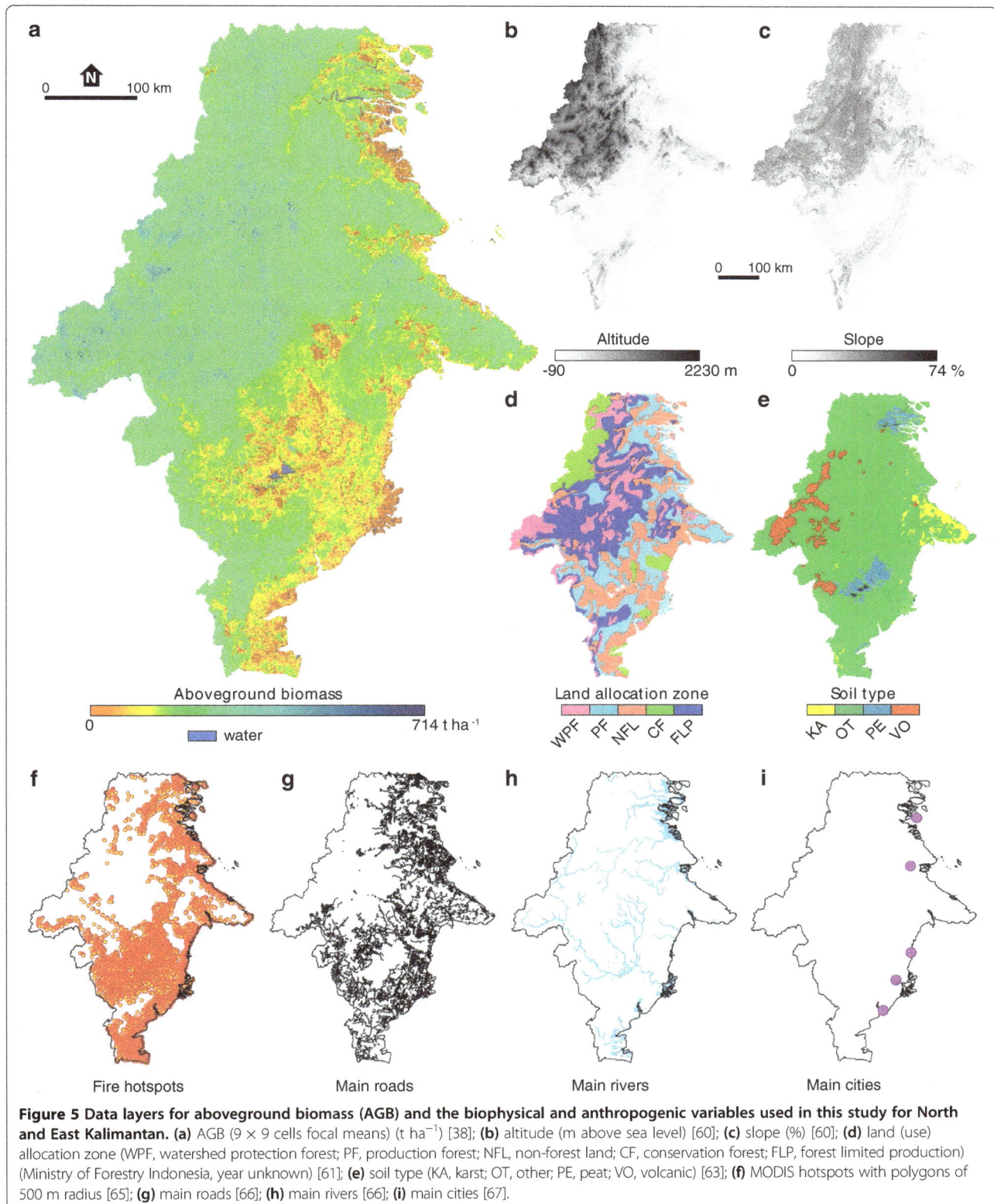

Figure 5 Data layers for aboveground biomass (AGB) and the biophysical and anthropogenic variables used in this study for North and East Kalimantan. (a) AGB (9 × 9 cells focal means) (t ha^{-1}) [38]; **(b)** altitude (m above sea level) [60]; **(c)** slope (%) [60]; **(d)** land (use) allocation zone (WPF, watershed protection forest; PF, production forest; NFL, non-forest land; CF, conservation forest; FLP, forest limited production) (Ministry of Forestry Indonesia, year unknown) [61] **(e)** soil type (KA, karst; OT, other; PE, peat; VO, volcanic) [63]; **(f)** MODIS hotspots with polygons of 500 m radius [65]; **(g)** main roads [66]; **(h)** main rivers [66]; **(i)** main cities [67].

linear regression and GWR was 'distance to the nearest fire'. For the fire data MODIS hotspot data from 2000 to 2008 were used [65], because of its high accuracy recording. According to NASA, each MODIS fire hotspot represents the centre point of a ~ 1 km pixel that contains one or more fires, rather than the exact location of a fire.

To overcome this uncertainty, a buffer of 500 m radius surrounding each location was created to define fire hotspot polygons. Additionally, for the ANOVA the fire variable was categorised as burned ($\leq$500 m from a hotspot; i.e. the area within a fire polygon), and non-burned areas (>500 m from a hotspot, i.e. the area outside a fire polygon) (see Appendix S9, in Additional file 1).

Both *roads* (Figure 5g) and *rivers* (Figure 5h) are the primary means of transportation in North and East Kalimantan, and improve accessibility from *cities* (Figure 5i) to forest frontier areas. Therefore, a relationship is expected between AGB and the proxy distance to nearest main road [66], the nearest main river [66] and the nearest main city [67].

Data pre-processing

In order to reduce the high local-scale variation in AGB caused by natural local variation and by the effect of speckle noise, 9 x 9 cell focal mean statistics was applied to the AGB map in ArcGIS, according to the results generated by Hoekman and Quiñones [68]. All shapefiles were rasterised, and the proximity variables were individually processed by means of the Euclidean Distance tool of the ArcGIS Spatial analyst. For optimal processing, a sample of 500 data points was selected randomly in the data layers with a minimum distance of 1,000 m from one another to minimise the effects spatial autocorrelation ([69]. The data layers were combined and all data queries were exported to a database in SPSS 20. Rows with missing values were deleted, resulting in a dataset of 465 data points.

The continuous explanatory variables showing a skewed distribution were transformed to attain normality. Natural logarithmic (ln) data transformation was in all cases the most suitable of a series of transformations tested for attaining a linear relationship between AGB and the explanatory variables.

Statistical analyses

Using the Pearson's correlation coefficient, the strength and direction of the predictive relationship between AGB and each of the continuous explanatory variables were defined. We conducted One-way ANOVA to analyse whether mean AGB among soil types and land allocation zones, among different altitudinal ranges, and between burned and unburned areas was significantly different.

Non-spatial backward multiple linear regression was conducted, and with every step non-significant ($p \geq 0.05$) variables were removed one-by-one. The categorical variables land allocation zoning and soil type were included as dummy variables with, respectively, 'non-forest land', and 'other' as the reference categories. Because land is allocated by the Ministry of Forestry of Indonesia based on climate, slope and soil type, tests for an interaction effect between land allocation zoning and altitude were carried out, by inclusion of product terms in the multiple linear regression [70].

To verify whether the output met the assumptions underlying multiple linear regression, tests for normality and multicollinearity were carried out. To test for normality, we plotted a histogram, a normal PP plot and a normal QQ plot of the standardised residuals. We tested for the presence of significant strong multicollinearity by examining the Tolerance.

Analysing ecological spatial data by multiple linear regression is challenging (e.g. [54,71], because of the possible existence of spatial autocorrelation and spatial non-stationarity, the latter being the variation in relationships and processes over space. Spatial non-stationarity was tested for by conducting the Breusch-Pagan test on random coefficients. Although often ignored, spatial autocorrelation or the spatial clustering of ecological conditions and processes is a natural, and thus widespread phenomenon [69]. Bini *et al.* [72] indicate that this can cause an unexplained shift in the regression coefficients of global or non-spatial models. To test for the presence of spatial autocorrelation, the Moran's I Index, the z-score and the p-value for the standardised residuals were calculated. If spatial autocorrelation was present, we additionally conducted GWR in ArcGIS. GWR is a spatial and local form of multiple linear regression that considers and models the spatially varying relationships between explanatory variables and the response variable [53-55]. The explanatory variables that showed multicollinearity were excluded from the model.

Additional file

Additional file 1: Figure S1. Frequency distribution of aboveground biomass in the sample. **Table S2.** Descriptive statistics of the data. **Table S3.** Correlation matrix for the combination of all continuous variables. **Figure S4.** Interaction effects between altitude and land allocation zoning. **Figure S5.** Frequency distribution, PP plot and QQ plot of the standardised residuals of the multiple linear regression. **Figure S6.** Frequency distribution of the standardised residuals of the GWR. **Table S7.** Overview of the variables and the data selected. **Appendix S8.** Data sources. **Appendix S9.** Categorisation of the variables.

Competing interests
The authors declare that they have no competing interests.

Authors' contributions
Authors: (CL), (PV), (MQ), (AF). The authors have made substantial contributions to conception and design: CL, PV or acquisition of data: CL, MQ or analyses: CL interpretation of data: CL, PV have been involved in drafting the manuscript: CL revising it critically for important intellectual content: PV, AF, MQ have given final approval of the version to be published: CL, PV, MQ, AF agree to be accountable for all aspects of the work in ensuring that questions related to the accuracy or integrity of any part of the work are appropriately investigated and resolved: CL, PV, MQ, AF. All authors read and approved the final manuscript.

Acknowledgements

The authors are grateful to Dr. Dirk Hoekman (Wageningen University) and Niels Wielaard (SarVision) for the AGB map and valuable discussions. The ALOS PALSAR data used in this paper have been provided by JAXA EORC within the framework of the JAXA Kyoto & Carbon Initiative. ALOS PALSAR data courtesy ALOS K&C © JAXA/METI. We thank Inge Kies, Rianne Visser, Leah Sazer-Krebbers and late Nina Schellinx for the AGB plot data collection. We are very grateful to Universitas Gadjah Mada, Universitas Mulawarman, The World Agroforestry Centre (ICRAF) in Bogor, and World Wide Fund for Nature in Jakarta and Samarinda for valuable support; Maarten Zeylmans for GIS-support; Derek Karssenberg for advice on the spatial analyses; Paul Westers and Mara Ding for statistical advice; and Ton Markus for optimizing the lay out of the figures. This project was funded by the Agriculture beyond Food Programme of NWO (Netherlands Organization for Scientific Research)/ WOTRO (Science for Global Development) and KNAW (Royal Netherlands Academy of Arts and Sciences).

Author details

[1]Copernicus Institute of Sustainable Development, Group Energy and Resources, Utrecht University, Heidelberglaan 2, 3584 CS Utrecht, The Netherlands. [2]SarVision BV, Agro Business Park 10, 6708 PW Wageningen, The Netherlands.

References

1. Van der Werf GR, Morton DC, DeFries RS, Olivier JGJ, Kasibhatla PS, Jackson RB, Collatz GJ, Randerson JT: **CO2 emissions from forest loss.** *Nat Geosci* 2009, **2**:737–738.
2. Gaveau DLA, Sloan S, Molidena E, Yaen H, Sheil D, Abram NK, Ancrenaz M, Nasi R, Quinones M, Wielaard N, Meijaard E: **Four Decades of Forest Persistence, Clearance and Logging on Borneo.** *PLoS One* 2014, **9**:e101654.
3. East Kalimantan's Green Growth Journey, *Presentation at the GCF Taskforce Annual Meeting*, Palangka Raya, Central Kalimantan, Indonesia, Sept. 20-22, 2011. Available upon request at the Governors' Climate and Forests Task Force Indonesia (http://www.gcftaskforce.org/events).
4. *About the UN-REDD Programme, UN-REDD*; [www.un-redd.org/AboutREDD/ tabid/102614/Default.aspx] Accessed: January 2014.
5. Wicke B, Sikkema R, Dornburg V, Faaij A: **Exploring land use changes and the role of palm oil production in Indonesia and Malaysia.** *Land Use Policy* 2011, **28**:193–206.
6. Smit HH, Meijaard E, Van der Laan C, Mantel S, Budiman A, Verweij PA: **Breaking the Link between Environmental Degradation and Oil Palm Expansion: A Method for Enabling Sustainable Oil Palm Expansion.** *PLoS One* 2013, **8**:e68610.
7. Chave J, Riera B, Dubois MA: **Estimation of biomass in a neotropical forest of French Guiana: spatial and temporal variability.** *J Trop Ecol* 2001, **17**:79–96.
8. Chave J, Condit R, Lao S, Caspersen JP, Foster RB, Hubbell SP: **Spatial and temporal variation of biomass in a tropical forest: results from a large census plot in Panama.** *J Ecol* 2003, **91**:240–252.
9. Baker T, Phillips O, Malhi Y: **Variation in wood density determines spatial patterns in Amazonian forest biomass.** *Glob Chang Biol* 2004, **10**:545–562.
10. Houghton R: **Aboveground Forest Biomass and the Global Carbon Balance.** *Glob Chang Biol* 2005, **11**:945–958.
11. De Castilho CV, Magnusson WE, de Araújo RNO, Luizão RCC, Luizão FJ, Lima AP, Higuchi N: **Variation in aboveground tree live biomass in a central Amazonian Forest: Effects of soil and topography.** *For Ecol Manag* 2006, **234**:85–96.
12. Page S, Siegert F, Rieley J, Boehm H: **The amount of carbon released from peat and forest fires in Indonesia during 1997.** *Nature* 1999, **2002**:61–65.
13. Slik JWF, Bernard CS, Van Beek M, Breman FC, Eichhorn KAO: **Tree diversity, composition, forest structure and aboveground biomass dynamics after single and repeated fire in a Bornean rain forest.** *Oecologia* 2008, **158**:579–588.
14. Berry NJ, Phillips OL, Lewis SL, Hill JK, Edwards DP, Tawatao NB, Ahmad N, Magintan D, Khen CV, Maryati M, Ong RC, Hamer KC: **The high value of logged tropical forests: lessons from northern Borneo.** *Biodivers Conserv* 2010, **19**:985–997.
15. Kronseder K, Ballhorn U, Böhm V, Siegert F: **Above ground biomass estimation across forest types at different degradation levels in Central Kalimantan using LiDAR data.** *Int J Appl Earth Observation Geoinformation* 2012, **18**:37–48.
16. Siegert F, Ruecker G, Hinrichs A, Hoffmann AA: **Increased damage from fires in logged forests during droughts caused by El Niño.** *Nature* 2001, **414**:437–440.
17. Slik JWF, Verburg RW, Keßler PJA: **Effects of fire and selective logging on the tree species composition of lowland dipterocarp forest in East Kalimantan, Indonesia.** *Biodiversity & Conservation* 2002, **11**:85–98.
18. Toma T, Ishida A, Matius P: **Long-term monitoring of post-fire aboveground biomass recovery in a lowland dipterocarp forest in East Kalimantan, Indonesia.** *Nutr Cycl Agroecosyst* 2005, **71**:63–72.
19. Matricardi EAT, Skole DL, Pedlowski MA, Chomentowski W, Fernandes LC: **Assessment of tropical forest degradation by selective logging and fire using Landsat imagery.** *Remote Sens Environ* 2010, **114**:1117–1129.
20. Hashimotio T, Kojima K, Tange T, Sasaki S: **Changes in carbon storage in fallow forests in the tropical lowlands of Borneo.** *For Ecol Manag* 2000, **126**:331–337.
21. Brown S: **Estimating Biomass and Biomass Change of Tropical Forests: A Primer.** *FAO Forestry Paper* 1997, **134**:55.
22. Feldpausch TR, Lloyd J, Lewis SL, Brienen RJW, Gloor M, Monteagudo Mendoza A, Lopez-Gonzalez G, Banin L, Abu Salim K, Affum-Baffoe K, Alexiades M, Almeida S, Amaral I, Andrade A, Aragão LEOC, Araujo Murakami A, Arets EJMM, Arroyo L, Aymard CGA, Baker TR, Bánki OS, Berry NJ, Cardozo N, Chave J, Comiskey JA, Alvarez E, de Oliveira A, Di Fiore A, Djagbletey G, Domingues TF, *et al*: **Tree height integrated into pantropical forest biomass estimates.** *Biogeosciences* 2012, **9**:3381–3403.
23. Laurance WF, Fearnside PM, Laurance SG, Delamonica P, Lovejoy TE, Rankin-de Merona JM, Chambers JQ, Gascon C: **Relationship between soils and Amazon forest biomass: a landscape-scale study.** *For Ecol Manag* 1999, **118**:127–138.
24. Paoli GD, Curran LM, Slik JWF: **Soil nutrients affect spatial patterns of aboveground biomass and emergent tree density in southwestern Borneo.** *Oecologia* 2008, **155**:287–299.
25. Whittaker R, Bormann F: **The Hubbard Brook ecosystem study: forest biomass and production.** *Ecol Monogr* 1974, **44**:233–254.
26. Asner GP, Flint Hughes R, Varga TA, Knapp DE, Kennedy-Bowdoin T: **Environmental and Biotic Controls over Aboveground Biomass Throughout a Tropical Rain Forest.** *Ecosystems* 2008, **12**:261–278.
27. Alves LF, Vieira SA, Scaranello MA, Camargo PB, Santos FAM, Joly CA, Martinelli LA: **Forest structure and live aboveground biomass variation along an elevational gradient of tropical Atlantic moist forest (Brazil).** *For Ecol Manag* 2010, **260**:679–691.
28. Ferry B, Morneau F, Bontemps J-D, Blanc L, Freycon V: **Higher treefall rates on slopes and waterlogged soils result in lower stand biomass and productivity in a tropical rain forest.** *J Ecol* 2010, **98**:106–116.
29. Van Nieuwstadt M, Van Sheil D: **Drought, fire and tree survival in a Borneo rain forest, East Kalimantan, Indonesia.** *J Ecol* 2005, **93**:191–201.
30. Dennis RA, Colfer CP: **Impacts of land use and fire on the loss and degradation of lowland forest in 1983–2000 in East Kutai District, East Kalimantan, Indonesia.** *Singap J Trop Geogr* 2006, **27**:30–48.
31. Broich M, Hansen M, Stolle F, Potapov P, Margono BA, Adusei B: **Remotely sensed forest cover loss shows high spatial and temporal variation across Sumatera and Kalimantan, Indonesia 2000–2008.** *Environ Res Lett* 2011, **6**:1–9.
32. Brown S, Iverson LR, Prasad A, Liu D: **Geographical distributions of carbon in biomass and soils of tropical Asian forests.** *Geocarto Int* 1993, **4**:45–59.
33. Gibbs HK, Brown S, Niles JO, Foley JA: **Monitoring and estimating tropical forest carbon stocks: making REDD a reality.** *Environ Res Lett* 2007, **2**:045023.
34. Yang C, Huang H, Wang S: **Estimation of tropical forest biomass using Landsat TM imagery and permanent plot data in Xishuangbanna, China.** *Int J Remote Sens* 2011, **32**:5741–5756.
35. Mitchard ETA, Saatchi SS, White LJT, Abernethy KA, Jeffery KJ, Lewis SL, Collins M, Lefsky MA, Leal ME, Woodhouse IH, Meir P: **Mapping tropical forest biomass with radar and spaceborne LiDAR: overcoming problems of high biomass and persistent cloud.** *Biogeosci Discuss* 2011, **8**:8781–8815.
36. Morel AC, Saatchi SS, Malhi Y, Berry NJ, Banin L, Burslem D, Nilus R, Ong RC: **Estimating aboveground biomass in forest and oil palm plantation in Sabah, Malaysian Borneo using ALOS PALSAR data.** *For Ecol Manag* 2011, **262**:1786–1798.

37. Englhart S, Keuck V, Siegert F: **Aboveground biomass retrieval in tropical forests — The potential of combined X- and L-band SAR data use.** *Remote Sens Environ* 2011, **115**:1260–1271.

38. Quiñones MJ, Schut V, Wielaard N, Hoekman D: **Above Ground Biomass map Kalimantan 2008 - Final report.** *SarVision Wageningen* 2011, 1-34. Downloaded on 15-09-2012 from: [http://www.globallandusechange.org/tl_files/glcfiles/wwf/SuLu%20Reports/Sarvision%20Report%20to%20Kalimantan%20AGB%20Maps.pdf]

39. Saatchi SS, Harris NL, Brown S, Lefsky M, Mitchard ETA, Salas W, Zutta BR, Buermann W, Lewis SL, Hagen S, Petrova S, White L, Silman M, Morel A: **Benchmark map of forest carbon stocks in tropical regions across three continents.** *Proc Natl Acad Sci U S A* 2011, **108**:9899–9904.

40. Houghton R, Skole DL, Nobre CA, Hackler JL, Lawrence KT, Chomentowski WH: **Annual fluxes of carbon from deforestation and regrowth in the Brazilian Amazon.** *Nature* 2000, **403**:301–304.

41. DeFries RS, Houghton RA, Hansen MC, Field CB, Skole D, Townshend J: **Carbon emissions from tropical deforestation and regrowth based on satellite observations for the 1980s and 1990s.** *Proc Natl Acad Sci U S A* 2002, **99**:14256–14261.

42. Carlson KM, Curran LM, Ratnasari D, Pittman AM, Soares-Filho BS, Asner GP, Trigg SN, Gaveau DA, Lawrence D, Rodrigues HO: **Committed carbon emissions, deforestation, and community land conversion from oil palm plantation expansion in West Kalimantan, Indonesia.** *Proc Natl Acad Sci U S A* 2012, **109**:7559–7564.

43. Morel AC, Fisher JB, Malhi Y: **Evaluating the potential to monitor aboveground biomass in forest and oil palm in Sabah, Malaysia, for 2000–2008 with Landsat ETM + and ALOS-PALSAR.** *Int J Remote Sens* 2012, **33**:3614–3639.

44. Saatchi SS, Houghton RA, Dos Santos Alvalá RC, Soares JV, Yu Y: **Distribution of aboveground live biomass in the Amazon basin.** *Glob Chang Biol* 2007, **13**:816–837.

45. Asner GP, Powell GVN, Mascaro J, Knapp DE, Clark JK, Jacobson J, Kennedy-Bowdoin T, Balaji A, Paez-Acosta G, Victoria E, Secada L, Valqui M, Hughes RF: **High-resolution forest carbon stocks and emissions in the Amazon.** *Proc Natl Acad Sci U S A* 2010, **107**:16738–16742.

46. Mitchard ETA, Saatchi SS, Lewis SL, Feldpausch TR, Woodhouse IH, Sonké B, Rowland C, Meir P: **Measuring biomass changes due to woody encroachment and deforestation/degradation in a forest–savanna boundary region of central Africa using multi-temporal L-band radar backscatter.** *Remote Sens Environ* 2011, **115**:2861–2873.

47. Langner A, Samejima H, Ong RC, Titin J, Kitayama K: **Integration of carbon conservation into sustainable forest management using high resolution satellite imagery: A case study in Sabah, Malaysian Borneo.** *Int J Appl Earth Observation Geoinformation* 2012, **18**:305–312.

48. Murdiyarso D, Adiningsih ES: **Climate anomalies, Indonesian vegetation fires and terrestrial carbon emissions.** *Mitig Adapt Strateg Glob Chang* 2006, **12**:101–112.

49. Fuller DO, Meijaard EM, Christy L, Jessup TC: **Spatial assessment of threats to biodiversity within East Kalimantan, Indonesia.** *Appl Geogr* 2010, **30**:416–425.

50. Budiharta S, Slik F, Raes N, Meijaard E, Erskine PD, Wilson KA: **Estimating the Aboveground Biomass of Bornean Forest.** *Biotropica* 2014, **0**:1–5.

51. Fargione J, Hill J, Tilman D, Polasky S, Hawthorne P: **Land clearing and the biofuel carbon debt.** *Science* 2008, **319**:1235–1238.

52. PAGE SE, RIELEY JO, BANKS CJ: **Global and regional importance of the tropical peatland carbon pool.** *Glob Chang Biol* 2011, **17**:798–818.

53. Wang Q, Ni J, Tenhunen J: **Application of a geographically-weighted regression analysis to estimate net primary production of Chinese forest ecosystems.** *Glob Ecol Biogeogr* 2005, **14**:379–393.

54. Foody GM: **Geographical weighting as a further refinement to regression modelling: An example focused on the NDVI–rainfall relationship.** *Remote Sens Environ* 2003, **88**:283–293.

55. Propastin P: **Modifying geographically weighted regression for estimating aboveground biomass in tropical rainforests by multispectral remote sensing data.** *Int J Appl Earth Observation Geoinformation* 2012, **18**:82–90.

56. Siegert F, Hoffmann AA: **The 1998 Forest Fires in East Kalimantan (Indonesia): A Quantitative Evaluation Using High Resolution, Multitemporal ERS-2 SAR Images and NOAA-AVHRR Hotspot Data.** *Remote Sens Environ* 2000, **72**:64-77.

57. Priadjati A: **Dipterocarpaceae: Forest Fires and Forest Recovery.** In *Wageningen University. Tropenbos-Kalimantan series 8.* Wageningen, The Netherlands: Tropenbos International; 2002:1–214.

58. Hoffmann AA, Hinrichs A, Siegert F: **Fire damage in East Kalimantan in 1997/98 related to land use and vegetation classes: Satellite radar inventory results and proposals for further actions.** *IFFM/SFMP GTZ* 1999, 1-26.

59. Mitchard ETA, Saatchi SS, Woodhouse IH, Nangendo G, Ribeiro NS, Williams M, Ryan CM, Lewis SL, Feldpausch TR, Meir P: **Using satellite radar backscatter to predict above-ground woody biomass: A consistent relationship across four different African landscapes.** *Geophys Res Lett* 2009, **36**:L23401.

60. Shuttle Radar Topography Mission *NASA*; 2012. [NASA: http://www2.jpl.nasa.gov/srtm/]

61. Ministry of Forestry Indonesia (year unknown): *Kawasan Hutan (Forest estate) land use maps, General Direktorat of Planning, Ministry of Forestry*; downloaded from http://appgis.dephut.go.id/appgis/kml.aspx. Processed and provided by Greenpeace. Prepared by the World Resources Institute (2012).

62. Brockhaus M, Obidzinski K, Dermawan A, Laumonier Y, Luttrell C: **An overview of forest and land allocation policies in Indonesia: Is the current framework sufficient to meet the needs of REDD +?** *Forest Policy Econ* 2012, **18**:30–37.

63. Consortium to Revise the HCV Toolkit for Indonesia: *Toolkit for identification of high conservation values in Indonesia.* Jakarta, Indonesia: Digital Appendix 12. Ecosystem proxy shapefiles for Kalimantan ver 1.0; 2008.

64. Cochrane MA: **Fire science for rainforests.** *Nature* 2003, **421**:913–919.

65. NASA/LANCE – FIRMS: *MODIS Hotspot / Active Fire Detections.* Data set; 2011. Acquired on 17-09-2012 online http://earthdata.nasa.gov/data/nrtdata/firms.

66. Bakosurtanal: Bakosurtanal, the Indonesian National Coordinating Agency for Surveys and Mapping (http://www.bakosurtanal.go.id). Data available in: Minnemeyer S, Boisrobert L, Stolle F, Muliastra YIKD, Hansen M, Arunarwati B, Prawijiwuri G, Purwanto J, Awaliyan R: Interactive Atlas of Indonesia's Forests CD-ROM 2009. Washington, DC: World Resources Institute; 2009.

67. CIESIN: *Center for International Earth Science Information Network, Columbia University; International Food Policy Research Institute (IPFRI); The World Bank; Centro Internacional de Agricultura Tropical (CIAT)*; 2012. http://sedac.ciesin.columbia.edu/gpw/.

68. Hoekman DH, Quiñones MJ: **Land cover type and biomass classification using AirSAR data for evaluation of monitoring scenarios in the Colombian Amazon.** *IEEE Trans Geosci Remote Sens* 2000, **38**:685–696.

69. Koenig W: **Spatial autocorrelation of ecological phenomena.** *Trends Ecol Evol* 1999, **14**:22–26.

70. Allison P: **Testing for interaction in multiple regression.** *Am J Sociol* 1977, **83**:144–153.

71. Graham MH: **Confronting multicollinearity in ecological multiple regression.** *Ecology* 2003, **84**:2809–2815.

72. Bini LM, Diniz-Filho JAF, Rangel TFLVB, Akre TSB, Albaladejo RG, Albuquerque FS, Aparicio A, Araújo MB, Baselga A, Beck J, Isabel Bellocq M, Böhning-Gaese K, Borges PAV, Castro-Parga I, Khen Chey V, Chown SL, de Marco P Jr, Dobkin DS, Ferrer-Castán D, Field R, Filloy J, Fleishman E, Gómez JF, Hortal J, Iverson JB, Kerr JT, Daniel Kissling W, Kitching IJ, León-Cortés JL, Lobo JM, *et al*: **Coefficient shifts in geographical ecology: an empirical evaluation of spatial and non-spatial regression.** *Ecography* 2009, **32**:193–204.

Choice of satellite imagery and attribution of changes to disturbance type strongly affects forest carbon balance estimates

Vanessa S. Mascorro[1]*, Nicholas C. Coops[1], Werner A. Kurz[2] and Marcela Olguín[3]

Abstract

Background: Remote sensing products can provide regular and consistent observations of the Earth´s surface to monitor and understand the condition and change of forest ecosystems and to inform estimates of terrestrial carbon dynamics. Yet, challenges remain to select the appropriate satellite data source for ecosystem carbon monitoring. In this study we examine the impacts of three attributes of four remote sensing products derived from Landsat, Landsat-SPOT, and MODIS satellite imagery on estimates of greenhouse gas emissions and removals: (1) the spatial resolution (30 vs. 250 m), (2) the temporal resolution (annual vs. multi-year observations), and (3) the attribution of forest cover changes to disturbance types using supplementary data.

Results: With a spatially-explicit version of the Carbon Budget Model of the Canadian Forest Sector (CBM-CFS3), we produced annual estimates of carbon fluxes from 2002 to 2010 over a 3.2 million ha forested region in the Yucatan Peninsula, Mexico. The cumulative carbon balance for the 9-year period differed by 30.7 million MgC (112.5 million Mg CO_{2e}) among the four remote sensing products used. The cumulative difference between scenarios with and without attribution of disturbance types was over 5 million Mg C for a single Landsat scene.

Conclusions: Uncertainty arising from activity data (rates of land-cover changes) can be reduced by, in order of priority, increasing spatial resolution from 250 to 30 m, obtaining annual observations of forest disturbances, and by attributing land-cover changes by disturbance type. Even missing a single year in the land-cover observations can lead to substantial errors in ecosystems with rapid forest regrowth, such as the Yucatan Peninsula.

Keywords: Carbon modeling, CBM-CFS3, Forest disturbance, MRV, Mexico, REDD+, Activity data

Background

Monitoring forest cover and change is essential to quantify the amount of carbon that is stored in the vegetation and soils, and the corresponding greenhouse gas (GHG) emissions and removals [1, 2]. Forest ecosystems can mitigate the impacts of climate change by absorbing significant amounts of atmospheric carbon through plant photosynthesis [3, 4]. However, degradation of forests and conversion to non-forested lands is the second largest anthropogenic cause of carbon dioxide (CO_2)

emissions into the atmosphere [5–8] which has led to initiatives such as Reducing Emissions from Deforestation and Forest Degradation (REDD+) [9].

Natural and anthropogenic disturbances are one of the main drivers that alter the forest structure over time, and understanding their impacts is therefore critical for quantifying carbon stock changes and associated emissions into and removals from the atmosphere [10–13]. Disturbances can range from fire, hurricane, insects and diseases, to timber harvesting, settlement expansion and other human activities. Since each disturbance type impacts the terrestrial carbon cycle in unique ways, annual observations of forest changes by disturbance type are necessary to accurately estimate carbon

*Correspondence: vanessa.mascorro@alumni.ubc.ca
[1] Faculty of Forestry, Department of Forest Resources Management, University of British Columbia, 2424 Main Mall, Vancouver, BC V6T 1Z4, Canada
Full list of author information is available at the end of the article

dynamics in the year of disturbance and the years after the disturbance [5, 10, 14].

Given the complexity of measuring and monitoring the terrestrial carbon cycle, a number of approaches have been developed [7]. Some techniques utilize data collected from ground-plot measurements and allometric equations that relate the physical attributes of trees to aboveground biomass in order to quantify forest carbon stocks (e.g., [15–17]). Other approaches use carbon budget models (e.g., [18]), remote sensing data (e.g. [19, 20]), or a combination of both with field measurements (e.g., [21–24]). Carbon budget modelling is a well-established approach to estimate carbon fluxes at regional to national-scales by integrating data from different spatial and temporal scales [25]. For example, Canada's National Forest Carbon Monitoring Accounting and Reporting System (NFCMARS) uses the Carbon Budget Model of the Canadian Forest Sector (CBM-CFS3) as the framework that integrates data from many sources [22, 26]. The CBM-CFS3 is also used as a decision support tool for forests managers to quantify the ecosystem carbon dynamics at the landscape level.

In compliance with the Intergovernmental Panel on Climate Change (IPCC) guidelines, the CBM-CFS3 quantifies carbon transfers among the five terrestrial carbon pools: above and belowground biomass, litter, dead wood and soil organic carbon, including atmospheric releases of CO_2 and non-CO_2 greenhouse gasses and transfers to the forest product sector [10]. The CBM-CFS3 offers the advantage of modeling the extent to which tree biomass and dead organic matter are affected by different disturbance types [27]. Following disturbance, it simulates subsequent forest transitions and successional dynamics and represents vegetation transfers to litter, coarse woody debris and soil organic carbon pools, and the decomposition of these pools. Based on the stage of stand development and ecological characteristics of the forest, the model accounts for growth and mortality using yield curves.

Carbon models, such as CBM-CFS3 require detailed spatio-temporal information about forest dynamics, including forest disturbance events, to accurately simulate carbon transfers, emissions and removals. Remote sensing (RS) is the primary source of land-cover data for forest monitoring due to its ability to monitor the Earth's surface on a regular and continuous basis, including areas otherwise difficult to access [28–30]. Recently, an increasing number of regional and global-scale RS studies and products have been developed specifically for forest carbon monitoring and REDD+ at different spatial and temporal scales [20, 31–33]. In a global study, Potter et al. [23] integrated large-scale Moderate Resolution Imaging Spectroradiometer (MODIS) satellite observations with the Carnegie Ames Stanford Approach (CASA), predicting a total of 0.51 Pg C $year^{-1}$ emissions from forest disturbance and biomass burning from 2000 to 2009. At a regional-scale, Masek and Collatz [21] integrated the CASA model with forest inventory data and higher spatial resolution satellite imagery to predict C fluxes from 1973 to 1999. They used Landsat time-series analysis to assess the effects of land-clearing disturbances (such as logging, harvesting and urbanization) on ecosystem productivity and quantified carbon emissions from biomass losses, decomposition and decay.

Significant variability among carbon stock estimates contributes to uncertainty in estimates of emissions and removals. Such variability is potentially due to differences inherent to the methods used, the spatial and temporal scales involved, and the definition of the components included [34]. A recent study by Achard et al. [33] estimated carbon emissions from deforestation in tropical countries for two decades (1990–2000, 2000–2010) by combining forest cover change maps derived from Landsat imagery and pan-tropical biomass maps. Contrasting their results with those obtained by Hansen et al. [35], they found discrepancies that can be explained by different approaches used in the satellite image analysis and their definitions of forest. A global study for tropical countries by Saatchi et al. [31] integrated inventory data and satellite light detection and ranging (LiDAR) data to quantify the carbon stored in the living biomass (247 Pg C) circa 2000. In a similar study, Baccini et al. [32] estimated carbon emissions from pan-tropical deforestation and land-use from 2000 to 2010, combining aboveground biomass with regional deforestation rates derived from LiDAR and MODIS, respectively. Comparing their results with the Forest Resource Assessment (FRA), they found their estimates (228.7 Pg C) were 21 % higher [32]. Likewise, Mitchard et al. [36] compared estimates from Saatchi et al. [31] and Baccini et al. [32] across the Amazon forest to a unique dataset of ground-plot data, and found that the regional differences over or underestimated the forest carbon estimates obtained from the ground data by >25 %.

It is critical to understand the potential role and implications of different RS observations for estimating ecosystem carbon dynamics following disturbance, in order to reduce uncertainties in measuring and monitoring carbon emissions and removals [37, 38]. In REDD+ studies, land-cover change observations are one of the main components (also called activity data) required for Monitoring Reporting and Verification (MRV) systems [39]. These activity data can be observations of changes in forest areas or land conversions between categories [40]. The design of efficient MRV systems requires an understanding of the implications of different attributes of remote sensing products on estimates of GHG emissions and removals.

This study assesses the impacts of using different RS products on estimates of terrestrial carbon dynamics. We examined the impacts of three attributes of RS products on the estimates of GHG emissions and removals: (1) the spatial resolution (30 vs. 250 m), (2) the temporal resolution (annual observations vs. multi-year observations), and (3) the attribution of forest cover changes to disturbance types. To do so we utilize four contrasting RS datasets—two land-cover change maps and two thematic maps—to compile activity data that we then use as spatially-explicit inputs of annual disturbance events to the CBM-CFS3. The RS data were derived from MODIS, Landsat, and a combination of Landsat-SPOT satellite imagery developed using different approaches, spatial and temporal resolutions. We derived two spatially-explicit layers for each of the RS products, one with land-cover change area, and the second attributing the observed change to its underlying disturbance cause. We then provided these activity data as inputs to CBM-CFS3 and estimated annual and cumulative greenhouse gas fluxes from 2002 to 2010 in the tropical dry forests of the Yucatan Peninsula, Mexico. We conclude with a discussion of key findings and implications of different activity data sources for future carbon budget analysis.

Methods and data
Study area
Our study focused on a ~3.2 million ha area located in the northwest side of the Yucatan Peninsula (YP), Mexico, covering parts mainly from Yucatan and Campeche states and a small area of Quintana Roo (see Fig. 1). The region's vegetation is dominated by secondary lowland dry tropical forests [41] of mixed ages that have regenerated after cycles of shifting agriculture and land abandonment. The region has a tropical climate, with rainy summers from June to October and dry winters from November to May. The mean average temperature is 26 °C with precipitation levels ranging from 900 to 1400 mm year^{-1} [42]. The regional topography is characterized by flat limestone areas and low moderate hills, with an average slope of 7 % and a mean elevation of 116 m [18]. The main soil types are well-drained rendzinas and shallow rocky lithosols [43].

Remote sensing data
Remote sensing observations are essential to track forest disturbance impacts and changes in land cover and forest

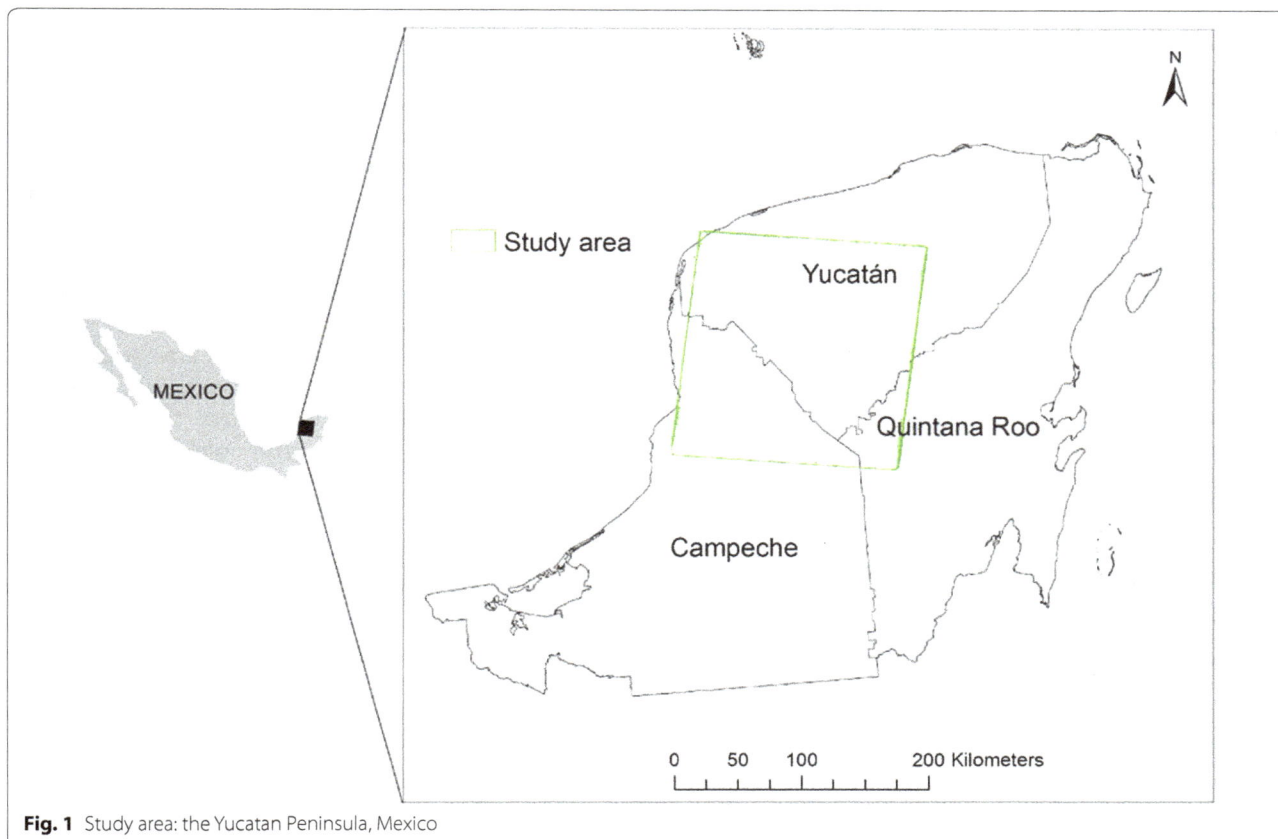

Fig. 1 Study area: the Yucatan Peninsula, Mexico

carbon at a range of spatial scales [44, 45]. Land-cover changes are those remotely sensed observations of the Earth's surface that changed from one point in time over two or more time periods because of any disturbance [7], including both natural and human-induced causes. We compiled four remote sensing disturbance products from Landsat, MODIS and a combination of Landsat-SPOT imagery—two classification and two pixel-based maps—derived using a range of approaches to compare their impacts on carbon budget estimates (see Table 1). We use the term pixel-based approach to differentiate the two maps that are labeled pixel by pixel as change/no change, from the thematic or classification maps that are comprised by various pixels grouped together into larger areas sharing similar characteristics (land cover class).

Landsat: VCT

The vegetation change tracker (VCT) is an automated algorithm that implements a disturbance index (DI, [46]) to reconstruct the history of the landscape disturbance on an annual or bi-annual basis (depending on data quality and availability) using Landsat 30 m time-series analysis [47]. This layer was generated and provided by the NASA Goddard Space Flight Center Biospheric Sciences Research branch. Each pixel of the spatial dataset was labeled according to the year in which the forest cover change was identified, from 1985 to 2010 (represented by 16 classes). From this map, we extracted disturbed areas for the years 2003, 2004, 2005, 2007, 2008 and 2010. Change data were unavailable due to cloud cover in 2002, 2006 and 2009.

Landsat: Hansen

Annual land-cover change data from a global forest cover change map derived from Landsat imagery at 30 m spatial resolution were also available from Hansen et al. [35]. The coverage provided direct estimates of forest cover loss and gain from 2000 to 2012, including detailed information of annual forest loss. Assessments of forest cover loss included pixels that completely lost tree canopy cover from a stand-replacing disturbance. Here, we obtained annual cover loss areas for our study period (2002–2010).

Landsat-SPOT: INEGI

We also retrieved three classification maps from the National Institute of Statistics and Geography of Mexico (in Spanish: INEGI): the vegetation and land-use cartography series developed for 2002, 2007 and 2011 at a scale 1:250,000 with approximately 70 classes (INEGI, SIII, SIV, and SV). The vegetation maps were generated using manual interpretation of Landsat and SPOT satellite images, validated with ground-plot data. In addition, a digital elevation model and ancillary datasets that describe the climate, geology, hydrology and topography of the Mexican ecosystems were used in their construction. These maps are a conventional source of information in Mexico for assessing the status and condition of the forested land base. The minimum mapping unit in all three vegetation and land-use cartography series is 25 ha [48]. We used the INEGI maps to generate two land-cover change maps: (1) changes detected from period observation one to two (2002–2007); and 2) changes detected between observation two and three (2007–2011; [49–51]). To simplify this process, we grouped the INEGI vegetation classes into twelve classes established by the National Commission for Knowledge and Use of Biodiversity of Mexico (CONABIO) in the Monitoring Activity Data for the Mexican REDD+ program (MAD-MEX; [48]) and assessed the change from one class to another.

Modis

Finally, we retrieved six classification maps developed by CONABIO in the North American Land Change Monitoring System (NALCMS) project using MODIS satellite imagery at 250 m spatial resolution [52]. Annual spatially-explicit maps were developed from 2005 to 2011, using monthly composites of MODIS, multiple classifications (ensemble classifier) with decision trees and attributes from auxiliary datasets to characterize 15 land-cover classes in Mexico [53]. These attributes include a digital elevation map, aspect, slope, mean temperature, maximum precipitation, sampling data, and aerial photography at high-spatial resolution. From these maps, we assessed the annual transition from one land-cover class to another from 2005 to 2010.

Table 1 Remote sensing products used as activity data inputs for carbon modeling with the CBM-CFS3 in the Yucatan Peninsula from 2002 to 2010

Map alias	Source	Type	Satellite	Spatial resolution	Temporality
VCT	NASA	Vegetation Change Tracker map	LANDSAT	30 m	2003, 2004, 2005, 2007, 2008 and 2010
Hansen	University of Maryland	Forest cover loss map	LANDSAT	30 m	2002, 2003... 2010
INEGI	INEGI	Classification map	LANDSAT-SPOT	30 m (25 ha MMU)	2002–2005, 2005–2007, 2007–2011
MODIS	NALCMS/CONABIO	Classification map	MODIS	250 m	2005, 2006... 2011

MMU minimum mapping unit

Accuracy assessment of forest cover change estimates

We performed an independent assessment of the remote sensing disturbance datasets using ground-plot data obtained from the Mexican National Forest Inventory (NFI). The NFI is the primary data source of information for forest and ecosystem studies in Mexico [54]. Data are collected in a two-stage sampling scheme comprised of primary plots of one-hectare size and four sub-sampling sites of 400 m^2 each. More than 150 variables are measured on a 5-year cycle, including attributes of the overstory and understory vegetation, the soil, and environmental characteristics of the landscape. For more details concerning the variables, methods and sampling design, readers are referred to the NFI sampling manual [54].We obtained forest change data over the period 2004–2012 of basal area loss/no-loss per hectare computed by Mascorro et al. [55] for 647 plots that were located across the study area. We intersected these plots with the RS derived land-cover change areas to identify and compare the number of plots with forest loss to the proportion of changed areas detected by each RS product using a standard error matrix. The matrix performs a cross-tabulation between the numbers of observations mapped with RS that agree with what is observed on the ground to assess the level of accuracy of the mapped products [56, 57].

Disturbance attribution

We attributed the land-cover change estimates obtained from the RS products using a Multi-Scale, Multi-Source Disturbance (MS-D) approach developed by Mascorro et al. [55]. This approach integrates RS data, forest inventory and ancillary datasets to attribute the land-cover change observations to the most likely disturbance type (natural or anthropogenic). Using NFI data and historical records of forest disturbances in tabular format (both considered ground-truth data), Mascorro et al. [55] derived annual spatially-explicit layers of major forest disturbance types, and obtained the forest change observed in the ground-plots. A regression tree analysis was undertaken to identify which of the constraining variables (forest disturbance types) best explained the observed forest loss in the field. Once the most likely cause of change was identified, remote sensing maps were used to obtain pixels with land-cover change and overlaid these with the spatially-explicit layers of disturbances characterized by type. Finally, the change was attributed to the most likely disturbance cause according to the relevance resulting from the regression tree analysis.

We retrieved ancillary forest disturbance layers compiled by Mascorro et al. [55] and overlaid them with the VCT, Hansen, MODIS and INEGI RS observations. The disturbance datasets retrieved included annual, spatially-explicit, natural and anthropogenic disturbances from 2005 to 2010: fires, hurricanes, and forest management areas. To characterize the permanent conversion of forestland to non-forestland caused by settlement, we used road coverages retrieved from INEGI [58]. In addition, land-cover classes that changed to urban areas in the classification maps were attributed as settlement.

Pre-processing data

To prepare the input data and provide the parameters required by the CBM-CFS3 for each of the simulation runs, we used a software tool called "Recliner" developed and provided by the carbon accounting team of the Canadian Forest Service. The tool facilitates the processing of large volumes of spatially-explicit data and assists users with the preparation of required input data for each simulation. We customized Recliner and matched the RS disturbance types with pre-defined disturbance matrices stored in the CBM-CFS3. Disturbance matrices define the impacts on the carbon pools for each disturbance type and specify the amount of carbon that is transferred from the biomass to the dead organic matter pools or is released into the atmosphere during a particular disturbance type [10, 27]. Disturbance matrices also define the amount of carbon transferred from forest ecosystems to the forest product sector by harvesting activities, and the amount of carbon released to the atmosphere as methane (CH_4), carbon monoxide (CO) and carbon dioxide (CO_2). Fire disturbances were matched to "wildfires", settlement to "deforestation" and harvest to "clear-cut with slash-burn".

The CBM-CFS3 can simulate stand-replacing disturbance by restoring the stand age to the initiation stage, or simulate partial mortality. "Generic mortality" disturbance matrices ranging from 5 to 95 % of impact are among the default choices [27]. Since the CBM-CFS3 does not contain a specific matrix for hurricanes, we matched these events with a generic disturbance matrix of "10 % mortality", estimated by dividing the mean basal area loss explained by hurricanes in Mascorro et al. [55] studies, over the mean basal area estimated across the Yucatan Peninsula (1.71 m^{2}ha^{-1}/15.54 m^{2}ha^{-1}). With Recliner, we assigned the year of disturbance impact to the corresponding land-cover change detection year. In cases where annual data were not available, we distributed equally the disturbance events among the number of years contained in the period of detection (e.g. VCT disturbances detected in 2007 where assigned in equal proportions to 2006 and 2007, since no observations were available for 2006).

Carbon budget modeling

We provided annual disturbance events to the CBM-CFS3 to quantify carbon exchange from tree biomass mortality, plant detritus decay, soil organic carbon and forest regrowth after disturbance. Additional parameters and data required by the CBM-CFS3 for the simulations were provided by the carbon accounting team of the Canadian Forest Service, and researchers from the Mexican Forest Service. These parameters were kept constant in all simulations so that only the activity data differed among the eight scenarios.

Ancillary datasets included detailed information on forest inventory, forest type, forest status, ecological boundaries, age-class structure, and yield curves [59]. Stand age and yield curve data, in combination with the ecological characteristics of the site, are used by the CBM-CFS3 to simulate forest growth and carbon accumulation [10]. The stage of stand development is also used to simulate litterfall and decomposition rates for dead organic matter and soil carbon. For forest status, we used protected and non-protected conservations areas, which the model employs as a classifier to differentiate the conditions on the landscape. We also used as a classifier the ecoregions level-I retrieved from the Commission for Environmental Cooperation (CEC) [60]: tropical humid forests and tropical dry forests. Estimates of carbon sinks and sources were then generated with the CBM-CFS3 for each of the two spatially-explicit layers of the four RS products, i.e. with and without attribution of cover loss to disturbance types. For simulations without attribution to specific disturbance types we assumed that all forest cover changes were caused by stand-replacing harvest with slash-burning.

Results
Forest change detection
The spatial variation in the location and extent of changes in forest area detected by the VCT, Hansen, MODIS and INEGI maps can be observed in Figs. 2 and 3. Overall, MODIS products detected much less changes across the landscape than the Landsat products, likely due to its lower spatial resolution. As can be seen in Fig. 2b, d, landscape changes between the VCT and Hansen products exhibited similar patterns. With a 30 m spatial resolution, these products detected finer disturbance events throughout the study area. Undisturbed forest areas were primarily located in the center of the region, expanding to the northwest and southeast of the area. In contrast, the INEGI maps generated with a classification approach detected larger contiguous areas of change from land-cover class transitions, due to the various 25 ha MMU's of similar characteristics grouped together into a class (Fig. 3e, f).

The cumulative area of change detected in the study area by each RS product across the analysis period is shown in Fig. 4. It is apparent that the Landsat products

detected the most change in all of the remote sensing approaches. However, the amount detected was less in the pixel-based approach than that derived from the classification maps (INEGI).

The range between the maximum (13 % by INEGI) and minimum (1 % by MODIS) cumulative percentages of area change detected, highlights the uncertainties resulting from the differences in spatial and temporal characteristics inherent in each RS product. Temporal trajectories of the land-cover changes detected in annual time-steps provided additional information on the inter-annual variation of disturbance events as shown in Fig. 5.

Forest change accuracy assessment
Table 2, summarizes the results of intersecting the NFI plots describing forest cover loss, with the VCT, Hansen, and INEGI land-cover change products in an error matrix. Of the 647 NFI plots located in the study area, 119 showed a forest cover loss, and 528 had no evidence of any basal area reduction in the plot. The first row of each table contains the number of mapping units that each RS product detected as forest cover loss compared to what was observed on the ground with the NFI plots (the total sum of the first column). As can be seen in Table 2a and b, the overall accuracy of the non-classification maps, the VCT (83 %) and Hansen (82 %), maps produced using pixel-based image processing approaches was similar, whereas the INEGI accuracy was 79 % (Table 2c). However, we can observe that all three maps presented a low producer's accuracy, detecting only a small fraction of the changes observed in the NFI plots likely due to forest degradation. The largest uncertainties were found in the MODIS maps, as there was no correspondence between the MODIS changed pixels and the NFI plots.

The impact of activity data on estimates of carbon fluxes
We used the CBM-CFS3 model to simulate the effects of activity data on emission and removal estimates during, and following disturbances from 2002 to 2010 (Fig. 6). Results show the annual net changes in ecosystem carbon stocks, with negative numbers indicating a decrease in carbon stocks (an emission of carbon to the atmosphere) and positive numbers an increase in carbon stocks (carbon uptake from the atmosphere). The estimates show both differences in the magnitude of changes and differences in trends due to differences in activity data.

The values from these simulations should not be considered as absolute values as research continues to improve yield curves and other ecological and modelling parameters that will likely affect future estimates of fluxes. What matters here is the impact of the activity

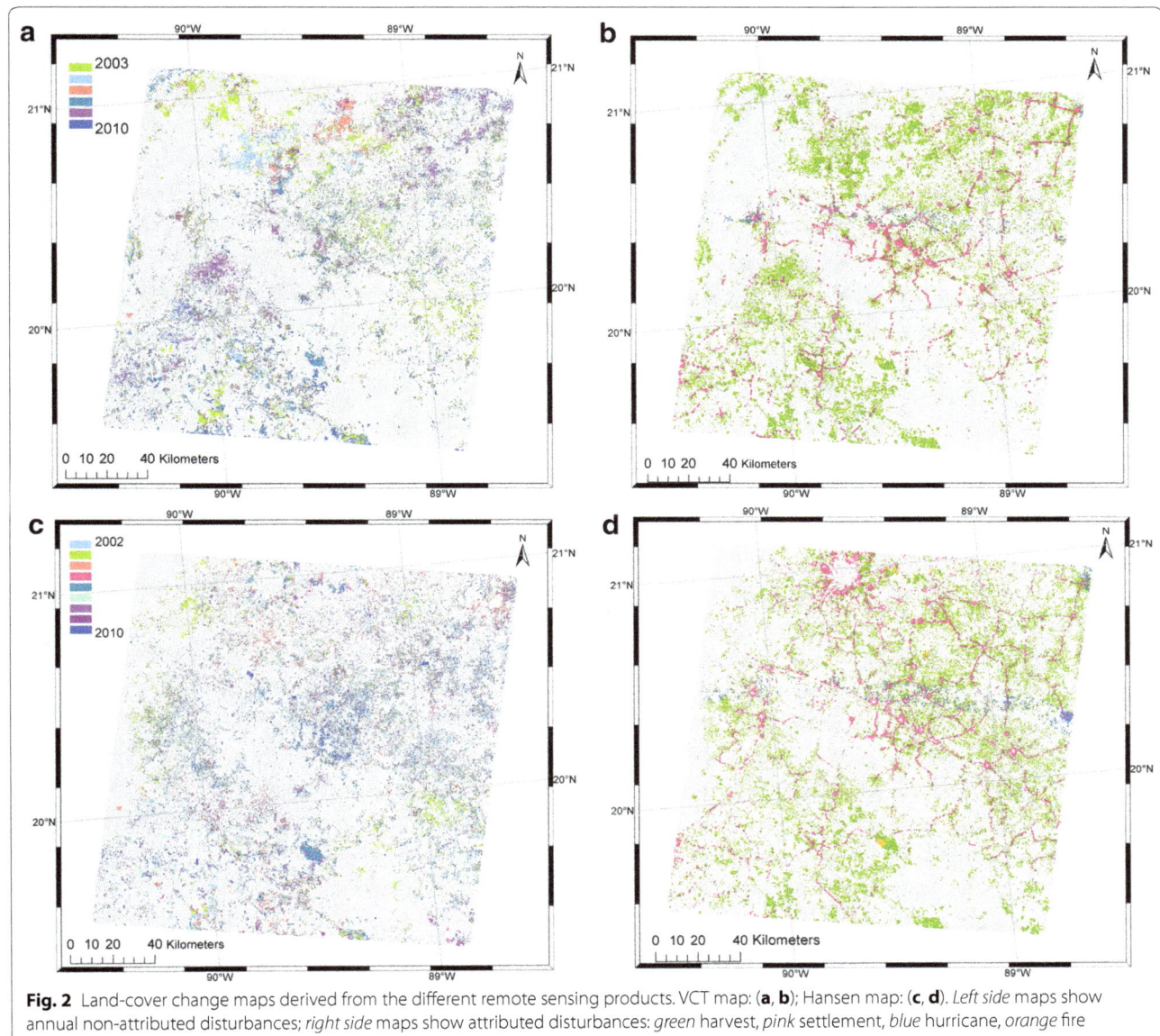

Fig. 2 Land-cover change maps derived from the different remote sensing products. VCT map: (**a**, **b**); Hansen map: (**c**, **d**). *Left side* maps show annual non-attributed disturbances; *right side* maps show attributed disturbances: *green* harvest, *pink* settlement, *blue* hurricane, *orange* fire

data on the differences in the annual and cumulative estimates of carbon fluxes (Fig. 6).

The cumulative difference in the carbon balance estimates resulting from the four RS products for the 9-year period (2002–2010) was over 30.7 million Mg C (112.5 million Mg CO_{2e}) see Fig. 7, with MODIS-derived estimates suggesting a large sink while INEGI-derived estimates remained a source for the whole period (Fig. 6).

Since the ecological parameters and additional data inputs used in each of the simulations were kept constant, the observed differences in the estimates reflect only the effects of the spatial and temporal resolutions inherent to each of the RS observations, and the impacts

of attributing cover changes to disturbance types with different impacts on carbon stocks.

Over the simulation period we observe the overriding trend of the MODIS disturbance detection underestimating cover changes and the associated carbon losses compared to the Landsat RS products. MODIS-based simulations shifted from a source to a sink in 2005 and remained a sink for the remainder of the simulations. Landsat-derived estimates in the other hand, followed a similar trend in 2002 and 2003, and began diverging in 2004. From 2005 to 2007, simulations based on VCT and Hansen predicted increasing carbon stocks. Consistent with the amount of land-cover change detected by each product

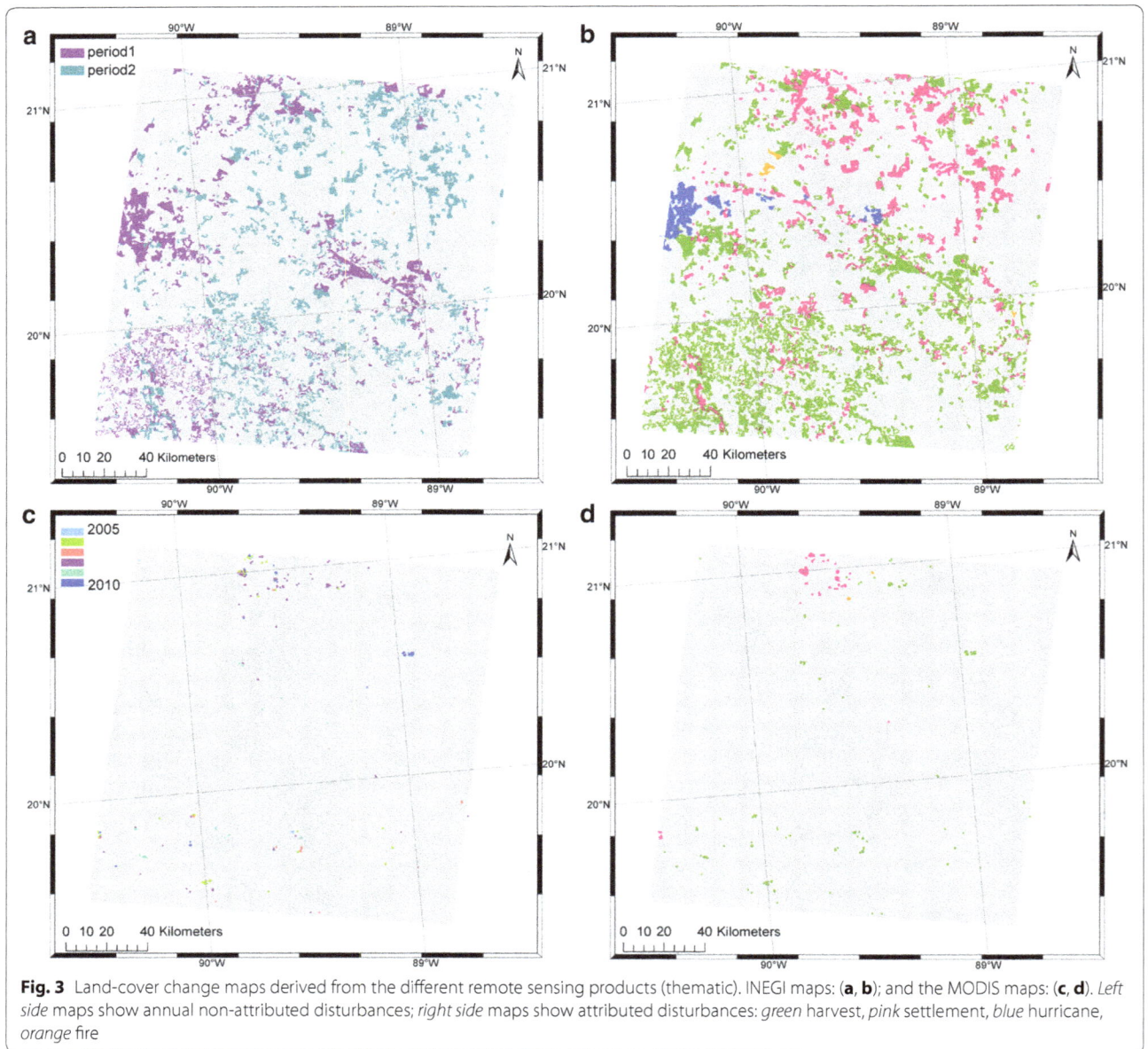

Fig. 3 Land-cover change maps derived from the different remote sensing products (thematic). INEGI maps: (**a**, **b**); and the MODIS maps: (**c**, **d**). *Left side* maps show annual non-attributed disturbances; *right side* maps show attributed disturbances: *green* harvest, *pink* settlement, *blue* hurricane, *orange* fire

(Fig. 5), the rate of increase in carbon uptake estimated with VCT from 2005 to 2007 was higher than that simulated with Hansen. In 2007, Hansen removals were 234,188 Mg C, whereas the VCT estimates were 1,033,608 Mg C (Fig. 6). While some disturbance events were detected in 2005–2007, the amount of area affected was not enough to bring the sink back to a source (e.g., estimates in 2009 from Hansen). Fewer disturbances were observed by VCT in this period, which translated to fewer carbon emissions and more carbon accumulation due to forest growth without disturbance. Nevertheless, in 2008 estimates from these two data sources went in opposite directions.

In the Hansen simulation, the forest continued to sequester carbon in 2008 as the amount of affected area decreased by 37 % compared to 2007 (Fig. 5). This was translated by the CBM-CFS as less carbon emissions and more living biomass accumulation, sequestering 3 times more carbon than in 2007 (234,188 Mg C in 2007 vs 729,215 Mg C in 2008). Conversely, with the VCT, the amount of change area detected in 2008 was more than double compared to 2007 (10,409 vs 28,278 ha), releasing considerable amounts of carbon (1,033,964 Mg C) to the atmosphere from forest disturbance events that killed the vegetation and redistributed the

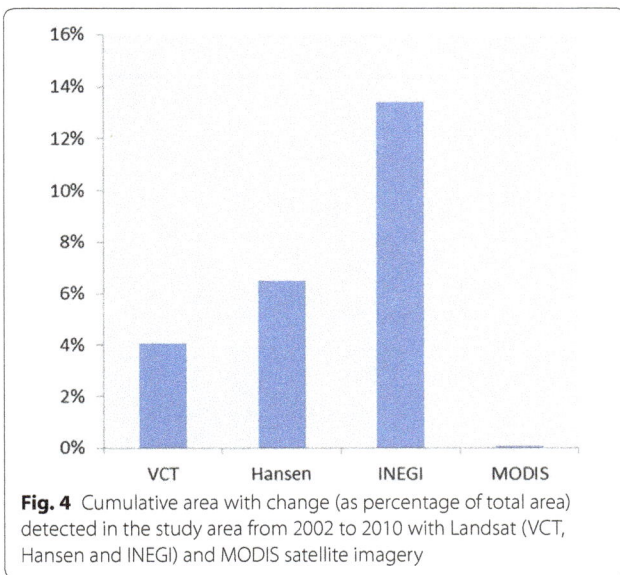

Fig. 4 Cumulative area with change (as percentage of total area) detected in the study area from 2002 to 2010 with Landsat (VCT, Hansen and INEGI) and MODIS satellite imagery

carbon in the dead organic matter pools during that year.

Moreover, Mascorro et al. [55] found that 2009 was an intense year of disturbances in the Yucatan Peninsula. Accordingly, the impact of the disturbance events detected by Hansen, resulted in significant amounts of carbon emissions into the atmosphere (stocks dropped by 2,774,453 Mg C; Fig. 6). MODIS-derived estimates also showed a decrease in carbon content in 2009, but by much smaller amount due to its limited capacity to detect changes in the landscape. However, with the VCT no land-cover change observations were included in 2009 due to limitations of image acquisition and cloud coverage, and therefore the simulations show no decrease in carbon emissions in 2009, despite the fact that the change observed in 2010 was equally distributed to 2009 and 2010. It is likely that some of the 2009 disturbance events were not detected by the VCT in 2010 due to the fast regrowth rate of the tropical forests of the Yucatan Peninsula.

Overall, estimates from non-attributed changes presented more carbon emissions than the attributed estimates in all the simulations. In the INEGI simulations this difference was more evident. In the second period, emissions from the non-attributed disturbances (7 million Mg C) were more than double than the attributed ones (3.2 million Mg C). When converting the additional non-CO_2 GHG emissions from fires (see Fig. 8a–c) in the form of carbon monoxide (CO), methane (CH_4) and nitrous oxide (N_2O) with their global warming potential (GWP) to 100 years (1, 25 and 298 respectively), we can see the additional CO_2 equivalent (CO_2e) emissions contributed by non-CO_2 greenhouse gases (i.e. 3 million Mg CO_2e more; see Fig. 8d).

Discussion

This study analyzed the contribution of three main attributes from RS activity data on estimates of forest carbon dynamics: (1) the spatial resolution (30 vs. 250 m), (2) the temporal resolution (annual vs. multi-year observations), and (3) the attribution of forest cover changes to disturbance types. Annual carbon fluxes were simulated with the CBM-CFS3 from 2002 to 2010 over a 180×180 km study area of dry tropical forests in the Yucatan Peninsula, Mexico.

Spatial resolution

The large difference in annual and cumulative carbon estimates generated with MODIS and Landsat-derived disturbance observations suggests that increased spatial resolution should be the first priority when deriving

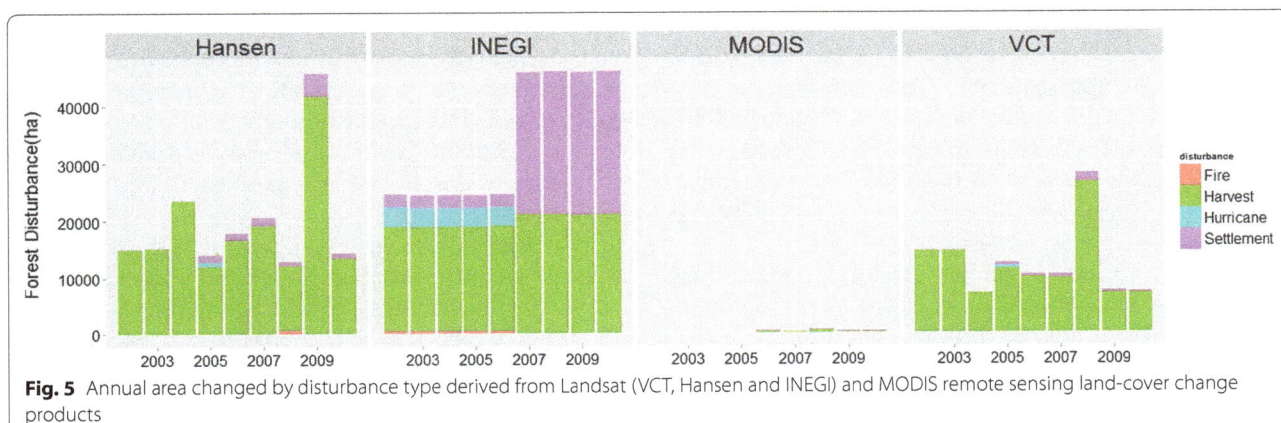

Fig. 5 Annual area changed by disturbance type derived from Landsat (VCT, Hansen and INEGI) and MODIS remote sensing land-cover change products

Table 2 Error matrix of landsat maps VCT (a), Hansen (b) and INEGI (c) identification of change and no change areas compared to the ground-plot data

(a) VCT

	Change	No change	Sum	Producer's
NFI plots				
Change	18	101	119	15 %
No change	11	517	528	98 %
Sum	29	618	647	
User's	62 %	84 %		
Overall accuracy = 83 %				

(b) Hansen

	Change	No change	Sum	Producer's
NFI plots				
Change	27	92	119	23 %
No change	22	506	528	96 %
Sum	49	598	647	
User's	55 %	85 %		
Overall accuracy = 82 %				

(c) INEGI

	Change	No change	Sum	Producer's
NFI plots				
Change	26	93	119	22 %
No change	44	484	528	92 %
Sum	70	577	647	
User's	37 %	84 %		
Overall accuracy = 79 %				

activity data for carbon modeling. Carbon flux estimates derived from Landsat at higher spatial resolution reflected the impacts of more accurate activity data. In contrast, the coarse spatial resolution of MODIS-derived products results in a substantial underestimate of change events and thus overestimates forest carbon sinks over the observation period. The latter occurred despite having MODIS observations on an annual basis from 2005 to 2010. While Mascorro et al. [55] were able to characterize major stand-replacing disturbances for the entire Yucatan Peninsula with MODIS, results from this study showed that this 250 m resolution satellite imagery has limitations for providing activity data for carbon modeling at finer spatial scales. Typically, satellite imagery from MODIS has focused on broad scale studies (e.g., [23, 29]).

Moreover, we found that the pixel-based image processing approach using VCT and Hansen was more accurate than change inferred from classification maps (i.e., INEGI, MODIS) (see Table 2). While the INEGI simulations of the first period observations (2002–2006) show a similar trend to VCT and Hansen, change observations for the period 2007–2011 are much higher than any of

the others. This is likely due to the fact that INEGI maps were generated with a classification approach, detecting larger areas of change formed by clusters of pixels that share similar characteristics and attributes. Therefore the larger disturbances observed over the second period, resulted in higher estimates of change likely due to larger areas that transitioned from one class to another (as opposed to single-pixel changes). Some of the predicted changes using thematic maps can also result from inaccuracies inherent to the classification method [61, 62], and omission and commission errors [63]. Nevertheless, post-classification analysis—a change detection technique that compares the initial and final class condition between two classification maps in time- has become a popular method for change detection [64], that helped to better identify settlement expansion embedded in the land-cover classification approach as urban areas. Yet, RS products generated with a pixel-based approach presented some limitations in attributing the changes to the underlying disturbance type due to the lack of spatially-explicit ancillary datasets. Further research would be required to increase the accuracy of land-cover changes

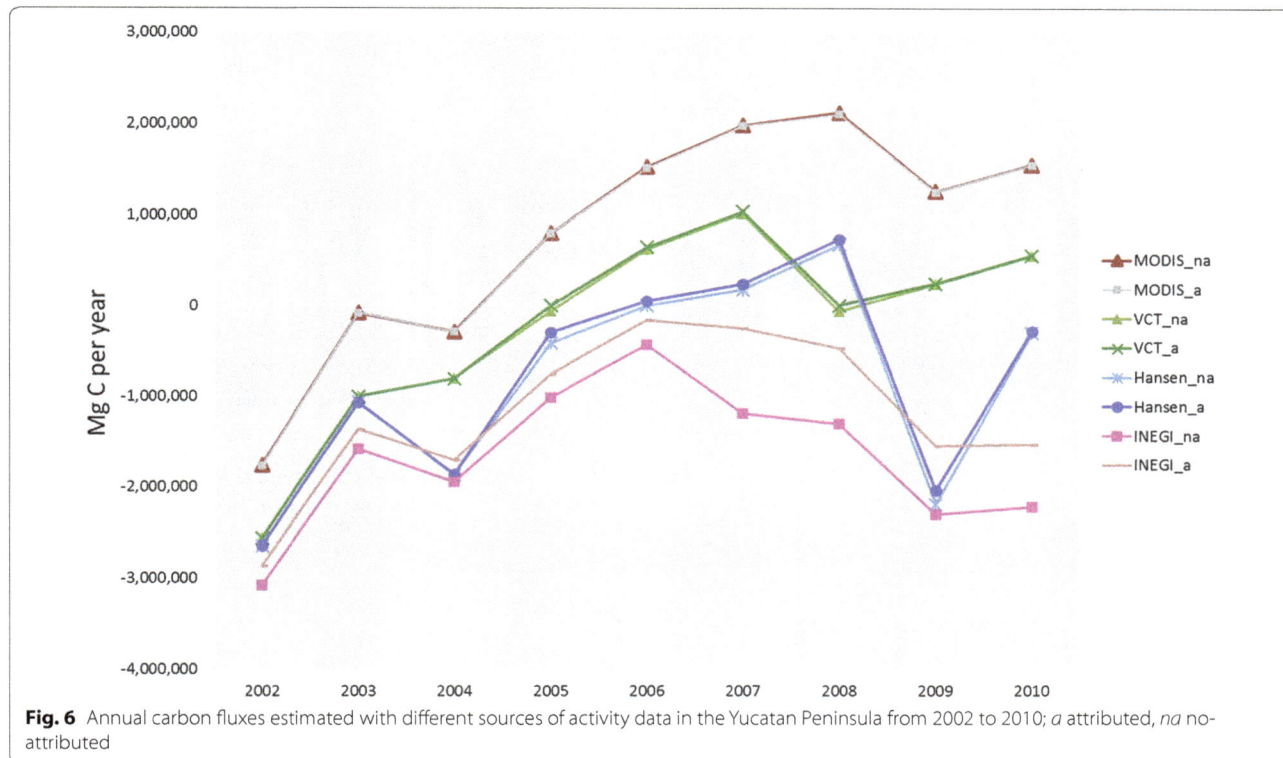

Fig. 6 Annual carbon fluxes estimated with different sources of activity data in the Yucatan Peninsula from 2002 to 2010; *a* attributed, *na* no-attributed

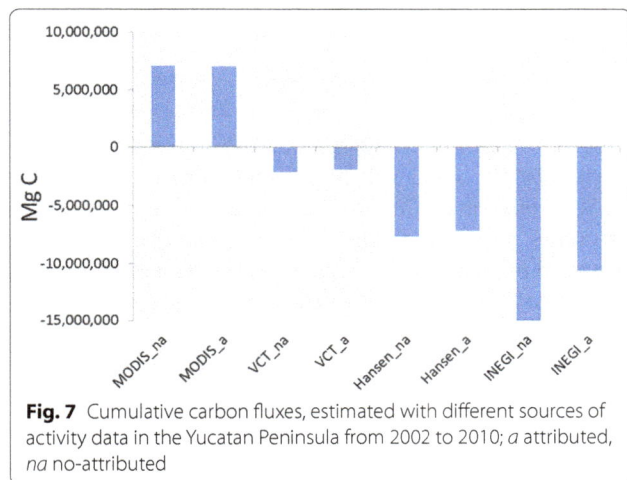

Fig. 7 Cumulative carbon fluxes, estimated with different sources of activity data in the Yucatan Peninsula from 2002 to 2010; *a* attributed, *na* no-attributed

characterized by disturbance type using additional datasets.

Temporal resolution

Reliable estimates of carbon dynamics require detailed observations of drivers of change on annual basis [14]. Our results showed that higher temporal resolution RS products improved the carbon dynamics estimates and captured the inter-annual variability in the forest. Consistent with the large disturbances detected in 2009, carbon fluxes predicted with the Hansen and MODIS products showed a decrease in total ecosystem carbon content. However, due to the lack of disturbance events detected with the VCT in 2009, corresponding decreases in the carbon stocks from the disturbance events were not reflected in the CBM-CFS3 estimates. This occurred despite the fact that the disturbance events observed over the period 2009–2010 were equally distributed in 2009 and 2010. This result showed that even missing a single year in the land-cover observations can lead to substantial errors; especially in ecosystems with rapid-regrowth forests, such as the Yucatan Peninsula.

Attribution of land-cover changes by disturbance type

Natural and human-induced disturbances modify the stand age and succession dynamics of the landscape in a unique way, changing the live biomass and dead organic matter stocks and subsequent turnover [65, 66]. Results from this study showed the relevance of identifying the cause of change in the landscape. The cumulative difference between scenarios with and without attribution of disturbance types was over 5 million Mg C for a single Landsat scene over the 9-year-period. INEGI estimates over the second period (2007–2011) from the

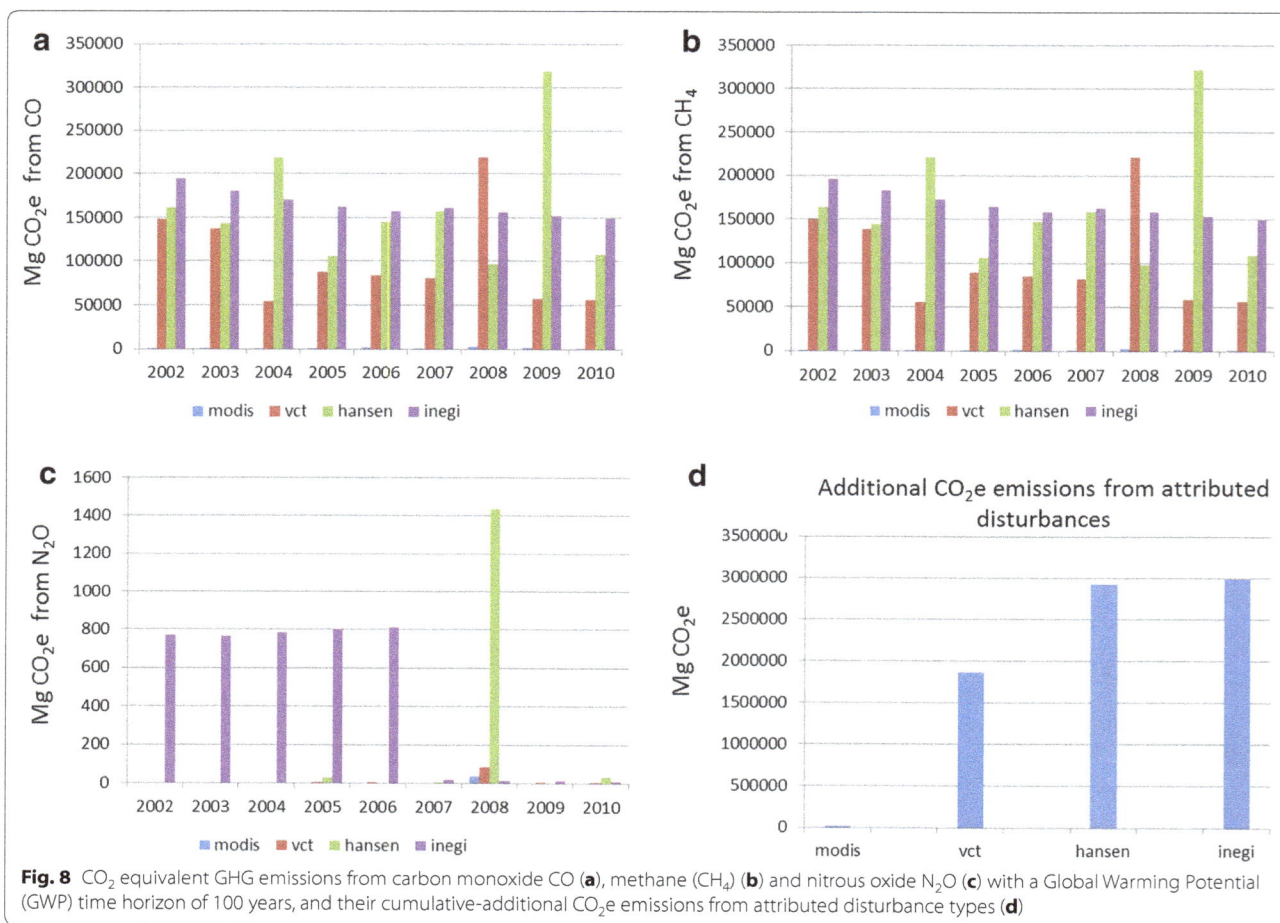

Fig. 8 CO_2 equivalent GHG emissions from carbon monoxide CO (**a**), methane (CH_4) (**b**) and nitrous oxide N_2O (**c**) with a Global Warming Potential (GWP) time horizon of 100 years, and their cumulative-additional CO_2e emissions from attributed disturbance types (**d**)

non-attributed disturbances, for example, dropped 84 % more than the attributed ones yielding higher emissions (3.2 million Mg C more). Since the default disturbance type associated with harvesting was clear-cut with slash-burn, it generated more emissions because it involved burning. Depending on the disturbance type, the CBM-CFS3 disturbance matrices determines how much biomass gets killed, re-distributing the carbon in the litter and dead organic matter pools accordingly; affecting the carbon changes more adequately. These differences between with and without attribution were even more pronounced when reported as CO_2 equivalent emissions (as required in REDD+ projects, see Fig. 8a–c), because fires cause additional non-CO_2 GHG emissions (in the form of CH_4 and N_2O), with higher global warming potentials than CO_2 (see Fig. 8d).

The carbon balance estimates obtained for this study are based on CBM-CFS3 simulations using preliminary inventory, age distribution and yield data for the study area [59]. Ongoing research as part of bigger collaboration of members from CONAFOR´s project "Reinforcing REDD+ and South–South Cooperation", SilvaCarbon

program, Colegio de Postgraduados, and research supported by the CEC will refine and improve the CBM-CFS3 input data. The absolute values of the numerical estimates obtained in this study may thus change in the future, but the general conclusions about the impacts on carbon balance estimates of the spatial and temporal resolution of RS products, and the attribution of land-cover change to causes of disturbance will not be affected by future changes to other input data of the model.

Conclusions

Systematic forest monitoring, reporting and verification (MRV) systems are required to aid in the successful development of national and regional strategies for REDD+, and to ensure long-term commitments to preserve forests [25, 40, 67]. This can only be achieved by implementing RS observations that will allow monitoring of large areas of land in a regular, consistent, and cost-efficient way and by developing modelling tools that can translate activity data derived from RS products into policy-relevant estimates of GHG emissions and removals. Results from this study provide an improved understanding of the role of RS disturbance observations on ecosystem carbon dynamics and the range of variability of carbon flux estimates following disturbance. Developing climate change mitigation scenarios, priorities, and initiatives requires further knowledge of the drivers of deforestation and forest degradation that can have a significant impact in the reduction of GHG emissions [7, 25, 68].

We conclude that spatial scale represents a serious mapping constraint when attempting to encompass large forest areas for ecosystem carbon accounting. Therefore, carbon monitoring decisions should consider direct trade-offs between the spatial detail of finer resolution products (e.g., Landsat, LiDAR, RapidEye) with more precision and less area covered per scene, versus moderate and broad spatial resolution observations derived at larger scales (e.g., MODIS). Our results support findings from the Global Forest Observations Initiative (GFOI) regarding MODIS imagery as "too large to be used for generating REDD+ activity data", suggesting it could be used in complimentary applications (e.g., monitoring near-real time forest change indicators) [69].

This research documents cumulative differences from four different RS products in estimates of carbon emissions of 30.7 million Mg C (112.5 million Mg CO_2e) over a 9-year period in a single Landsat scene in Mexico. While we did not attempt to extrapolate this to the national-scale implications, it is evident that with 135 Landsat scenes required to cover all of Mexico [48], the choice of satellite data sources to derive activity data can have a large impact on uncertainties in national estimates of forest carbon budgets. The magnitude of these differences highlights that efforts to improve the activity data can yield substantial reductions in uncertainty of GHG estimates at the regional or national scale. Uncertainties arising from activity data can be reduced by, in order of priority, increasing spatial resolution from 250 to 30 m, obtaining annual observations of forest disturbances, and by attributing land-cover changes by disturbance type. Results from this study can help justify the monitoring expenses required to reduce uncertainties in GHG emissions and removals estimates for countries like Mexico, with a variety of small scale-low impact disturbances and a strong interest in participating in REDD+ initiatives. Improved estimates also help meet the requirements of the Intergovernmental Panel on Climate Change (IPCC) to identify, quantify and reduce sources of uncertainty in estimates of GHG emissions and removals as far as is practicable.

Abbreviations
CEC: Commission for Environmental Cooperation; CBM-CFS3: carbon budget model of the Canadian Forest Sector; CONABIO: National Commission for Knowledge and Use of Biodiversity of Mexico; GHG: greenhouse gas; INEGI: National Institute of Statistics and Geography of Mexico; LiDAR: light detection and ranging; MODIS: moderate resolution imaging spectroradiometer; NFI: Mexican National Forest Inventory; REDD+: reducing emissions from deforestation and forest degradation; RS: remote sensing; VCT: vegetation change tracker algorithm; YP: Yucatan Peninsula.

Authors' contributions
All the authors have made substantial contribution towards the successful completion of this manuscript. They all have been involved in designing the study, drafting the manuscript and engaging in critical discussion. VSM, contributed with the processing, data analysis and write up. NCC and WAK contributed to the interpretation, quality control and revisions of the manuscript. MO contributed with the analysis of forest status, age distribution and yield data. All authors read and approved the final manuscript.

Author details
[1] Faculty of Forestry, Department of Forest Resources Management, University of British Columbia, 2424 Main Mall, Vancouver, BC V6T 1Z4, Canada. [2] Canadian Forest Service, Natural Resources Canada, 506 West Burnside Road, Victoria, BC V8Z 1M5, Canada. [3] Proyecto Mexico-Noruega, Comision Nacional Forestal, Coyoacán, D.F. Col. del Carmen Coyoacán, CP 04100 Mexico, Mexico.

Acknowledgements
This research was undertaken as part of the "Integrated Modeling and Assessment of North American Forest Carbon Dynamics and Climate Change Mitigation Options" project funded by the Commission for Environmental Cooperation (Montreal, Canada) representing Canada, Mexico and the USA. Additional funding was provided by the Secretary of Public Education of Mexico to Mascorro and the Natural Sciences and Engineering Research Council of Canada to Coops. We thank Shannon Franks and Jeff Masek from NASA for the provision of the VCT land cover time series and Rene Colditz from CONABIO for providing the MODIS maps. We thank Scott Morken and Max Fellows of the Canadian Forest Service for software development, training and support. We also thank Craig Wayson and Gregorio Ángeles for their help with developing the growth and age information that was used to run the model. Statements in this paper are by the authors and do not necessarily represent the positions of the sponsors or their governments.

Competing interests
The authors declare that they have no competing interests.

References
1. Bosworth D, Birdsey R, Joyce L, Millar C. Climate change and the nation's forests: challenges and opportunities. J For. 2008;106:214–21.
2. Turner M. Disturbance and landscape dynamics in a changing world. Ecology. 2010;91:2833–49.
3. Chapin FS III, McFarland J, David McGuire A, Euskirchen ES, Ruess RW, Kielland K. The changing global carbon cycle: linking plant-soil carbon dynamics to global consequences. J Ecol. 2009;97:840–50.
4. Pan Y, Birdsey RA, Fang J, Houghton R, Kauppi PE, Kurz WA, Phillips OL, Shvidenko A, Lewis SL, Canadell JG, Ciais P, Jackson RB, Pacala SW, McGuire AD, Piao S, Rautiainen A, Sitch S, Hayes D. A large and persistent carbon sink in the world's forests. Science. 2011;333:988–93.
5. Spalding D. The role of forests in global carbon budgeting. Forests an. In: Tyrrell ML, Ashton MS, Spalding D, Gentry B, Editors. Yale School of Forestry & Environmental Studies; 2009. p 223–53.
6. Hansen MC, Stehman SV, Potapov PV. Quantification of global gross forest cover loss. Proc Natl Acad Sci USA. 2010;107:8650–5.
7. Houghton RA, House JI, Pongratz J, van der Werf GR, DeFries RS, Hansen MC, Le Quéré C, Ramankutty N. Carbon emissions from land use and land-cover change. Biogeosciences. 2012;9:5125–42.
8. Le Quéré C, Moriarty R, Andrew RM, Peters GP, Ciais P, Friedlingstein P, Jones SD, Sitch S, Tans P, Arneth A, Boden TA, Bopp L, Bozec Y, Canadell JG, Chini LP, Chevallier F, Cosca CE, Harris I, Hoppema M, Houghton RA, House JI, Jain AK, Johannessen T, Kato E, Keeling RF, Kitidis V, Klein

Goldewijk K, Koven C, Landa CS, Landschützer P, et al. Global carbon budget 2014. Earth Syst Sci Data. 2015;7:47–85.

9. UNFCCC. Outcome of the work of the ad hoc working group on long-term cooperative action under the convention—policy approaches and positive incentives on issues relating to reducing emissions from deforestation and forest degradation in developing countries. United Nations Framework Convention on Climate Change COP 16; 2010.

10. Kurz WA, Dymond CC, White TM, Stinson G, Shaw CH, Rampley GJ, Smyth C, Simpson BN, Neilson ET, Trofymow JA, Metsaranta J, Apps MJ. CBM-CFS3: a model of carbon-dynamics in forestry and land-use change implementing IPCC standards. Ecol Modell. 2009;220:480–504.

11. Birdsey R, Pan Y, Houghton R. Sustainable landscapes in a world of change: tropical forests, land use and implementation of REDD+: Part II. Carbon Manag. 2013;4:567–9.

12. Pickell P, Gergel S, Coops N, Andison D. Monitoring forest change in landscapes under-going rapid energy development: challenges and new perspectives. Land. 2014;3:617–38.

13. Lorenz K, Lal R. Effects of disturbance, succession and management on carbon sequestration. In: Carbon sequestration in forest ecosystems. Dordrecht: Springer Netherlands; 2010. p. 103–57.

14. Kurz WA. An ecosystem context for global gross forest cover loss estimates. Proc Natl Acad Sci USA. 2010;107:9025–6.

15. Vargas R, Allen MF, Allen EB. Biomass and carbon accumulation in a fire chronosequence of a seasonally dry tropical forest. Glob Chang Biol. 2007;14:109–24.

16. Jaramillo VJ, Kauffman JB, Renteria-Rodriguez L, Cummings DL, Ellingson LJ. Biomass, carbon, and nitrogen pools in Mexican tropical dry forest landscapes. Ecosystems. 2003;6:609–29.

17. Orihuela-Belmonte DE, de Jong BHJ, Mendoza-Vega J, Van der Wal J, Paz-Pellat F, Soto-Pinto L, Flamenco-Sandoval A. Carbon stocks and accumulation rates in tropical secondary forests at the scale of community, landscape and forest type. Agric Ecosyst Environ. 2013;171:72–84.

18. Dai Z, Birdsey RA, Johnson KD, Dupuy JM, Hernandez-Stefanoni JL, Richardson K. Modeling carbon stocks in a secondary tropical dry forest in the Yucatan Peninsula, Mexico. Water Air Soil Pollut. 1925;2014:225.

19. De Fries RS, Houghton RA, Hansen MC, Field CB, Skole D, Townshend J. Carbon emissions from tropical deforestation and regrowth based on satellite observations for the 1980s and 1990s. Proc Natl Acad Sci USA. 2002;99:14256–61.

20. Asner GP, Powell GVN, Mascaro J, Knapp DE, Clark JK, Jacobson J, Kennedy-Bowdoin T, Balaji A, Paez-Acosta G, Victoria E, Secada L, Valqui M, Hughes RF. High-resolution forest carbon stocks and emissions in the Amazon. Proc Natl Acad Sci USA. 2010;107:16738–42.

21. Masek JG, Collatz GJ. Estimating forest carbon fluxes in a disturbed southeastern landscape: Integration of remote sensing, forest inventory, and biogeochemical modeling. J Geophys Res. 2006;111:G01006.

22. Stinson G, Kurz WA, Smyth CE, Neilson ET, Dymond CC, Metsaranta JM, Boisvenue C, Rampley GJ, Li Q, White TM, Blain D. An inventory-based analysis of Canada's managed forest carbon dynamics, 1990 to 2008. Glob Chang Biol. 2011;17:2227–44.

23. Potter C. Terrestrial ecosystem carbon fluxes predicted from MODIS satellite data and large-scale disturbance modeling. Int J Geosci. 2012;03:469–79.

24. Espírito-Santo FDB, Gloor M, Keller M, Malhi Y, Saatchi S, Nelson B, Junior RCO, Pereira C, Lloyd J, Frolking S, Palace M, Shimabukuro YE, Duarte V, Mendoza AM, López-González G, Baker TR, Feldpausch TR, Brienen RJW, Asner GP, Boyd DS, Phillips OL. Size and frequency of natural forest disturbances and the Amazon forest carbon balance. Nat Commun. 2014;5:3434.

25. Birdsey R, Angeles-Perez G, Kurz WA, Lister A, Olguin M, Pan Y, Wayson C, Wilson B, Johnson K. Approaches to monitoring changes in carbon stocks for REDD+. 2013:519–37.

26. Kurz WA, Apps MJ. Developing Canada's national forest carbon monitoring, accounting and reporting system to meet the reporting requirements of the Kyoto Protocol. Mitig Adapt Strateg Glob Chang. 2006;11:33–43.

27. Kull S, Kurz WA, Rampley GJ, Banfield G, Schivatcheva R, Apps M. Operational-Scale Carbon Budget Model of the Canadian Forest Sector (CBM-CFS3): Version 1.0, User's Guide. Edmonton, Alberta; 2011.

28. Coops NC, Wulder MA, White JC: Identifying and describing forest disturbance and spatial pattern: data selection issues and methodological implications; 2006. p 31–62.

29. Wulder MA, White JC, Gillis MD, Walsworth N, Hansen MC, Potapov P. Multiscale satellite and spatial information and analysis framework in support of a large-area forest monitoring and inventory update. Environ Monit Assess. 2010;170:417–33.

30. Hermosilla T, Wulder MA, White JC, Coops NC, Hobart GW. Remote sensing of environment an integrated landsat time series protocol for change detection and generation of annual gap-free surface reflectance composites. Remote Sens Environ. 2015;158:220–34.

31. Saatchi SS, Harris NL, Brown S, Lefsky M, Mitchard ETA, Salas W, Zutta BR, Buermann W, Lewis SL, Hagen S, Petrova S, White L, Silman M, Morel A. Benchmark map of forest carbon stocks in tropical regions across three continents. Proc Natl Acad Sci USA. 2011;108:9899–904.

32. Baccini A, Goetz SJ, Walker WS, Laporte NT, Sun M, Sulla-Menashe D, Hackler J, Beck PSA, Dubayah R, Friedl MA, Samanta S, Houghton RA. Estimated carbon dioxide emissions from tropical deforestation improved by carbon-density maps. Nat Clim Chang. 2012;2:182–5.

33. Achard F, Beuchle R, Mayaux P, Stibig H-J, Bodart C, Brink A, Carboni S, Desclée B, Donnay F, Eva HD, Lupi A, Raši R, Seliger R, Simonetti D. Determination of tropical deforestation rates and related carbon losses from 1990 to 2010. Glob Chang Biol. 2014;20:2540–54.

34. Chapin FS, Woodwell GM, Randerson JT, Rastetter EB, Lovett GM, Baldocchi DD, Clark DA, Harmon ME, Schimel DS, Valentini R, Wirth C, Aber JD, Cole JJ, Goulden ML, Harden JW, Heimann M, Howarth RW, Matson PA, McGuire AD, Melillo JM, Mooney HA, Neff JC, Houghton RA, Pace ML, Ryan MG, Running SW, Sala OE, Schlesinger WH, Schulze E-D. Reconciling carbon-cycle concepts, terminology, and methods. Ecosystems. 2006;9:1041–50.

35. Hansen MC, Potapov PV, Moore R, Hancher M, Turubanova SA, Tyukavina A, Thau D, Stehman SV, Goetz SJ, Loveland TR, Kommareddy A, Egorov A, Chini L, Justice CO, Townshend JRG. High-resolution global maps of 21st-century forest cover change. Science. 2013;342:850–3.

36. Mitchard ET, Feldpausch TR, Brienen RJW, Lopez-Gonzalez G, Monteagudo A, Baker TR, Lewis SL, Lloyd J, Quesada CA, Gloor M, ter Steege H, Meir P, Alvarez E, Araujo-Murakami A, Aragão LEOC, Arroyo L, Aymard G, Banki O, Bonal D, Brown S, Brown FI, Cerón CE, Chama Moscoso V, Chave J, Comiskey JA, Cornejo F, Corrales Medina M, Da Costa L, Costa FRC, Difiore A, et al. Markedly divergent estimates of Amazon forest carbon density from ground plots and satellites. Glob Ecol Biogeogr. 2014;23:935–46.

37. DeFries R, Achard F, Brown S, Herold M, Murdiyarso D, Schlamadinger B, de Souza C. Earth observations for estimating greenhouse gas emissions from deforestation in developing countries. Environ Sci Policy. 2007;10:385–94.

38. De Sy V, Herold M, Achard F, Asner GP, Held A, Kellndorfer J, Verbesselt J. Synergies of multiple remote sensing data sources for REDD+ monitoring. Curr Opin Environ Sustain. 2012;4:696–706.

39. GOFC-GOLD. A sourcebook of methods and procedures for monitoring and reporting anthropogenic greenhouse gas emissions and removals caused by deforestation, gains and losses of carbon stocks in forests remaining forests, and forestation; 2010.

40. Hewson J, Steininger M, Pesmajoglou S: REDD+ measurement, reporting and verification (MRV) manual. USAID-supported Forest Carbon, Markets and Communities Program; 2013.

41. Read L, Lawrence D. Recovery of biomass following shifting cultivation in dry tropical forests of the Yucatan. Ecol Appl. 2003;13:85–97.

42. Vandecar KL, Lawrence D, Richards D, Schneider L, Rogan J, Schmook B, Wilbur H. High mortality for rare species following hurricane disturbance in the southern Yucatan. 2011; 43:676–84.

43. Urquiza-Haas T, Dolman PM, Peres CA. Regional scale variation in forest structure and biomass in the Yucatan Peninsula, Mexico: effects of forest disturbance. For Ecol Manage. 2007;247:80–90.

44. Cohen WB, Yang Z, Kennedy R. Detecting trends in forest disturbance and recovery using yearly Landsat time series: 2. TimeSync—Tools for calibration and validation. Remote Sens Environ. 2010;114:2911–24.

45. Powers RP, Hermosilla T, Coops NC, Chen G. Remote sensing and object-based techniques for mapping fine-scale industrial disturbances. Int J Appl Earth Obs Geoinf. 2015;34:51–7.

46. Healey S, Cohen W, Zhiqiang Y, Krankina O. Comparison of Tasseled Cap-based Landsat data structures for use in forest disturbance detection. Remote Sens Environ. 2005;97:301–10.

47. Huang C, Goward SN, Masek JG, Thomas N, Zhu Z, Vogelmann JE. An automated approach for reconstructing recent forest disturbance history using dense Landsat time series stacks. Remote Sens Environ. 2010;114:183–98.

48. Gebhardt S, Wehrmann T, Ruiz M, Maeda P, Bishop J, Schramm M, Kopeinig R, Cartus O, Kellndorfer J, Ressl R, Santos L, Schmidt M. MAD-MEX: automatic wall-to-wall land cover monitoring for the Mexican REDD-MRV program using all Landsat data. Remote Sens. 2014;6:3923–43.

49. INEGI. Conjunto de Datos Vectoriales de La Carta de Uso Del Suelo Y Vegetación, Escala 1:250,000, Serie III (CONTINUO NACIONAL). Aguascalientes, Mexico: Instituto Nacional de Estadística y Geografía; 2003.

50. INEGI. Conjunto de Datos Vectoriales de La Carta de Uso Del Suelo Y Vegetación, Escala 1:250,000, Serie IV (CONTINUO NACIONAL). Aguascalientes, Mexico: Instituto Nacional de Estadística y Geografía; 2007.

51. INEGI. Conjunto de Datos Vectoriales de La Carta de Uso Del Suelo Y Vegetación, Escala 1:250,000, Serie V (CONTINUO NACIONAL). Aguascalientes, Mexico: Instituto Nacional de Estadística y Geografía; 2011.

52. Latifovic R, Homer C, Ressl R, Pouliot D, Hossain SN, Colditz RR, Giri OC, Victoria A. North american land change monitoring system; 2010.

53. Colditz RR, López Saldaña G, Maeda P, Espinoza JA, Tovar CM, Hernández AV, Benítez CZ, Cruz López I, Ressl R. Generation and analysis of the 2005 land cover map for Mexico using 250 m MODIS data. Remote Sens Environ. 2012;123:541–52.

54. INFyS. Inventario Nacional Forestal Y de Suelos Informe 2004–2009 (National Forest and Soils Inventory Report). Zapopan, Jalisco: Comision Nacional Forestal de Mexico; 2012.

55. Mascorro VS, Coops NC, Kurz WA, Olguín M. Attributing changes in land cover using independent disturbance datasets: a case study of the Yucatan Peninsula, Mexico. Reg Environ Chang. 2014.

56. Congalton RG. A review of assessing the accuracy of classifications of remotely sensed data. 1991;46(October 1990):35–46.

57. Olofsson P, Foody GM, Herold M, Stehman SV, Woodcock CE, Wulder MA. Good practices for estimating area and assessing accuracy of land change. Remote Sens Environ. 2014;148:42–57.

58. INEGI. Conjunto de Datos Vectoriales de Carreteras y Vialidades Urbanas Edición 1.0. 2013.

59. Olguin M, Wayson C, Kurz. W, Fellows M, Fellows, Ángeles G, Maldonado V, Carrillo O, López D. Input Data Improvements for Version 2 of State-Level Carbon Dynamics Runs Using the CBM-CFS3 Model in Mexico. Mexico City: Mexico-Norway Project of the National Forestry Commission of Mexico; 2014.

60. CEC. Ecological Regions of North America: Towards a Common Perspective. Montreal, Quebec: Commission for Environmental Cooperation; 1997.

61. Dal XL, Khorram S. Remotely sensed change detection based on artificial neural networks. Photogramm Eng Remote Sens. 1999;65:1187–94.

62. Fuller R, Smith G, Devereux B. The characterisation and measurement of land cover change through remote sensing: problems in operational applications? Int J Appl Earth Obs Geoinf. 2003;4:243–53.

63. Olofsson P, Foody GM, Herold M, Stehman SV, Woodcock CE, Wulder MA. Remote sensing of environment good practices for estimating area and assessing accuracy of land change. Remote Sens Environ. 2014;148:42–57.

64. Lu D, Mausel P, Brondízio E, Moran E. Change detection techniques. Int J Remote Sens. 2004;25:2365–401.

65. Lorenz K, Lal R. The Natural Dynamic of Carbon in Forest Ecosystems. 2010(C).

66. Franklin J, Mitchell R, Palik B. Natural disturbance and stand development principles for ecological forestry. USDA Forest Service 2007. p 44.

67. Herold M, Skutsch M. Monitoring, reporting and verification for national REDD+ programmes: two proposals. Environ Res Lett. 2011;6:014002.

68. Hosonuma N, Herold M, De Sy V, De Fries RS, Brockhaus M, Verchot L, Angelsen A, Romijn E. An assessment of deforestation and forest degradation drivers in developing countries. Environ Res Lett. 2012;7:044009.

69. GFOI. Integrating remote-sensing and ground-based observations for estimation of emissions and removals of greenhouse gases in forests. Geneva: Group on Earth Observations; 2014.

EU mitigation potential of harvested wood products

Roberto Pilli[*], Giulia Fiorese and Giacomo Grassi

Abstract

Background: The new rules for the Land Use, Land Use Change and Forestry sector under the Kyoto Protocol recognized the importance of Harvested Wood Products (HWP) in climate change mitigation. We used the Tier 2 method proposed in the 2013 IPCC KP Supplement to estimate emissions and removals from HWP from 1990 to 2030 in EU-28 countries with three future harvest scenarios (constant historical average, and +/−20% in 2030).

Results: For the historical period (2000–2012) our results are consistent with other studies, indicating a HWP sink equal on average to −44.0 Mt CO_2 yr^{-1} (about 10% of the sink by forest pools). Assuming a constant historical harvest scenario and future distribution of the total harvest among each commodity, the HWP sink decreases to −22.9 Mt CO_2 yr^{-1} in 2030. The increasing and decreasing harvest scenarios produced a HWP sink of −43.2 and −9.0 Mt CO_2 yr^{-1} in 2030, respectively. Other factors may play an important role on HWP sink, including: (i) the relative share of different wood products, and (ii) the combined effect of production, import and export on the domestic production of each commodity.

Conclusions: Maintaining a constant historical harvest, the HWP sink will slowly tend to saturate, i.e. to approach zero in the long term. The current HWP sink will be maintained only by further increasing the current harvest; however, this will tend to reduce the current sink in forest biomass, at least in the short term. Overall, our results suggest that: (i) there is limited potential for additional HWP sink in the EU; (ii) the HWP mitigation potential should be analyzed in conjunction with other mitigation components (e.g. sink in forest biomass, energy and material substitution by wood).

Keywords: Harvested wood products (HWP); Carbon; FAOSTAT; LULUCF

Background

Forests and the forest sector play a relevant role in the carbon (C) cycle and can significantly contribute to the mitigation of global climate change [1,2]. Specifically, forest-related mitigation options include a change in C stocks - which reflects emissions or removals of CO_2 from the atmosphere - and a substitution effect. Changes in C stocks can happen both within the forest pools (living biomass, dead wood, litter and soil) and in the harvested wood products (HWP) pool [3]. Substitution effects can occur when wood products replace materials or energy (e.g., concrete or fossil fuels) [4,5]. Given the trade-offs between the forest sink at large scale, wood products and bioenergy (see for example [6-9]), the most effective forest mitigation strategy is the one that maximizes the sum of various mitigation components.

Accounting approaches for HWP

The role of HWP in mitigating GHG emissions has been recognized only recently by the Kyoto Protocol (KP). For the first KP commitment period (2008–2012) it was assumed that the annual amount of C leaving the HWP pool equals the annual C inflow to the pool. This means that all C in the harvested biomass is oxidized at the time of harvest. In reality, wood-based materials may emit C over a long time frame. Depending on the balance between C inflow and outflow, and the corresponding C stock change, the HWP pool may indeed act as a sink or as a source of CO_2. For this reason, for the second KP commitment period (2013–2020) accounting rules have been changed to include explicitly C stock changes in the HWP pool [10].

* Correspondence: roberto.pilli@jrc.ec.europa.eu
European Commission, Joint Research Centre, Institute for Environment and Sustainability, Via E. Fermi 2749, I-21027 Ispra, VA, Italy

Carbon stock changes in HWP depend on several factors such as the amount of harvest, the final products and their end use, the service life of products, and the disposal/recycling or use as fuel at the end of service life [11]. Different approaches exist to account for C stock changes in the HWP pool [3,12,13]. In particular, the IPCC production accounting approach [3] has been applied to many individual countries or regions such as Northern US [14] and Ireland [13]. More recently, the 2013 Revised Supplementary Methods and Good Practice Guidance Arising from the Kyoto Protocol (2013 IPCC KP Supplement in the following) defined the methods, named Tiers 1 to 3, to be used under the KP according to the level of detail and of accuracy of the available data [15]. The principles behind this new method are the same as applied in Rüter [16] to estimate the current and future HWP emissions/removals in EU countries.

Future mitigation potential of forests

At the EU level (i.e. 28 Member States) forests cover about 165 Mha [17]. The EU forest area increased by 5% compared to 1990 and now equals about 37% of total EU area. Most of the EU forest area (83%) is available for wood supply [18] and EU is one of the main world producer of roundwood, with about 405 million m^3 in 2010. Nevertheless, on average only 64% of the EU forest annual increment is removed from the growing stock by fellings[a]. As a consequence, forests in the EU are a major carbon sink: between 1990 and 2012 the average annual forest sink was about 435 million tons of CO_2 [17]. This corresponds to about 9% of 2010 EU emissions in the same period. While this forest sink has been approximately stable in the past two decades, possible first signs of saturation have been suggested [19], also based on a reported decline in stem volume increment possibly related to forest aging.

In addition, several analyses suggest a significant increase in harvest removals at EU level for the next few decades, mainly due to increasing wood demand for renewable energy production. The EU Reference Scenario 2013 [20] expects an increase of harvest by 17% in 2030 compared to 2005, associated to a decline by about 30% of the forest sink. In the same Reference Scenario, based on the 2010 statistics from EUROSTAT and including the effect of the on-going economic downturn, it is expected that in 2030 forest wood used for energy will increase by 41% while forest used for wood products will increase only by 13%, with respect to 2005 [20]. Mantau et al. [21], for the EU (without Croatia), estimated an increase of the total demand for wood biomass (including biomass power plantations, pellets, etc.) from almost 800 million m^3 in 2010 to nearly 1400 million m^3 and 1200 million m^3 under the A1 and B2 IPCC 2000 scenario, respectively. Considering the B2 IPCC 2000 scenario (i.e. assuming a modest GDP growth rates in Europe), EFI-GTM models

projects from 2010 to 2030 a 15% increase of total consumption for wood products in UNECE countries in Europe, including Russia [5]. This trend appears mainly driven by the consumption of wood fuel, which is expected to increase by 35% from 2010 to 2030. Similar trends are suggested also by other studies, e.g. the EU blueprint for forest-based industries [22] and Böttcher et al. [7]. Overall, the expected increase in harvest rate at EU level will heavily influence both the forest sink and the carbon stock changes in the HWP pool.

Objectives of this study

In order to optimize the overall forest mitigation potential, tools are needed to estimate the specific mitigation potentials of forest management (e.g., [23]), energy uses (e.g., [24]) and wood products. The aim of this paper is to describe a tool for estimating the present and future C stock changes in the HWP pool at EU level, as part of a comprehensive modelling framework for the forest sector [25]. Specifically, following the methods in the 2013 IPCC KP Supplement [15], we estimated HWP emissions/removals for EU countries (with the exception of Malta and Cyprus) (i) for the historical period (from 1990 to 2012) and (ii) until 2030. In this second case we assessed the impact of different harvest amounts on the HWP mitigation potential. Three different scenarios for future total harvest (constant historical average, and +/−20% in 2030) were analyzed. Furthermore, for the constant harvest scenario, the impact of three different distributions of future harvest between each commodity was explored.

Accounting method and activity data in this study

According to the 2013 IPCC KP Supplement [15], for the second commitment period of the KP, countries have to account the C stock change on the HWP pool from domestic harvest (i.e., the trees harvested in the reporting countries) following one of these methods: (i) instantaneous oxidation (Tier 1); (ii) a default method proposed in the same supplement (Tier 2); (iii) country-specific methods (Tier 3).

The first approach ignores the changes of the HWP stock, with the consequent assumption that all wood is instantaneously burned. This method has been used in the first commitment period of KP. The Tier 2 method applies first order decay functions based on default half-lives numbers distinguished between the main semi-finished wood products (i.e. sawn wood, wood panels and paper) and defined by the international classification system of forestry products. All the countries which proposed a "reference level" for forest management in the second commitment period of the KP (i.e., all the EU countries) have to use at least this approach. If more detailed data and methodologies are available, a country-specific method can be used (Tier 3 approach).

In order to be consistent across all EU countries, the Tier 2 method is used in the following analysis. From the FAOSTAT database[b] [26] we collected consistent, transparent and verifiable activity data on HWP (production, import and export) for each country. The activity data required for the Tier 2 method are:

- Roundwood removals[c]
- Industrial Roundwood (IRW) i.e., the portion of roundwood removals used for the production of wood commodities
- Sawn wood (SW), wood panels (WP) and paper and paperboard (PP), i.e. the three semi-finished wood products categories.

These categories can also be distinguished between coniferous and not-coniferous (i.e., broadleaves). The historical FAOSTAT data are complete from 1961 for 17 out of 28 countries and largely missing for 2 countries (Malta and Cyprus). For some countries, such as Luxembourg or Eastern EU countries, only data for the last 10 – 20 years are available (Table 1).

We compared FAOSTAT data with other available data sources, including: (i) the 2010 Forest Resource Assessment country's reports [27]; (ii) the 2013 National Inventory Reports [28] submitted by each country to the United Nations Framework Convention on Climate Change (UNFCCC); (iii) the last National Forest Inventories (NFIs, when public data were available) and (iv) the countries' submissions for Forest Management Reference Level [29]. These last documents, submitted to UNFCCC in 2011, generally provided additional information on the historical (i.e., until 2008) and future harvest rate, used to assess the FMRL for the second commitment period of the KP. All data were preliminary harmonized, taking into account over and under bark correction factors and other possible corrections due to the accounting of forest residues or under- overestimates reported by official statistics. A comparison of the different data sources and a summary of the corrections applied to original FAOSTAT data are reported in Table 1.

The 2013 IPCC KP Supplement Tier 2 method is basically a flux data method where estimates of net emissions are derived from a stock change calculation applied to products derived from domestic harvest, i.e., imported HWP are excluded. To implement this method, it is first necessary to estimate the annual fraction of the industrial roundwood (sawn wood and wood based panels) and wood pulp commodities coming from domestic harvest. According to IPCC, the C stock included in fuelwood is immediately released to the atmosphere. The main steps on this method are summarized in Figure 1.

We assume that all domestic harvest derives from forest management, thus we do not differentiate harvest coming from afforestation activities. Indeed, due to the low rate of afforestation and the young age of the afforested lands, the share of harvest potentially provided by this category is generally negligible [30], with few exceptions [31]. According to 2013 IPCC KP Supplement [15], the instantaneous oxidation method must be applied to harvest from deforestation. Due to the lack of reliable data, in this study we did not consider any harvest from deforestation and thus did not assess the related C emissions. Moreover, due to the relative small amount of area affected by deforestation in EU, equal on average to 109 kha yr^{-1} [32] for the period 2008 – 2012 over a total FM area equal to 140,030 kha (as considered in this study), we can assume that the total amount of harvest provided by deforestation is negligible compared with that of the forest management area. Finally, because this work is part of a more comprehensive modelling framework for the forest sector [25], we plan to include deforestation in future developments.

The share of domestically produced SW, WP and PP for the domestic production (DP) is computed (step 1 of Figure 1) considering the production, import and export of the feedstock commodities, IRW and Pulp. When the amount of DP in each commodity has been estimated (step 2 of Figure 1), it is then possible to calculate the associated flows of carbon (step 3 of Figure 1).

The C stock of each HWP category, in each year, is estimated by applying a first order decay function taking into account the C outflow and inflow from and to each category, as reported in the Methods (see the subsection on First Order decay functions).

To complete the assessment, we account also for the inherited emissions, i.e. the emissions that occur during the second commitment period from HWP removed from forests prior to the beginning of the second commitment period [15]. We estimate the accumulation of the historic inflow, starting from 1900, assuming that the C inflow until 1961 (or until the first available year, see Table 1) is constant and equal to the average of the first available five years (e.g., generally, 1961–1965), as in [16].

Finally (step 4, Figure 1), the total C stock and C stock change in HWP can be calculated for each country by summing up all the C stocks and C stock changes of all the commodities[d] (e.g., $SW_C + SW_{NC} + WP + PP$). This approach was applied to estimate for each EU country the historical C stock change in HWP until 2012, i.e., the last year reported by 2013 FAO statistics.

Future trends

To establish the future mitigation potential of HWP, we first need to estimate how the harvest demand will evolve. To this aim, we used three different harvest scenarios up to 2030:

Table 1 Activity data analysis: for 28 EU countries

Countries	A. FAOSTAT	B. Specific data sources in comparison with FAOSTAT				C. Possible explanations for differences between A and B	D. Corr. factors
	Available since	FRA CR	NIR	NFI	FMRL		
Austria	1961	=	↑	X	↑	Bark fraction	1.15
Belgium	2000	↓	↓		↓	Accounting methods	-
Bulgaria	1961	=			=		-
Croatia	1992	X			↑	Forest residues & bark	1.10
Cyprus	N. A.	X					-
Czech Rep	1993	=	X		↑	Forest residues & bark	1.10
Denmark	1961	=			X	Forest residues	-
Estonia	1992	X	=[1]		=[1]	Forest residues, bark & other	1.10
Finland	1961	↑	↑		↑	Forest residues, bark & other	1.10
France	1961	=	↓		=		-
Germany	1961	↑	=		↑	Forest residues, bark & other	1.44[2]
Greece	2007	↑			↑	Bark fraction	1.15
Hungary	1961	X	↑		↑	Bark fraction & forest residues	1.20[3]
Ireland	1961	X			↑	Bark fraction	1.10
Italy	1961	=		X	↑	Forest residues, bark & other	1.57[4]
Latvia	1992	X	↓		↓	Bark fraction	1.12
Lithuania	1992	X			↑	Bark fraction	1.12
Luxemb.	2000	=			↑		-
Malta	N. A.	X					
Netherlands	1961	X			↑	Bark fraction	1.15-1.18
Poland	1961	↑			=	Bark fraction	1.20
Portugal	1961	↑			=	Bark fraction	1.25-1.18
Romania	1961	=	↑		↑	General CF	1.23[5]
Slovakia	1993	X	=		=	Bark fraction	1.10-1.12
Slovenia	1993	X			↑	Bark fraction	1.17-1.13
Spain	1961	↑			↑		1.10
Sweden	1961	↑			↑	Bark fraction & forest residues	1.14
UK	1961	X			↑	Bark fraction	1.14-1.12

The table reports: A. The first year from which FAOSTAT data are available; B. the additional data sources considered by this study, including: the 2010 Forest Resource Assessment country's report (FRA CR), the 2013 National Inventory Reports (NIR), the last National Forest Inventory (NFI, when public available) and the Submission for Forest Management Reference Level (FMRL).
Symbols highlight if the amount of harvest reported by these specific data sources are, on average: equal (=), higher (↑), lower (↓) or not comparable (X, because of different time scales or other reasons) as compared to the FAOSTAT data. C. Possible differences between FAOSTAT and the other specific data sources. D. The correction factors applied to the original FAOSTAT data, mostly based on a correction for bark. The bark's correction factor (based on data from the literature, when available at country level) was applied when, comparing FAOSTAT data with other sources (mainly the 2010 FRA Country Report), we argued that the volume reported by original FAOSTAT data were under-bark.
[1] the NIR 2013 reports the same values reported by FAOSTAT since 2003.
[2] average general correction factor (accounting for bark and other corrections) applied to original FAOSTAT data from 2000 to 2012; the CF varied year by year, assuming that the figures reported by the Submission for FMRL represent the correct estimates (Joachim Rock, pers. com).
[3] bark's CF applied only to the industrial roundwood compartment.
[4] average general correction factor (accounting for bark, forest residues and other corrections, suggested by [28] and by [23].
[5] average general correction factor suggested by NIR 2013 [28].

1. A constant harvest scenario (CH) equal to the average historical harvest (2000–2012);
2. An increasing harvest scenario (CH+) assuming +20% with respect to the CH scenario in 2030 and a linear increase from 2013 to 2030;
3. A decreasing harvest scenario (CH-) assuming −20% with respect to the CH scenario in 2030 and a linear decrease from 2013 to 2030.

For all these scenarios, the future harvest demand (Figure 2) was defined on the basis of the historical

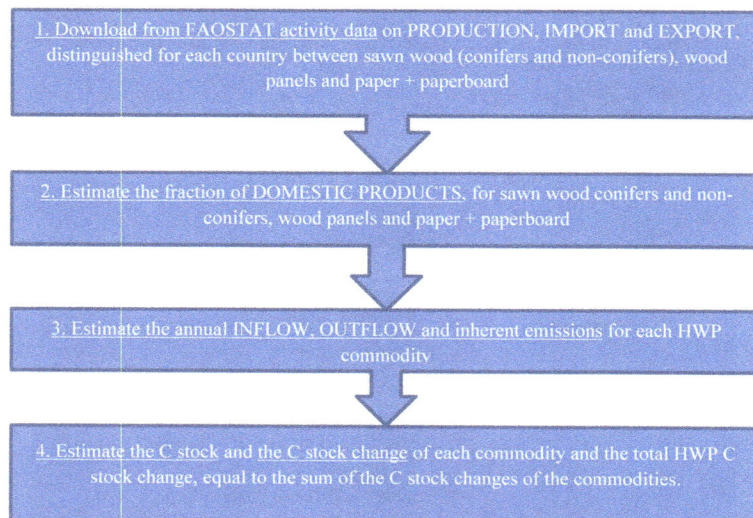

Figure 1 IPCC Tier 2 steps: main steps applied to the HWP pool to estimate the total C stock and C stock change according to the Tier 2 method [15].

amount of harvest reported by the FAOSTAT data, corrected according to the analysis described above (see Table 1). Of course, we are not considering the fact that harvest projections provided by countries are driven either by projected potential supply of forest biomass and projected demand of forest biomass for the subsequent production of wood products and wood fuel.

The share of domestic feedstock for the production of particular HWP category originating from domestic forests is not directly correlated to the total harvest because of the production processes. For example there

may be recycled products, recovered paper, slashes, wood-chips used in wood-based panel production, etc.

To apply the method described above to each harvest scenario, we first calculated the 2008–2012 average C inflow (e.g., the C inflow of the last five years) and then we applied a constant, increasing (+20%) and decreasing (−20%) variation rate to this average, according to each harvest scenario. This implicitly assumes to use the historical distribution of the total harvest between each commodity and to vary this distribution proportionally to the future harvest scenario. Because this is the same approach used in Rüter [16] and applied

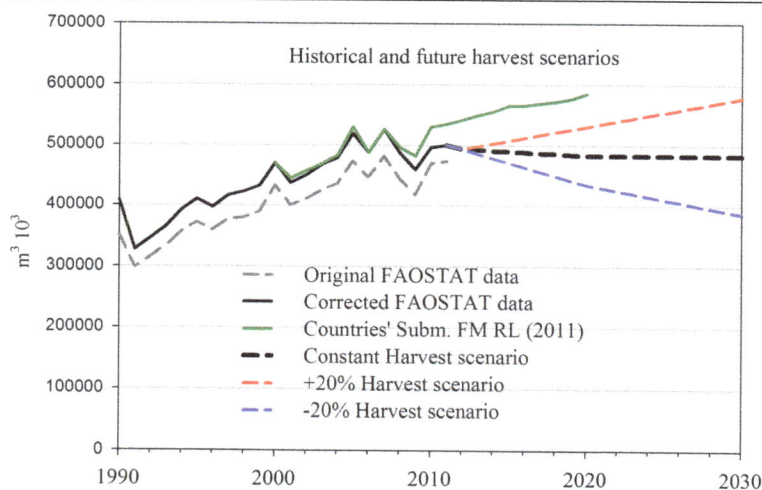

Figure 2 Total harvest demand: total harvest demand (in m³ 10³) for 28 EU countries, based on the historical FAO statistics (until 2012, Original FAOSTAT data) corrected to account for possible under-/over-estimates (Corrected FAOSTAT data used in this paper, see Table 1) and three future harvest scenarios: constant harvest, increasing harvest (+20%) and decreasing harvest (−20%) up to 2030. A comparison with the harvest provided by countries's Submission FMRL is also reported.

by the majority of EU countries for the submissions for FMRL, we considered this method as Approach 0 (AP0).

For the constant harvest scenario, we also explored two other possible approaches (AP1 and AP2). The aim of these two approaches is to explore, through statistical correlations among the variables requested by the default Tier 2 method, the impact of deviating from the AP0 assumption that the future distribution of the total harvest among each commodity is constant and equal to the historical distribution. All these approaches follow the IPCC 2013 Tier 2 method.

Approach 1 (AP1) starts at step 1 described in the Methods (Figure 1). With AP1 we estimate the future distribution of the total harvest between each commodity (i.e., SW, WP and PP) and the amount of IRW and pulp production, import and export (this is only aimed to calculate the annual fraction of domestic production).

Approach 2 (AP2) starts at step 2 described in the Methods (i.e., the estimate of the domestic production of the 4 commodities).

To this aim, for each country, we first looked for possible statistical correlations between the available information on the total harvest or the time (e.g., years) and the SW, WP, and PP production (for AP1) or domestic production (for AP2). The rationale is that the correlations that best describe the past data could then be used to estimate the future evolution of each variable, applying the default Tier 2 Method.

Results and discussions
Historical and future harvest rate
The total amount of harvest at EU level, reported by FAOSTAT and corrected to account for bark's fractions or other possible over- or under-estimations, is shown in Figure 2.

Based on FAOSTAT data, more than 60% of the total wood harvest (measured as the 2000–2012 average) is provided by only five countries: Sweden (17% of the total), Germany (14%), Finland (12%), France (12%) and Poland (8%). Eight countries (Austria, Czech Republic, Romania, Spain, Latvia, Portugal, Italy, UK) contribute to another 25%, each with 2-4% share of the total EU harvest. The remaining 12% of wood is harvested in 13 countries, each contributing on average less than 1%. From these figures, it emerges that the estimates and data regarding the first five countries are extremely relevant with respect to any analysis of the forest sector in general and of the HWP sector in particular. For example, the input data for Germany extracted from the FAOSTAT database (based on production statistics) is different from other data sources, such as the NIR (derived from forest inventory).

Between 2000 and 2011 (for some countries data prior to 2000 are missing, see Table 1), FAOSTAT original data show the same trend reported by the country's submissions for FMRL [29]. However, the total amount of harvest

reported by the submissions is on average 10% higher than FAOSTAT. In most cases, this difference is due to the bark fraction, which sometimes is not accounted for in FAOSTAT. Furthermore, other specific issues that vary country-by-country can explain the difference (Table 1). When FAOSTAT data are corrected according to additional information from FRA 2010 Country Reports and to other available data (i.e., from the last NIRs), then the difference with data from FMRL decreases to less than 2%. If we look specifically at 2010 and 2011, the difference between the two data sources is higher (about 7%). This difference is related to the fact that in FMRL submissions the years 2010 and 2011 were already part of the "projected" estimates, i.e. based on future assumptions and not on statistics.

The 2010 harvest rate based on FAOSTAT corrected data is about 72% of the potential forest woody biomass resource (686 million m^3) estimated by Mantau et al. [21] according to the IPCC 2000 scenario A1, assuming a medium mobilization scenario. Reducing this amount by forest residues (118 million m^3, not accounted by FAOSTAT), the resulting amount (568 million m^3) is about 14% higher than our FAOSTAT corrected data. This difference appears related to the fact that Mantau et al. [21] did not consider the effect of the last economic crisis.

The 2030 harvest rate applied in our study is equal to 482 million m^3, 386 million m^3 and 579 million m^3, assuming respectively a constant, decreasing and increasing harvest rate. This last value is similar (about 6% lower) to the estimates used by the G4M and GLOBIOM models in the EU reference scenario [20].

Approach 1 (AP1)
A synthesis of the results based on AP1 is presented in Table 2, for each country and commodity. Applying the AP1 to the historical FAOSTAT data– corrected according to our analysis - shows that, in most countries, the production of *SW*, *WP* and *PP* is statistically correlated with the total roundwood production and with time. This means that, for some countries, (i) there is a statistical correlation between time (i.e., years) and/or total roundwood production and the SW, WP and PP production and (ii) in most cases, we highlighted a temporal trend (increasing or decreasing) on the data series (i.e., a correlation with time) above all for the PP and WP production.

For 42% (for SW) and 30% (for WP and PP) of the countries, the production of each commodity was estimated using the total RW production as main driver (i.e., as independent variable applied to the linear model described in Materials) and it was therefore directly related to the future harvest demand. In other countries, however, these data were correlated with time (34% and 50% of the countries for WP and PP, respectively) suggesting the existence of an increasing

Table 2 Results from AP1: For each key variable the table reports the independent variable (i.e., total roundwood (RW) or time (t)) applied to Eq. 6 followed by the coefficient of determination R^2 of the linear regression model

Country	To calculate the C stock and C stock change — PRODUCTION				To calculate the share of domestically produced IRW and Pulp — PRODUCTION			IMPORTS			EXPORTS		
	SWc	SWnc	WP	PP	IRWc	IRWnc	Pulp	IRWc	IRWnc	Pulp	IRWc	IRWnc	Pulp
Austria	f(RW) 0.79	Avg	f(RW) 0.82	f(RW) 0.76	f(RW) 0.95	IRW-IRWc	f(t) 0.85	f(t) 0.87	f(t) 0.72	f(t) 0.92	f(t) 0.79	Avg	f(t) 0.95
Belgium	Avg	Avg	f(RW) 0.78	f(RW) 0.75	f(RW) 0.94	IRW-IRWc	Avg	Avg	Avg	Avg	f(t) 0.64	Avg	Avg
Bulgaria	f(RW) 0.78	Avg	Avg	f(RW) 0.71	f(RW) 0.94	IRW-IRWc	f(RW) 0.69	Avg	f(t) 0.65	f(t) 0.95	f(t) 0.68	f(t) 0.67	f(t) 0.76
Croatia	Avg	Avg	Avg	f(RW) 0.89	IRWnc-IRW	f(RW) 0.97	Avg	Avg	Avg	Avg	Avg	Avg	Avg
Cyprus													
Czech Rep	f(RW) 0.86	Avg	f(RW) 0.86	f(RW) 0.81	f(RW) 0.99	IRW-IRWc	f(RW) 0.95	Avg	Avg	Avg	Avg	Avg	Avg
Denmark	Avg	Avg	Avg	f(t) 0.91	IRWnc-IRW	f(t) -0.82	Avg	f(t) 0.70	Avg	f(t) 0.97	Avg	Avg	f(t) 0.92
Estonia	Avg	f(t) 0.72	Avg	f(t) 0.73	f(RW) 0.99	IRW-IRWc	f(t) 0.88	Avg	Avg	f(t) 0.93	Avg	Avg	f(t) 0.90
Finland	f(RW) 0.76	Avg	f(t) 0.88	f(t) 0.96	f(RW) 0.93	IRW-IRWc	f(t) 0.69	Avg.	f(t) 0.72	f(t) 0.86	f(t) 0.72	Avg	f(t) 0.95
France	f(RW) 0.85	Avg	f(t) 0.85	f(RW) 0.80	f(RW) 0.83	IRW-IRWc	f(t) 0.95	Avg	Avg	f(t) 0.95	Avg	Avg	f(t) 0.91
Germany	f(RW) 0.77	f(t) 0.75	f(t) 0.91	f(t) 0.97	f(RW) 0.99	IRW-IRWc	f(t) 0.84	Avg.	Avg	f(t) 0.95	Avg	f(t) 0.66	f(t) 0.91
Greece	Avg	Avg	f(t) 0.82	f(t) 0.91	Avg	IRW-IRWc	Avg	f(t) 0.66	Avg	f(t) 0.88	Avg	Avg	f(t) 0.75
Hungary	Avg	Avg	f(t) 0.85	Avg	Diff.	Avg	Avg	Avg	Avg	f(t) 0.9	Avg	Avg	f(t) 0.93
Ireland	f(RW) 0.96	f(t) 0.74	f(RW) 0.81	Avg	f(RW) 0.99	IRW-IRWc	Avg	Avg	Avg.	f(t) 0.92	f(t) 0.72	Avg	Avg
Italy	Avg	f(t) 0.69	f(t) 0.88	f(t) 0.94	IRWnc-IRW	f(t) -0.81	Avg	Avg	Avg.	f(t) 0.93	Avg.	Avg.	f(t) 0.92
Latvia	f(RW) 0.92	Avg	Avg	f(t) 0.91	f(RW) 0.94	IRW-IRWc	Avg	Avg	Avg	f(t) 0.94	Avg.	Avg.	f(t) 0.77
Lithuania	Avg	f(RW) 0.70	f(t) 0.88	f(t) 0.94	f(RW) 0.80	IRW-IRWc	Avg	Avg	f(t) 0.70	f(t) 0.95	Avg	f(t) 0.66	f(t) 0.95
Luxemb.	Avg	Avg	f(t) 0.72	f(t) 0.66	IRWnc-IRW	f(RW) 0.96	Avg	Avg	Avg	Avg	Avg	Avg	Avg
Malta													
Netherl.	f(t) 0.70	Avg	f(t) 0.84	f(t) 0.92	f(RW) 0.83	IRW-IRWc	f(t) 0.81	Avg	Avg	f(t) 0.98	Avg	Avg	f(t) 0.91
Poland	f(t) 0.74	f(RW) 0.66	f(RW) 0.95	f(RW) 0.96	f(RW) 0.97	IRW-IRWc	Avg	Avg	Avg	Avg	Avg	Avg	f(t) 0.69
Portugal	Avg	Avg	f(RW) 0.85	f(RW) 0.69	Avg	f(RW) 0.85	f(t) 0.90	Avg	Avg	f(t) 0.87	Avg	Avg	f(t) 0.84
Romania	Avg	f(RW) 0.77	Avg	Avg	f(RW) 0.87	IRW-IRWc	Avg	Avg	Avg	f(t) 0.96	Avg	Avg	Avg
Slovakia	f(RW) 0.87	f(RW) 0.87	f(RW) 0.78	f(t) 0.66	f(RW) 0.95	IRW-IRWc	f(t) 0.83	Avg	Avg	f(t) 0.97	Avg	f(t) 0.66	f(t) 0.91
Slovenia	Avg	Avg	Avg	f(t) 0.74	f(RW) 0.94	IRW-IRWc	Avg	Avg	Avg	f(t) 0.9	Avg	Avg	f(t) 0.88
Spain	f(t) 0.78	Avg	f(t) 0.91	f(t) 0.98	IRWnc-IRW	f(t) 0.95	f(t) 0.97	Avg	Avg	f(t) 0.84	Avg	Avg	f(t) 0.79
Sweden	f(RW) 0.86	Avg	Avg	Avg	f(RW) 0.99	IRW-IRWc	Avg	Avg	Avg	f(t) 0.89	Avg	Avg	f(t) 0.98
UK	f(RW) 0.98	f(RW) 0.90	f(RW) 0.97	Avg	f(RW) 0.99	IRW-IRWc	Avg	Avg	Avg	f(t) 0.98	Avg	Avg	f(t) 0.73

Where the coefficient of determination r < |0.69|, the average historical values (Avg) was applied. Acronyms stand for: SW_C sawn wood coniferous; SW_{NC} sawn wood non-coniferous; WP wood based panels; PP paper and paper boards; IRW_C Industrial Roundwood coniferous; IRW_{NC} Industrial Roundwood non-coniferous; IRW Total Industrial Roundwood.

(this was for example the case of the PP production in Spain, and of many other countries) or decreasing (this was the case of the SW_{NC} production in Italy and in few other countries) trend over time. Where no correlation was detected (i.e., 30%, 34% and 20% of the countries for SW, WP and PP production, respectively), we used the constant average of the previous years. In these cases, the production is not statistically correlated to the total amount of harvest but is probably linked to other drivers. This reflects the fact that in some countries the domestic harvest amounts sufficiently supply the demand for producing subsequent products (for these countries we detected a correlation with RW), whereas other countries need to import the feedstock. Of course, this is a quite simplified approach that ignores any technical or economic correlations between forest biomass and production of semi-finished wood products.

Approach 2 (AP2)

A synthesis of the results based on AP2 is presented in Table 3, for each country and commodity. When using the AP2, the SW and WP domestic production were estimated using the IRW as independent variable for the linear regression model for 50% and 46% of the countries, respectively. In these cases the DP is estimated from the future amount of wood for non-energy use applied to each scenario. Only in few cases, 10% of the countries for SW and 15% for WP, the domestic production is statistically correlated with time (i.e., a temporal trend on the data

Table 3 Results from AP2: For each key variable the table reports (if Eq. 6 was applied) the independent variable (i.e., industrial roundwood (IRW) or time (t)) applied and the coefficient of determination R^2 of the linear regression model

COUNTRY	SWt Function	R^2	SWc Function	R^2	SWcn Function	R^2	WP Function	R^2	PP Function	R^2
Austria	f(IRW$_t$)	0.82	f(IRW$_t$)	0.83	SWt-SWc		f(IRW$_t$)	0.69	Average	
Belgium			Average		Average		Average		f(IRW$_t$)	0.64
Bulgaria	f(IRW$_t$)	0.68	SW$_t$-SW$_{nc}$		f(IRW$_t$)	0.82	Average		Average	
Croatia			Average		Average		f(IRW$_t$)	0.64	f(IRW$_t$)	0.86
Cyprus	na		na		na		na		na	
Czech Rep	f(IRW$_t$)	0.78	f(IRW$_t$)	0.75	SWt-SWc		f(IRW$_t$)	0.82	Average	
Denmark			Average		average		Average		Average	
Estonia	f(IRW$_t$)	0.66	f(IRW$_t$)	0.66	SWt-SWc		Average		Average	
Finland	f(IRW$_t$)		f(IRW$_t$)	0.87	SWt-SWc		Average		f(IRW$_t$)	0.7
France	Average		f(t)	0.72	SWt-SWc		f(t)	0.87	f(t)	0.87
Germany	f(IRW$_t$)	0.7	f(IRW$_t$)	0.75	SWt-SWc		f(IRW$_t$)	0.65	f(t)	0.97
Greece			Average		Average		f(IRW$_t$)	0.92	Average	
Hungary			Average		Average		f(IRW$_t$)	0.91	Average	
Ireland	f(IRW$_t$)	0.97	f(IRW$_t$)	0.97	SWt-SWc		f(IRW$_t$)	0.86	Average	
Italy	f(t)	0.76	SWt-SWnc		f(t)	0.81	f(t)	0.86	Average	
Latvia	f(IRW$_t$)	0.95	f(IRW$_t$)	0.95	SWt-SWc		f(t)	0.71	Average	
Lithuania	Average		SWt-SWnc		f(IRW$_t$)	0.70	f(t)	0.84	Average	
Luxemb.			Average		Average		Average		Average	
Malta	na		na		Na		Na		na	
Netherl.			Average		Average		Average		f(t)	0.74
Poland	f(t)	0.75	f(t)	0.75	SWt-SWc		f(IRW$_t$)	0.96	Average	
Portugal	Average		Average		SWt-SWc		f(IRW$_t$)	0.8	f(IRW$_t$)	0.8
Romania	Average		SWt-SWnc		f(IRW$_t$)	0.95	Average		Average	
Slovakia	f(IRW$_t$)	0.9	f(IRW$_t$)	0.92	SWt-SWc		f(IRW$_t$)	0.76	Average	
Slovenia			Average		Average		Average		Average	
Spain	f(IRW$_t$)	0.68	f(IRW$_t$)	0.61	SWt-SWc		f(IRW$_t$)	0.75	f(IRW$_t$)	0.88
Sweden	f(IRW$_t$)	0.86	f(IRW$_t$)	0.86	SWt-SWc		Average		f(t)	0.72
UK	f(IRW$_t$)	0.98	f(IRW$_t$)	0.95	SWt-SWc		f(IRW$_t$)	0.98	Average	

Acronyms stand for: SW_C sawn wood coniferous; SW_{NC} sawn wood non-coniferous; WP wood based panels; PP paper and paper boards.

series can be clearly highlighted). In the remaining countries, we calculated the average DP of the last years and kept it constant. For the majority of the countries (65%) we also calculated the average PP domestic production of the last years and kept it constant.

HWP mitigation potential

The historic domestic production of the three HWP commodities, using the IPCC Tier 2 method is shown in Figure 3.

Starting from 2013 and considering the constant harvest scenario, both the AP1 and the AP2 estimate a quite stable SW domestic production. This is due to the fact that: (i) in many countries (42% for AP1 and 50% for AP2) the SW production or DP were estimated using the RW (for AP1) or the IRW (for AP2) as main driver (i.e., a correlation between this commodity and the total harvest was detected); and (ii) where no statistical correlation was detected (i.e., 30% of the countries both for the AP1 and the AP2) a constant production or DP was assumed (of course, this is fully consistent with the constant harvest scenario). In some cases however, including Germany and Poland (i.e., two of the 5 most important EU countries detected by our preliminary analysis on harvest) the SW production or DP is statistically related to time, i.e., a temporal trend was detected, without any correlation with the total amount of harvest. This may also explain the increasing domestic production estimated with AP1 compared with AP2, even with a constant harvest scenario.

The same considerations may explain the increasing WP domestic production estimated with AP1 (+25% compared to the average 2000–2012 values). In this case, for 36% of the countries (including Sweden, Germany

and France), the WP production was not statistically correlated with the RW but with time. On the contrary, with AP2 we estimated a constant WP domestic production and only for 4 countries this variable was related to time. For the PP, the 2030 DP is 14% (with AP1) and 18% (with AP2) higher than the 2000–2012 average production. Indeed, with AP1 the PP production was mainly estimated using the time as independent variable and with AP2, we used the time as independent variable for Sweden, Germany and France.

Overall, these results suggest that due to the combined effect of IRW and pulp production, import and export (indirectly affecting the estimates of the DP), variations on the total harvest rate may have different effects on the DP of each commodity and each country. At EU level, the SW and WP DPs generally have a stronger correlation with the total harvest rate but for some important country we detected a temporal trend (i.e., a correlation with time) and no correlation with the total RW or IRW production. For many countries the PP commodity is not related to the total RW production but to other parameters i.e., economic and technical drivers (e.g., recycled paper) not directly considered by our analysis but indirectly included in the variable time. In these cases, the resulting DP is not correlated with the total amount of harvest. Of course, all commodities are not only correlated with the harvest amount but also with the prices development of woody biomass, which also impact the dynamics of import and export.

The historical net sink from HWP estimated by our analysis is reported in Figure 4 (see the upper panel). Until 2009 (i.e., the historical period considered by the submissions for FMRL), we estimated the same trend reported by the countries' submissions and by Rüter [16][e],

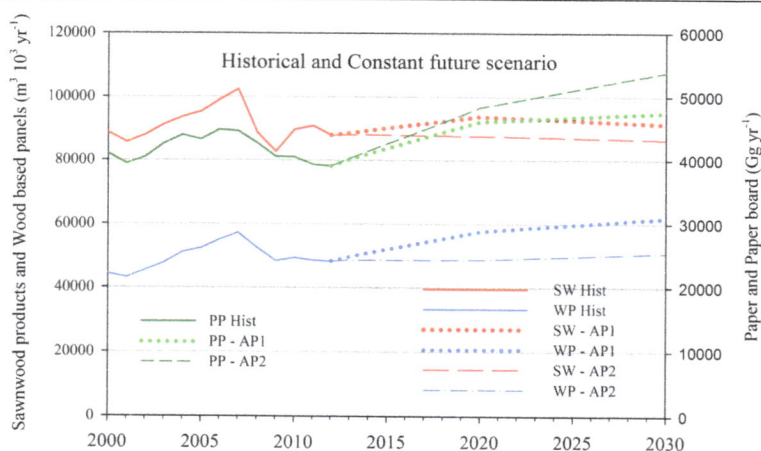

Figure 3 Total domestic production: total domestic production distinguished between sawn wood products (SW), wood based panels (WP, both in m^3 10^3 yr^{-1}, left axis) and paper and paper board (PP, Gg yr^{-1}, right axis), estimated applying the IPCC Tier 2 method.
Solid lines show historic data; dotted lines show future trends based on constant harvest scenarios, estimated with the Approach 1 (AP1) and the Approach 2 (AP2).

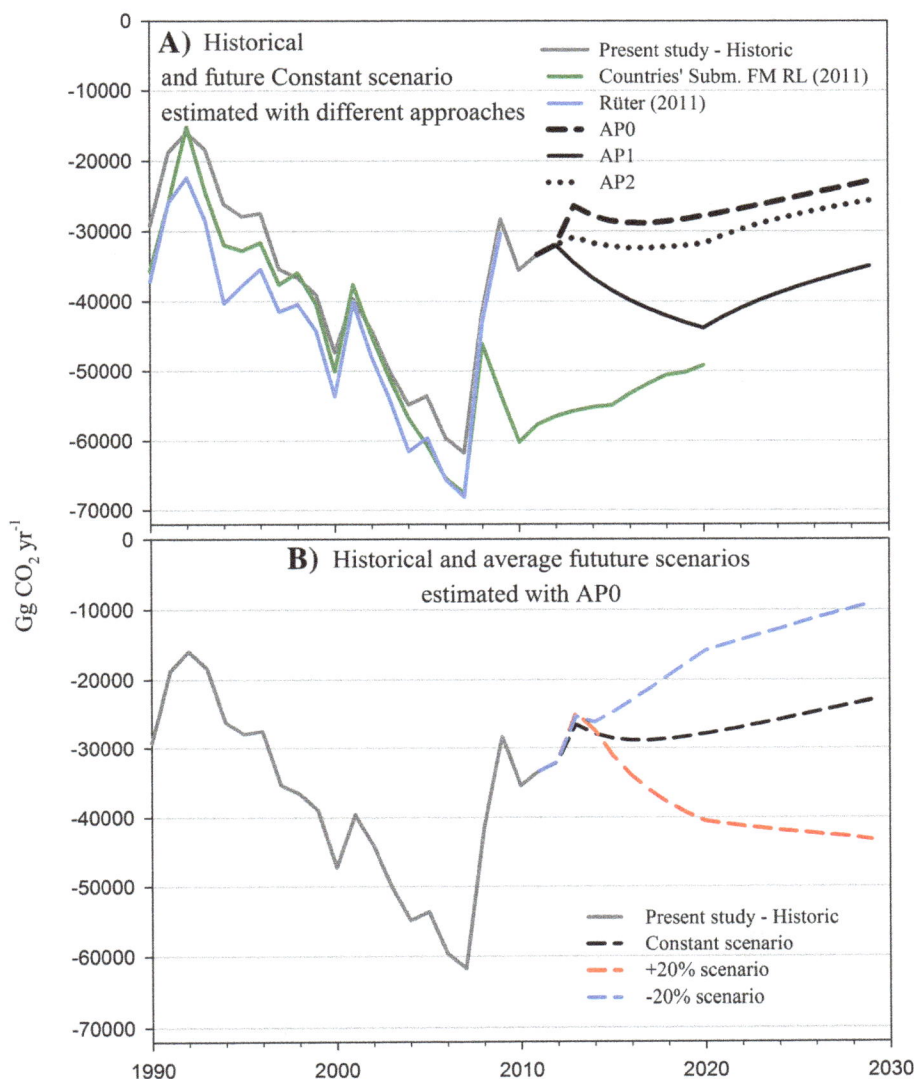

Figure 4 HWP sink: total sink from HWP (in Gg CO_2 yr^{-1}) for the historical period (until 2012) and projections until 2030. The upper panel **(A)** reports: (i) the estimates provided by our study for the historical period (based on FAOSTAT corrected data); (ii) a comparison with the estimates provided by Rüter [16] and by the country's Submission for FMRL (2011) and (iii) the future C sink estimated by our study using (a) the AP0, (b) the AP1 and (c) the AP2 approaches. The lower panel **(B)** reports the historical sink estimated by our study until 2012 and the future sink estimated by the AP0 for the constant, increasing (+20%) and decreasing (−20%) harvest scenarios. Please note that for some countries no data was available before 2000.

even if the sink estimated by our study is on average 11% (compared with the submissions for FMRL) and 21% (compared with Rüter's estimates) lower than these studies (considering the period 1990 – 2008). These differences may be due to (i) different data sources (e.g., [26]) used in Rüter [16] and the present study, respectively, (ii) the application of different carbon conversion factors, (iii) the total harvest rate of some countries and (iv) on the methods used by some countries (i.e., Finland, according to the country's submission for FMRL, did not consider the import and total production to estimate the fraction of domestic production).

For 2010, our estimates are 70% lower than the estimates reported by the submissions for FMRL. These documents, submitted to UNFCCC in 2011, were based on different assumptions on the future (after 2009) harvest rates. For the same reason, the submissions for FMRL estimated after 2009 a higher sink in 2020, equal to −49,162 Gg CO_2 yr^{-1}, due to the higher harvest rate as compared to our scenarios (see Figure 2).

Overall, for 2030 with a constant harvest scenario, the AP1 estimates higher removals (−34,901 Gg CO_2 yr^{-1}) than the AP2 (−25,652 Gg CO_2 yr^{-1}), mainly due to the higher SW and PP domestic production (see Figure 4, upper panel). With AP0 (i.e., the same approach proposed in Rüter [16] and applied by the majority of EU countries for the submissions for FMRL), the resulting sink in 2030 is equal to −22,942 Gg CO_2 yr^{-1} i.e., 12% and 50% lower than the sink estimated by AP2 and AP1, respectively. As discussed above, the differences are mainly due to the increasing SW (for AP1) and PP domestic production (for AP1 and AP2) estimated by our analysis even with a constant harvest rate (Figure 3). The pattern highlighted by all the approaches (with an increasing sink from 2012 to 2020) is due to the temporary increase of DP estimated in this period (see for example the slope of the future DP estimated with AP1 in Figure 3), due to the difference between the historical DP in 2012 and the production (with AP1), DP (with AP2) or Inflow (with AP0) estimated in 2020. Between 2012 and 2020 we applied a linear regression between the last historical value and 2020 for all the approaches. After this date, all our approaches report a decreasing sink, with a quite similar trend (i.e., the slope of the lines in the upper panel of Figure 4, from 2020 to 2030).

The average historical HWP sink from 2000 to 2012 is equal to −44,731 Gg CO_2 yr^{-1}. This is about 10% of the sink by EU forest pools and nearly 1% of the total EU GHG emissions in the same period. In 2030, with a constant harvest scenario, the future HWP sink was reduced to −22,942 Gg CO_2 yr^{-1} (with AP0), −34,901 Gg CO_2 yr^{-1} (with AP2) and −25,652 Gg CO_2 yr^{-1} (with AP2), i.e. -49% (with AP0), −22% (with AP1) and −43% (with AP2) compared with the historical average sink. This trend of decreasing HWP sink can be explained observing that, in the constant harvest scenario, the domestic production of each commodity (and the consequent inflow of C in the HWP pool) stabilizes and, as a consequence, the difference between the inflow and outflow tends to balance out. This means that with a constant harvest the HWP sink will eventually tend to zero, i.e. to "saturate".

The lower panel in Figure 4 reports the historical and the future HWP sinks estimated, for each scenario, with AP0. In the increasing harvest scenario, the final HWP sink in 2030 (−43,172 Gg CO_2 yr^{-1}) is almost equal to the historical average HWP sink (2000–2012). This can be explained by the fact that the rate of increase of harvest assumed in this scenario is similar to the one observed in the previous period (see Figure 2). This means that in order to keep a constant HWP sink the rate of increase in future harvest (assuming a constant distribution of harvest to the various commodities) should not be lower than the rate of increase observed in the past.

As expected, reducing by 20% the future harvest rate, the 2030 sink decreases to −9,078 Gg CO_2 yr^{-1} in 2030, i.e., −80% compared with the average historical sink. This is due to the cumulative effect of a reduced inflow to the HWP pool, to the annual decay rate affecting each commodity (i.e., the outflow) and to the quite strong reduction in the domestic production.

Despite our higher harvest scenario seems similar to the one followed in the EU Reference Scenario [20], results for HWP are not comparable due to different methodological assumptions. The main difference is that the EU Reference Scenario assumes that the HWP pool was in steady state in 2000 [33].

Conclusions

The contribution of the forest sector to climate change mitigation results from different and partly competing options, such as increasing the forest sink or maximizing the energy or material substitution by wood products [6-9]. In this context, a better understanding of potential future carbon stock changes in the Harvest Wood Products pool is essential to define an effective mitigation strategy, capable to maximize the sum of the contribution of different mitigation options.

In this paper we estimated the CO_2 emissions and removals in the HWP pool at EU level from 1990 to 2030, using the Tier 2 method from the 2013 IPCC KP Supplement and applying different scenarios of future harvest (and its distribution in different products).

The results of our study show that by assuming a constant historical harvest till 2030 the HWP sink at EU level will tend to decrease, for all the assumptions made on the distribution of harvest to different products. This is a consequence of the fact that, with a constant inflow of C in HWP pool, the HWP sink will sooner or later tend to saturate, i.e. to approach zero. A decreasing harvest will further speed up the tendency of HWP sink to approach zero. The current HWP sink will be maintained only by further increasing the current harvest in the future, as shown in our increasing harvest scenario. On the other hand, this latter scenario will tend to reduce the current sink in forest biomass, at least in the short term. Overall, our results suggest that there is limited potential for additional HWP sink in the EU.

Furthermore, our analysis suggests that other factors other than the total harvest may also play an important role in determining the future HWP emissions or removals. These factors include: (i) the relative share of different commodities such as furniture, plywood, paper and paper-like products, or energy; and (ii) the combined effect of production, import and export on the domestic production of each commodity. Depending on the specific country situation, in some case these factors may be even more important than the total amount of harvest in determining the future HWP emissions or removals. Therefore, when making projections on future HWP mitigation potential, the assumptions on (i) and (ii) above may play a crucial role.

Our results are based on possible correlations between harvest projections and subsequent HWP productions and do not consider technical and economic correlations between these variables. Looking to the material composition of each HWP commodity, a technical correlation between roundwood consumption (including production, import and export) used as feedstock for SW could be assessed.

Whereas in this paper we evaluated the HWP alone, from the analysis above it is clear that the HWP mitigation potential should be analyzed in conjunction with other mitigation components (e.g. sink in forest biomass, energy and material substitution by wood). To this aim, our future work will incorporate progressively the HWP into a broader modeling framework, including interactions with different options for forest management or the use of forest products for energy or material purposes (e.g., [6,25]).

Methods

The IPCC Tier 2 method described by the 2013 Revised Supplementary Methods and Good Practice Guidance Arising from the Kyoto Protocol [15] involves the following steps:

1. Estimate the fraction of domestically produced commodities, distinguished between sawn wood, wood based panels and paper and paperboards (Figure 5). The share of domestic IRW originating from domestic forests in the overall consumption of IRW in year i, is computed considering the production (IRW_P and $PULP_P$), imports (IRW_{IM} and $PULP_{IM}$) and exports (IRW_{EX} and $PULP_{EX}$) of industrial roundwood (IRW) and pulp (PULP), according to the following equations:

$$f_{IRW}(i) = \frac{IRW_P(i) - IRW_{EX}(i)}{IRW_P(i) + IRW_{IM}(i) - IRW_{EX}(i)} \tag{1}$$

$$f_{PULP}(i) = \frac{PULP_P(i) - PULP_{EX}(i)}{PULP_P(i) + PULP_{IM}(i) - PULP_{EX}(i)} \tag{2}$$

Where, f_{IRW} is the share of industrial roundwood for the domestic production of HWP originating from domestic forests in year i; f_{PULP} is the share of domestically produced pulp for the domestic production of paper and paperboard in year i. The term in the denominator of Eq. (1) and (2^f) equals the consumption.

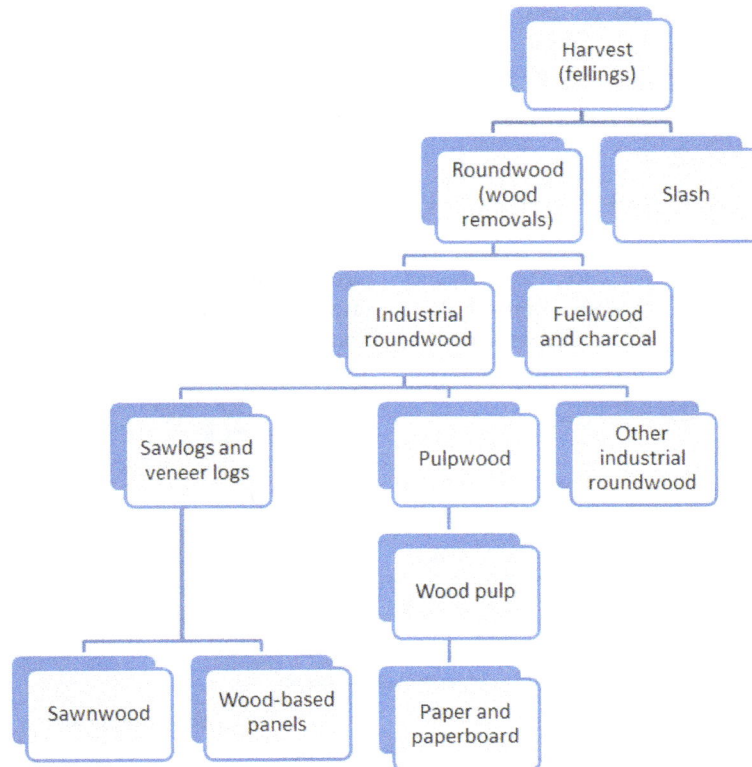

Figure 5 Wood products FAO classification: classification of wood products based on FAO forest products definitions, adapted from 2013 IPCC KP Supplement [15].

2. Estimate the annual *Inflow* of C since 1900, applying the 2013 IPCC KP Supplement default conversion factors to the domestically produced commodities;

3. Estimate the annual *Outflow* of C, applying a first order decay function with constant annual default decay factors for each commodity;

4. Estimate the total C stock and C stock change.

This method can be applied not only to historic data series, but also to projections in the future.

Equations (1) and (2) were estimated using the original FAO units (i.e., m^3 or Mg). Eq. (1) was applied to the IRW activity data (distinguished between conifers and non-conifers); Eq. (2) was applied to paper and paperboard.

The final annual HWP amount produced from domestic harvest is estimated as:

$$HWP(i) = \begin{cases} HWP_P(i) \times f_{IRW}(i) & Eq.\ (3a) \\ HWP_P(i) \times \overline{f_{IRW}(i)} \times f_{PULP}(i) & Eq.\ (3b) \end{cases}$$

Where: $HWP(i)$ is the harvested wood products amount produced from domestic harvest in year i for each commodity, in m^3 (or Mg for paper) yr^{-1}; $HWP_P(i)$ is the production of the particular HWP commodities (i.e., sawn wood, distinguished between conifers and non-conifers, wood based panels or paper and paper board) in year i, in m^3 (or Mg for paper) yr^{-1}. Eq. (3a) was applied to sawn wood and wood-based panels, Eq. (3b) to paper and paper board. In this last case we estimated an average share of domestic feedstock ($\overline{f_{IRW}(i)}$) for the production of this specific commodity, equal to:

$$f_{IRW}^{-}(i) = f_{IRW}(i)_{Con} \times w_{Con} + f_{IRW}(i)_{Non-Con} \times w_{Non-Con} \qquad (3)$$

Where, $f_{IRW}(i)_{Con}$ and $f_{IRW}(i)_{Non-Con}$ are the shares of IRW estimated by Eq.(1) for conifers and non-conifers, respectively; w_{Con} and $w_{Non-Con}$ are weighting factors derived by the total amount (i.e., production + import + export) of conifers and non-conifers, respectively.

First-order decay functions

Inflow–outflow methods estimate the changes in carbon stocks by counting the amount of wood products into and out of the stock (FCCC/tP/2003/7). Changes in carbon stocks in year i are estimated on the basis of information (i) on the inflow of wood products into the stock and

(ii) of assumed lifetimes and (iii) decay factors of these products (*Lifetime analysis*), according to the following equations (see Equations two-eight-five, 2013 IPCC KP Supplement and [34]):

$$C(i+1) = e^{-k}C(i) + \left[\frac{(1-e^{-k})}{k}\right] * Inflow(i) \qquad (4)$$

$C(i)$ = carbon stock in the particular HWP category at the beginning of year i, in Gg C (default conversion factors are in Table 4).

k = first-order decay constant for each HWP category, equal to $\ln(2)/HL$, where HL is the half-life of each HWP pool in years (Table 4).

Inflow(i) = inflow to the particular HWP category during the year i, in Gg C yr^{-1}.

The carbon stock change ($\Delta C(i)$ in Gg C yr^{-1}) of the HWP category during the year i, is equal to:

$$\Delta C = C(i+1) - C(i) \qquad (5)$$

Equations (4) and (5) were applied separately for each semi-finished wood products category (sawn wood coniferous and non-coniferous, wood-based panels and paper and paperboards). Finally (step 4, Figure 1), the total C stock and C stock change in HWP can be calculated for each country by summing up all the C stocks and C stock changes of all the commodities[d].

Approach 1 (AP1)

To apply the default Tier 2 IPCC method [15], the production of each commodity is needed. Moreover, to estimate the share of domestically produced industrial roundwood and pulp, import and export data are also requested. Therefore, in the first approach, we first looked for possible statistical correlations between the production, import and export quantities (i.e., the variables used in the first step described in Figure 1) and total roundwood (RW) and time (t, i.e., years) (Table 5). Based on a preliminary analysis of the results, the relations having a Pearson's coefficient of correlation $r > |0.69|$ were used to estimate the values of these parameters.

Table 2 reports the correlations that were used for the linear regression to extrapolate production quantities into the future.

Depending on the highest resulting correlation, the future production of sawn wood (conifer and non-conifer), wood based panels and paper and paperboards and the

Table 4 Default conversion factors for HWP categories (based on [15])

HWP categories	Sawn wood		Wood based panels	Paper and paper boards
	Coniferous	Non-coniferous		
Conversion factor per air dry density	0.225 Mg C m^{-3}	0.280 Mg C m^{-3}	0.269 Mg C m^{-3}	0.386 Mg C Mg^{-1}
Default half-lives	35 years		25 years	2 years

Table 5 AP1 summary table: we estimated the statistical correlations (highlighted by the black crosses) between these categories (dependent variables, y) and the following independent variables (x: total roundwood production (RW) and time (years)) using a simple linear model $y = a + b\,x$

Correlations	Dependent variables (y)									Independent variables (x)	
	PRODUCTION			IMPORT			EXPORT				
KEY INPUT DATA	Total	C	NC	Total	C	NC	Total	C	NC	Total RW	Time
IRW and Pulp		X	X		X	X		X	X	X	X
Sawn wood		X	X					X	X	X	X
Wood based panels	X									X	X
Paper & paperboard	X									X	X

import and export data requested by Eq. (1) and (2), were estimated with a simple linear model:

$$y = a + bx \qquad (6)$$

with y and x defined according to Table 5, and a and b defining the intercept and the slope of the function, respectively. The values of these parameters was estimated, for each country, using the *Proc Reg* procedure in the SAS® software, excluding possible outliers from the analysis of the distribution of the studentized residuals (i.e., the scaled version of residuals that are obtained by dividing each residual by its standard error) and evaluating the fitness of each model through the coefficient of determination R^2. The highest correlation coefficients were chosen for the regression, without taking into account the composition of feedstock of the particular HWP category. When no correlation could be established, an average of the previous 20 years was calculated and assumed to remain valid in future decades as well. For some countries, such as Belgium, Greece or Luxembourg

(Table 1), where no data was available for the last 20 years, a shorter period (generally 10 years) was considered.

Approach 2 (AP2)

With the second method, we check the statistical correlation between the domestic production of each commodity (i.e., the parameters used in the second step described above and reported in Figure 1), and both the industrial roundwood total production (IRW_t) and time. Depending on the highest resulting correlation, the future domestic production of sawn wood (conifer and non-conifer), wood based panels and paper and paperboards can be estimated with the linear model reported in Eq. (6).

Coniferous and non-coniferous sawn wood are estimated as a function either of IRW_t or time, depending on the highest coefficient of correlation. If the correlation with one of these two categories is too low (i.e., based on a preliminary analysis of the data, r < |0.66|), but the total sawn wood (SWt) is above the threshold, then the difference between the SWt and the other commodity was used. If the correlation is valid only for SWt, we estimate the

Table 6 AP2 summary table: SW_T was used as ancillary variable to estimate the SW_{NC} or the SW_C input data

Approach	(a)	(b)		(c)		(d)
	Eq. (6)	Eq. (6) + difference		Eq. (6) + Const. aver.		Constant average
Key data	$R^2 > |0.66|$ for IRW_T or T	$R^2 > |0.66|$ for IRW_T or T	$R^2 > |0.66|$ for IRW_T or T	$R^2 > |0.66|$ for IRW_T or T	$R^2 > |0.66|$ for IRW_T or T	$R^2 > |0.66|$ for IRW_T or T
SW_C	Yes: f(IRW_T or T)	Yes: f(IRW_T or T)	No: SW_T–SW_{NC}	No: Const. Average	No: SW_T–SW_{NC}	No: Const. Average
SW_T	Not used	Yes: f(IRW_T or T)	Yes: f(IRW_T or T)	Yes: f(IRW_T or T)	Yes: f(IRW_T or T)	Not used
SW_{NC}	Yes: f(IRW_T or T)	No: SW_T–SW_C	Yes: f(IRW_T or T)	No: SW_T–SW_C	No: Const. Average	No: Const. Average
WP	Yes: f(IRW_T or T)					No: Const. Average
PP	Yes: f(IRW_T or T)					No: Const. Average

If the coefficient of determination (R^2) is > |0.66| (Yes in the table), a linear regression is used with either IRW_T or time. If no significant correlation can be established, the average of the past 20 years is used for the future (Constant average in the table).

average amount of one commodity for the paste 1–2 decades (depending by the available country data), assume it remains valid for the future and then estimate the amount of the other commodity as the difference with SWt. If non correlation can be established, then the average amount of the sawn wood conifers (SWc) and sawn wood non-conifers ($SWnc$) for the past 10–20 years is applied as constant in the future as well.

The same approach is used for wood based panels (WP) and paper and paperboards (PP). Future productions are a function of either $IRWt$ or time, depending on the highest correlation. If no correlation can be established, then the average of the past 10–20 years is assumed as valid for the future as well. Table 6 shows a synthesis of the method.

Endnotes

[a]This is the standing volume of all trees live or dead that are felled during a certain period, including those parts of trees that are not removed from the forest (harvest removals are a subset of fellings [35]).

[b]For forest products definitions, see: http://faostat.fao.org/Portals/_Faostat/documents/pdf/FAOSTAT-Forestry-def-e.pdf.

[c]i.e., "wood in the rough" which includes all wood in its natural state, used for wood products or for energy production (FAO, 2000).

[d]To convert carbon to CO_2 multiply by (44/12).

[e]For many countries these data are the same.

[f]Please note that, according to GPG 2013 (Equation. two-eight-four) [15], f_{PULP} =0, if f_{PULP} <0. We assumed that f_{PULP} = 0 when ($PULP_P(i) - PULP_{EX}(i)$) = 0.

Abbreviations

ARD: Afforestation, reforestation and deforestation; DP: Domestic production; AP2: Approach 2; FM: Forest management; HWP: Harvested wood product; IRW: Industrial roundwood; LULUCF: Land use, land use change and forestry; KP: Kyoto protocol; PCA: Principal component analysis; AP1: Approach 1; PP: Paper and paperboards; RL: Reference level; RW: Roundwood; SWc: Coniferous sawnwood; SWnc: Non-coniferous sawnwood; SWt: Total sawnwood; WP: Wood based panels; FRA: Forest resources assessment; NIR: National inventory report.

Competing interests

The authors declare that they have no competing interests.

Authors' contributions

RP carried out the data analysis. GF and GG helped in the design of the study and together with RP wrote the manuscript. All authors read and approved the final manuscript.

Acknowledgments

We thank Sandro Federici and Richard Sikkema, for the useful comments and suggestions provided for improving this paper, and Jose V. Moris and Nuria Guerrero, for their help on collecting data for the analysis. We especially thank the two anonymous Reviewers, who provided useful comments and suggestions to improve the manuscript.
This paper was prepared in the context of the Contract n° 31502, Administrative arrangement 070307/2009/539525/AA/C5 between JRC and DG CLIMA.

The views expressed are purely those of the authors and may not in any circumstances be regarded as stating an official position of the European Commission.

References

1. Ciais P, Sabine C, Bala G, Bopp L, Brovkin V, Canadell J, et al. Carbon and Other Biogeochemical Cycles. In: Stocker TF, Qin D, Plattner G-K, Tignor M, Allen SK, Boschung J, Nauels A, Xia Y, Bex V, Midgley PM, editors. Climate Change 2013: The Physical Science Basis. Contribution of Working Group I to the Fifth Assessment Report of the Intergovernmental Panel on Climate Change. Cambridge, United Kingdom and New York, NY, USA: Cambridge University Press; 2013.
2. Settele J, Scholes R, Betts R, Bunn S, Leadley P, Nepstad D, et al. Terrestrial and inland water systems. In: Field CB, Barros VR, Dokken DJ, Mach KJ, Mastrandrea MD, Bilir TE, Chatterjee M, Ebi KL, Estrada YO, Genova RC, Girma B, Kissel ES, Levy AN, MacCracken S, Mastrandrea PR, White LL, editors. Climate Change, 2014: Impacts, Adaptation, and Vulnerability. Part A: Global and Sectorial Aspects. Contribution of Working Group II to the Fifth Assessment Report of the Intergovernmental Panel on Climate Change. Cambridge, United Kingdom and New York, NY, USA: Cambridge University Press; 2014.
3. IPCC. Guidelines for national greenhouse gas inventories. In: Eggleston S, Buendia L, Miwa K, Ngara T, Tanabe K, editors. Agriculture, forestry and other land use, vol. 4. Hayama, Japan: Institute for Global Environmental Strategies; 2006.
4. UNECE. The forest sector in the green economy. Geneva timber and forest discussion paper 54. 2010.
5. United Nations, UNECE, FAO. The European Forest Sector Outlook Study II. Geneva: United Nations, United Nations Economic Commission for Europe, Food and Agriculture Organization of the United Nations; 2011.
6. Smyth CE, Stinson G, Neilson E, Lemprière TC, Hafer M, Rampley GJ, et al. Quantifying the biophysical climate change mitigation potential of Canada's forest sector. Biogeosci Discuss. 2014;11:441–80.
7. Böttcher H, Verkerk PJ, Gusti M, Havlík P, Grassi G. Projection of the future EU forest CO_2 sink as affected by recent bioenergy policies using two advanced forest management models. GCB Bioenergy. 2012;4:773–83.
8. Marland G, Obersteiner M, Schlamadinger B. The Carbon Benefits of Fuels and Forests. Science. 2007;318(5853):1066–8.
9. Mitchell SR, Harmon ME, O'Connell KEB. Carbon debt and carbon sequestration parity in forest bioenergy production. GCB Bioenergy. 2012;4:818–27.
10. UNFCCC, United Nations Framework Convention on Climate Change, 2011. URL: http://unfccc.int/resource/docs/tp/tp0307.pdf.
11. Pingoud K, Pohjola J, Valsta L. Assessing the integrated climatic impacts of forestry and wood products. Silva Fennica. 2010;44(1):155–75.
12. Dias AC, Louro M, Arroja L, Capela I. Comparison of methods for estimating carbon in harvested wood products. Biomass Bioenergy. 2009;33(2):213–22.
13. Donlan J, Skog K, Byrne KA. Carbon storage in harvested wood products for Ireland 1961–2009. Biomass Bioenergy. 2012;46:731–8.
14. Stockmann KD, Anderson NM, Skog KE, Healey SP, Loeffler DR, Jones G, et al. Estimates of carbon stored in harvested wood products from the United States forest service northern region, 1906–2010. Carbon Balance Manag. 2012;7:1–16.
15. IPCC. In: Hiraishi T, Krug T, Tanabe K, Srivastava N, Baasansuren J, Fukuda M, Troxler TG, editors. 2013 Revised Supplementary Methods and Good Practice Guidance Arising from the Kyoto Protocol. Switzerland: IPCC; 2014.
16. Rüter S. Projections of Net-Emissions from Harvested Wood Products in European Countries. Hamburg: Johann Heinrich von Thünen-Institute (vTI), 63 p, Work Report of the Institute of Wood Technology and Wood Biology, Report No: 2011/1, 2011.
17. EEA. Annual European Union greenhouse gas inventory 1990–2012 and inventory report 2014 Submission to the UNFCCC Secretariat, Technical report No 09/2014, 2014. URL: http://unfccc.int/national_reports/annex_i_ghg_inventories/national_inventories_submissions/items/8108.php.
18. FOREST EUROPE, UNECE and FAO. State of Europe's Forests 2011. Status and Trends in Sustainable Forest Management in Europe, 2011. URL: http://www.foresteurope.org/full_SoEF.

19. Nabuurs G-J, Lindner M, Verkerk PJ, Gunia K, Deda P, Michalak R, et al. First signs of carbon sink saturation in European forest biomass. Nature Clim Change. 2013;3:792–6.

20. EC - European Commission. EU Reference scenario. 2013; 2013b.

21. Mantau U, Saal U, Prins K, Steierer F, Lindner M, Verkerk H, et al. Real potential for changes in growth and use of EU forests. Hamburg: EUwood, Methodology report; 2010.

22. EC - European Commission. A blueprint for the EU forest- Based industries (Staff working document – SWD (2013)343), 2013.

23. Pilli R, Grassi G, Kurz W, Smyth CE, Bluydea V. Application of the CBM-CFS3 model to estimate Italy's forest carbon budget, 1995 to 2020. Ecol Model. 2013;266:144–71.

24. Fiorese G, Gatto M, Guariso G. Optimization of combustion bioenergy in a farming district under different localisation strategies. Biomass Bioenergy. 2013;58:20–30.

25. Mubareka S, Jonsson R, Rinaldi F, Fiorese G, San Miguel J, Sallnas O, et al. An integrated modelling framework for the forest-Based bioeconomy. Earth-zine; 2014.

26. FAOSTAT data, 2013 URL (last access 10/12/2013): http://faostat3.fao.org/home/index.html#DOWNLOAD.

27. FRA 2010 Country Reports. URL (last access 10/12/2013). http://www.fao.org/forestry/fra/67090/en/.

28. NIR, National Inventory Reports, 2013. URL (last access 12/12/2013): http://unfccc.int/national_reports/annex_i_ghg_inventories/national_inventories_submissions/items/7383.php.

29. Submissions for Forest Management (FM). Reference Level (RM), 2011. URL (last access 12/07/2014): http://unfccc.int/bodies/awg-kp/items/5896.php.

30. Pilli R, Grassi G, Moris JV, Kurz WA. Assessing the carbon sink of afforestation with the Carbon Budget Model at the country level: an example for Italy, iForest. 2014.

31. Portugal: National Inventory Report on Greenhouse Gasses. Amadora, 2014. URL (last access October 2014): http://unfccc.int/national_reports/annex_i_ghg_inventories/national_inventories_submissions/items/8108.php.

32. Annual European Community greenhouse gas inventory 1990–2012 and inventory report 2014. Submission to the UNFCCC Secretariat. Technical report No 09/2014, European Environment Agency, 2014.

33. Böttcher H, Frank S, Havlík P, Valin H, Witzke P. Methodology for estimation and modelling of EU LULUCF greenhouse gas emissions and removals until 2050. Laxembourg: IIASA; 2013.

34. Pingoud K, Wagner F. Methane emissions from landfills and carbon dynamics of harvestedwood products: the first-order decay revisited, Mitigation and Adaptation Strategies for Global Change. 2006.

35. IPCC. Good practice guidance for land use, land-use change and forestry. In: Penman J, Gytarsky M, Hiraishi T, Krug T, Kruger D, Pipatti R, Buendia L, Miwa K, Ngara T, Tanabe K, Wagner F, editors. Hayama, Japan: Institute for Global Environmental Strategies; 2003.

PERMISSIONS

LIST OF CONTRIBUTORS

Oliver L Phillips and Jon C Lovett
School of Geography, University of Leeds, Leeds LS2 9JT, UK

Simon Willcock
School of Geography, University of Leeds, Leeds LS2 9JT, UK
School of Biological Sciences, University of Southampton, Southampton SO17 1BJ, UK

Philip J Platts and Rob Marchant
Environment Department, University of York, York YO10 5DD, UK

Andrew Balmford and Julian Bayliss
Department of Zoology, University of Cambridge, Cambridge CB2 3EJ, UK

Neil D Burgess
WWF US, Washington, USA
UNEP World Conservation Monitoring Centre, Cambridge CB3 0DL, UK

Antje Ahrends
Genetics and Conservation, Royal Botanic Garden Edinburgh, Edinburgh, UK

Nike Doggart
Tanzanian Forest Conservation Group, Dar es Salaam, Tanzania

Kathryn Doody
Frankfurt Zoological Society, Frankfurt D-60316, Germany

Eibleis Fanning
The Society for Environmental Exploration, London EC2A 3QP, UK

Jonathan MH Green
STEP Program, Princeton University, Princeton 08544, USA

Jaclyn Hall
Department of Geography, University of Florida, PO Box 117315, Gainesville, Florida, FL 32611, USA

Kim L Howell
The University of Dar es Salaam, Dar es Salaam, Tanzania

Andrew R Marshall
Environment Department, University of York, York YO10 5DD, UK
Centre for the Integration of Research, Conservation and Learning, Flamingo Land Ltd, Malton YO 17 6UX, UK

Boniface Mbilinyi and Pantaleon KT Munishi
Sokoine University of Agriculture, PO Box 3001, Morogoro, Tanzania

Nisha Owen
The Society for Environmental Exploration, London EC2A 3QP, UK
EDGE of Existence, Conservation Programmes, Zoological Society of London, London, UK

Ruth D Swetnam
Department of Geography, Staffordshire University, Stoke-on-Trent ST4 2DF, UK

Elmer J Topp-Jorgensen
Department of Bioscience, Aarhus University, Aarhus C DK-8000, Denmark

Simon L Lewis
School of Geography, University of Leeds, Leeds LS2 9JT, UK
Department of Geography, University College London, London WC1E 6BT, UK

Roberto Pilli, Giacomo Grassi, Raúl Abad Viñas and Nuria Hue Guerrero
European Commission, Joint Research Centre, Institute for Environment and Sustainability, Via E. Fermi 2749, 21027 Ispra, VA, Italy

Werner A. Kurz
Natural Resources Canada, Canadian Forest Service, Victoria, BC V8Z 1M5, Canada

Shengli Huang
ASRC Federal InuTeq, Contractor to the U.S. Geological Survey (USGS) Earth Resources Observation and Science (EROS) Center, 47914 252nd Street, Sioux Falls, SD 57198, USA

Shuguang Liu and Terry L Sohl
USGS EROS Center, 47914 252nd Street, Sioux Falls, SD 57198, USA

Ben Sleeter and Jinxun Liu
Contractor to USGS Western Geographic Science Center, 345 Middlefield Rd, Menlo Park, CA 94025, USA

Devendra Dahal and Brian Davis
Stinger Ghaffarian Technologies (SGT), Inc., Contractor to the USGS EROS Center, Sioux Falls, SD 57198, USA

Claudia Young
Innovate!, Inc. Contractor to the USGS EROS Center, Sioux Falls, SD 57198, USA

Todd J Hawbaker
U.S. Geological Survey, Denver, CO, USA

Zhiliang Zhu
U.S. Geological Survey, Reston, VA, USA

Terje Gobakken and Ørjan Totland
Department of Ecology and Natural Resource Management, Norwegian University of Life Sciences, P.O. Box 5003, 1432 Ås, Norway

Beatrice Tarimo
Department of Ecology and Natural Resource Management, Norwegian University of Life Sciences, P.O. Box 5003, 1432 Ås, Norway
Department of Geoinformatics, School of Geospatial Sciences and Technology, Ardhi University, P.O. Box 35176, Dar es Salaam, Tanzania

Øystein B Dick
Department of Mathematical Sciences and Technology, Norwegian University of Life Sciences, P.O. Box 5003, 1432 Ås, Norway

T. C. Hill
Department of Earth and Environmental Science, University of St Andrews, Irvine Building, North Street, St Andrews, UK
School of Geo Sciences, The University of Edinburgh, Edinburgh, UK
The NERC National Centre for Earth Observation, St Andrews, UK

C. M. Ryan
School of Geo Sciences, The University of Edinburgh, Edinburgh, UK

M. Williams
School of Geo Sciences, The University of Edinburgh, Edinburgh, UK
The NERC National Centre for Earth Observation, St Andrews, UK

Joseph Mascaro
American Association for the Advancement of Science, 1300 New York Ave NW, Washington, DC 20001, USA

Gregory P Asner
Department of Global Ecology, Carnegie Institution for Science, 260 Panama St., Stanford, CA 94305, USA

Stuart Davies
Forest Global Earth Observatory, Center for Tropical Forest Science, Smithsonian Tropical Research Institute, PO Box 37012, Washington, DC 20013, USA

Alex Dehgan
Conservation X Labs, 2380 Champlain St. NW #203, Washington, DC 20009, USA

Sassan Saatchi
Jet Propulsion Laboratory, California Institute of Technology, 4800 Oak Grove Drive, Pasadena, CA 91109, USA

Florian Reimer
Center for Development Research (ZEF), Group Börner, Rheinische Friedrich-Wilhelm University, Walter-Flex-Str. 3, 53113 Bonn, Germany

Gregory P Asner
Department of Global Ecology, Carnegie Institution for Science, 260 Panama Street, Stanford 94305CA, USA

Shijo Joseph
Forest and Environment Program, Center for International Forestry Research, Jalan CIFOR, Bogor 16115, Indonesia

Rajesh Bahadur Thapa, Takeshi Motohka, Manabu Watanabe and Masanobu Shimada
Earth Observation Research Center, Japan Aerospace Exploration Agency (JAXA), 2-1-1 Sengen, Tsukuba, Ibaraki 305-8505, Japan

Tarquinio Mateus Magalhães
Departamento de Engenharia Florestal, Universidade Eduardo Mondlane, Campus Universitário, Edifício no.1, 257, Maputo, Mozambique

David P Turner, William D Ritts and Zhiqiang Yang
Department of Forest Ecosystems and Society, Oregon State University, 97331 Corvallis, OR, USA

Robert E Kennedy
College of Earth, Ocean, and Atmospheric Sciences, Oregon State University, 97331 Corvallis, OR, USA

Andrew N Gray
USDA Forest Service, Pacific Northwest Station, 97331 Corvallis, OR, USA

Kristofer D Johnson, Richard Birdsey and Craig Wayson
USDA Forest Service, Northern Research Station, Newtown Square, Pennsylvania, USA

Andrew O Finley
Departments of Forestry and Geography, Michigan State University, East Lansing, Michigan, USA

Anu Swantaran and Ralph Dubayah
Department of Geographical Sciences, University of Maryland, College Park, Maryland, USA

Rachel Riemann
USDA Forest Service, Northern Research Station, Troy, New York, USA

Gregory P. Asner, Sinan Sousan, David E. Knapp and Roberta E. Martin
Department of Global Ecology, Carnegie Institution for Science, 260 Panama St, Stanford, CA 94305, USA

Paul C. Selmants
Department of Natural Resources and Environmental Management, University of Hawaii at Manoa, 1910 East–West Rd., Honolulu, HI 96822, USA

R. Flint Hughes and Christian P. Giardina
USDA Forest Service, Pacific Southwest Research Station, Institute of Pacific Islands Forestry, 60 Nowelo Street, Hilo, HI 96720, USA

Da B Tran and Tho V Hoang
The Vietnam Forestry University, Hanoi, Vietnam

Paul Dargusch
School of Geography, Planning and Environmental Management, The University of Queensland, Brisbane, QLD, Australia

Carina Van der Laan, Pita A Verweij and André PC Faaij
Copernicus Institute of Sustainable Development, Group Energy and Resources, Utrecht University, Heidelberglaan 2, 3584 CS Utrecht, The Netherlands

Marcela J Quiñones
SarVision BV, Agro Business Park 10, 6708 PW Wageningen, The Netherlands

Vanessa S. Mascorro and Nicholas C. Coops
Faculty of Forestry, Department of Forest Resources Management, University of British Columbia, 2424 Main Mall, Vancouver, BC V6T 1Z4, Canada

Werner A. Kurz
Canadian Forest Service, Natural Resources Canada, 506 West Burnside Road, Victoria, BC V8Z 1M5, Canada

Marcela Olguín
Proyecto Mexico-Noruega, Comision Nacional Forestal, Coyoacán, D.F. Col. del Carmen Coyoacán, CP 04100 Mexico, Mexico

Roberto Pilli, Giulia Fiorese and Giacomo Grassi
European Commission, Joint Research Centre, Institute for Environment and Sustainability, Via E. Fermi 2749, I-21027 Ispra, VA, Italy

Index

A

Aboveground Biomass, 2, 17, 31, 71, 74, 77, 89, 91, 100-101, 103, 114, 129-130, 138-139, 146, 151-152, 167, 169-172, 175-178, 180

Activity Data, 179-182, 184, 187-191, 195-197, 206

Additivity, 102, 105, 107, 110, 113, 115-116

Anchorage, 102-104, 109-110, 112, 115

Anthropogenic Variables, 5, 167-168, 171-173, 175

B

Belowground Biomass Allocation Patterns, 102, 105

Biomass, 2, 9, 11-12, 15-22, 24-28, 30-31, 35, 42-46, 48-49, 51-52, 59, 64-69, 71-74, 76-77, 89-91, 98, 100-105, 107-116, 118, 121, 125, 128-139, 143, 146-147, 149, 151-152, 154, 163-173, 175-178, 180, 183, 186, 189, 192, 194-195, 198, 201-202, 204-205, 208-209

Biophysical Variables, 168

Borneo, 88, 167, 169, 172, 177-178

Burned Area, 6-7, 33, 51-52, 54-55, 58-61, 63-65, 120, 176

C

Carbon, 1-19, 21, 26-27, 30, 34-38, 40-52, 64, 66, 72, 74-78, 80-81, 84, 87-91, 93-97, 100-103, 110, 113-118, 120-130, 132-134, 138-148, 150-154, 156-159, 161, 163-169, 173, 177-196, 203-204, 206, 208-209

Carbon Budget Model, 18-19, 21, 30, 34-35, 179-180, 192, 209

Carbon Dioxide (CO2), 179, 183

Carbon Sequestration, 6-8, 12-13, 36, 45, 47-49, 52, 84, 103, 117-118, 123, 127, 151, 153, 161, 164, 192, 208

Carbon Stocks, 3, 15, 17, 37, 47, 50-52, 64, 75-78, 88-90, 94-97, 100, 110, 114-118, 123, 125, 127-128, 140, 142-143, 146, 148, 151-154, 156, 159, 161, 163-169, 173, 177-178, 184-185, 189, 192

Carnegie Airborne Observatory, 140, 151-152

Carnegie Ames Stanford Approach (CASA), 180

Chocó, 78-79, 85

Climate Change, 2, 15, 17-19, 30, 34, 36-38, 45-50, 63-64, 78, 81, 84, 87-88, 101-102, 117, 122, 124, 127-128, 140-141, 152-153, 159, 161, 164-165, 167, 179-180, 191-192, 194, 196, 208

Colombia, 64, 78-79, 85, 87-88

D

Deforestation, 1-2, 10, 14-16, 21, 31-32, 52, 64, 66-68, 71, 74, 76-90, 94-97, 100-101, 140, 152, 167, 174, 178-180, 183, 191-193, 196, 208

Degradation, 1-2, 9-10, 14-16, 52, 64, 66-68, 71, 74, 76, 78-79, 81, 83, 85, 87-90, 92, 98, 100-101, 140, 152, 161, 167-168, 174, 177-179, 184, 191-192

D

Disturbance, 1, 5-6, 10, 22, 29, 32-33, 36-37, 46-48, 50, 59, 66-75, 77, 85, 117-118, 120-128, 141, 154, 161-162, 167-168, 173, 179-193

E

Eastern Arc Mountains, 1, 3, 14, 16-17

Ecosystem Service, 1, 14-15, 47

Eu Mitigation, 35

European Countries, 18, 21, 30, 208

F

Faostat Data, 32, 35, 196-199, 209

Fire History, 51-52, 55, 64-65

Forest, 1-2, 4-5, 8-11, 13-21, 25-27, 29-38, 40-43, 45-50, 52, 63-68, 71, 73-98, 100-103, 112, 114-120, 123-136, 138-184, 186-199, 201, 204-205, 208-209

Forest Cover, 67, 78-86, 88, 90, 94, 98, 100-101, 143-144, 146-147, 168, 172, 177, 179-184, 187, 192

Forest Inventory, 1, 15, 40, 42, 49, 126-129, 135, 138-141, 143, 151-152, 180, 183, 191-192, 197, 199

Forest Management, 15-16, 18-19, 29, 31-32, 34, 37, 49, 52, 64-65, 97, 118, 140, 178, 195-197, 205, 208-209

Frequency-size Distribution, 51, 59

G

Geographically Weighted Regression, 167, 169, 171, 178

Greenhouse Gas (GHG), 18, 80

Greenhouse Gas Inventories, 15, 18, 34, 101, 208

H

Harvest, 16, 18-20, 22, 25-34, 117-127, 136-137, 183-186, 194-199, 201-204, 206, 208

Harvested Wood Products (HWP), 19, 21, 194

I

Intensity, 16, 57, 60, 64, 66-71, 73-74, 118, 120-121, 123-125, 174

Inter-comparison, 129

Invasive Species, 49, 140-141, 147, 152

Ipcc Tier 3, 1

J

Jungle, 75, 152

L

Land Cover, 2, 4-5, 8-10, 13-16, 30, 35-36, 38, 49, 55, 64-67, 83, 85, 88, 91, 98, 101, 119, 125, 127-128, 131, 138, 152, 159, 165, 167, 174, 178, 181-182, 191, 193

Land Use, 14-16, 19-20, 22, 27, 34, 36-37, 40-41, 47-50, 64-65, 82-85, 89, 91, 94, 98, 101, 117, 124, 126-127, 140-141, 143, 165, 167-168, 174, 177-178, 191-192, 194, 208-209

Landsat, 51-56, 60-61, 63-65, 78-80, 82-88, 100, 117-118, 124-127, 138, 141, 146, 148, 150, 162, 168, 177-182, 184-185, 187-189, 191-193

Laser Ranging Technology, 75

Lidar, 2, 5, 75-77, 89-92, 97-98, 100-101, 129-138, 140-143, 145, 148-152, 174, 177, 180, 191

Lulucf, 19, 29, 34-35, 101, 194, 208-209

M

Mecrusse, 102-104, 110, 112, 114-115

Melaleuca, 153-166

Mexico, 179, 181-183, 187, 191-193

Miombo Woodland, 51, 53-55, 59-60, 64-65, 73, 115

Modis, 5, 12, 17, 38, 41-42, 51-52, 54-57, 59-61, 63-65, 128, 138, 170, 175, 178-189, 191-193

Multiple Regression, 167, 171, 173, 178

N

Natural Disturbances, 18-19, 21, 26-27, 29-30, 32, 34, 50

Net CO2 Emissions, 18-22, 24-25, 27-29

Net Ecosystem Carbon Balance, 117, 124, 126-127

Net Ecosystem Exchange, 117-118, 124, 126, 128

Net Ecosystem Production, 117-118, 123-124, 126-128

P

Palsar, 69, 73, 85, 88-93, 98, 100-101, 168, 174, 177-178

R

Random Forest Machine Learning, 140, 142-143, 149, 151-152

Reducing Emissions from Degradation and Deforestation (REDD+), 2

Reference Emissions, 78

Regional Spatial Variation, 167, 173

Remote Sensing, 16, 48-49, 64-67, 72, 74, 77-79, 87-88, 90, 92, 96-98, 101, 114-115, 117-118, 123-124, 127, 129-130, 138, 140-141, 145, 151-152, 174, 178-187, 191-193

Remote Sensing Observations, 181

Riau, 89-90, 93, 98, 101, 167

Root Components, 102-104, 108, 110, 112-113

S

Satellite, 15-17, 51-56, 60, 64, 66-67, 73-74, 76, 78-79, 83-85, 87-88, 90, 118, 125, 128, 140-142, 146, 148-152, 168, 178-180, 182, 187-188, 191-192

Satellite Imagery, 16, 78, 83, 141, 148, 178-180, 187-188

Sodicity, 153-154, 159-161, 165-166

Surface Fires, 51

T

Tanzania, 1, 3, 10, 14-17, 51-60, 63-65, 100, 114

Tropical Forest Landscapes, 89, 167-168, 174

Tropical Forests, 2, 14-17, 75-78, 84, 88, 90, 97, 100-101, 151, 156, 164, 166, 177-178, 187, 192

U

United Nations Framework Convention On Climate Change (UNFCCC), 18

V

Verified Carbon Standard (VCS), 78

W

West Cascades Ecoregion, 117-118, 123-124

Wildfires, 36-37, 41, 45-48, 183

Y

Yellowstone, 36-38, 44, 46, 48-50